Algebra 2

Instruction Manual

By Steven P. Demme

1·888·854·MATH(6284) - MathUSee.com
Sales@MathUSee.com

Algebra 2 Instruction Manual

©2010 Math-U-See, Inc.
Published and distributed by Demme Learning

www.MathUSee.com

1-888-854-6284 or +1 717-283-1448 | www.demmelearning.com
Lancaster, Pennsylvania USA

ISBN 978-1-60826-039-3
Revision Code 0616

Printed in the United States of America by Bindery Associates LLC

For information regarding CPSIA on this printed material call: 1-888-854-6284
and provide reference #0616-07272016

Algebra 2

4

Curriculum Sequence

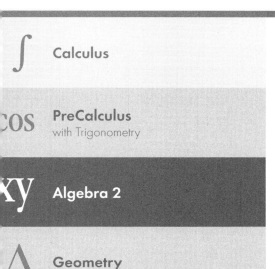

∫ **Calculus**

cos **PreCalculus**
with Trigonometry

xy **Algebra 2**

Δ **Geometry**

x^2 **Algebra 1**

X **Pre-Algebra**

ζ **Zeta**
Decimals and Percents

ε **Epsilon**
Fractions

δ **Delta**
Division

γ **Gamma**
Multiplication

β **Beta**
Multiple-Digit Addition and Subtraction

α **Alpha**
Single-Digit Addition and Subtraction

P **Primer**
Introducing Math

Math-U-See is a complete, K-12 math curriculum that uses manipulatives to illustrate and teach math concepts.
We strive toward "Building Understanding" by using a mastery-based approach suitable for all levels and learning preferences. While each book concentrates on a specific theme, other math topics are introduced where appropriate. Subsequent books continuously review and integrate topics and concepts presented in previous levels.

Where to Start
Because Math-U-See is mastery-based, students may start at any level. We use the Greek alphabet to show the sequence of concepts taught rather than the grade level. Go to MathUSee.com for more placement help.

Each level builds on previously learned skills to prepare a solid foundation so the student is then ready to apply these concepts to algebra and other upper-level courses.

Major concepts and skills for Algebra 2:
- Scientific notation
- Operations with radicals
- Factoring polynomials
- Fractional exponents
- Imaginary and complex numbers
- Binomial theorem
- Quadratic formula
- Discriminants
- Ratios and unit multipliers
- Graphs of lines
- Conic sections: circles, ellipses, parabolas, and hyperbolas
- Solving systems of equations to solve problems

Find more information and products at MathUSee.com

HONORS TOPICS

Here are the topics for the special challenge lessons included in the student text. You will find one honors page after the last systematic review page for each regular lesson. Instructions for the honors pages are included in the student text.

LESSON TOPIC

HOW TO USE MATH·U·SEE

Welcome to *Algebra 2*. I believe you will have a positive experience with the unique Math·U·See approach to teaching math. These first few pages explain the essence of this methodology which has worked for thousands of students and teachers. I hope you will take five minutes and read through these steps carefully.

If you are using the program properly and still need additional help, you may call 888-854-6284 or visit us online at MathUSee.com/support. —**Steve Demme**

The Goal of Math-U-See

The underlying assumption or premise of Math·U·See is that the reason we study math is to apply math in everyday situations. Our goal is to help produce confident problem solvers who enjoy the study of math. These are students who learn their math facts, rules, and formulas *and* are able to use this knowledge in solving word problems and real-life applications. Therefore, the study of math is much more than simply committing to memory a list of facts. It includes memorization, but it also encompasses learning underlying concepts that are critical to problem solving.

More than Memorization

Many people confuse memorization with understanding. Once while I was teaching seven junior high students, I asked how many pieces they would each receive if there were fourteen pieces. The students' response was, "What do we do: add, subtract, multiply, or divide?" Knowing *how* to divide is important; understanding *when* to divide is equally important.

THE SUGGESTED 4-STEP MATH·U·SEE APPROACH

In order to train students to be confident problem solvers, here are the four steps that I suggest you use to get the most from the Math·U·See curriculum:

Step 1. Preparation for the lesson.
Step 2. Presentation of the new topic.
Step 3. Practice for mastery.
Step 4. Progression after mastery.

Step 1. Preparation for the lesson.

This course assumes a knowledge of Algebra 1. The first few lessons include some review. If you need more review, study the "Basic Algebra Review." It is found before lesson 1 in the instruction manual, and near the end of the student text. Watch the DVD to learn the concept. Study the written explanations and examples in the instruction manual. Many students watch the DVD along with their instructor. Students in the secondary level who have taken responsibility to study math themselves will do well to watch the DVD and read through the instruction manual.

Step 2. Presentation of the new topic.

Now that you have studied the new topic, choose problems from the instruction manual to present the new concept to your students.

a. Write: Record the step-by-step solutions on paper as you work them.
b. Say: Explain the "why" and "what" of the math as you work the problems.

Do as many problems as you feel are necessary until the student is comfortable with the new material. One of the joys of teaching is hearing a student say, *"Now I get it!"* or *"Now I see it!"*

Step 3. Practice for mastery.

Using the examples in the instruction manual and the lesson practice problems from the student text, have the students practice the new concept until they understand it. It is one thing for students to watch someone else do a problem, it is quite another to do the same problem themselves. Do enough examples together so that they can do them without assistance.

Do as many of the lesson practice pages as necessary (not all pages may be needed) until the students understand the new material. Give special attention to the word problems, which are designed to apply the concept being taught in the lesson. Additional lesson practice pages are available for download at MathUSee. com/downloads.php if your student needs more practice.

Step 4. Progression after mastery.

Once mastery of the new concept is demonstrated, proceed into the systematic review pages for that lesson. Mastery can be demonstrated by having each student teach the new material back to you. The goal is not to fill in worksheets, but to be able to teach back what has been learned.

The systematic review worksheets review the new material as well as provide practice of the math concepts previously studied. Remediate missed problems as they arise to ensure continued mastery.

After the last systematic review page in each lesson, you will find an "honors" lesson. These are optional, but highly recommended for students who will be taking advanced math or science courses. These challenging problems are a good way for all students to hone their problem-solving skills.

Proceed to the lesson tests. These were designed to be an assessment tool to help determine mastery, but they may also be used as extra worksheets. Your students will be ready for the next lesson only after demonstrating mastery of the new concept and continued mastery of concepts found in the systematic review worksheets.

Confucius is reputed to have said, "Tell me, I forget; show me, I understand; let me do it, I will remember." To which we add, **"Let me teach it and I will have achieved mastery!"**

Length of a Lesson

So how long should a lesson take? This will vary from student to student and from topic to topic. You may spend a day on a new topic, or you may spend several days. There are so many factors that influence this process that it is impossible to predict the length of time from one lesson to another. I have spent three days on a lesson, and I have also invested three weeks in a lesson. This occurred in the same book with the same student. If you move from lesson to lesson too quickly without the student demonstrating mastery, he will become overwhelmed and discouraged as he is exposed to more new material without having learned the previous topics. But if you move too slowly, your student may become bored and lose interest in math. I believe that as you regularly spend time working along with your student, you will sense when is the right time to take the lesson test and progress through the book.

By following the four steps outlined above, you will have a much greater opportunity to succeed. Math must be taught sequentially, as it builds line upon line and precept upon precept on previously learned material. I hope you will try this methodology and move at your student's pace. As you do, I think you will be helping to create a confident problem solver who enjoys the study of math.

Note for Chemistry Students

Some students may be taking *Algebra 2* and chemistry at the same time. If that is the case, you may want to consider changing the order of some of the lessons.

Begin with lessons 1, 2, and 3 as usual. They review and teach algebra skills you will need in chemistry. Then move ahead and do lessons 14, 15, and 16 for topics applicable to chemistry. The concepts in these lessons use basic skills, so there should be no difficulty in using this sequence. Then move to lesson 28 to prepare for chemical mixture problems in your chemistry course. This lesson requires solving simple systems of equations as was taught in *Algebra 1*.

If you choose this order, simply skip any review questions that cover concepts you passed over, and do them when you go back to the normal sequence.

ONGOING SUPPORT
AND ADDITIONAL RESOURCES

Welcome to the Math·U·See Family!

Many of our customer service representatives have been with us for over 10 years. What makes them unique is their desire to serve and their expertise. They are able to answer your questions, place your student(s) in the appropriate level, and provide knowledgeable support throughout the school year.

Come to your local curriculum fair where you can meet us face-to-face, see the latest products, attend a workshop, meet other MUS users at the booth, and be refreshed. We are at most curriculum fairs and events. To find the fair nearest you, click on "Events" under "E-sources."

The **Website**, at MathUSee.com, is continually being updated and improved. It has many excellent tools to enhance your teaching and provide more practice for your student(s).

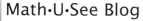

Math·U·See Blog

Interesting insights and up-to-date information appear regularly on the Math·U·See Blog. It features updates, rep highlights, fun pictures, and stories from other users. Visit us to get the latest scoop on what is happening .

Email Newsletter

For the latest news and practical teaching tips, sign up online for the free Math·U·See e-mail newsletter. Each month you will receive an e-mail with a teaching tip from Steve as well as the latest news from the website. It's short, beneficial, and fun. Sign up today at MathUSee.com!

Online Support

You will find a variety of helpful tools on our website, including corrections lists, placement tests, answers to questions, and support options.

For Specific Math Help

When you have watched the DVD instruction and read the instruction manual and still have a question, we are here to help. Call us or click the support link. Our trained staff are available to answer a question or walk you through a specific lesson.

Feedback

Send us an e-mail by clicking the feedback link. We are here to serve you and help you teach math. Ask a question, leave a comment, or tell us how you and your student are doing with Math·U·See.

Our hope and prayer is that you and your students will be equipped to have a successful experience with math!

Blessings,

Steve Demme

Basic Algebra Review
for Math-U-See Algebra 2

Math-U-See *Algebra 2* assumes a solid understanding of math concepts taught in *Algebra 1* and *Pre-Algebra*. However, we recognize that students may need to be reminded of what they have learned.

The following summary is designed to help a student "warm up" for *Algebra 2*. Hopefully each concept will be familiar to the student. If a concept listed here is confusing or unfamiliar, we strongly recommend reviewing it before beginning *Algebra 2*. You may use the "Secondary Levels Master Index" near the end of your instruction manual to find which previous level introduced a particular concept.

Some concepts, such as graphing and coin problems, are re-taught in detail in *Algebra 2*, and so are not included in this review.

Absolute Value

Absolute value lines make the value between them positive.

$$|3 + 5| = |8| = 8$$
$$|3 - 5| = |-2| = 2$$

Associative Property

When adding or multiplying, numbers may be grouped differently without affecting the answer. This property does not work for subtraction or division.

$$(3 + 4) + 2 = 3 + (4 + 2) = 9$$
$$(3 \times A) \times B = 3 \times (A \times B) = 3AB$$

Commutative Property

When adding or multiplying, the order of numbers may be changed without affecting the answer. This property does not work for subtraction or division.

$$2A + 5A = 5A + 2A = 7A$$
$$B \times A \times 6 = 6 \times A \times B = (6)(A)(B) = 6AB$$

Distributive Property

A common factor may be multiplied across all the terms of an expression.

$$4(3 + 4X - 2Y) = (4)(3) + (4)(4X) - (4)(2Y) = 12 + 16X - 8Y$$
$$-X(2 - 6Y) = (-X)(2) - (-X)(6Y) = -2X + 6XY$$

Exponents

An exponent indicates fast multiplying of the same number. Operations with exponents are reviewed in lesson 1.

$$3^2 = 3 \times 3 = 9$$

$$(3X)^3 = (3X)(3X)(3X) = 27X^3$$

$$\left(\frac{2}{3}\right)^2 = \frac{4}{9}$$

Fractions

Fractions with unknowns are handled according to the same rules as numerical fractions. Keep in mind that a fraction is another way of writing a division problem.

Greatest Common Factor (GCF)

This is the largest number that will divide evenly into two or more factors.

The GCF of 28 and 35 is 7.
The GCF of $4X$ and $2X^2Y$ is $2X$.

Least Common Multiple (LCM)

This is the smallest number that is a multiple of two or more other numbers.

The LCM of 10 and 100 is 100.
The LCM of 5Y, 10Y, and 25 is 50Y.

Negative Numbers

Operations with negative numbers follow these rules.

Addition	$(-4) + (-5) = -9$	$(+4) + (-5) = -1$
		$(-4) + (+5) = 1$
Subtraction	$(-9) - (+5) = -14$	$9 - 5 = 4$
		$(-9) - (-5) = (-9) + 5 = -4$
Multiplication	$(-3)(4) = -12$	$(-3)(-4) = 12$
Division	$(-12) \div (-3) = 4$	$(-12) \div (3) = -4$

Order of Operations

When simplifying an equation, operations must be done in the following order: parentheses, exponents, multiplication and division, addition and subtraction. Multiplication and division are done together left to right across the problem as they occur. The same is true for addition and subtraction.

Radicals

The radical sign indicates the square root of a number. A perfect square has a whole number square root. The square roots of many other numbers are irrational numbers. Operations with radicals are reviewed in lesson 4.

$$\sqrt{4} = 2$$

$$\sqrt{x^2} = x$$

$$\sqrt{2} = 1.4142\ldots$$

Solving for an Unknown

There is often more than one way to solve an equation. The basic principle is that any operation may be done to an equation as long as the same thing is done to both sides.

Terms

The terms in an algebra problem are separated by addition or subtraction signs. They may not be combined unless they have the same value. Values within terms are being multiplied by each other. Study the examples.

$2A + 4A = 6A$ Note that $6A = (6)(A) = 6 \times A = 6 \cdot A$.

$2X + 4Y = 2X + 4Y$ X and Y are different values, so the terms cannot be combined.

$5X^3 + 3X^3 = 8X^3$ X^3 and X^3 are the same value, so the terms can be combined.

$5AX^2 + 3AX^2 = 8AX^2$ AX^2 and AX^2 are the same value, so the terms can be combined.

$5AX^2 + 3AX^3 = 5AX^2 + 3AX^3$ AX^2 and AX^3 are different values, so the terms cannot be combined.

Exponents

We'll begin this lesson with a review of what we know about **exponents**. If any of this is new, spend what time you need to learn the new material and do all of the practice problems. Once you feel comfortable with what you have learned, move on to the next topic. Even though we'll review this material again on the worksheets, it won't be taught again. So get it down now, as it will be assumed from this point on that you understand these concepts. If this is all pure review, do a few problems in each sub-section, and then proceed to the next lesson.

Negative Exponents

There are two options for placing a number or variable when writing a fraction; either put it in the numerator or in the denominator. Similarly, there are two signs to use when describing a number—positive or negative. When you put these two concepts together, you have everything you need to understand negative exponents. The key phrase is: when you change the place for a number or variable, you change the sign at the same time. Another way to state this is: opposite place, opposite sign. Closely observe the following examples, do the practice problems, and compare your work with the solutions.

Make the exponent positive, and then simplify.

Example 1

$$9^{-2} = \frac{1}{9^2} = \frac{1}{81}$$

Example 2

$$x^{-3} = \frac{1}{x^3}$$

Move the term with the exponent to the numerator, and then simplify. It is important to note that the unknown cannot have any value that will cause the denominator to be zero. This is because we cannot divide by zero.

Example 3

$$\frac{1}{A^4} = A^{-4}$$ (A ≠ 0, because 0 to the fourth power is zero.)

Example 4

$$\frac{1}{10^{-3}} = 10^3 = 1000$$

Practice Problems 1

1. $5^{-2} =$

2. $X^{-5} =$

3. $Y^{-4} =$

4. $3^{-3} =$

5. $\dfrac{1}{Y^6} =$

6. $\dfrac{1}{2^{-3}} =$

7. $7^2 =$

8. $\dfrac{1}{10^{-4}} =$

Solutions 1

1. $5^{-2} = \dfrac{1}{5^2} = \dfrac{1}{25}$

2. $X^{-5} = \dfrac{1}{X^5}$

3. $Y^{-4} = \dfrac{1}{Y^4}$

4. $3^{-3} = \dfrac{1}{3^3} = \dfrac{1}{27}$

5. $\dfrac{1}{Y^6} = Y^{-6}$

6. $\dfrac{1}{2^{-3}} = 2^3 = 8$

7. $7^2 = 49$

8. $\dfrac{1}{10^{-4}} = 10^4 = 10,000$

Multiplying Numbers with the Same Base

If a number is multiplied by another number with the same base, you may *add the exponents*. The same holds true for variables with exponents. Study the examples and observe this relationship being worked out.

Example 5

$$2^3 \cdot 2^4 = \left(2^1 \cdot 2^1 \cdot 2^1\right)\left(2^1 \cdot 2^1 \cdot 2^1 \cdot 2^1\right) = 2^7 = 128$$

or

$$2^3 \cdot 2^4 = (8)(16) = 128$$

Example 6

$$4^2 \cdot 4^3 = \left(4^1 \cdot 4^1\right)\left(4^1 \cdot 4^1 \cdot 4^1\right) = 4^5 = 1{,}024$$

or

$$4^2 \cdot 4^3 = (16)(64) = 1{,}024$$

Practice Problems 2

1. $X^2 \cdot X^3 \cdot X^4 =$

2. $4^2 \cdot 4^1 \cdot 4^{-1} =$

3. $X^A \cdot X^{2B} =$

4. $X^8 \cdot X^3 \cdot X^A =$

5. $Y^{-1} \cdot Y^5 \cdot Y^2 =$

6. $5^{-2} \cdot 5^6 \cdot 5^0 =$

7. $Y^A \cdot Y^{-B} =$

8. $7^{2X} \cdot 7^3 \cdot 7^X =$

Solutions 2

1. $X^2 \cdot X^3 \cdot X^4 = X^9$

2. $4^2 \cdot 4^1 \cdot 4^{-1} = 4^2 = 16$

3. $X^A \cdot X^{2B} = X^{A+2B}$

4. $X^8 \cdot X^3 \cdot X^A = X^{11+A}$

5. $Y^{-1} \cdot Y^5 \cdot Y^2 = Y^6$

6. $5^{-2} \cdot 5^6 \cdot 5^0 = 5^4 = 625$

7. $Y^A \cdot Y^{-B} = Y^{A-B}$

8. $7^{2X} \cdot 7^3 \cdot 7^X = 7^{3X+3}$

Dividing Numbers with the Same Base

If a number is divided by another number with the same base, you can *subtract the exponents*. The same holds true for variables with exponents. Study the examples and observe this relationship as it is being worked out.

Example 7

$$2^5 \div 2^2 = \frac{\cancel{2} \cdot \cancel{2} \cdot 2 \cdot 2 \cdot 2}{\cancel{2} \cdot \cancel{2}} = 2^3 = 8$$

or

$$2^5 \div 2^2 = 2^{5-2} = 2^3 = 8$$

Example 8

$$2^1 \div 2^5 = \frac{\cancel{2}}{\cancel{2} \cdot 2 \cdot 2 \cdot 2 \cdot 2} = \frac{1}{2^4} = \frac{1}{16}$$

or

$$2^1 \div 2^5 = 2^{1-5} = 2^{-4} = \frac{1}{2^4} = \frac{1}{16}$$

Practice Problems 3

1. $X^5 \div X^2 =$

2. $Y^8 \div Y^{-1} =$

3. $\dfrac{X^3}{X^7} =$

4. $\dfrac{3^{-4}}{3^{-1}} =$

5. $Y^4 \div Y^{-6} =$

6. $6^{-2} \div 6^{-5} =$

7. $4^{-5} \div 4^{-2} =$

8. $\dfrac{9^4}{9^8} =$

Solutions 3

1. $X^5 \div X^2 = X^3$

2. $Y^8 \div Y^{-1} = Y^9$

3. $\dfrac{X^3}{X^7} = X^{-4}$

4. $\dfrac{3^{-4}}{3^{-1}} = 3^{-3} = \dfrac{1}{3^3} = \dfrac{1}{27}$

5. $Y^4 \div Y^{-6} = Y^{10}$

6. $6^{-2} \div 6^{-5} = 6^3 = 216$

7. $4^{-5} \div 4^{-2} = 4^{-3} = \dfrac{1}{4^3} = \dfrac{1}{64}$

8. $\dfrac{9^4}{9^8} = 9^{-4} = \dfrac{1}{9^4} = \dfrac{1}{6{,}561}$

Zero as an Exponent

Using what we just covered about adding and subtracting exponents when multiplying and dividing numbers with the same base, we'll show that anything with a zero exponent equals one.

Example 9

$$\frac{10^2}{10^2} = 10^{2-2} = 10^0$$

$$1 = \frac{100}{100} = \frac{10^2}{10^2} = 10^2 \cdot 10^{-2} = 10^0 \quad \text{so} \quad 10^0 = 1$$

Example 10

$$\frac{X^3}{X^3} = X^{3-3} = X^0$$

$$\frac{X^3}{X^3} = \frac{X \cdot X \cdot X}{X \cdot X \cdot X} = 1 \quad \text{so} \quad X^0 = 1$$

Raising a Power to a Power

When raising an exponent to another *power*, or exponent, you fast add or *multiply the exponents.* In this example, we either add $7 + 7 + 7 = 21$ or multiply $7 \times 3 = 21$. In this case, the result of multiplying 6 times itself 21 times would be a very large number, so we leave it as shown below.

Example 11

$$\left(6^7\right)^3 = \left(6^7\right)\left(6^7\right)\left(6^7\right) = \left(6^{7+7+7}\right) = \left(6^{(7)(3)}\right) = \left(6^{21}\right)$$

Notice in example 12 that both the number and the letter inside the parentheses are raised to the power outside the parentheses.

Example 12

$$\left(2x^3\right)^3 = \left(2x^3\right)\left(2x^3\right)\left(2x^3\right) = \left(8x^{3+3+3}\right) = 8x^9$$

Practice Problems 4

1. $\left(2Y^3\right)^5 =$

2. $\left(4^{-2}\right)^9 =$

3. $\left(8^0\right)^{-3} =$

4. $\left(B^{-2}\right)^{-4} =$

5. $\left(A^7\right)^{-3} =$

6. $\left(x^{-3}\right)^0 =$

7. $\left(10^4\right)^{-6} =$

8. $\left[\left(Y^7\right)^3\right]^3 =$

Solutions 4

1. $\left(2Y^3\right)^5 = (2)^5\left(Y^3\right)^5 = 2^5 Y^{15}$ or $32Y^{15}$

2. $\left(4^{-2}\right)^9 = \left(4^{(-2)(9)}\right) = 4^{-18}$

3. $\left(8^0\right)^{-3} = \left(8^{(0)(-3)}\right) = \left(8^0\right) = 1$

4. $\left(B^{-2}\right)^{-4} = \left(B^{(-2)(-4)}\right) = B^8$

5. $\left(A^7\right)^{-3} = \left(A^{(7)(-3)}\right) = A^{-21}$

6. $\left(x^{-3}\right)^0 = \left(x^{(-3)(0)}\right) = \left(x^0\right) = 1$

7. $\left(10^4\right)^{-6} = \left(10^{(4)(-6)}\right) = 10^{-24}$

8. $\left[\left(Y^7\right)^3\right]^3 = \left(Y^{(7)(3)(3)}\right) = Y^{63}$

Simplifying an Exponential Expression

There are two techniques that may be employed in simplifying an exponential expression. The first is to have everything in the numerator, and the second is to make all the exponents positive. In the examples, the problems will be done both ways for your examination.

Although we will not state it for every problem, we will assume that the unknowns do not have values that would make the denominator of the problem equal to zero.

Example 13

Put everything in the numerator.

$$\frac{X^3 X^{-2}}{X^1 X^4} = \frac{X^3 X^{-2} X^{-1} X^{-4}}{1} = X^{-4}$$

Make all the exponents positive.

$$\frac{X^3 X^{-2}}{X^1 X^4} = \frac{X^3}{X^1 X^4 X^2} = \frac{X^3}{X^7} = \frac{1}{X^{7-3}} = \frac{1}{X^4}$$

Example 14

Put everything in the numerator.

$$\frac{X^{-2} Y^{-3} Y^1 X^4}{X^1 Y^2} = \frac{X^{-2} X^{-1} X^4 Y^{-3} Y^1 Y^{-2}}{1} = X^1 Y^{-4}$$

Make all the exponents positive.

$$\frac{X^{-2} Y^{-3} Y^1 X^4}{X^1 Y^2} = \frac{Y^1 X^4}{X^2 X^1 Y^3 Y^2} = \frac{Y^1 X^4}{Y^5 X^3} = \frac{X^4 X^{-3}}{Y^5 Y^{-1}} = \frac{X^{4-3}}{Y^{5-1}} = \frac{X^1}{Y^4}$$

Practice Problems 5

Use either method to solve.

1. $\dfrac{A^2 B^2}{A^{-6} B^5} =$

2. $\dfrac{B^{-6} C^4}{B^1 C^9} =$

3. $\dfrac{H^4 N^7}{H^{-1} N^3} =$

4. $\dfrac{C^{-2} C^{-3} D^1}{D^1 D^2 C^4} =$

5. $\dfrac{A^{-2} B^6 B^{-8}}{A^6 B^{-1} A^{-5}} =$

6. $\dfrac{P^2 Q^{-2} P^1}{Q^{-4} Q^2 P^3} =$

Solutions 5

1. $\dfrac{A^2 B^2}{A^{-6} B^5} = A^8 B^{-3}$ or $\dfrac{A^8}{B^3}$

2. $\dfrac{B^{-6} C^4}{B^1 C^9} = B^{-7} C^{-5}$ or $\dfrac{1}{B^7 C^5}$

3. $\dfrac{H^4 N^7}{H^{-1} N^3} = H^5 N^4$

4. $\dfrac{C^{-2} C^{-3} D^1}{D^1 D^2 C^4} = C^{-9} D^{-2}$ or $\dfrac{1}{C^9 D^2}$

5. $\dfrac{A^{-2} B^6 B^{-8}}{A^6 B^{-1} A^{-5}} = A^{-3} B^{-1}$ or $\dfrac{1}{A^3 B^1}$

6. $\dfrac{P^2 Q^{-2} P^1}{Q^{-4} Q^2 P^3} = 1$

Rational Expressions

A *rational expression* is a fancy word for algebra in fractions, or fractions with letters for unknowns and variables as well as numbers. The same rules or concepts that apply to fractions apply to rational expressions, except that if there is a letter in the denominator, it cannot make the denominator equal to zero. In the fraction X/A, you must say that $A \neq 0$. Let's review the two key concepts of fractions.

Concept #1
You can only compare or combine (add or subtract) two fractions that are the same "kind," i.e. that have the same denominator.

Concept #2
Multiplying or dividing by one does not change the value of a fraction.

It is tricky when there is more than one symbol in the numerator and/or denominator. It is helpful to remember that the denominator tells what kind and that the numerator tells how many.

Example 1

$$\frac{2}{5} + \frac{X}{5} = \frac{X+2}{5} \quad \text{The converse is also true.} \quad \frac{X+2}{5} = \frac{2}{5} + \frac{X}{5}$$

Since they are the same kind (have the same denominator), they can be combined. But numbers and letters are not the same kind, so we have to leave X + 2 as it is.

Example 2

$$\frac{3}{X+1} \neq \frac{2}{X} + \frac{1}{1}$$

You cannot separate X + 1 in the denominator, since this is one value.

You can combine the following fractions because they have the same denominator.

$$\frac{3}{X+1} = \frac{1}{X+1} + \frac{2}{X+1}$$

Let's replace X with 7 in both of the examples to verify our conclusions.

$$\frac{3}{7+1} \neq \frac{2}{7} + \frac{1}{1} \qquad \rightarrow \qquad \frac{3}{8} \neq 1\frac{2}{7}$$

$$\frac{3}{7+1} = \frac{1}{7+1} + \frac{2}{7+1} \qquad \rightarrow \qquad \frac{3}{8} = \frac{1}{8} + \frac{2}{8}$$

Our conclusions were correct. When in doubt, replace variables with numbers to see whether the problem makes sense. We have learned that we need to treat (X + 1) as a single term. Using parentheses is a big help in keeping this concept straight. The rule of thumb is that numbers or variables in the numerator may be separated because the numerator tells how many. The term in the denominator (the divisor) must maintain its integrity because it tells what kind. Here are examples for more clarification:

Example 3

$$\frac{X^2 + 3X + 7}{X^2 + 4X + 3} = \frac{X^2}{X^2 + 4X + 3} + \frac{3X}{X^2 + 4X + 3} + \frac{7}{X^2 + 4X + 3} \qquad \text{Correct}$$

Example 4

$$\frac{X^2 + 3X + 7}{X^2 + 4X + 3} = \frac{X^2}{X^2} + \frac{3X}{4X} + \frac{7}{3} \qquad \text{Incorrect}$$

Example 5

$$\frac{5}{X-5} = \frac{3}{X-5} + \frac{1}{X-5} + \frac{1}{X-5} \quad \text{Correct}$$

Example 6

$$\frac{5}{X-5} = \frac{7}{X} - \frac{2}{5} \quad \text{Incorrect}$$

Example 7

$$\frac{X}{X} + \frac{3}{X} = \frac{X+3}{X} \quad \text{Correct}$$

Example 8

$$\frac{X+3}{X} = \frac{X}{X} + \frac{3}{X} = 1 + \frac{3}{X} \quad \text{Correct}$$

Example 9

$$\frac{4X^2 - X}{X} = \frac{4X^2}{X} - \frac{X}{X} = \frac{4X}{1} - \frac{1}{1} = 4X - 1$$

or

$$\frac{4X^2 - X}{X} = \frac{X(4X-1)}{X} = 4X - 1 \quad \text{Both are correct}$$

Example 10

$$\frac{2}{X+1} + \frac{3}{X} = \frac{2}{X+1} + \left(\frac{3+1}{X+1}\right) \quad \text{Incorrect}$$

Adding one is not acceptable. You may multiply the numerator and denominator by one, but you can't add one to each of them without changing the nature of the problem. Replace X with a number to verify this. See example 11 for the correct solution.

Example 11

$$\frac{2}{X+1} + \frac{3}{X} = \frac{2(X)}{(X+1)(X)} + \frac{3(X+1)}{(X)(X+1)} =$$

$$\frac{2(X)+3(X+1)}{(X)(X+1)} = \frac{5X+3}{(X)(X+1)} = \frac{5X+3}{X^2+X} \qquad \text{Correct}$$

This is the correct way to add rational expressions. It is the same as finding the common denominator as we do when we are adding simple fractions with different denominators.

Practice Problems 1

Simplify by factoring first. Then use canceling if possible.

1. $\dfrac{3X^2+X}{X} =$

2. $\dfrac{YX+Y^2X}{YX} =$

3. $\dfrac{25Y-15}{5} =$

4. $\dfrac{16X^2+24X}{8X} =$

Find the same denominator and add.

5. $\dfrac{2}{X+1} + \dfrac{3X}{X-1} =$

6. $\dfrac{3}{Y} - \dfrac{2}{X} =$

7. $\dfrac{4}{X} + \dfrac{5}{7} =$

8. $\dfrac{Y}{X} + \dfrac{4Y}{X+2} =$

Solutions 1

1. $\dfrac{3X^2 + X}{X} = \dfrac{X(3X+1)}{X} = 3X + 1$

2. $\dfrac{YX + Y^2X}{YX} = \dfrac{YX(1+Y)}{YX} = 1 + Y$

3. $\dfrac{25Y - 15}{5} = \dfrac{5(5Y-3)}{5} = 5Y - 3$

4. $\dfrac{16X^2 + 24X}{8X} = \dfrac{8X(2X+3)}{8X} = 2X + 3$

5. $\dfrac{2}{X+1} + \dfrac{3X}{X-1} = \dfrac{2(X-1)}{(X+1)(X-1)} + \dfrac{3X(X+1)}{(X-1)(X+1)} =$

   $\dfrac{2X - 2 + 3X^2 + 3X}{(X-1)(X+1)} = \dfrac{3X^2 + 5X - 2}{(X^2 - 1)}$

6. $\dfrac{3}{Y} - \dfrac{2}{X} = \dfrac{3X}{YX} - \dfrac{2Y}{XY} = \dfrac{3X - 2Y}{YX}$

7. $\dfrac{4}{X} + \dfrac{5}{7} = \dfrac{4(7)}{7X} + \dfrac{5X}{7X} = \dfrac{28 + 5X}{7X}$

8. $\dfrac{Y}{X} + \dfrac{4Y}{X+2} = \dfrac{Y(X+2)}{X(X+2)} + \dfrac{4YX}{X(X+2)} = \dfrac{4YX + YX + 2Y}{X(X+2)} = \dfrac{5YX + 2Y}{X^2 + 2X}$

Scientific Notation; Combining Like Terms

Scientific notation is used in science to solve equations with very large and/or very small numbers. If we were asked to compute 200 times the distance from the earth to the sun (93,000,000 miles) using the normal method of multiplying, it would require a good deal of paper and pencil and have lots of zeroes.

$$
\begin{array}{r}
93{,}000{,}000 \\
\times \qquad 200 \\
\hline
\end{array}
$$

Scientific notation provides an easier and more efficient method. It is closely related to exponential notation.

93,000,000 in exponential notation is $9 \times 10^7 + 3 \times 10^6$. In scientific notation, you keep the 9 and the 3 together instead of separating them, and place the decimal point so the 9 is in the units place. Then choose the exponent figured from the number in the units place. The 3 is ignored when choosing the exponent.

Example 1
Multiply 93,000,000 times 200.

93,000,000 in scientific notation is 9.3×10^7.
200 in scientific notation is 2×10^2.

To solve our original problem of 200 times the distance from the sun to the earth, multiply the numbers without exponents times the other numbers without exponents and the numbers with exponents times the other numbers with exponents.

$$\left(9.3 \times 10^7\right)\left(2 \times 10^2\right) = \left(9.3 \times 2\right)\left(10^7 \times 10^2\right) = \left(18.6\right)\left(10^9\right)$$

$$= \left(1.86 \times 10^1\right)\left(10^9\right) = 1.86 \times 10^{10}$$

Example 2
Multiply: 1,900 x 50

1,900 = 1.9 x 10^3 in scientific notation.
50 = 5 x 10^1 in scientific notation.

Multiply the numbers times the numbers and the exponential terms times the exponential terms.

$$\left(1.9 \times 10^3\right)\left(5 \times 10^1\right) = (1.9 \times 5)\left(10^3 \times 10^1\right) = (9.5)\left(10^4\right)$$

You can also show very small decimal numbers with scientific notation. Remember that you may only have one number in the units place and the rest as decimals. Study the examples:

Example 3
Change .000054 to scientific notation.

$$.00005\underset{\uparrow}{4} = 5.4 \times 10^{-5} = 5.4 \times \frac{1}{10^5} \text{ or } 5.4 \times \frac{1}{100,000}$$

Example 4
Multiply: 30,000,000 x .000023

30,000,000 = 3.0 x 10^7 in scientific notation.
.000023 = 2.3 x 10^{-5} in scientific notation.

$$\left(3 \times 10^7\right)\left(2.3 \times 10^{-5}\right) = (3 \times 2.3)\left(10^7 \times 10^{-5}\right) = (6.9)\left(10^2\right) = 6.9 \times 10^2$$

Example 5
Divide: 500,000 ÷ 8,000

500,000 = 5.0 x 10^5 in scientific notation.
8,000 = 8.0 x 10^3 in scientific notation.

$$\left(5 \times 10^5\right) \div \left(8 \times 10^3\right) = (5 \div 8)\left(10^5 \div 10^3\right) = (.625)\left(10^2\right)$$
$$= \left(6.25 \times 10^{-1}\right) \times 10^2 = 6.25 \times 10^1$$

Practice Problems 1

1. 93,000,000 x .000054 =

2. 18,000 x .007 =

3. 640,000 x .92 =

4. 12,400 ÷ .04 =

5. 40,000 x 3,000 ÷ 60 =

6. .00058 x .0023 =

Solutions 1

1. $(9.3 \times 10^7)(5.4 \times 10^{-5}) = (9.3 \times 5.4)(10^7 \times 10^{-5}) = 50.22 \times 10^2$
$$= (5.022 \times 10^1) \times 10^2 = 5.022 \times 10^3$$

2. $(1.8 \times 10^4)(7 \times 10^{-3}) = (1.8 \times 7)(10^4 \times 10^{-3}) = 12.6 \times 10^1$
$$= (1.26 \times 10^1) \times 10^1 = 1.26 \times 10^2$$

3. $(6.4 \times 10^5)(9.2 \times 10^{-1}) = (6.4 \times 9.2)(10^8 \times 10^{-1}) = 58.88 \times 10^4$
$$= (5.888 \times 10^1) \times 10^4 = 5.888 \times 10^5$$

4. $(1.24 \times 10^4) \div (4 \times 10^{-2}) = (1.24 \div 4)(10^4 \div 10^{-2}) = .31 \times 10^6$
$$= (3.1 \times 10^{-1}) \times 10^6 = 3.1 \times 10^5$$

5. $(4 \times 10^4)(3 \times 10^3) \div (6 \times 10^1) = (4 \times 3 \div 6)(10^4 \times 10^3 \div 10^1) = 2 \times 10^6$

6. $(5.8 \times 10^{-4})(2.3 \times 10^{-3}) = (5.8 \times 2.3)(10^{-4} \times 10^{-3}) = 13.34 \times 10^{-7}$
$$= (1.334 \times 10^1) \times 10^{-7} = 1.334 \times 10^{-6}$$

Even though in proper form the digit must be in the units place, there are times when doing larger problems that it is advantageous to leave the numbers larger and make the necessary corrections in the exponents. This allows for reducing or dividing by a common factor (sometimes called canceling) and saves steps in problem solving. Example 6 employs this technique. You can do the long method to compare the results if you wish.

Example 6

$$\frac{(2,700)(3,500)}{9,000,000} = \frac{(27 \times 10^2)(35 \times 10^2)}{(90 \times 10^5)}$$

$$= \frac{^3(\cancel{27} \times 10^2)\ ^7(\cancel{35} \times 10^2)}{_2\cancel{10}(\cancel{90} \times 10^5)}$$

You can divide 27 and 90 by 9. Then divide 35 and 10 by 5.

$$= \frac{(3 \times 10^2)(7 \times 10^2)}{(2 \times 10^5)}$$

$$= \frac{(3 \times 7)(10^2 \times 10^2)}{(2 \times 10^5)}$$

$$= (21)(10^4) \div (2 \times 10^5)$$

$$= (21 \div 2)(10^{4-5})$$

$$= (10.5)(10^{-1})$$

$$= (1.05 \times 10^1)(10^{-1})$$

$$= 1.05 \text{ or } 1.05 \times 10^0$$

Combining Like Terms

One of the key concepts we have focused on since the beginning is that numbers tell us how many, and place value tells what kind. Building on this, we found that you can only combine or compare things that are the same kind.

The terms and expressions in *Algebra 2* are becoming more complex, but the concepts remain the same. What we need to learn in this lesson is how to identify which terms are the same kind, and then combine them. The strategies to be employed are not new either. When the terms are rational expressions, the first step is to reduce or simplify as much as possible. When exponents are present, either make them all positive or put all of them on the same line by changing the signs of the exponent (opposite sign, opposite place). Some of the problems look tricky, but after patiently applying these strategies, we can distinguish between the apples and the oranges and combine the apples with the apples and the oranges with the oranges. Let's do a few examples.

Example 7

$$\frac{30X^{-1}}{10} + 2^2Y^2 - \frac{2}{X} + \frac{2Y}{Y^{-1}}$$

Step #1 — Simplify what you can.

$$3X^{-1} + 4Y^2 - \frac{2}{X} + \frac{2Y}{Y^{-1}}$$

Step #2 — Make all exponents positive,
and when possible combine like terms.

$$\frac{3}{X} + 4Y^2 - \frac{2}{X} + 2Y^2 =$$

$$\frac{3}{X} - \frac{2}{X} + 4Y^2 + 2Y^2 = \frac{1}{X} + 6Y^2$$

Or put all the variables on one line, and when possible combine like terms.

$$3X^{-1} + 4Y^2 - 2X^{-1} + 2Y^2 =$$

$$3X^{-1} - 2X^{-1} + 4Y^2 + 2Y^2 = X^{-1} + 6Y^2$$

Example 8

$$3XY^{-2} - 4X^2Y^2 + \frac{7X^2}{XY^2}$$

Step #1 — Simplify what you can.

$$\frac{3X}{Y^2} - 4X^2Y^2 + \frac{7X}{Y^2}$$

Step #2 — Combine like terms.

$$\frac{3X}{Y^2} - 4X^2Y^2 + \frac{7X}{Y^2} = \frac{10X}{Y^2} - 4X^2Y^2$$

Step #3 — Keep all exponents positive,
or put all the variables on one line.

$$\frac{10X}{Y^2} - 4X^2Y^2 = 10XY^{-2} - 4X^2Y^2$$

Practice Problems 2

1. $\dfrac{5B^{-1}}{A^{-1}} - \dfrac{7B^2A^2}{B^3A^1} + \dfrac{3B^3B^0}{A^1A^{-2}} =$

2. $\dfrac{5X^4X^{-1}}{X^2Y^2} - 2X^2Y^{-2} + \dfrac{6X^4Y^2X^{-1}}{XY^4} =$

3. $8XXY - YXXY + \dfrac{2XY}{X^{-1}} =$

4. $4X^2X^{-1} - \dfrac{12X^2}{X^3} + 8X =$

5. $6A^2B^{-2}A - \dfrac{4AB^2B^{-1}}{A^{-1}B^3} + \dfrac{9B^3B^{-3}}{A^{-3}AB^2} =$

6. $9X^{-2}A^{-2}X + X^3AA^{-1} + \dfrac{7A^3X^0}{A^2X^3} =$

Solutions 2

1. $\dfrac{5B^{-1}}{A^{-1}} - \dfrac{7B^2A^2}{B^3A^1} + \dfrac{3B^3B^0}{A^1A^{-2}} = 5B^{-1}A - 7B^2B^{-3}A^2A^{-1} + 3B^3B^0A^{-1}A^2 =$

   $5AB^{-1} - 7AB^{-1} + 3AB^3 = -2AB^{-1} + 3AB^3$ or $-\dfrac{2A}{B} + 3AB^3$ *

2. $\dfrac{5X^4X^{-1}}{X^2Y^2} - 2X^2Y^{-2} + \dfrac{6X^4Y^2X^{-1}}{XY^4} =$

   $5X^4X^{-1}X^{-2}Y^{-2} - 2X^2Y^{-2} + 6X^4Y^2X^{-1}X^{-1}Y^{-4} =$

   $5XY^{-2} - 2X^2Y^{-2} + 6X^2Y^{-2} = 5XY^{-2} + 4X^2Y^{-2}$ or $\dfrac{5X + 4X^2}{Y^2}$

3. $8XXY - YXXY + \dfrac{2XY}{X^{-1}} = 8X^2Y - X^2Y^2 + 2X^2Y = 10X^2Y - X^2Y^2$

4. $4X^2X^{-1} - \dfrac{12X^2}{X^3} + 8X = 4X - 12X^{-1} + 8X = 12X - 12X^{-1}$ or $12X - \dfrac{12}{X}$

5. $6A^2B^{-2}A - \dfrac{4AB^2B^{-1}}{A^{-1}B^3} + \dfrac{9B^3B^{-3}}{A^{-3}AB^2} =$

   $6A^3B^{-2} - 4A^2B^{-2} + 9A^2B^{-2} =$

   $6A^3B^{-2} + 5A^2B^{-2}$ or $\dfrac{6A^3}{B^2} + \dfrac{5A^2}{B^2} = \dfrac{6A^3 + 5A^2}{B^2}$

6. $9X^{-2}A^{-2}X + X^3AA^{-1} + \dfrac{7A^3X^0}{A^2X^3} =$

   $9A^{-2}X^{-1} + X^3 + 7AX^{-3}$ or $\dfrac{9}{A^2X} + X^3 + \dfrac{7A}{X^3}$

*It is customary to put letters in alphabetical order within terms.

Radicals, Basic Operations, and Simplifying

Add and Subtract Radicals

You can only add or combine two things if they are the same kind. $\sqrt{9}$ is the same as 3 because the square root of 9 is 3. But $\sqrt{3}$ is not a whole number; it is a *radical*. (Technically, the symbol in front of the number is the radical, but for convenience, we call the whole expression a radical.) You may add a radical to a radical or a whole number to a whole number, but you can't combine a whole number and a radical.

We know $2X + 5X = 7X$, $3 + 8 = 11$, and $2X + 5 = 2X + 5$. In the same way, $4\sqrt{3} + \sqrt{3} = 5\sqrt{3}$ and $\sqrt{3} + 8 = \sqrt{3} + 8$. You can add $2 + \sqrt{9}$, because you can change the square root of 9 to the whole number 3, and $2 + 3 = 5$.

Example 1

$$5\sqrt{2} + 3\sqrt{2} = 8\sqrt{2}$$

Example 2

$$6\sqrt{5} - \sqrt{5} = 5\sqrt{5}$$

Just as X is the same as 1X, $\sqrt{5}$ is the same as $1\sqrt{5}$.

Example 3

$$11\sqrt{5} + 4\sqrt{7} = 11\sqrt{5} + 4\sqrt{7}$$

You can't add or combine terms if they are not the same kind.

Practice Problems 1

1. $4\sqrt{3} + 5\sqrt{3} =$

2. $9\sqrt{7} - 3\sqrt{8} =$

3. $11\sqrt{X} + 8\sqrt{X} =$

4. $6\sqrt{5} - \sqrt{5} =$

5. $\sqrt{15} + 4\sqrt{10} =$

6. $8\sqrt{6} - 2\sqrt{6} =$

Solutions 1

1. $4\sqrt{3} + 5\sqrt{3} = 9\sqrt{3}$

2. $9\sqrt{7} - 3\sqrt{8} = 9\sqrt{7} - 3\sqrt{8}$

3. $11\sqrt{X} + 8\sqrt{X} = 19\sqrt{X}$

4. $6\sqrt{5} - \sqrt{5} = 5\sqrt{5}$

5. $\sqrt{15} + 4\sqrt{10} = \sqrt{15} + 4\sqrt{10}$

6. $8\sqrt{6} - 2\sqrt{6} = 6\sqrt{6}$

Multiply and Divide Radicals

When we multiply, we multiply whole numbers times whole numbers and radicals times radicals. The same holds true for division.

Example 4

$$\sqrt{7} \times \sqrt{6} = \sqrt{42}$$

Example 5

$$2\sqrt{3} \times 4\sqrt{5} = 8\sqrt{15}$$

Example 6

$$\frac{\sqrt{21}}{\sqrt{3}} = \sqrt{7}$$ This is true because $\sqrt{7} \times \sqrt{3} = \sqrt{21}$.

Practice Problems 2

1. $\left(5\sqrt{6}\right)\left(2\sqrt{10}\right) =$

2. $\dfrac{2\sqrt{30}}{\sqrt{5}} =$

3. $\left(9\sqrt{2}\right)\left(4\sqrt{2}\right) =$

4. $\dfrac{2\sqrt{28}}{\sqrt{7}} =$

5. $\left(3\sqrt{5}\right)\left(6\sqrt{7}\right) =$

6. $\dfrac{8\sqrt{12}}{2\sqrt{6}} =$

7. $\left(7\sqrt{11}\right)\left(8\sqrt{12}\right) =$

8. $\dfrac{9\sqrt{200}}{12\sqrt{2}} =$

Solutions 2

1. $\left(5\sqrt{6}\right)\left(2\sqrt{10}\right) = 10\sqrt{60}$

2. $\dfrac{2\sqrt{30}}{\sqrt{5}} = 2\sqrt{6}$

3. $\left(9\sqrt{2}\right)\left(4\sqrt{2}\right) = 36\sqrt{4} = 36 \times 2 = 72$

4. $\dfrac{2\sqrt{28}}{\sqrt{7}} = 2\sqrt{4} = 2 \times 2 = 4$

5. $\left(3\sqrt{5}\right)\left(6\sqrt{7}\right) = 18\sqrt{35}$

6. $\dfrac{8\sqrt{12}}{2\sqrt{6}} = 4\sqrt{2}$

7. $\left(7\sqrt{11}\right)\left(8\sqrt{12}\right) = 56\sqrt{132}$

8. $\dfrac{9\sqrt{200}}{12\sqrt{2}} = \dfrac{3}{4}\sqrt{100} = \dfrac{3}{4} \times 10 = \dfrac{15}{2}$

Simplifying Radicals

We can separate a radical into factors. The key is to choose a factor that is a perfect square, such as 4, 9, 16, 25, etc. Perfect squares are the only factors that may be transformed into whole numbers instead of being left as radicals. In example 7, there are other possible factors, but only $\sqrt{4}$ will become a whole number.

Example 7

$\sqrt{12} = \sqrt{6} \times \sqrt{2} = \sqrt{12}$ This hasn't been simplified.

$\sqrt{12} = \sqrt{4} \times \sqrt{3} = 2\sqrt{3}$ This has been simplified.

Example 8

$\sqrt{18} = \sqrt{3} \times \sqrt{6} = \sqrt{18}$ This hasn't been simplified.

$\sqrt{18} = \sqrt{9} \times \sqrt{2} = 3\sqrt{2}$ This has been simplified.

For further proof, estimate your answers by using your calculator.
The symbol ≈ means "approximately equal to."

$\sqrt{12} = 2\sqrt{3}$
$3.46 \approx 2.00 \times 1.73$
$3.46 \approx 3.46$

$\sqrt{18} = 3\sqrt{2}$
$4.24 \approx 3.00 \times 1.414$
$4.24 \approx 4.23$

Practice Problems 3

1. $\sqrt{24} =$

2. $\sqrt{50} =$

3. $\sqrt{200} =$

4. $\sqrt{27} =$

5. $2\sqrt{90} =$

6. $3\sqrt{28} =$

7. $10\sqrt{125} =$

8. $4\sqrt{72} =$

Solutions 3

1. $\sqrt{24} = \sqrt{4}\sqrt{6} = 2\sqrt{6}$

2. $\sqrt{50} = \sqrt{25}\sqrt{2} = 5\sqrt{2}$

3. $\sqrt{200} = \sqrt{100}\sqrt{2} = 10\sqrt{2}$

4. $\sqrt{27} = \sqrt{9}\sqrt{3} = 3\sqrt{3}$

5. $2\sqrt{90} = 2\sqrt{9}\sqrt{10} = 6\sqrt{10}$

6. $3\sqrt{28} = 3\sqrt{4}\sqrt{7} = 6\sqrt{7}$

7. $10\sqrt{125} = 10\sqrt{25}\sqrt{5} = 50\sqrt{5}$

8. $4\sqrt{72} = 4\sqrt{36}\sqrt{2} = 24\sqrt{2}$

Radicals in the Denominator

Up to this point, we've been dealing with normal radicals. But there are radical radicals that live in places they shouldn't, namely in the denominator. Only whole numbers are permitted in the denominator. In the example $7/\sqrt{2}$, we need to multiply $\sqrt{2}$ by something to make it a whole number. The easiest factor to choose is $\sqrt{2}$ itself. But we can't randomly multiply the denominator alone because it would change the value of the expression. If we multiply the numerator by the same factor, then we are multiplying by $\sqrt{2}/\sqrt{2}$, which equals one. Now the radical is in the numerator, which is acceptable, and the denominator is occupied by a whole number, which is also acceptable.

Example 9

$$\frac{7}{\sqrt{2}} \times \frac{\sqrt{2}}{\sqrt{2}} = \frac{7\sqrt{2}}{\sqrt{4}} = \frac{7\sqrt{2}}{2}$$ Look this over carefully.

Example 10

$$\frac{3}{\sqrt{5}} \times \frac{\sqrt{5}}{\sqrt{5}} = \frac{3\sqrt{5}}{\sqrt{25}} = \frac{3\sqrt{5}}{5}$$

Example 11

You can do this problem in either of the following ways:

$$\frac{4}{\sqrt{8}} \times \frac{\sqrt{2}}{\sqrt{2}} = \frac{4\sqrt{2}}{\sqrt{16}} = \frac{4\sqrt{2}}{4} = \sqrt{2}$$

or

$$\frac{4}{\sqrt{8}} \times \frac{\sqrt{8}}{\sqrt{8}} = \frac{4\sqrt{8}}{\sqrt{64}} = \frac{4 \times 2\sqrt{2}}{8} = \frac{8\sqrt{2}}{8} = \sqrt{2}$$

Practice Problems 4

1. $\dfrac{5}{\sqrt{13}} =$

2. $\dfrac{7}{\sqrt{11}} =$

3. $\dfrac{4}{\sqrt{12}} =$

4. $\dfrac{6\sqrt{5}}{\sqrt{7}} =$

5. $\dfrac{8\sqrt{2}}{\sqrt{6}} =$

6. $\dfrac{9\sqrt{7}}{\sqrt{8}} =$

Solutions 4

1. $\dfrac{5}{\sqrt{13}} \times \dfrac{\sqrt{13}}{\sqrt{13}} = \dfrac{5\sqrt{13}}{13}$

2. $\dfrac{7}{\sqrt{11}} \times \dfrac{\sqrt{11}}{\sqrt{11}} = \dfrac{7\sqrt{11}}{11}$

3. $\dfrac{4}{\sqrt{12}} \times \dfrac{\sqrt{3}}{\sqrt{3}} = \dfrac{2\sqrt{3}}{3}$

4. $\dfrac{6\sqrt{5}}{\sqrt{7}} \times \dfrac{\sqrt{7}}{\sqrt{7}} = \dfrac{6\sqrt{35}}{7}$

5. $\dfrac{8\sqrt{2}}{\sqrt{6}} \times \dfrac{\sqrt{6}}{\sqrt{6}} = \dfrac{8\sqrt{12}}{6} = \dfrac{8\sqrt{3}}{3}$

6. $\dfrac{9\sqrt{7}}{\sqrt{8}} \times \dfrac{\sqrt{2}}{\sqrt{2}} = \dfrac{9\sqrt{14}}{4}$

Now we can simplify radicals, find the common denominator, and add the terms.

Example 12

$$\dfrac{3}{\sqrt{2}} + \dfrac{5}{\sqrt{3}} = \dfrac{3}{\sqrt{2}} \times \dfrac{\sqrt{2}}{\sqrt{2}} + \dfrac{5}{\sqrt{3}} \times \dfrac{\sqrt{3}}{\sqrt{3}} = \dfrac{3\sqrt{2}}{2} + \dfrac{5\sqrt{3}}{3}$$

$$\dfrac{3 \times 3\sqrt{2}}{3 \times 2} + \dfrac{2 \times 5\sqrt{3}}{2 \times 3} = \dfrac{9\sqrt{2}}{6} + \dfrac{10\sqrt{3}}{6} = \dfrac{9\sqrt{2} + 10\sqrt{3}}{6}$$

Practice Problems 5

1. $\dfrac{8}{\sqrt{5}} + \dfrac{3}{\sqrt{6}} =$

2. $\dfrac{2}{\sqrt{3}} - \dfrac{9}{\sqrt{7}} =$

Solutions 5

1. $\dfrac{8}{\sqrt{5}} + \dfrac{3}{\sqrt{6}} = \dfrac{8}{\sqrt{5}} \times \dfrac{\sqrt{5}}{\sqrt{5}} + \dfrac{3}{\sqrt{6}} \times \dfrac{\sqrt{6}}{\sqrt{6}} =$

$\dfrac{8\sqrt{5}}{5} + \dfrac{3\sqrt{6}}{6} = \dfrac{6 \times 8\sqrt{5}}{6 \times 5} + \dfrac{5 \times 3\sqrt{6}}{5 \times 6} = \dfrac{48\sqrt{5} + 15\sqrt{6}}{30} = \dfrac{16\sqrt{5} + 5\sqrt{6}}{10}$

2. $\dfrac{2}{\sqrt{3}} - \dfrac{9}{\sqrt{7}} = \dfrac{2}{\sqrt{3}} \times \dfrac{\sqrt{3}}{\sqrt{3}} - \dfrac{9}{\sqrt{7}} \times \dfrac{\sqrt{7}}{\sqrt{7}} =$

$\dfrac{2\sqrt{3}}{3} - \dfrac{9\sqrt{7}}{7} = \dfrac{7 \times 2\sqrt{3}}{7 \times 3} - \dfrac{3 \times 9\sqrt{7}}{3 \times 7} = \dfrac{14\sqrt{3} - 27\sqrt{7}}{21}$

Factoring Polynomials; Rational Expressions

Factoring Trinomials with Positive Numbers

Here is how to find the factors of $X^2 + 7X + 12$ using the blocks with the algebra inserts snapped into the back. First $X^2 + 7X + 12$ is built. This is the product that is given. Next a rectangle is built using all the blocks. The factors can be found by reading the lengths of the "over" dimension and the "up" dimension.

Example 1

Find the factors of $X^2 + 7X + 12$.

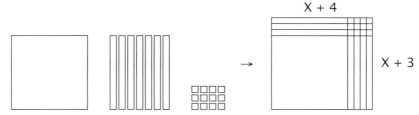

The factors are $(X + 4)(X + 3)$.

Example 2

Find the factors of $X^2 + 8X + 12$.

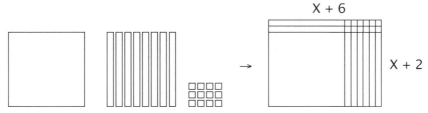

The factors are $(X + 6)(X + 2)$.

Notice the relationship between the last term (12), the middle term (7X or 8X), and the factors.

$$x^2 + 7X + 12 = (X + 4)(X + 3)$$

The last term is found by **multiplying** 3 x 4.
The middle term is found by **adding** 3X + 4X.

$$x^2 + 8X + 12 = (X + 6)(X + 2)$$

The last term is found by **multiplying** 6 x 2.
The middle term is found by **adding** 6X + 2X.

Summary: If the coefficient of x^2 is one, then the factors of the last term are the addends of the middle term.

Practice Problems 1
Find the factors.

1. $x^2 + 5X + 6$

2. $x^2 + 13X + 12$

3. $x^2 + 13X + 42$

4. $x^2 + 9X + 20$

5. $x^2 + 6X + 9$

Solutions 1

1. $x^2 + 5X + 6 =$
   $(X + 2)(X + 3)$

2. $x^2 + 13X + 12 =$
   $(X + 12)(X + 1)$

3. $x^2 + 13X + 42 =$
   $(X + 6)(X + 7)$

4. $x^2 + 9X + 20 =$
   $(X + 4)(X + 5)$

5. $x^2 + 6X + 9 = (X + 3)(X + 3)$

Factoring Trinomials with Negative Numbers

The shaded bars are –X (negative X). They are represented by the gray algebra inserts snapped into the back of the blue ten bars. The shaded unit bars are unit pieces upside down (the hollow side showing), representing negative numbers. $X^2 - X - 6$ looks like this:

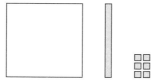

To find the factors, we must build a rectangle. However, it is not possible to build a rectangle with this group of manipulatives, so we have to add something to them. We know that zero plus anything is still the same thing. So, we will add zero (nothing) to the blocks by adding a positive X and a negative X, or a gray insert and a blue insert. We still don't have enough pieces to build a rectangle, so we'll add zero again (+X – X). This works! We still have $X^2 - X - 6$, just in a different form: $X^2 - 3X + 2X - 6$. The factors are $(X + -3)(X + 2)$ or $(X - 3)(X + 2)$.

X – 3

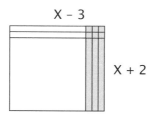

X + 2

For clarity, pick up all the negative pieces and place them on top of the positive pieces. Think of the negative pieces and the positive pieces directly beneath them as making zero. See the factors in the remaining rectangle.

zero

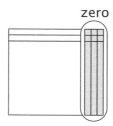

Example 3

Find the factors of $X^2 - 2X - 8$.

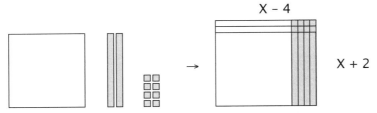

The factors are $(X - 4)(X + 2)$.

Practice Problems 2

Find the factors.

1. $X^2 + 2X - 15$

2. $X^2 - 4X - 12$

3. $X^2 + 6X - 7$

4. $X^2 - 6X + 8$

5. $X^2 - 9X + 20$

Solutions 2

1. $X^2 + 2X - 15 = (X - 3)(X + 5)$ 2. $X^2 - 4X - 12 = (X + 2)(X - 6)$

3. $X^2 + 6X - 7 = (X + 7)(X - 1)$ 4. $X^2 - 6X + 8 = (X - 2)(X - 4)$

5. $X^2 - 9X + 20 = (X - 4)(X - 5)$

The Difference of Two Squares

When factoring, we are given the product, and we have to find the factors. Let's begin this type of problem by going in reverse. Build a rectangle to find the product of $(X - 3)(X + 3)$. Notice the middle term especially. It is equal to zero.

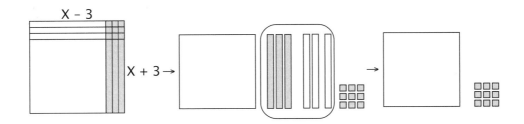

The first term (X^2) and the third term (3^2 or 9) are both squares. So when asked to factor the *difference of two squares*, the answer is the square root of the first term plus the square root of the third term, times the square root of the first term minus the square root of the third term. Algebraically, we may represent this pattern as $A^2 - B^2 = (A + B)(A - B)$. In the problem shown on the previous page, this is $X^2 - 9 = (X + 3)(X - 3)$.

Example 4
Find the factors of $X^2 - 4$.

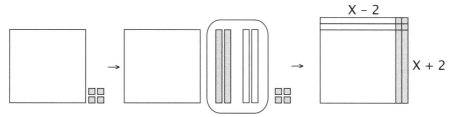

The factors of $X^2 - 4$ are $(X + 2)(X - 2)$.

Practice Problems 3
Find the factors.

1. $X^2 - 36$ 2. $X^2 - A^2$

3. $X^2 - 1$ 4. $4X^2 - 9$

5. $9X^2 - 25$

Solutions 3

1. $X^2 - 36 = (X - 6)(X + 6)$ 2. $X^2 - A^2 = (X + A)(X - A)$

3. $X^2 - 1 = (X + 1)(X - 1)$ 4. $4X^2 - 9 = (2X - 3)(2X + 3)$

5. $9X^2 - 25 = (3X - 5)(3X + 5)$

Factoring Trinomials with Coefficients

In a polynomial, *coefficients* are the numbers and *variables* (X^3, X^2, X^1, etc.) are the place values. The coefficient tells "how many," and the variable tells "what kind." In $2X^2$, the first 2 is the coefficient. In $7X$, the 7 is the coefficient. When factoring a trinomial where the coefficient of X^2 is 1, the factors of the last term are the addends of the middle term. When the coefficient is not 1, the factoring process becomes more difficult. Now instead of finding the factors of the third term alone, you also have factors of the first term to contend with. If the sign is negative, there are still more options. You can always work out all the possibilities systematically until you find the solution, which can be tedious. After a lot of practice, you will begin to see the patterns that reveal the solutions more quickly.

Example 5
Find the factors of $2X^2 + 7X + 6$.

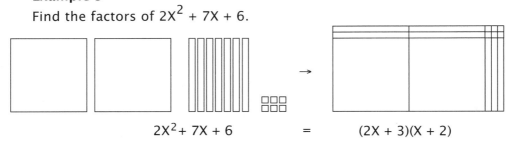

$$2X^2 + 7X + 6 \qquad = \qquad (2X + 3)(X + 2)$$

$(2X + 3)(X + 2)$ may be multiplied to check the factors vertically (below left), or linearly, using the distributive property.

$$
\begin{array}{r}
2X + 3 \\
X + 2 \\
\hline
4X + 6 \\
2X^2 + 3X \\
\hline
2X^2 + 7X + 6
\end{array}
$$

$(X + 2)(2X + 3) = (X)(2X + 3) + (2)(2X + 3) =$
$2X^2 + 3X + 4X + 6$

Another way that multiplication of binomials has been taught is with the FOIL method. FOIL represents the four products, or combinations, of the factors: F - first, O - outside, I - inside, L - last.

F   In (X + 2)(2X + 3), X · 2X is the
    First term times the **First** term. $2X^2$

O   In (X + 2)(2X + 3), X · 3 is the
    Outside term times the **O**utside term. $3X$

I   In (X + 2)(2X + 3), 2 · 2X is the
    Inside term times the **I**nside term. $4X$

L   In (X + 2)(2X + 3), 2 · 3 is the
    Last term times the **L**ast term. 6

$$2X^2 + 3X + 4X + 6$$
$$2X^2 + \quad 7X \quad + 6$$

Practice Problems 4
Find the factors.

1. $2X^2 + 9X + 4$ 2. $3X^2 + 17X + 10$

3. $4X^2 + 8X + 3$ 4. $6X^2 + 19X + 15$

5. $3X^2 - X - 4$ 6. $5X^2 + 13X - 6$

7. $6X^2 + X - 2$ 8. $12X^2 - 11X + 2$

Solutions 4

1. $2X^2 + 9X + 4 = (2X + 1)(X + 4)$

2. $3X^2 + 17X + 10 = (3X + 2)(X + 5)$

3. $4X^2 + 8X + 3 = (2X + 3)(2X + 1)$

4. $6X^2 + 19X + 15 = (2X + 3)(3X + 5)$

5. $3X^2 - X - 4 = (3X - 4)(X + 1)$

6. $5X^2 + 13X - 6 = (5X - 2)(X + 3)$

7. $6X^2 + X - 2 = (2X - 1)(3X + 2)$

8. $12X^2 - 11X + 2 = (3X - 2)(4X - 1)$

Finding the Greatest Common factor

Before factoring, look for a *greatest common factor* (GCF) that you can divide out first. This makes a much easier problem. A GCF may contain coefficients, variables, or both.

Example 6
Find the factors of $2X^2 - 18$.

Notice that both have a common factor of 2,
    so we can make this $2\left(X^2 - 9\right)$.

$\left(X^2 - 9\right)$ is the difference of two squares, $X^2 - 3^2$,
    so our factors are $2(X + 3)(X - 3)$.

Example 7

Find the factors of $3X^3 + 9X^2 - 54X$.

Notice that each term has common factors of 3 and X.

We can factor 3X out of each term, yielding $3X(X^2 + 3X - 18)$.
Factor again to get a final result of $3X(X + 6)(X - 3)$.

Practice Problems 5

Find the GCF and then the factors.

1. $2X^3 + 12X^2 + 18X$ 2. $3X^2 + 9X + 6$

3. $4X^4 - 25X^2$ 4. $20X^4 + 10X^3 - 30X^2$

Solutions 5

1. $2X^3 + 12X^2 + 18X = 2X(X^2 + 6X + 9) = 2X(X + 3)(X + 3)$

2. $3X^2 - 9X + 6 = 3(X^2 - 3X + 2) = 3(X - 1)(X - 2)$

3. $4X^4 - 25X^2 = X^2(4X^2 - 25) = X^2(2X + 5)(2X - 5)$

4. $20X^4 + 10X^3 - 30X^2 = 10X^2(2X^2 + X - 3) = 10X^2(2X + 3)(X - 1)$

Repeated Factoring

$X^4 - 16$ is the same as $(X^2)(X^2) - (4)(4)$ or $(X^2)^2 - (4)^2$. This is the difference of two squares. The resultant factors will be $(X^2 - 4)(X^2 + 4)$.

Notice that we are not done yet, as $X^2 - 4$ is also the difference of two squares $(X^2 - 2^2)$, and its factors are $(X - 2)(X + 2)$. Putting it all together:

$$x^4 - 16 = \left(x^2\right)^2 - (4)^2$$
$$= \left(x^2 + 4\right)\left(x^2 - 4\right)$$
$$= \left(x^2 + 4\right)(x - 2)(x + 2)$$

Note: $(X^2 + 4)$ cannot be factored using the difference of two squares or any other method we have learned so far, so we leave it alone.

Example 8

Find the factors of $X^4 - 81$.

$$x^4 - 81 = \left(x^2\right)^2 - (9)^2 = \left(x^2 + 9\right)\left(x^2 - 9\right) = \left(x^2 + 9\right)(x - 3)(x + 3)$$

Practice Problems 6

Find the factors.

1. $X^4 - 1$

2. $X^4 - 104X^2 + 400$

Solutions 6

1. $\left(x^2 + 1\right)\left(x^2 - 1\right) = \left(x^2 + 1\right)(x + 1)(x - 1)$

2. $\left(x^2 - 4\right)\left(x^2 - 100\right) = (x + 2)(x - 2)(x + 10)(x - 10)$

After learning all of these skills about factoring, you can apply them when solving equations.

Using Factoring to Solve Equations

Example 9

Find the solution of $X^2 + 5X + 6 = 20$.

$$X^2 + 5X + 6 = 20 \qquad \text{First subtract 20 from both sides.}$$
$$X^2 + 5X - 14 = 0 \qquad \text{Now find the factors.}$$
$$(X + 7)(X - 2) = 0 \qquad \text{Think of these two factors as A and B.}$$

If A x B = 0, then either A or B or both have to be equal to zero.

The options are:

$\quad$ 0 x B = 0, or A x 0 = 0, or 0 x 0 = 0.

So either $(X + 7) = 0$ or $(X - 2) = 0$.

$\quad$ If $X + 7 = 0$, then $X = -7$.

$\quad$ If $X - 2 = 0$, then $X = 2$.

The solutions are $X = -7$ or $X = 2$.

Let's check these values of X in the original equation.

$$(-7)^2 + 5(-7) + 6 = 20 \quad \text{and} \quad (2)^2 + 5(2) + 6 = 20.$$
$$49 \quad - \quad 35 \quad + 6 = 20 \qquad\qquad 4 \quad + \quad 10 + 6 = 20$$

They both work, thus validating our solutions.

Practice Problems 7

Find the solution(s) for X. In other words, "What values of X will satisfy the equation?" Then check your answers by substituting them into the original equation to see whether they work.

1. $2X^2 + 3X = 2$ $\qquad\qquad$ 2. $3X^3 = 27X$

Solutions 7

1. $2X^2 + 3X = 2$

   A. Set the equation equal to zero: $2X^2 + 3X - 2 = 0$

   B. Find the factors

       and set them equal to zero: $(2X - 1)(X + 2) = 0$

                                                   $2X - 1 = 0 \quad X + 2 = 0$

   C. Solve the equations: $X = 1/2 \quad\quad X = -2$

   D. Check by substituting the solutions:

   $$2(1/2)^2 + 3(1/2) = 2 \quad\quad 2(-2)^2 + 3(-2) = 2$$
   $$2 = 2 \quad\quad\quad\quad\quad\quad\quad 2 = 2$$
   $$\text{Yes} \quad\quad\quad\quad\quad\quad\quad\quad \text{Yes}$$

2. $3X^3 = 27X$

   A. Set the equation equal to zero: $3X^3 - 27X = 0$

   B. Find the factors $3X(X^2 - 9) = 0$

       and set them equal to zero: $3X(X + 3)(X - 3) = 0$

                                        $3X = 0 \quad X + 3 = 0 \quad X - 3 = 0$

   C. Solve the equations: $X = 0 \quad\quad X = -3 \quad\quad X = 3$

   D. Check by substituting the solutions:

   $$3(0)^3 = 27(0) \quad\quad 3(-3)^3 = 27(-3) \quad\quad 3(3)^3 = 27(3)$$
   $$0 = 0 \quad\quad\quad\quad -81 = -81 \quad\quad\quad\quad 81 = 81$$
   $$\text{Yes} \quad\quad\quad\quad\quad \text{Yes} \quad\quad\quad\quad\quad\quad \text{Yes}$$

More Rational Expressions

When we put what we know about polynomials with what we've been learning about rational expressions, we will have some interesting equations, like puzzles, to solve. In the process, remember that you can never have zero as a denominator. The denominator is the divider, or divisor, and you can't divide by zero.

Example 10

If $\dfrac{6}{2} = 3$, then 2 x 3 = 6. This is true.

If $\dfrac{18}{X} = 9$, then X = 2 because 2 x 9 = 18.

What if you were to solve this equation?

What is X? What times 0 equals 18? $\quad \dfrac{18}{X} = 0$

This is the same equation written differently.

What is X here? What times 0 equals 18? $\quad \dfrac{18}{0} = X$

There is no such solution to either of these, and so we say that the solutions are *undefined*. This applies to polynomials because you often have variables in the denominator. Look at the example below and notice the denominators. What are the two values that X cannot be? X cannot be either 1 or −3, because then one of the denominators would be zero, and the solution would be undefined.

$$\frac{X}{X-1} + \frac{X}{X+3} = 5 \qquad \frac{X}{1-1} + \frac{X}{-3+3} = 5 \qquad \frac{X}{0} + \frac{X}{0} = 5$$

In this example, we qualify the answer by saying: $X \neq 1$ or $X \neq -3$. This states that X can be any number except 1 or −3. Whenever we have a possibility of the denominator being zero, we must qualify the answer.

Let's solve some more difficult equations with rational expressions. Keep in mind that the denominator cannot be zero in any of the steps.

Example 11
Combine the rational expressions shown below.

The common denominator is (X + 5)(X − 5), or $X^2 - 25$, and X ≠ 5, −5.

$$\frac{2}{X+5} + \frac{5}{X-5} - \frac{10}{X^2-25} \rightarrow \frac{2}{X+5}\frac{(X-5)}{(X-5)} + \frac{5}{X-5}\frac{(X+5)}{(X+5)} - \frac{10}{X^2-25} =$$

$$\frac{2X-10+5X+25-10}{X^2-25} = \frac{7X+5}{X^2-25}$$

What if the expression was the same except for the denominator in the second term?

$$\frac{2}{X+5} + \frac{5}{5-X} - \frac{10}{X^2-25}$$

There are three ways of showing that a fraction is negative.

$$-\frac{1}{2} \quad \text{or} \quad \frac{-1}{2} \quad \text{or} \quad \frac{1}{-2}$$

In the example, we can use a double negative since this will still have the same value, just a different form. For example, $+2 = -(-2)$.

$$\frac{5}{5-X} = \left[-\left(-\frac{5}{5-X} \right) \right] \quad \text{and} \quad -\frac{5}{5-X} = \frac{5}{-5+X} = \frac{5}{X-5}$$

$$\text{so} \quad -\left[-\frac{5}{5-X} \right] = -\left(\frac{5}{X-5} \right)$$

To transform this equation, we introduced a double negative, so that we can use the same denominator as in the original problem.

$$\frac{2}{X+5} + \frac{5}{5-X} - \frac{10}{X^2-25} = \frac{2}{X+5} - \frac{5}{X-5} - \frac{10}{X^2-25}$$

Example 12
Combine the rational expressions shown below.

The common denominator is $(X + 2)(X - 2)$, or $X^2 - 4$, and $X \neq 2, -2$.

$$\frac{X-3}{X-2} + \frac{X+3}{X+2} + \frac{4X+3}{X^2-4} \rightarrow \frac{(X-3)(X+2)}{(X-2)(X+2)} + \frac{(X+3)(X-2)}{(X+2)(X-2)} + \frac{4X+3}{X^2-4} \rightarrow$$

$$\frac{X^2-X-6+X^2+X-6+4X+3}{X^2-4} \rightarrow \frac{2X^2+4X-9}{X^2-4}$$

Example 13
Combine the rational expressions shown below.

The common denominator is $(X + 4)(X - 3)$ and $X \neq 3, -4$.

$$\frac{X}{X+4} + \frac{X}{X-3} \rightarrow \frac{(X)(X-3)}{(X+4)(X-3)} + \frac{(X)(X+4)}{(X-3)(X+4)} \rightarrow$$

$$\frac{X^2-3X+X^2+4X}{(X-3)(X+4)} \rightarrow \frac{2X^2+X}{(X-3)(X+4)}$$

Practice Problems 8

1. $\dfrac{4}{X+1} + \dfrac{7}{X}$

2. $\dfrac{X+1}{X+3} + \dfrac{X+1}{X+2} - \dfrac{2X}{X^2+5X+6}$.

3. $\dfrac{X}{X+5} + \dfrac{3X}{X-2}$

Solutions 8

The common denominator is $(X)(X+1)$ and $X \neq 0, -1$.

1. $\dfrac{4}{X+1} + \dfrac{7}{X} \rightarrow \dfrac{(4)(X)}{(X+1)(X)} + \dfrac{(7)(X+1)}{(X)(X+1)} \rightarrow \dfrac{4X+7X+7}{(X)(X+1)} \rightarrow \dfrac{11X+7}{(X)(X+1)}$

The common denominator is $(X+2)(X+3)$, which is $X^2 + 5X + 6$, and $X \neq -2, -3$.

2. $\dfrac{X+1}{X+3} + \dfrac{X+1}{X+2} - \dfrac{2X}{X^2+5X+6} \rightarrow$

$\dfrac{(X+1)(X+2)}{(X+3)(X+2)} + \dfrac{(X+1)(X+3)}{(X+2)(X+3)} - \dfrac{2X}{X^2+5X+6} \rightarrow$

$\dfrac{X^2+3X+2+X^2+4X+3-2X}{X^2+5X+6} \rightarrow \dfrac{2X^2+5X+5}{X^2+5X+6}$

The common denominator is $(X+5)(X-2)$, and $X \neq -5, 2$.

3. $\dfrac{X}{X+5} + \dfrac{3X}{X-2} \rightarrow \dfrac{(X)(X-2)}{(X+5)(X-2)} + \dfrac{(3X)(X+5)}{(X-2)(X+5)} \rightarrow$

$\dfrac{X^2-2X+3X^2+15X}{(X+5)(X-2)} \rightarrow \dfrac{4X^2+13X}{(X+5)(X-2)}$

A Fraction Divided by a Fraction

Another new concept is a fraction or rational expression divided by another fraction or rational expression. We know that a fraction divided by a fraction is the same as a fraction times its reciprocal. We'll show this below. Then we will show another way to simplify this process by multiplying by one.

$$\dfrac{\dfrac{1}{2}}{\dfrac{1}{8}} \text{ is the same as: } \dfrac{1}{2} \div \dfrac{1}{8} = \dfrac{1}{2} \times \dfrac{8}{1} = \dfrac{4}{1} \text{ or } \dfrac{\dfrac{1}{2}}{\dfrac{1}{8}} \times \dfrac{\dfrac{8}{1}}{\dfrac{8}{1}} = \dfrac{4}{1}$$

In the second method, we multiply the denominator 1/8 by its reciprocal 8/1. Then the denominator becomes one. But we can't multiply the denominator by 8/1 without also multiplying the numerator by 8/1, so that we are multiplying the whole fraction by one, changing its form without affecting its value. Let's do some examples. Note that X cannot be any value that would make the denominator zero in any of the steps of the solution.

Example 14
Simplify.

$$\dfrac{\dfrac{2}{X}}{\dfrac{3}{X+1}} \rightarrow \dfrac{\dfrac{2}{X}}{\dfrac{3}{X+1}} \cdot \dfrac{\dfrac{X+1}{3}}{\dfrac{X+1}{3}} = \dfrac{2}{X} \cdot \dfrac{X+1}{3} = \dfrac{2X+2}{3X} \qquad X \neq 0 \text{ or } -1$$

Example 15
Simplify.

$$\dfrac{4 + \dfrac{1}{2}}{2 - \dfrac{2}{3}} \rightarrow \dfrac{\dfrac{9}{2}}{\dfrac{4}{3}} \rightarrow \dfrac{\dfrac{9}{2}}{\dfrac{4}{3}} \cdot \dfrac{\dfrac{3}{4}}{\dfrac{3}{4}} = \dfrac{9}{2} \cdot \dfrac{3}{4} = \dfrac{27}{8}$$

Example 16

Simplify.

$$\dfrac{1+\dfrac{2}{X}}{1+\dfrac{3}{X+1}} \to \dfrac{\dfrac{X}{X}+\dfrac{2}{X}}{\dfrac{X+1}{X+1}+\dfrac{3}{X+1}} \to \dfrac{\dfrac{X+2}{X}}{\dfrac{X+4}{X+1}} \to \dfrac{\dfrac{X+2}{X}}{\dfrac{X+4}{X+1}} \cdot \dfrac{\dfrac{X+1}{X+4}}{\dfrac{X+1}{X+4}} = \dfrac{X^2+3X+2}{X^2+4X}$$

$X \neq -4, 0,$ or -1

Example 17

Simplify.

$$\dfrac{\dfrac{X^2+4X+3}{X^2+6X+8}}{\dfrac{X^2-X-2}{X^2+3X-4}} \to \dfrac{\dfrac{(X+1)(X+3)}{(X+4)(X+2)}}{\dfrac{(X+1)(X-2)}{(X+4)(X-1)}} \to \dfrac{(X+1)(X+3)}{(X+4)(X+2)} \cdot \dfrac{(X+4)(X-1)}{(X+1)(X-2)} \to$$

$$\dfrac{\cancel{(X+1)}(X+3)\cdot\cancel{(X+4)}(X-1)}{\cancel{(X+4)}(X+2)\cdot\cancel{(X+1)}(X-2)} \to \dfrac{X^2+2X-3}{X^2-4} \qquad X \neq -4, -2, -1, 1, 2$$

Practice Problems 9

1. $\dfrac{\dfrac{4}{X}}{\dfrac{X+1}{2X}}$

2. $\dfrac{2+\dfrac{1}{4}}{5-\dfrac{5}{8}}$

3. $\dfrac{1-\dfrac{5}{A}}{1+\dfrac{3}{A+2}}$

4. $\dfrac{\dfrac{X^2-4}{X^2+7X+12}}{\dfrac{X^2+3X-10}{X^2+6X+9}}$

Solutions 9

1. $\dfrac{\frac{4}{X}}{\frac{X+1}{2X}} \rightarrow \dfrac{\frac{4}{X}}{\frac{X+1}{2X}} \cdot \dfrac{\frac{2X}{X+1}}{\frac{2X}{X+1}} = \dfrac{4}{X} \cdot \dfrac{2X}{X+1} = \dfrac{8}{X+1}$ $X \neq -1 \text{ or } 0$

2. $\dfrac{2+\frac{1}{4}}{5-\frac{5}{8}} \rightarrow \dfrac{\frac{9}{4}}{\frac{35}{8}} \rightarrow \dfrac{\frac{9}{4}}{\frac{35}{8}} \cdot \dfrac{\frac{8}{35}}{\frac{8}{35}} = \dfrac{9}{4} \cdot \dfrac{\overset{2}{8}}{35} = \dfrac{18}{35}$

3. $\dfrac{1-\frac{5}{A}}{1+\frac{3}{A+2}} \rightarrow \dfrac{\frac{A}{A}-\frac{5}{A}}{\frac{A+2}{A+2}+\frac{3}{A+2}} \rightarrow \dfrac{\frac{A-5}{A}}{\frac{A+5}{A+2}} \cdot \dfrac{\frac{A+2}{A+5}}{\frac{A+2}{A+5}} =$

$\dfrac{A^2-3A-10}{A^2+5A}$ $A \neq 0, -2, \text{ or } -5$

4. $\dfrac{\frac{X^2-4}{X^2+7X+12}}{\frac{X^2+3X-10}{X^2+6X+9}} \rightarrow \dfrac{\frac{(X+2)(X-2)}{(X+4)(X+3)}}{\frac{(X+5)(X-2)}{(X+3)(X+3)}} \rightarrow$

$\dfrac{(X+2)(X-2)}{(X+4)(X+3)} \cdot \dfrac{(X+3)(X+3)}{(X+5)(X-2)} \rightarrow$
$\dfrac{(X+5)(X-2)}{(X+3)(X+3)}$

$\dfrac{(X+2)\cancel{(X-2)} \cdot (X+3)\cancel{(X+3)}}{(X+4)\cancel{(X+3)} \cdot (X+5)\cancel{(X-2)}} \rightarrow \dfrac{(X+2)(X+3)}{(X+4)(X+5)} \rightarrow \dfrac{X^2+5X+6}{X^2+9X+20}$

$X \neq 2, -3, -4, \text{ or } -5$

LESSON 6

Fractional Exponents

As we learned in *Algebra 1*, square roots and cube roots can also be written as exponents. Instead of $\sqrt{5}$, we can write $5^{1/2}$. The cube root of 7 can be written as $\sqrt[3]{7}$ or $7^{1/3}$.

Two or more operations can be expressed with one ***fractional exponent***. The square root of 9 raised to the third power is written as: $\left(\sqrt{9}\right)^3 = \left(9^{1/2}\right)^3 = 9^{3/2} = 27$. The square root of 9 is 3, and 3 to the third power is 27.

Example 1

$$\left(\sqrt{4}\right)^5 = \left(4^{1/2}\right)^5 = 4^{5/2} = 32 \text{ or } \left(\sqrt{4}\right)^5 = (2)^5 = 32$$

The square root of 4 is 2, and 2 to the fifth power is 32.

Example 2

$$\left(\sqrt[3]{27}\right)^2 = \left(27^{1/3}\right)^2 = 27^{2/3} = 9 \text{ or } \left(\sqrt[3]{27}\right)^2 = (3)^2 = 9$$

The cube root of 27 is 3, and 3 to the second power is 9.

The denominator of the fractional exponent tells what root, and the numerator tells to what power the expression is raised.

Example 3

$$\left(\sqrt{\sqrt{5}}\right) = \left(5^{1/2}\right)^{1/2} = 5^{1/4}$$

Example 4

$$\left(\sqrt{\sqrt{81}}\right) = \left(81^{1/2}\right)^{1/2} = 81^{1/4} = 3 \text{ or } \left(\sqrt{\sqrt{81}}\right) = \sqrt{9} = 3$$

The square root of 81 is 9, and the square root of 9 is 3. So the square root of the square root of 81 is 3.

As you can see, there may be more than one way to approach these problems. Don't be afraid to experiment in order to find the most efficient way to simplify expressions with radicals or fractional exponents.

Practice Problems 1

1. $\left(8^{1/3}\right)^2 =$

2. $\left(81^{1/4}\right)^3 =$

3. $\left(32^{1/5}\right)^3 =$

4. $\left(3^4\right)^{1/2} =$

Rewrite using fractional exponents, and then solve.

5. $\left(\sqrt[4]{16}\right)^3 =$

6. $\left(\sqrt{3^4}\right) =$

7. $\left(\sqrt{4}\right)^3 =$

8. $\left(\sqrt{25}\right)^3 =$

9. $\left(\sqrt{\sqrt[3]{X}}\right) =$

10. $\left(\sqrt{\sqrt{Y}}\right) =$

11. $\left(\sqrt[3]{Q^4}\right) =$

12. $\left(\sqrt[3]{A^2}\right) =$

Solutions 1

1. $\left(8^{1/3}\right)^2 = 2^2 = 4$

2. $\left(81^{1/4}\right)^3 = (3)^3 = 27$

3. $\left(32^{1/5}\right)^3 = 32^{3/5} = 8$
$\qquad = (2)^3 = 8$

4. $\left(3^4\right)^{1/2} = 3^{4/2} = 3^2 = 9$
$\qquad = 81^{1/2} = 9$

5. $\left(\sqrt[4]{16}\right)^3 = \left(16^{1/4}\right)^3 = 2^3 = 8$

6. $\left(\sqrt{3^4}\right) = \left(3^4\right)^{1/2} = 3^{4/2} = 3^2 = 9$

7. $\left(\sqrt{4}\right)^3 = \left(4^{1/2}\right)^3 = 2^3 = 8$

8. $\left(\sqrt{25}\right)^3 = \left(25^{1/2}\right)^3 = 5^3 = 125$

9. $\left(\sqrt{\sqrt[3]{X}}\right) = \left(X^{1/3}\right)^{1/2} = X^{1/6}$

10. $\left(\sqrt{\sqrt{Y}}\right) = \left(Y^{1/2}\right)^{1/2} = Y^{1/4}$

11. $\left(\sqrt[3]{Q^4}\right) = \left(Q^4\right)^{1/3} = Q^{4/3}$

12. $\left(\sqrt[3]{A^2}\right) = \left(A^2\right)^{1/3} = A^{2/3}$

LESSON 7

Imaginary and Complex Numbers

So far in algebra, we have encountered positive numbers and negative numbers. Both of these are real. There is another option that arises when working with squares and square roots. We know that $(+3) \times (+3) = +9$ and $(+3)^2 = +9$. Also $(-3) \times (-3) = +9$ and $(-3)^2 = +9$. The converse of squaring a number is finding the square root of a number. $\sqrt{9}$ is either $+3$ or -3, and we can write this as ± 3, read "plus or minus 3." So whether we are squaring a positive number or a negative number, the answer is a positive number. The expressions $(-3)^2$ and $(+3)^2$ both equal 9.

What happens when we encounter $\sqrt{-9}$? I like to separate this using what we know about multiplying radicals: $\sqrt{-9} = \sqrt{9}\sqrt{-1}$. Now we have $\sqrt{9}$, which we can solve. The new concept involves $\sqrt{-1}$. Since there is no real number that when squared equals negative one, we call this an ***imaginary number*** or imaginary unit and refer to it as *i*. By definition, i^2 equals -1. (In electrical engineering, capital *I* represents current, so you may see *j* used instead of *i* to represent imaginary numbers.)

What is interesting is that even though *i* is imaginary, i^2 is not. Remember that $\sqrt{4}\sqrt{4} = \sqrt{16} = 4$ so $\sqrt{4}\sqrt{4} = 4$ and $\sqrt{-7}\sqrt{-7} = -7$ and $\sqrt{X}\sqrt{X} = X$, so $\sqrt{-1}\sqrt{-1} = -1$.

Warning: Simply multiplying the square roots of negative numbers can lead to incorrect results in some cases. For that reason, we always factor out the square root of -1 first, and change it to *i*. Study example 8 on the next page.

A ***complex number*** is a combination of a real number and an imaginary number. It is similar to a mixed number, which is a number and a fraction. Some examples of complex numbers are 4 + 3i and 27 − 9i. Imaginary numbers can be used in all the basic operations. Treat *i* as you would a radical or a variable.

Example 1

$3i + 5i = 8i$ or $= 3\sqrt{-1} + 5\sqrt{-1} = 8\sqrt{-1} = 8i$

Example 2

$7i - 4i = 3i$

Example 3

$7i + 5i = 12i$, but $7i + 5 = 7i + 5$

Remember that you can only combine or compare two numbers that are the same kind or value. You cannot combine 7i and 5 because they are not the same kind. The expression 7i is an imaginary number, while 5 is a real number.

Example 4

$(4i)(3i) = 12i^2 = 12(-1) = -12$ Remember $i^2 = -1$.

Another way to write this is:

$(4i)(3i) = (4\sqrt{-1})(3\sqrt{-1}) = 12(\sqrt{-1})^2 = 12(-1) = -12$

Example 5

$\sqrt{-121} = \sqrt{121}\sqrt{-1} = 11i$

Example 6

$\sqrt{-64} + \sqrt{-49} = \sqrt{64}\sqrt{-1} + \sqrt{49}\sqrt{-1} = 8i + 7i = 15i$

Example 7

$i \cdot i \cdot i \cdot i = i^2 \cdot i^2 = (-1)(-1) = 1$

Example 8

$(2\sqrt{-3})(8\sqrt{-3}) = (2\sqrt{3}\sqrt{-1})(8\sqrt{3}\sqrt{-1}) =$
$(2)(8)(\sqrt{3})(\sqrt{3})(\sqrt{-1})(\sqrt{-1}) = 16(3)(i^2) = 48i^2 = -48$

Practice Problems 1

Simplify each in terms of *i*.

1. $\sqrt{-36}$

2. $\sqrt{-169}$

3. $\sqrt{-225}$

4. $\sqrt{-9/49}$

5. $\sqrt{-81X^2Y^2}$

6. $\sqrt{-27X^5}$

Simplify each, and then combine like terms.

7. $\sqrt{-16} + \sqrt{-25} =$

8. $\sqrt{-4} + \sqrt{-144} =$

9. $\sqrt{-169} - 2\sqrt{-25} =$

10. $3\sqrt{-27} - 4\sqrt{-8} =$

11. $\sqrt{-100} - 3\sqrt{-16} =$

12. $4\sqrt{196} + 3\sqrt{289} =$

13. $i \cdot i \cdot i \cdot i \cdot i \cdot i \cdot i \cdot i =$

14. $\sqrt{-7} \cdot \sqrt{-14} =$

15. $\left(i^2\right)^3 =$

16. $i^3 =$

17. $(9i)\sqrt{-64} =$

18. $\sqrt{-11}\sqrt{-11} =$

19. $(-8i)(5i) =$

20. $\left(7\sqrt{-5}\right)\left(4\sqrt{-5}\right) =$

Solutions 1

1. $\sqrt{-36} = \sqrt{36}\sqrt{-1} = 6i$

2. $\sqrt{-169} = \sqrt{169}\sqrt{-1} = 13i$

3. $\sqrt{-225} = \sqrt{225}\sqrt{-1} = 15i$

4. $\sqrt{-9/49} = \sqrt{9/49}\sqrt{-1} = \frac{3}{7}i$

5. $\sqrt{-81X^2Y^2} =$
   $\sqrt{81}\sqrt{-1}\sqrt{X^2}\sqrt{Y^2} = 9XYi$

6. $\sqrt{-27X^5} = \sqrt{27}\sqrt{X^4}\sqrt{X}\sqrt{-1} = 3X^2i\sqrt{3X}$
   (You will often see this written as $3X^2\sqrt{3X}$ i.
   Either order has the same value.)

7. $\sqrt{-16} + \sqrt{-25} = 4i + 5i = 9i$

8. $\sqrt{-4} + \sqrt{-144} = 2i + 12i = 14i$

9. $\sqrt{-169} - 2\sqrt{-25} =$
   $13i - 2(5i) = 3i$

10. $3\sqrt{-27} - 4\sqrt{-8} =$
    $3(3\sqrt{3})i - 4(2\sqrt{2})i = 9i\sqrt{3} - 8i\sqrt{2}$

11. $\sqrt{-100} - 3\sqrt{-16} =$
    $10i - 3(4i) = 10i - 12i = -2i$

12. $4\sqrt{196} + 3\sqrt{289} =$
    $4 \times 14 + 3 \times 17 = 107$

13. $i \cdot i \cdot i \cdot i \cdot i \cdot i \cdot i \cdot i = i^2 \cdot i^2 \cdot i^2 \cdot i^2 =$
    $(-1)(-1)(-1)(-1) = 1$

14. $\sqrt{-7} \cdot \sqrt{-14} = i\sqrt{7} \cdot i\sqrt{14} =$
    $i^2\sqrt{98} = -7\sqrt{2}$

15. $\left(i^2\right)^3 = (-1)^3 = -1$

16. $i^3 = i^2 \cdot i = (-1) \cdot i = -i$

17. $(9i)\sqrt{-64} = 9i \cdot 8i = 72i^2 = -72$

18. $\sqrt{-11}\sqrt{-11} = i\sqrt{11}\, i\sqrt{11} = -11$

19. $(-8i)(5i) = -40\left(i^2\right) =$
    $-40(-1) = 40$

20. $\left(7\sqrt{-5}\right)\left(4\sqrt{-5}\right) = \left(7i\sqrt{5}\right)\left(4i\sqrt{5}\right) =$
    $28\left(i^2\right)(5) = -140$

Conjugate Numbers

When given one term squared minus another term squared, the factors are the first term plus the second term, times the first term minus the second term. This sounds confusing, but it is the easiest and most concise method of factoring. In the example shown below, the first term is X and the second term is 3.

$$X^2 - 9 \text{ or } (X)^2 - (3)^2 = (X + 3)(X - 3)$$

We can check this by multiplying $(X + 3)(X - 3)$ to find the product.

$$
\begin{array}{r}
X + 3 \\
X - 3 \\
\hline
-3X - 9 \\
X^2 + 3X \\
\hline
X^2 - 9
\end{array}
$$

This particular operation, which we recognize from factoring in lesson 5, is referred to as the ***difference of two squares***.

$$Y^2 - 25 \text{ or } (Y)^2 - (5)^2 = (Y + 5)(Y - 5)$$

The difference of two squares is very useful in eliminating radicals and complex numbers from their position in the denominator of a fraction or a rational expression. In this scenario, we will be looking for a factor which, when multiplied by the existing factor, gives us the difference of two squares. This missing factor, which produces a difference of two squares, is called a *conjugate*. Some examples are in order to make this clear.

Example 1

Given: X + 7. What factor can we multiply times X + 7 that will produce the difference of two squares?

Given X + 7, the conjugate is X – 7 and X + 7 times X – 7 equals $X^2 - (7)^2$ or $X^2 - 49$.

Example 2

Given: 2X – 5. What factor can we multiply times 2X – 5 that will produce the difference of two squares?

Given 2X – 5, the conjugate is 2X + 5 and 2X + 5 times 2X - 5 equals $(2X)^2 - (5)^2$ or $4X^2 - 25$.

You will see how helpful this is in making sure there are no imaginary numbers or radicals in the denominator of a rational expression.

Example 3

Find the conjugate of $\left(4 - \sqrt{3}\right)$.

The conjugate is $\left(4 + \sqrt{3}\right)$ and $\left(4 - \sqrt{3}\right)\left(4 + \sqrt{3}\right) = (4)^2 - \left(\sqrt{3}\right)^2 = 16 - 3 = 13$.

Example 4

Find the conjugate of (9 + 2i).

The conjugate is (9 – 2i) and $(9 + 2i)(9 - 2i) = 9^2 - 4i^2 = 81 - (-4) = 81 + 4 = 85$.

Practice Problems 1

1. Find the conjugate of (X + A).

2. Find the conjugate of (Y – B).

3. Find the conjugate of (3 – 2X).

4. Find the conjugate of $\left(11 + \sqrt{5}\right)$.

5. Find the conjugate of (3X + 5).

6. Find the conjugate of (4 – 3i).

Solutions 1

1. The conjugate is (X – A).   2. The conjugate is (Y + B).

3. The conjugate is (3 + 2X).   4. The conjugate is $\left(11 - \sqrt{5}\right)$.

5. The conjugate is (3X – 5).   6. The conjugate is (4 + 3i).

Example 5
Simplify the expression so there are no imaginary numbers in the denominator. The key is finding the conjugate of 6 + i.

$$\frac{5}{6+i} \times \frac{(6-i)}{(6-i)} = \frac{5(6-i)}{(6+i)(6-i)} = \frac{30-5i}{6^2-i^2} = \frac{30-5i}{36-(-1)} = \frac{30-5i}{37}$$

Example 6
Simplify the expression so there are no radicals in the denominator. The key is finding the conjugate of $3 + \sqrt{2}$.

$$\frac{4}{3+\sqrt{2}} \times \frac{3-\sqrt{2}}{3-\sqrt{2}} = \frac{4\left(3-\sqrt{2}\right)}{3^2-\left(\sqrt{2}\right)^2} = \frac{12-4\sqrt{2}}{9-2} = \frac{12-4\sqrt{2}}{7}$$

Multiplying by the conjugate yields "squares," and we know that a radical or an imaginary number squared is a whole number. When there is no radical or imaginary number in the denominator, we say that the expression is in *standard form.*

Practice Problems 2
Use the conjugate to simplify each expression (put it in standard form).

1. $\dfrac{X}{9+2i} =$

2. $\dfrac{-A}{4+\sqrt{10}} =$

3. $\dfrac{12}{7-3i} =$

4. $\dfrac{4Q}{5-\sqrt{11}} =$

5. $\dfrac{2i}{8+5i} =$

6. $\dfrac{18}{13-4\sqrt{5}} =$

7. $\dfrac{Y}{1-6i} =$

8. $\dfrac{-9}{15+2\sqrt{3}} =$

Solutions 2

1. $\dfrac{X}{9+2i} \times \dfrac{(9-2i)}{(9-2i)} = \dfrac{X(9-2i)}{(9+2i)(9-2i)} = \dfrac{9X-2Xi}{81-(-4)} = \dfrac{9X-2Xi}{85}$

2. $\dfrac{-A}{4+\sqrt{10}} \times \dfrac{4-\sqrt{10}}{4-\sqrt{10}} = \dfrac{-A(4-\sqrt{10})}{4^2-(\sqrt{10})^2} = \dfrac{-4A+A\sqrt{10}}{16-10} = \dfrac{-4A+A\sqrt{10}}{6}$

3. $\dfrac{12}{7-3i} \times \dfrac{(7+3i)}{(7+3i)} = \dfrac{12(7+3i)}{(7-3i)(7+3i)} = \dfrac{84+36i}{49-(-9)} = \dfrac{84+36i}{58} = \dfrac{42+18i}{29}$

4. $\dfrac{4Q}{5-\sqrt{11}} \times \dfrac{5+\sqrt{11}}{5+\sqrt{11}} = \dfrac{4Q(5+\sqrt{11})}{5^2-(\sqrt{11})^2} = \dfrac{20Q+4Q\sqrt{11}}{25-11} =$

   $\dfrac{20Q+4Q\sqrt{11}}{14} = \dfrac{10Q+2Q\sqrt{11}}{7}$

5. $\dfrac{2i}{8+5i} \times \dfrac{(8-5i)}{(8-5i)} = \dfrac{2i(8-5i)}{(8+5i)(8-5i)} = \dfrac{16i-10i^2}{64-(-25)} = \dfrac{16i+10}{89}$

6. $\dfrac{18}{13-4\sqrt{5}} \times \dfrac{13+4\sqrt{5}}{13+4\sqrt{5}} = \dfrac{18(13+4\sqrt{5})}{13^2-(4\sqrt{5})^2} = \dfrac{234+72\sqrt{5}}{169-80} = \dfrac{234+72\sqrt{5}}{89}$

7. $\dfrac{Y}{1-6i} \times \dfrac{(1+6i)}{(1+6i)} = \dfrac{Y(1+6i)}{(1-6i)(1+6i)} = \dfrac{Y+6Yi}{1^2-(6i)^2} = \dfrac{Y+6Yi}{1-(-36)} = \dfrac{Y+6Yi}{37}$

8. $\dfrac{-9}{15+2\sqrt{3}} \times \dfrac{15-2\sqrt{3}}{15-2\sqrt{3}} = \dfrac{-9(15-2\sqrt{3})}{15^2-(2\sqrt{3})^2} =$

   $\dfrac{-135+18\sqrt{3}}{225-12} = \dfrac{-135+18\sqrt{3}}{213}$

Squares, Cubes, and Pascal's Triangle

Before we start cubing binomials, we'll look for a pattern as we square binomials. This will help us square and factor them more quickly. By now you should be getting the modus operandi under your belt. We'll do several examples of real problems to reveal the pattern, and then we'll move to an algebraic equation to formalize our observations and give us a formula. When multiplying these binomials, there are always two options. Either distribute them using the FOIL method, or multiply them vertically as you have multiplied numbers for much of your life. I'll do most examples horizontally but a few vertically. Choose whichever method you are comfortable with.

Example 1
$$(X + 4)^2 = (X + 4)(X + 4) = X^2 + 4X + 4X + 16 = X^2 + 8X + 16$$

Example 2
$$(A + 5)^2 = (A + 5)(A + 5) = A^2 + 5A + 5A + 25 = A^2 + 10A + 25$$

Example 3
$$(Y + 7)^2 = (Y + 7)(Y + 7) = Y^2 + 7Y + 7Y + 49 = Y^2 + 14Y + 49$$

In examples 4 and 5, the signs are different.

Example 4
$$(X - 3)^2 = (X - 3)(X - 3) = X^2 - 3X - 3X + 9 = X^2 - 6X + 9$$

Example 5
$$(B - 6)^2 = (B - 6)(B - 6) = B^2 - 6B - 6B + 36 = B^2 - 12B + 36$$

Did you notice the pattern? See whether you can predict what $(A + B)^2$ and $(A - B)^2$ will be, and then compare your answers with the solutions below.

Formula 1
$$(A + B)^2 = (A + B)(A + B) = A^2 + AB + AB + B^2 = A^2 + 2AB + B^2$$

The first term (A) squared **plus** 2 times the middle term (AB) plus the last term (B) squared.

Formula 2
$$(A - B)^2 = (A - B)(A - B) = A^2 - AB - AB + B^2 = A^2 - 2AB + B^2$$

The first term (A) squared **minus** 2 times the middle term (AB) plus the last term (B) squared.

Let's use the new formulas to solve the next two examples. Try $(X + 3)^2$ and $(Y - 9)^2$ first, and then compare your work.

Example 6
$$(X + 3)^2 = (X)^2 + 2(3X) + (3)^2 = X^2 + 6X + 9$$

Example 7
$$(Y - 9)^2 = (Y)^2 - 2(9Y) + (9)^2 = Y^2 - 18Y + 81$$

The converse is also true. If we see a product with the first and last terms squared and the middle term twice the product of the first and last terms, then we know it is a binomial squared, and we can deduce the factors.

Example 8
Find the binomial root of the following trinomial: $X^2 + 14X + 49$.

The square root of X^2 is X, and the square root of 49 is 7.
Now if the middle term is twice the product of 7X, or 14X, then we are in business. The square root of $X^2 + 14X + 49$ is (X+7).

Example 9

Find the binomial root of the following trinomial: $X^2 - 16X + 64$.

The square root of X^2 is X, and the square root of 64 is 8.
Now if the middle term is twice the product of 8 times X, or 16X,
then we are in business. Because it is a negative middle term, then the
binomial must have a negative sign. The square root is (X – 8).

Practice Problems 1

Find the product of each binomial squared.

1. $(A + 7)^2$ 2. $(X - 1)^2$

3. $(B + 2)^2$ 4. $(X - 5)^2$

5. $(2A - 1)^2$ 6. $(2B - 3)^2$

7. $(5Y - 2)^2$ 8. $(3X + 4)^2$

Find the square root of each trinomial.

9. $X^2 + 10X + 25$ 10. $4A^2 + 12A + 9$

11. $B^2 + 12B + 36$ 12. $Y^2 - 12Y + 36$

Solutions 1

1. $A^2 + 14A + 49$

2. $X^2 - 2X + 1$

3. $B^2 + 4B + 4$

4. $X^2 - 10X + 25$

5. $4A^2 - 4A + 1$

6. $4B^2 - 12B + 9$

7. $25Y^2 - 20Y + 4$

8. $9X^2 + 24X + 16$

9. $(X + 5)^2$

10. $(2A + 3)^2$

11. $(B + 6)^2$

12. $(Y - 6)^2$

Cubes

Now that you are experts on raising a binomial to a power of two, what about a power of three? Instead of multiplying $(A + B)$ times itself three times, which can be intimidating, let's multiply $(A + B)$ times $(A + B)^2$, which gives us $(A + B)^3$. When a number is raised to the third power, we often refer to it as "cubed." When we raise it to the second power, we say "squared." So $(A + B)$ times $(A + B)$ squared is $(A + B)$ **cubed**. We'll do it vertically and horizontally.

Formula 3

$$(A + B)^3 = (A + B)^1 (A + B)^2 = (A + B)^1 \left(A^2 + 2AB + B^2 \right) =$$
$$A^3 + 2A^2B + AB^2 + A^2B + 2AB^2 + B^3 = A^3 + 3A^2B + 3AB^2 + B^3$$

$$
\begin{array}{r}
A^2 + 2AB + B^2 \\
\times \quad A + B \\
\hline
BA^2 + 2AB^2 + B^3 \\
A^3 + 2BA^2 + AB^2 \quad\quad \\
\hline
A^3 + 3BA^2 + 3AB^2 + B^3
\end{array}
$$

Now let's find the formula for $(A - B)^3$.

Formula 4

$(A - B)^3 = (A - B)^1 (A - B)^2 = (A - B)^1 (A^2 - 2AB + B^2) =$
$A^3 - 2A^2B + AB^2 - A^2B + 2AB^2 - B^3 = A^3 - 3A^2B + 3AB^2 - B^3$

$$
\begin{array}{r}
A^2 - 2AB + B^2 \\
\times \quad A - B \\
\hline
-BA^2 + 2AB^2 - B^3 \\
A^3 - 2BA^2 + AB^2 \\
\hline
A^3 - 3BA^2 + 3AB^2 - B^3
\end{array}
$$

Example 10

Solve $(X + 5)^3$ by multiplying $(X + 5)(X + 5)^2$ and then by using the formula.

$(X + 5)^3 = (X + 5)(X + 5)^2 = (X + 5)(X^2 + 10X + 25) =$
$X^3 + 10X^2 + 25X + 5X^2 + 50X + 125 = X^3 + 15X^2 + 75X + 125$

$(X + 5)^3 = (X)^3 + 3(X)^2(5) + 3(X)(5)^2 + (5)^3 = X^3 + 15X^2 + 75X + 125$

Practice Problems 2

1. $(A + 2)^3 =$

2. $(X + 10)^3 =$

3. $(2A + 4)^3 =$

4. $(2X + 2Y)^3 =$

5. $(B - 2)^3 =$

6. $(Y - 10)^3 =$

7. $(3B - 1)^3$

8. $(R - 1/2)^3 =$

Solutions 2

1. $(A + 2)^3 = (A)^3 + 3(A)^2(2) + 3(A)(2)^2 + (2)^3 = A^3 + 6A^2 + 12A + 8$

2. $(X + 10)^3 = (X)^3 + 3(X)^2(10) + 3(X)(10)^2 + (10)^3 =$
   $X^3 + 30X^2 + 300X + 1,000$

3. $(2A + 4)^3 = (2A)^3 + 3(2A)^2(4) + 3(2A)(4)^2 + (4)^3 =$
   $8A^3 + 48A^2 + 96A + 64$

4. $(2X + 2Y)^3 = (2X)^3 + 3(2X)^2(2Y) + 3(2X)(2Y)^2 + (2Y)^3 =$
   $8X^3 + 24X^2Y + 24XY^2 + 8Y^3$

5. $(B - 2)^3 = (B)^3 - 3(B)^2(2) + 3(B)(2)^2 - (2)^3 = B^3 - 6B^2 + 12B - 8$

6. $(Y - 10)^3 = (Y)^3 - 3(Y)^2(10) + 3(Y)(10)^2 - (10)^3 =$
   $Y^3 - 30Y^2 + 300Y - 1,000$

7. $(3B - 1)^3 = (3B)^3 - 3(3B)^2(1) + 3(3B)(1)^2 - (1)^3 =$
   $27B^3 - 27B^2 + 9B - 1$

8. $(R - 1/2)^3 = (R)^3 - 3(R)^2(1/2) + 3(R)(1/2)^2 - (1/2)^3 =$
   $R^3 - 3/2\,R^2 + 3/4\,R - 1/8$

Pascal's Triangle

We know how to raise a binomial to powers of zero, one, two, and three. We can keep going and, with a good bit of ink, figure out how to raise it to the fourth power, and on up. But a pattern has been emerging in the coefficients, as well as in the variables. Next we'll raise $(A + B)$ to the power of four, and then review and look for the pattern. A young man named Blaise Pascal is credited with discovering this pattern. Let's follow in his footsteps.

Example 11

$$A^3 + 3A^2B + 3AB^2 + B^3$$
$$\times \quad A + B$$
$$\overline{A^3B + 3A^2B^2 + 3AB^3 + B^4}$$
$$A^4 + 3A^3B + 3A^2B^2 + AB^3$$
$$\overline{A^4 + 4A^3B + 6A^2B^2 + 4AB^3 + B^4}$$

Notice the coefficients in the following sequence.

With coefficients, variables and exponents:

$(A + B)^0$	1
$(A + B)^1$	$1A^1 + 1B^1$
$(A + B)^2$	$1A^2 + 2A^1B^1 + 1B^2$
$(A + B)^3$	$1A^3 + 3A^2B^1 + 3A^1B^2 + 1B^3$
$(A + B)^4$	$1A^4 + 4A^3B^1 + 6A^2B^2 + 4A^1B^3 + 1B^4$

With coefficients and variables:

$(A + B)^0$	1
$(A + B)^1$	$1A + 1B$
$(A + B)^2$	$1A + 2AB + 1B$
$(A + B)^3$	$1A + 3AB + 3AB + 1B$
$(A + B)^4$	$1A + 4AB + 6AB + 4AB + 1B$

With just the coefficients:

$(A + B)^0$	1
$(A + B)^1$	1 1
$(A + B)^2$	1 2 1
$(A + B)^3$	1 3 3 1
$(A + B)^4$	1 4 6 4 1

While studying the triangle with just the coefficients, see whether you notice the pattern, and then try and predict what the coefficients would be for $(A + B)$ to the fifth power. After you've done this, read the explanation on the next page.

With coefficients:

```
                    1
              1           1        Add 1 + 1 in line two, to get 2 in line three.

          1       2       1        Add 1 + 2 in line three, to get 3 in line four.

      1       3       3       1    Add 1 + 3 in line four, to get 4 in line five,
                                   and add 3 + 3 to get 6 in line five.
    1       4       6       4       1
  1     5       10      10      5       1    This is what the next line should be.
```

Practice Problems 3

Find the coefficients for the next three rows representing the sixth, seventh, and eighth powers.

Solutions 3

```
    1   6   15  20  15   6   1
  1   7   21  35  35  21   7   1
1   8  28  56  70  56  28   8   1
```

Binomial Theorem

So far we've seen the pattern for the coefficients in the triangle. Now take a look at the variables and the exponents. The first study will be the variables. In $(A + B)^2$, the first term in the product has an A but no B. The last term has a B but no A. The middle term has one of each letter. It will be easier to observe the pattern if we put one of each variable in all the terms. We can do this by adding a zero exponent in the first and last terms. Notice also that the number of terms is one more than the exponent of the original problem.

$$1A^2 + 2A^1B^1 + 1B^2 = 1A^2B^0 + 2A^1B^1 + 1A^0B^2$$

Notice how the exponents of A begin with the same exponent to which we are raising the binomial, which is 2. Then the exponents decrease by one each time. For the Bs, the exponents begin at 0 and increase by one until they get to 2. If you add the exponents in each term, $2 + 0$, $1 + 1$, and $0 + 2$, they all add up to 2. With this knowledge and Pascal's triangle, we can predict the product of a binomial raised to a power.

```
              1
            1   1
          1   2   1
        1   3   3   1
      1   4   6   4   1
    1   5  10  10   5   1
  1   6  15  20  15   6   1
1   7  21  35  35  21   7   1
```

Example 1

Expand $(A + B)^5$.

From the triangle, we get the coefficients: 1, 5, 10, 10, 5, 1. Since we are raising to the fifth power we start with $A^5 B^0$ and go from there.

$$1A^5B^0 + 5A^4B^1 + 10A^3B^2 + 10A^2B^3 + 5A^1B^4 + 1A^0B^5$$

Example 2

Expand $(A + B)^7$.

From the triangle, we get the coefficients: 1, 7, 21, 35, 35, 21, 7, 1. Since we are raising to the seventh power we start with $A^7 B^0$ and go from there.

$$1A^7B^0 + 7A^6B^1 + 21A^5B^2 + 35A^4B^3 + 35A^3B^4 + 21A^2B^5 + 7A^1B^6 + 1A^0B^7$$

Practice Problems 1

Tell how many terms there will be and expand.

1. $(A + B)^6$

2. $(A + B)^4$

3. $(X + 2)^5$

1. 7 terms (6 + 1)
   $1A^6B^0 + 6A^5B^1 + 15A^4B^2 + 20A^3B^3 + 15A^2B^4 + 6A^1B^5 + 1A^0B^6$

2. 5 terms (4 + 1)
   $1A^4B^0 + 4A^3B^1 + 6A^2B^2 + 4A^1B^3 + 1A^0B^4$

3. 6 terms (5 + 1)

   $1X^5 2^0 + 5X^4 2^1 + 10X^3 2^2 + 10X^2 2^3 + 5X^1 2^4 + 1X^0 2^5 =$
   $X^5 + 10X^4 + 40X^3 + 80X^2 + 80X + 32$

The *binomial theorem* is a way of predicting what the product of a binomial raised to any power will be without using Pascal's triangle. The key is expressing the formula in algebraic terms. The triangle can be expressed using factors and fractions. Here are four rows of the triangle. Predict the fifth and sixth rows before checking below.

$$1$$
$$1 \qquad \frac{1}{1}$$
$$1 \qquad \frac{2}{1} \qquad \frac{2 \cdot 1}{1 \cdot 2}$$
$$1 \qquad \frac{3}{1} \qquad \frac{3 \cdot 2}{1 \cdot 2} \qquad \frac{3 \cdot 2 \cdot 1}{1 \cdot 2 \cdot 3}$$

Notice how the fractions reduce to come up with the same numbers that we had in the triangle.

$$1 \qquad \frac{4}{1} \qquad \frac{4 \cdot 3}{1 \cdot 2} \qquad \frac{4 \cdot 3 \cdot 2}{1 \cdot 2 \cdot 3} \qquad \frac{4 \cdot 3 \cdot 2 \cdot 1}{1 \cdot 2 \cdot 3 \cdot 4}$$
$$1 \qquad \frac{5}{1} \qquad \frac{5 \cdot 4}{1 \cdot 2} \qquad \frac{5 \cdot 4 \cdot 3}{1 \cdot 2 \cdot 3} \qquad \frac{5 \cdot 4 \cdot 3 \cdot 2}{1 \cdot 2 \cdot 3 \cdot 4} \qquad \frac{5 \cdot 4 \cdot 3 \cdot 2 \cdot 1}{1 \cdot 2 \cdot 3 \cdot 4 \cdot 5}$$

Here is the algebra to predict the variables. It looks intimidating, but that is because of the increasing and decreasing of the exponents. In the theorem, N represents any positive integer or positive whole number.

The binomial theorem is:

$$(A+B)^N = \underbrace{A^N B^0}_{\substack{\text{First} \\ \text{Term}}} + \underbrace{\frac{N}{1} A^{N-1} B^1}_{\substack{\text{Second} \\ \text{Term}}} + \underbrace{\frac{N(N-1)}{1 \cdot 2} A^{N-2} B^2}_{\substack{\text{Third} \\ \text{Term}}} + \underbrace{\frac{N(N-1)(N-2)}{1 \cdot 2 \cdot 3} A^{N-3} B^3}_{\substack{\text{Fourth} \\ \text{Term}}} \dots \underbrace{A^0 B^N}_{\substack{\text{Last} \\ \text{Term}}}$$

This is useful in finding what a specific term will be without expanding the whole product. Look at the third term. There are two factors in the coefficient. In the fourth term there are three. So there is always one less factor in the coefficient than the number of the term. In the seventh term there would be six factors. You will notice that the exponent for B is 2, which is also one less than the number of the term. The exponent for A, when added to the exponent for B, will always add up to N. Let's do some examples.

Example 3
Find the fourth term of $(X + Y)^6$.
The coefficient will have three factors or (4 − 1). $\frac{6 \cdot 5 \cdot 4}{1 \cdot 2 \cdot 3} = 20$

The exponent of Y will be 3 or (4 − 1),
and the exponent for X will also be 3, since 3 + 3 = 6. $X^3 Y^3$

Putting it all together: $20 X^3 Y^3$

Example 4
Find the third term of $(X + 2)^5$.
The coefficient will have two factors or (3 − 1). $\frac{5 \cdot 4}{1 \cdot 2} = 10$

The exponent of 2 will be 2 or (3 − 1),
and the exponent for X will be 3 since 2 + 3 = 5. $X^3 2^2$

Putting it all together: $10 X^3 2^2 = 10 X^3 (4) = 40 X^3$

Practice Problems 2

For #1–3, tell how many terms and expand.

1. $(A + 3)^5$

2. $(3X + 4)^3$

3. $(X - Y)^4$

   [Hint: $X - Y = X + (-Y)$]

4. What is the third term in $(P + Q)^6$?

5. What is the fourth term in $(3A + B)^5$?

6. What is the third term in $(2C - D)^4$?

Solutions 2

1. 6 terms (5 + 1)

   $1A^5 3^0 + 5A^4 3^1 + 10A^3 3^2 + 10A^2 3^3 + 5A^1 3^4 + 1A^0 3^5 =$
   $A^5 + 15A^4 + 90A^3 + 270A^2 + 405A + 243$

2. 4 terms (3 + 1)

   $1(3X)^3 4^0 + 3(3X)^2 4^1 + 3(3X)^1 4^2 + 1(3X)^0 4^3 =$
   $27X^3 + 108X^2 + 144X + 64$

3. 5 terms (4 + 1)

   $1X^4(-Y)^0 + 4X^3(-Y)^1 + 6X^2(-Y)^2 + 4X^1(-Y)^3 + 1X^0(-Y)^4 =$
   $X^4 - 4X^3Y + 6X^2Y^2 - 4XY^3 + Y^4$

4. $\dfrac{6 \cdot 5}{1 \cdot 2} P^4 Q^2 = 15 P^4 Q^2$

5. $\dfrac{5 \cdot 4 \cdot 3}{1 \cdot 2 \cdot 3} (3A)^2 B^3 = 90 A^2 B^3$

6. $\dfrac{4 \cdot 3}{1 \cdot 2} (2C)^2 (-D)^2 = 24 C^2 D^2$

Completing the Square

In lesson 9, while working on squaring a binomial, we observed the product as the result of squaring the first term, adding (or subtracting) two times the middle term, and squaring the last term. We also learned to recognize the square root of a binomial squared. Let's begin by reviewing this, and then find out what is needed to make a perfect square.

Example 1
Square the binomial: $(X + 7)^2$

The first term squared is X^2, and the last term squared is 49.
The middle term is the first term times the second term times two, or $(7X) \cdot 2 = 14X$.

$$X^2 + 14X + 49$$

Example 2
Square the binomial: $(X - 6)^2$

The first term squared is X^2, and the last term squared is 36.
The middle term is the first term times the second term times two, or $(-6X) \cdot 2 = -12X$.

$$X^2 - 12X + 36$$

Example 3

Square the binomial: $(X + 2/3)^2$

The first term squared is X^2 and the last term squared is 4/9.
The middle term is the first term times the second term times two, or
$(2/3 \ X) \cdot 2 = 4/3 \ X$.

$$X^2 + 4/3 \ X + 4/9$$

What is needed to make a perfect square if you have only the first and middle terms? Looking at the examples, we can see that the last term is half of the coefficient of the middle term squared. So if you are given the first and middle, take half of the coefficient of the middle term and square it. This is called *completing the square.*

Example 4

Complete the square by finding the last term.

$$X^2 + 26X + \underline{\hspace{1cm}}$$

Half of 26 is 13, which when squared equals 169.

$$X^2 + 26X + 169 = (X + 13)^2$$

Example 5

Complete the square by finding the last term.

$$X^2 - 2X + \underline{\hspace{1cm}}$$

Half of (–2) is (–1), which when squared equals 1.

$$X^2 - 2X + 1 = (X - 1)^2$$

Example 6

Complete the square by finding the last term.

$$X^2 + 5X + \underline{\hspace{1cm}}$$

Half of 5 is 5/2, which when squared equals 25/4.

$$X^2 + 5X + 25/4 = (X + 5/2)^2$$

Example 7
Complete the square by finding the last term.

$$X^2 + 3/5\ X + \underline{\hspace{2cm}}$$

Half of 3/5 is 3/10, which when squared equals 9/100.

$$X^2 + 3/5\ X + 9/100 = (X + 3/10)^2$$

Practice Problems 1
Complete the square by finding the last term.

1. $X^2 + 18X + \underline{\hspace{1.5cm}}$ 2. $X^2 - 4X + \underline{\hspace{1.5cm}}$

3. $X^2 + 7X + \underline{\hspace{1.5cm}}$ 4. $X^2 - 11X + \underline{\hspace{1.5cm}}$

5. $X^2 + X + \underline{\hspace{1.5cm}}$ 6. $X^2 - 1/2\ X + \underline{\hspace{1.5cm}}$

7. $X^2 + 3/4\ X + \underline{\hspace{1.5cm}}$ 8. $X^2 - 5/3\ X + \underline{\hspace{1.5cm}}$

Solutions 1

1. $X^2 + 18X + 81$ 2. $X^2 - 4X + 4$

3. $X^2 + 7X + 49/4$ 4. $X^2 - 11X + 121/4$

5. $X^2 + X + 1/4$ 6. $X^2 - 1/2\ X + 1/16$

7. $X^2 + 3/4\ X + 9/64$ 8. $X^2 - 5/3\ X + 25/36$

What is needed to make a perfect square if you have only the first and last terms? Looking at the examples, we can see that the last term is half of the coefficient of the middle term squared. So if you are given the last term, take the square root of it and then double that to find the coefficient of the middle term.

Example 8
Complete the square by finding the middle term.

$$X^2 + \underline{} + 100$$

The square root of 100 is 10, and 10 doubled is 20.

$$X^2 + 20X + 100 = (X + 10)^2$$

Example 9
Complete the square by finding the middle term.

$$X^2 - \underline{} + 64$$

The square root of 64 is 8, and 8 doubled is 16.
The negative sign makes the square root (X – 8).

$$X^2 - 16X + 64 = (X - 8)^2$$

Example 10
Complete the square by finding the middle term.

$$X^2 + \underline{} + 1/9$$

The square root of 1/9 is 1/3, and 1/3 doubled is 2/3.

$$X^2 + 2/3\, X + 1/9 = (X + 1/3)^2$$

Practice Problems 2

Complete the square by finding the middle term.

1. $x^2 +$ _____ $+ 144$

2. $x^2 -$ _____ $+ 169$

3. $x^2 +$ _____ $+ 121$

4. $x^2 -$ _____ $+ 49$

5. $x^2 +$ _____ $+ 36$

6. $x^2 -$ _____ $+ 16/25$

7. $x^2 -$ _____ $+ 9/64$

8. $x^2 +$ _____ $+ 1/4$

Solutions 2

1. $x^2 + 24x + 144$
   $(x + 12)^2$

2. $x^2 - 26x + 169$
   $(x - 13)^2$

3. $x^2 + 22x + 121$
   $(x + 11)^2$

4. $x^2 - 14x + 49$
   $(x - 7)^2$

5. $x^2 + 12x + 36$
   $(x + 6)^2$

6. $x^2 - 8/5\ x + 16/25$
   $(x - 4/5)^2$

7. $x^2 - 3/4\ x + 9/64$
   $(x - 3/8)^2$

8. $x^2 + 1\ x + 1/4$
   $(x + 1/2)^2$

Now that you are an expert at completing the square, you can apply this knowledge to solve equations that need to be factored but have no whole-number solutions.

Example 11

Solve for X by completing the square: $X^2 + 8X - 10 = 0$.

Step 1 — Add 10 to both sides. $X^2 + 8X = 10$

Step 2 — Complete the square and add to both sides.

$$X^2 + 8X + 16 = 10 + 16$$

Step 3 — Rewrite as a square.

$$(X + 4)^2 = 26$$

Step 4 — Take the square root of both sides.

$$\sqrt{(X + 4)^2} = \sqrt{26}$$

Step 5 — Solve for X. $X + 4 = \sqrt{26}$

$$X = -4 \pm \sqrt{26}$$

The solutions, or roots, are $-4 + \sqrt{26}$ and $-4 - \sqrt{26}$.

Put these into the original equation to see whether they check. They do.

$$\left(-4 + \sqrt{26}\right)^2 + 8\left(-4 + \sqrt{26}\right) - 10 = 0$$
$$16 - 8\sqrt{26} + 26 - 32 + 8\sqrt{26} - 10 = 0$$
$$16 + 26 - 32 - 10 = 0$$

$$\left(-4 - \sqrt{26}\right)^2 + 8\left(-4 - \sqrt{26}\right) - 10 = 0$$
$$16 + 8\sqrt{26} + 26 - 32 - 8\sqrt{26} - 10 = 0$$
$$42 - 32 - 10 = 0$$

Example 12

Solve for X by completing the square: $X^2 - 6X + 4 = 0$.

Step 1 — Add –4 to both sides. $X^2 - 6X = -4$

Step 2 — Complete the square and add to both sides.
$$X^2 - 6X + 9 = -4 + 9$$

Step 3 — Rewrite as a square. $(X - 3)^2 = 5$

Step 4 — Take the square root of both sides.
$$\sqrt{(X-3)^2} = \pm\sqrt{5}$$

Step 5 — Solve for X. $X - 3 = \pm\sqrt{5}$

$$X = 3 \pm \sqrt{5}$$

The solutions, or roots, are $3 + \sqrt{5}$ and $3 - \sqrt{5}$.

Put these into the original equation to see whether they check. They do.

$$\left(3+\sqrt{5}\right)^2 - 6\left(3+\sqrt{5}\right) + 4 = 0$$
$$9 + 6\sqrt{5} + 5 - 18 - 6\sqrt{5} + 4 = 0$$
$$14 - 18 + 4 = 0$$

$$\left(3-\sqrt{5}\right)^2 - 6\left(3-\sqrt{5}\right) + 4 = 0$$
$$9 - 6\sqrt{5} + 5 - 18 + 6\sqrt{5} + 4 = 0$$
$$14 - 18 + 4 = 0$$

Example 13

Solve for X by completing the square. If there is a coefficient before the first term, divide through by that number first. It is easier to complete the square when the coefficient of X^2 is 1.

$$2X^2 + 10X + 4 = 0$$

Step A — Divide everything by 2. $X^2 + 5X + 2 = 0.$

Step 1 — Add –2 to both sides. $X^2 + 5X = -2$

Step 2 — Complete the square and add to both sides.

$$X^2 + 5X + \frac{25}{4} = \frac{-8}{4} + \frac{25}{4}$$

Step 3 — Rewrite as a square. $\left(X + \frac{5}{2}\right)^2 = \frac{17}{4}$

Step 4 — Take the square root of both sides.

$$\sqrt{\left(X + \frac{5}{2}\right)^2} = \pm\sqrt{\frac{17}{4}}$$

Step 5 — Solve for X.

$$X + \frac{5}{2} = \pm\sqrt{\frac{17}{4}}$$

$$X = -\frac{5}{2} \pm \frac{\sqrt{17}}{2}$$

The solutions, or roots, are $-\frac{5}{2} + \frac{\sqrt{17}}{2}$ and $-\frac{5}{2} - \frac{\sqrt{17}}{2}$.

Put these into the original equation to see whether they check.

$$\left(-\frac{5}{2} + \frac{\sqrt{17}}{2}\right)^2 + 5\left(-\frac{5}{2} + \frac{\sqrt{17}}{2}\right) + 2 = 0$$

$$\frac{25}{4} - \frac{10\sqrt{17}}{4} + \frac{17}{4} - \frac{25}{2} + \frac{5\sqrt{17}}{2} + 2 = 0$$

$$\frac{25}{4} - \frac{5\sqrt{17}}{2} + \frac{17}{4} - \frac{50}{4} + \frac{5\sqrt{17}}{2} + \frac{8}{4} = 0$$

$$0 = 0$$

$$\left(-\frac{5}{2} - \frac{\sqrt{17}}{2}\right)^2 + 5\left(-\frac{5}{2} - \frac{\sqrt{17}}{2}\right) + 2 = 0$$

$$\frac{25}{4} + \frac{10\sqrt{17}}{4} + \frac{17}{4} - \frac{25}{2} - \frac{5\sqrt{17}}{2} + 2 = 0$$

$$\frac{25}{4} + \frac{5\sqrt{17}}{2} + \frac{17}{4} - \frac{50}{4} - \frac{5\sqrt{17}}{2} + \frac{8}{4} = 0$$

$$0 = 0$$

They do.

Practice Problems 3

Solve for X by completing the square. Check your work.

1. $X^2 + 8X + 1 = 0$

2. $X^2 - 6X - 2 = 0$

3. $2X^2 + 4X + 10 = 0$

4. $3X^2 - 36X - 36 = 0$

5. $X^2 + 3X + 1 = 0$

6. $X^2 - 7X + 5 = 0$

Solutions 3

1. $X^2 + 8X + 1 = 0$

$X^2 + 8X = -1$

$X^2 + 8X + 16 = -1 + 16$

$(X + 4)^2 = 15$

$\sqrt{(X+4)^2} = \sqrt{15}$

$X + 4 = \pm\sqrt{15}$

$X = -4 \pm \sqrt{15}$

$\left(-4 + \sqrt{15}\right)^2 + 8\left(-4 + \sqrt{15}\right) + 1 = 0$

$16 - 8\sqrt{15} + 15 - 32 + 8\sqrt{15} + 1 = 0$

$31 - 32 + 1 = 0$

$\left(-4 - \sqrt{15}\right)^2 + 8\left(-4 - \sqrt{15}\right) + 1 = 0$

$16 + 8\sqrt{15} + 15 - 32 - 8\sqrt{15} + 1 = 0$

$31 - 32 + 1 = 0$

2. $X^2 - 6X - 2 = 0$

$X^2 - 6X = 2$

$X^2 - 6X + 9 = 2 + 9$

$(X - 3)^2 = 11$

$\sqrt{(X-3)^2} = \sqrt{11}$

$X - 3 = \pm\sqrt{11}$

$X = 3 \pm \sqrt{11}$

$\left(3 + \sqrt{11}\right)^2 - 6\left(3 + \sqrt{11}\right) - 2 = 0$

$9 + 6\sqrt{11} + 11 - 18 - 6\sqrt{11} - 2 = 0$

$20 - 18 - 2 = 0$

$\left(3 - \sqrt{11}\right)^2 - 6\left(3 - \sqrt{11}\right) - 2 = 0$

$9 - 6\sqrt{11} + 11 - 18 + 6\sqrt{11} - 2 = 0$

$20 - 18 - 2 = 0$

3. $2X^2 + 4X + 10 = 0$

$X^2 + 2X + 5 = 0$

$X^2 + 2X = -5$

$X^2 + 2X + 1 = -5 + 1$

$(X + 1)^2 = -4$

$\sqrt{(X+1)^2} = \sqrt{-4}$

$X + 1 = \pm\sqrt{-4}$

$X = -1 \pm 2i$

$(-1 + 2i)^2 + 2(-1 + 2i) + 5 = 0$

$1 - 4i - 4 - 2 + 4i + 5 = 0$

$6 - 6 = 0$

$(-1 - 2i)^2 + 2(-1 - 2i) + 5 = 0$

$1 + 4i - 4 - 2 - 4i + 5 = 0$

$6 - 6 = 0$

4. $3X^2 - 36X - 36 = 0$
$X^2 - 12X - 12 = 0$
$X^2 - 12X = 12$
$X^2 - 12X + 36 = 12 + 36$
$(X - 6)^2 = 48$
$\sqrt{(X - 6)^2} = \sqrt{48}$
$X - 6 = \pm\sqrt{48}$
$X = 6 \pm 4\sqrt{3}$

$(6 + 4\sqrt{3})^2 - 12(6 + 4\sqrt{3}) - 12 = 0$
$36 + 48\sqrt{3} + 16 \cdot 3 - 72 - 48\sqrt{3} - 12 = 0$
$84 - 84 = 0$

$(6 - 4\sqrt{3})^2 - 12(6 - 4\sqrt{3}) - 12 = 0$
$36 - 48\sqrt{3} + 16 \cdot 3 - 72 + 48\sqrt{3} - 12 = 0$
$84 - 84 = 0$

5. $X^2 + 3X + 1 = 0$
$X^2 + 3X = -1$
$X^2 + 3X + \dfrac{9}{4} = -1 + \dfrac{9}{4}$
$\left(X + \dfrac{3}{2}\right)^2 = \dfrac{5}{4}$
$\sqrt{\left(X + \dfrac{3}{2}\right)^2} = \sqrt{\dfrac{5}{4}}$
$X + \dfrac{3}{2} = \pm\sqrt{\dfrac{5}{4}}$
$X = -\dfrac{3}{2} \pm \dfrac{\sqrt{5}}{2}$

$\left(-\dfrac{3}{2} + \dfrac{\sqrt{5}}{2}\right)^2 + 3\left(-\dfrac{3}{2} + \dfrac{\sqrt{5}}{2}\right) + 1 = 0$
$\dfrac{9}{4} - \dfrac{6\sqrt{5}}{4} + \dfrac{5}{4} - \dfrac{9}{2} + \dfrac{3\sqrt{5}}{2} + 1 = 0$
$\dfrac{9}{4} - \dfrac{3\sqrt{5}}{2} + \dfrac{5}{4} - \dfrac{18}{4} + \dfrac{3\sqrt{5}}{2} + \dfrac{4}{4} = 0$

$\left(-\dfrac{3}{2} - \dfrac{\sqrt{5}}{2}\right)^2 + 3\left(-\dfrac{3}{2} - \dfrac{\sqrt{5}}{2}\right) + 1 = 0$
$\dfrac{9}{4} + \dfrac{6\sqrt{5}}{4} + \dfrac{5}{4} - \dfrac{9}{2} - \dfrac{3\sqrt{5}}{2} + 1 = 0$
$\dfrac{9}{4} + \dfrac{3\sqrt{5}}{2} + \dfrac{5}{4} - \dfrac{18}{4} - \dfrac{3\sqrt{5}}{2} + \dfrac{4}{4} = 0$

6. $X^2 - 7X + 5 = 0$

       $X^2 - 7X = -5$

   $X^2 - 7X + \dfrac{49}{4} = -5 + \dfrac{49}{4}$

$$\left(X - \dfrac{7}{2}\right)^2 = \dfrac{29}{4}$$

$$\sqrt{\left(X - \dfrac{7}{2}\right)^2} = \sqrt{\dfrac{29}{4}}$$

$$X - \dfrac{7}{2} = \pm\sqrt{\dfrac{29}{4}}$$

$$X = \dfrac{7}{2} \pm \dfrac{\sqrt{29}}{2}$$

$$\left(\dfrac{7}{2} + \dfrac{\sqrt{29}}{2}\right)^2 - 7\left(\dfrac{7}{2} + \dfrac{\sqrt{29}}{2}\right) + 5 = 0$$

$$\dfrac{49}{4} + \dfrac{14\sqrt{29}}{4} + \dfrac{29}{4} - \dfrac{49}{2} - \dfrac{7\sqrt{29}}{2} + 5 = 0$$

$$\dfrac{49}{4} + \dfrac{7\sqrt{29}}{2} + \dfrac{29}{4} - \dfrac{98}{4} - \dfrac{7\sqrt{29}}{2} + \dfrac{20}{4} = 0$$

$$\left(\dfrac{7}{2} - \dfrac{\sqrt{29}}{2}\right)^2 - 7\left(\dfrac{7}{2} - \dfrac{\sqrt{29}}{2}\right) + 5 = 0$$

$$\dfrac{49}{4} - \dfrac{14\sqrt{29}}{4} + \dfrac{29}{4} - \dfrac{49}{2} + \dfrac{7\sqrt{29}}{2} + 5 = 0$$

$$\dfrac{49}{4} - \dfrac{7\sqrt{29}}{2} + \dfrac{29}{4} - \dfrac{98}{4} + \dfrac{7\sqrt{29}}{2} + \dfrac{20}{4} = 0$$

Quadratic Formula

A *quadratic* is an equation that has an unknown or variable raised to the second power, as in Y^2 or A^2. In factoring and in completing the square, we have been dealing exclusively with quadratic equations. So far, we can find the solution to a quadratic equation by factoring it, or if this fails, by completing the square. In this lesson we are going to complete the square with variables in order to discover a formula to solve all quadratics. If you've mastered the previous lesson, try solving the following equation by completing the square, and then compare your solution with mine.

$$AX^2 + BX + C = 0$$

Divide by the coefficient of X^2.

$$\frac{AX^2}{A} + \frac{BX}{A} + \frac{C}{A} = 0$$
$$X^2 + \frac{BX}{A} + \frac{C}{A} = 0$$

Add the opposite of the third term to both sides.

$$X^2 + \frac{BX}{A} = -\frac{C}{A}$$

Take one-half of the coefficient of the middle term, square it, and add the result to both sides.

$$X^2 + \frac{BX}{A} + \left(\frac{B}{2A}\right)^2 = -\frac{C}{A} + \left(\frac{B}{2A}\right)^2$$

Factor the left side.

$$\left(X + \frac{B}{2A}\right)^2 = -\frac{C}{A} + \frac{B^2}{4A^2}$$

Combine terms on the right.

$$\left(X + \frac{B}{2A}\right)^2 = -\frac{4AC}{4A^2} + \frac{B^2}{4A^2}$$

Take the square root of both sides.

$$X + \frac{B}{2A} = \sqrt{-\frac{4AC}{4A^2} + \frac{B^2}{4A^2}} = \pm\frac{\sqrt{-4AC + B^2}}{2A}$$

Subtract B/2A from both sides, and combine.

$$X = -\frac{B}{2A} \pm \frac{\sqrt{-4AC + B^2}}{2A}$$

The *quadratic formula*! This is the form in which it is usually written.

$$X = \frac{-B \pm \sqrt{B^2 - 4AC}}{2A}$$

Example 1

Let's try an equation that we can answer by factoring, and "plug in" the values for A, B, and C. Remember that to find A, B, and C, the equation must be in the form $AX^2 + BX + C = 0$.

$$X^2 + 5X + 6 = 0$$

$$A = 1, B = 5, \text{ and } C = 6$$

$$X = \frac{-B \pm \sqrt{B^2 - 4AC}}{2A}$$

$$X = \frac{-5 \pm \sqrt{5^2 - 4 \cdot 1 \cdot 6}}{2 \cdot 1}$$

$$X = \frac{-5 \pm \sqrt{25 - 24}}{2} = \frac{-5 \pm \sqrt{1}}{2}$$

$$X = \frac{-5 \pm 1}{2} = \frac{-4}{2} \text{ or } \frac{-6}{2} = -2 \text{ or } -3$$

We can also solve $X^2 + 5X + 6 = 0$ by factoring.

$$X^2 + 5X + 6 = 0$$
$$(X + 2)(X + 3) = 0$$
$$X + 2 = 0 \quad X + 3 = 0$$
$$X = -2 \qquad X = -3$$

For this problem, it would have much easier to solve by factoring. Try factoring first, and if it doesn't work, use the quadratic formula. Here is another problem to try.

Example 2

Find the factors of $2X^2 = -7X - 4$.

To find A, B, and C, the equation must be in the form $AX^2 + BX + C = 0$.

$$2X^2 + 7X + 4 = 0$$

$$A = 2, B = 7, \text{ and } C = 4$$

$$X = \frac{-B \pm \sqrt{B^2 - 4AC}}{2A}$$

$$X = \frac{-7 \pm \sqrt{7^2 - 4 \cdot 2 \cdot 4}}{2 \cdot 2}$$

$$X = \frac{-7 \pm \sqrt{49 - 32}}{4} = \frac{-7 \pm \sqrt{17}}{4}$$

$$X = \frac{-7 \pm \sqrt{17}}{4}$$

$$X = \frac{-7 + \sqrt{17}}{4} \text{ or } \frac{-7 - \sqrt{17}}{4}$$

Practice Problems 1

Solve for X. Try factoring first, and then use the quadratic formula if necessary.

1. $X^2 - 25 = 0$

2. $X^2 - 18X = -81$

3. $2X^2 + 7X + 6 = 0$

4. $3X^2 + X - 4 = 0$

5. $4A^2 - 36 = 0$

6. $X^2 + 5 = -3X$

7. $7X^2 = -2X + 1$

8. $2X^2 + 2X - 5 = 0$

9. $\dfrac{5}{X+3} + \dfrac{2}{X-3} = 5$ $(X \neq \pm 3)$

10. $4X^2 = 9$

11. $4X^2 + 20X = -25$

12. $3Q^2 = -4Q - 2$

Solutions 1

1. $(X+5)(X-5) = 0$

   $X+5=0 \quad X-5=0$

   $\quad X=-5 \quad\quad X=5$

2. $(X-9)(X-9) = 0$

   $X-9=0 \quad X-9=0$

   $\quad X=9 \quad\quad X=9$

3. $(2X+3)(X+2) = 0$

   $2X+3=0 \qquad X+2=0$

   $2X=-3$

   $\quad X=-3/2 \quad\quad X=-2$

4. $(3X+4)(X-1) = 0$

   $3X+4=0 \qquad X-1=0$

   $3X=-4$

   $\quad X=-4/3 \quad\quad X=1$

5. $(2A-6)(2A+6) = 0$

   $2A-6=0 \qquad 2A+6=0$

   $2A=6 \qquad\quad 2A=-6$

   $\quad A=6/2 \qquad\quad A=-6/2$

   $\quad A=3 \qquad\qquad A=-3$

6. $X = \dfrac{-3 \pm \sqrt{3^2 - 4 \cdot 1 \cdot 5}}{2 \cdot 1}$

   $X = \dfrac{-3 \pm \sqrt{-11}}{2} = \dfrac{-3 + i\sqrt{11}}{2}$ or $X = \dfrac{-3 - i\sqrt{11}}{2}$

7. $X = \dfrac{-2 \pm \sqrt{2^2 - 4 \cdot 7 \cdot -1}}{2 \cdot 7} = \dfrac{-2 \pm 4\sqrt{2}}{14} = \dfrac{-1 + 2\sqrt{2}}{7}$ or $X = \dfrac{-1 - 2\sqrt{2}}{7}$

8. $X = \dfrac{-2 \pm \sqrt{2^2 - 4 \cdot 2 \cdot -5}}{2 \cdot 2} = \dfrac{-2 \pm 2\sqrt{11}}{4} = \dfrac{-1 + \sqrt{11}}{2}$ or $X = \dfrac{-1 - \sqrt{11}}{2}$

9. $\left(\dfrac{5}{X+3} + \dfrac{2}{X-3} \right) = 5$ $\qquad$ $X = \dfrac{-(-7) \pm \sqrt{(-7)^2 - 4 \cdot 5 \cdot -36}}{2 \cdot 5}$

   $5(X-3) + 2(X+3) = 5\left(X^2 - 9\right)$ $\qquad$ $X = \dfrac{7 \pm \sqrt{769}}{10}$

   $\qquad\qquad 7X - 9 = 5X^2 - 45$ $\qquad$ $X = \dfrac{7 + \sqrt{769}}{10}$ or $X = \dfrac{7 - \sqrt{769}}{10}$

   $\qquad 5X^2 - 7X - 36 = 0$

10. $(2X - 3)(2X + 3) = 0$ $\qquad$ 11. $(2X + 5)(2X + 5) = 0$

$\qquad 2X - 3 = 0 \qquad 2X + 3 = 0 \qquad\qquad 2X + 5 = 0 \qquad 2X + 5 = 0$

$\qquad\quad 2X = 3 \qquad\quad 2X = -3 \qquad\qquad\quad 2X = -5 \qquad\quad 2X = -5$

$\qquad\qquad X = 3/2 \qquad\quad X = -3/2 \qquad\qquad\quad X = -5/2 \qquad\quad X = -5/2$

12. $3Q^2 + 4Q + 2 = 0$ $\qquad$ $X = \dfrac{-(4) \pm \sqrt{(4)^2 - 4 \cdot 3 \cdot 2}}{2 \cdot 3}$

$\qquad\qquad\qquad\qquad X = \dfrac{-4 \pm \sqrt{16 - 24}}{2 \cdot 3} = \dfrac{-4 \pm \sqrt{-8}}{2 \cdot 3}$

$\qquad\qquad\qquad\qquad X = \dfrac{-4 \pm i\sqrt{2 \cdot 4}}{2 \cdot 3} = \dfrac{-4 \pm 2i\sqrt{2}}{2 \cdot 3}$

$\qquad\qquad\qquad\qquad X = \dfrac{-2 + i\sqrt{2}}{3}$ or $\dfrac{-2 - i\sqrt{2}}{3}$

Discriminants

In the quadratic formula, what goes on underneath the radical sign indicates the kinds of roots, or what the nature of the solutions will be. Solutions fall into one of three categories: real and rational (as 3, 1/4, or .75), imaginary (as 4 + 7i or 3i), or real and irrational (as $\sqrt{2}$).

The part of the quadratic formula that helps us discriminate what kind of roots we will have is $B^2 - 4AC$, which is the part under the radical sign. For example, if this expression is a negative number, the solution will obviously be imaginary because you are being asked to find the square root of a negative number. This key part of the quadratic formula is called the ***discriminant*** for its ability to discriminate the type of root.

In the following examples, keep an eye on the $B^2 - 4AC$ and see how it affects the final solutions.

Example 1

$X^2 + 7X + 10 = 0$

A = 1, B = 7, and C = 10

$$X = \frac{-B \pm \sqrt{B^2 - 4AC}}{2A} = \frac{-7 \pm \sqrt{(7)^2 - 4 \cdot 1 \cdot 10}}{2 \cdot 1}$$

$$X = \frac{-7 \pm \sqrt{49 - 40}}{2} = \frac{-7 \pm \sqrt{9}}{2} = \frac{-7 + 3}{2} \text{ or } \frac{-7 - 3}{2}$$

$$X = \frac{-4}{2}, \frac{-10}{2} = -2, -5$$

$B^2 - 4AC$ equals 9 in example 1, and the roots (–2 and –5) are real and rational. A rational number is a number that can be expressed as a ratio or fraction. Notice that 9 is a perfect square. If the discriminant is a perfect square, then your roots will be real, not imaginary.

Example 2

$$X^2 + 3X - 3 = 0$$
$$A = 1, B = 3, \text{ and } C = -3$$

$$X = \frac{-B \pm \sqrt{B^2 - 4AC}}{2A} = \frac{-(3) \pm \sqrt{(3)^2 - 4 \cdot 1 \cdot (-3)}}{2 \cdot 1}$$

$$X = \frac{-3 \pm \sqrt{9 + 12}}{2} = \frac{-3 \pm \sqrt{21}}{2} = \frac{-3 + \sqrt{21}}{2}, \frac{-3 - \sqrt{21}}{2}$$

$B^2 - 4AC$ equals 21 in example 2, and the roots are real and irrational. The square root of 21 is an irrational number. It cannot be expressed as a ratio or a fraction, but it is real.

Example 3

$$X^2 - 10X + 25 = 0$$
$$A = 1, B = -10, \text{ and } C = 25$$

$$X = \frac{-B \pm \sqrt{B^2 - 4AC}}{2A} = \frac{-(-10) \pm \sqrt{(-10)^2 - 4 \cdot 1 \cdot 25}}{2 \cdot 1}$$

$$X = \frac{10 \pm \sqrt{100 - 100}}{2} = \frac{10}{2} = 5$$

$B^2 - 4AC$ equals zero in example 3. The roots are real, rational, and equal. If this polynomial, which is a perfect square, had been factored, there would have been two identical solutions (5 and 5). This is called a double root. When the discriminant is zero, you can expect a double root.

$$X^2 - 10X + 25 = 0$$
$$(X - 5)(X - 5) = 0$$
$$X - 5 = 0 \qquad X - 5 = 0$$
$$X = 5 \qquad X = 5$$

Example 4

$$X^2 - 5X + 8 = 0$$

$$A = 1, B = -5, \text{ and } C = 8$$

$$X = \frac{-B \pm \sqrt{B^2 - 4AC}}{2A}$$

$$X = \frac{-(-5) \pm \sqrt{(-5)^2 - 4 \cdot 1 \cdot 8}}{2 \cdot 1}$$

$$X = \frac{5 \pm \sqrt{25 - 32}}{2}$$

$$X = \frac{5 \pm \sqrt{-7}}{2} = \frac{5 + i\sqrt{7}}{2} \text{ or } \frac{5 - i\sqrt{7}}{2}$$

$B^2 - 4AC$ equals -7 in example 4. The roots include *i*, so they are imaginary.

In summary:

1. If the discriminant is a perfect square, the roots are **real**, **rational**, and **unequal**.

2. If the discriminant is positive (greater than zero), and not a perfect square, the roots are **real**, **irrational**, and **unequal**.

3. If the discriminant is equal to zero, the roots are **real**, **rational**, and **equal** (a double root).

4. If the discriminant is negative (less than zero), the roots are **imaginary**.

Practice Problems 1
Predict the nature of the solutions. Solve #1-3 to find the exact roots. Factor when possible.

1. $5X^2 + 2X = 2X + 45$

2. $X^2 + 16 = -8X$

3. $X^2 = 2X - 5$

4. $X^2 - 2/3 \, X = 4/3$

5. $2X^2 = X + 3$

Solutions 1

1.
$$5X^2 + 2X = 2X + 45$$
$$5X^2 + 2X - 2X - 45 = 0$$
$$5X^2 - 45 = 0$$
$$X^2 - 9 = 0$$

$A = 1, \; B = 0, \; C = -9$

$$B^2 - 4AC = 0^2 - (4)(1)(-9) = 36$$
Disc. $= 36$, a perfect square
real, rational, unequal
$$X^2 - 9 = 0$$
$$(X - 3)(X + 3) = 0$$
$$X - 3 = 0 \quad X + 3 = 0$$
$$X = 3 \qquad X = -3$$
$$(3, \, -3)$$

2.
$$X^2 + 16 = -8X$$
$$X^2 + 8X + 16 = 0$$

$A = 1, \; B = 8, \; C = 16$

$$B^2 - 4AC = (8)^2 - (4)(1)(16) = 0$$
Disc. $= 0$, double root
real, rational, equal
$$X^2 + 8X + 16 = 0$$
$$(X + 4)(X + 4) = 0$$
$$X + 4 = 0 \quad X + 4 = 0$$
$$X = -4 \qquad X = -4$$
$$(-4, -4)$$

3.
$$X^2 = 2X - 5$$
$$X^2 - 2X + 5 = 0$$

$A = 1, \; B = -2, \; C = 5$

$$B^2 - 4AC = (-2)^2 - (4)(1)(5) = -16$$
Disc. < 0
imaginary roots
$$\frac{-(-2) \pm \sqrt{(-2)^2 - 4(1)(5)}}{2(1)}$$
$$\frac{2 \pm \sqrt{4 - 20}}{2} = \frac{2 \pm \sqrt{-16}}{2} = \frac{2 \pm 4i}{2}$$
$$(1 + 2i, \; 1 - 2i)$$

4. $X^2 - \dfrac{2}{3}X - \dfrac{4}{3} = 0$

$$B^2 - 4AC = \left(-\frac{2}{3}\right)^2 - 4(1)\left(-\frac{4}{3}\right) = \frac{52}{9}$$

Disc. > 0
real, irrational, unequal

5. $2X^2 - X - 3$

$$B^2 - 4AC = (-1)^2 - 4(2)(-3) = 25$$
Disc. $= 25$, a perfect square
real, rational, unequal

Applications Using Percent

Language is very important when solving problems concerning price markups and price discounts. Notice particularly whether the problem is discussing a percentage of the final cost or a percentage of the original cost. This can be very tricky. In order to compare or combine, you must use the same kind. You must be talking from the same perspective, comparing apples to apples and oranges to oranges. Recently I was explaining this concept to a contractor. He saw the percentage as one thing, and the customer saw it as something else. A disagreement ensued, yet they both were correct. This is what happened:

The wholesale cost of a light was $10.00, and the contractor charged $15.00, with the explanation that he had marked it up 1/3 or 33 1/3%. The customer disagreed and said it had been marked up 1/2 or 50%. Who was right?

The contractor was taking 33 1/3%, or 1/3, of the final cost, and 1/3 of $15.00 is $5.00, which was the amount of the markup. The customer agreed that the profit was $5.00 but added that this was 1/2 or 50% of the original cost. Both were correct, but they began at different points, taking a percentage of two different costs, the original and the final. The final cost is the original cost plus the markup. In economic terms, the original cost is referred to as the *wholesale price*. This is what a store pays for products. The retail store then adds the markup, which is its *profit*, to the wholesale price. The final price, which the store charges, is called the *retail price*. So the wholesale price plus the profit equals the retail price.

Example 1

What percent of the retail price did you save if you bought a swimsuit for $27.00 that was priced at $36.00?

Solution A—You saved $9.00. The equation is:
WP (what percentage) x (of) 36.00 (retail price) = (is) 9.00 (savings)

$$WP \times 36 = 9$$

$$\frac{WP \times 36}{36} = \frac{9}{36}$$ Divide both sides by 36.

$$WP = 1/4 \text{ or } 25\%$$ Change the fraction to a percent.

Solution B — Another way to solve this is based on the word percent, which means per cent or per hundred. When writing the equation, write WP as WP/100.

$$\frac{WP}{100} \times 36 = 9$$

$$100 \times \frac{WP}{100} \times 36 = 9 \times 100$$ Multiply both sides by 100.

$$WP \times 36 = 900$$
$$\frac{WP \times 36}{36} = \frac{900}{36}$$ Divide both sides by 36.
$$WP = 25\%$$

Example 2

What percent of the final cost is the markup of a basketball if the wholesale price is $25.00, and the retail price is $35.00?

$$WP \qquad x \qquad C_f \qquad = \qquad M$$
(what percentage)(of)(final cost)(is)(markup)

Solution A — The Markup is $10.00. ($35.00 – $25.00)

$$WP \times 35 = 10$$

$$WP = \frac{10}{35}$$

$$WP = 2/7 \text{ or } 28.6\%$$

Solution B

$$\frac{WP}{100} \times 35 = 10$$

$$100 \times \frac{WP}{100} \times 35 = 10 \times 100$$

$$WP \times 35 = 1,000$$

$$WP \times \frac{35}{35} = \frac{1,000}{35}$$

$$WP = 28.6\%$$

A variable can be identified further with a smaller letter beside it. This is called a ***subscript***, with "sub" meaning under, and "script" meaning write. So a subscript is "written under." In examples 2 and 3, we chose C to represent cost. To further distinguish C, we can use the subscripts o for original and f for final.

Example 3

What percent of the original wholesale cost is the markup of the basketball in example 2?

$$WP \quad \times \quad C_O \quad = \quad M$$
$$\text{(what percentage)(of)(original cost)(is)(markup)}$$

Solution A

$$WP \times 25 = 10$$

$$WP = 2/5 \text{ or } 40\%$$

Solution B

$$\frac{WP}{100} \times 25 = 10$$

$$100 \times \frac{WP}{100} \times 25 = 10 \times 100$$

$$WP \times 25 = 1{,}000$$

$$WP \times \frac{25}{25} = \frac{1{,}000}{25}$$

$$WP = 40\%$$

Practice Problems 1

1. The wholesale price of the golf clubs is $254.00, and the retail price is $299.00. What percentage of the wholesale price is the profit, or markup?

2. The wholesale price of the golf clubs is $254.00, and the retail price is $299.00. What percentage of the retail price is the profit, or markup?

3. The final retail price was 22% above the original wholesale cost, which was $34.00. What is the final price?

4. The used car sold for $1,750.00. The car dealer made a profit of 12% of the original cost. What was the original cost?

Solutions 1

1. $WP \times C_w = M$

    Solution A

    $WP \times \dfrac{254}{254} = \dfrac{45}{254}$

    $WP = .177 = 17.7\%$

    Solution B

    $\dfrac{WP}{100} \times 254 = 45$

    $WP = \dfrac{4500}{254}$

    $WP = 17.7\%$

2. $WP \times C_r = M$

    Solution A

    $WP \times \dfrac{299}{299} = \dfrac{45}{299}$

    $WP = .15 = 15\%$

    Solution B

    $\dfrac{WP}{100} \times 299 = 45$

    $WP = \dfrac{4500}{299}$

    $WP = 15\%$

3. $P_f = P_o + M$

    Final price = Original price + Profit

    $P_f = P_o + 22\%(P_o)$

    $P_f = 34.00 + 22\%(34.00)$

    $P_f = 34.00 + 7.48 = \$41.48$

4. $P_f = P_o + M$

    Final price = Original price + Profit

    $P_f = P_o + 12\%(P_o)$

    $1{,}750 = P_o(1 + .12)$

    $\dfrac{1{,}750}{1.12} = P_o \dfrac{(1.12)}{1.12}$

    $\$1{,}562.50 = P_o$

Percentages are also used in computing taxes, tips at restaurants, chemistry, and in many other applications. The key is how you verbalize the problem and create the equation.

Example 4

I collected $142.08 in sales tax for the fourth quarter. Pennsylvania state sales tax is 6%. What were my gross sales during this quarter?

$$Sales \times 6\% = 142.08$$
$$.06S = 142.08$$
$$\frac{.06S}{.06} = \frac{142.08}{.06}$$
$$S = \$2,368.00$$

Example 5

The final bill at the restaurant was $35.09. This amount included a 15% tip and 6% sales tax. What was the cost of the food, the tip, and the tax? Round to the nearest cent.

$$C + 15\%C + 6\%C = 35.09$$
$$C + .15C + .06C = 35.09$$
$$\frac{1.21C}{1.21} = \frac{35.09}{1.21}$$
$$C = \$29.00$$

Tax = 6% of $29.00
Tax = $1.74

Tip = 15% of $29.00
Tip = $4.35

Percents of Elements in Chemical Compounds

A chemical compound is made up of elements. Each element has an atomic weight. When you add up the atomic weights for each of the elements in a compound, you get the molecular weight. The percentage of each element tells us how the mass or weight of the compound is distributed. Atomic weights for selected elements are given in a table on the next page and under "Symbols and Tables" in the back of your student book. The weights have been rounded to whole numbers for ease of use in the problems. For a complete guide to the elements, consult a periodic table.

Symbol	Element	Atomic Weight	Symbol	Element	Atomic Weight
H	Hydrogen	1	Mg	Magnesium	24
Li	Lithium	7	Si	Silicon	28
Be	Beryllium	9	P	Phosphorus	31
B	Boron	11	S	Sulfur	32
C	Carbon	12	Cl	Chlorine	35
N	Nitrogen	14	K	Potassium	39
O	Oxygen	16	Ca	Calcium	40
F	Fluorine	19	Cr	Chromium	52
Na	Sodium	23	Fe	Iron	56

Example 1

Find the percentage of hydrogen in water.

H_2O* is the chemical compound for water.

This means there are two atoms of hydrogen and one atom of oxygen:
    H + H + O.
The atomic weight of the whole compound is:
    1 + 1 + 16 = 18.
The weight of the hydrogen is:
    1 + 1 = 2.

The percentage of hydrogen is 2 of 18 or 2/18 = 1/9 = 11%.
The percentage of oxygen is 16 of 18 or 16/18 = 8/9 = 89%.
(The answers are rounded to the nearest whole percent.)

*In this compound, the 2 tells how many hydrogen atoms.

Example 2

Find the percentage of carbon in C_2H_2.

C_2H_2 is the chemical compound.

This means there are two atoms of carbon and two atoms of hydrogen:

C + C + H + H.

The atomic weight of the whole compound is:

12 + 12 + 1 + 1 = 26.

The weight of the hydrogen is:

1 + 1 = 2.

The weight of the carbon is:

12 + 12 = 24.

The percentage of carbon is 24 of 26 or 24/26 = 12/13 = 92%.
The percentage of hydrogen is 2 of 26 or 2/26 = 1/13 = 8%.

Practice Problems 2

Round your answers to the nearest whole percent.

1. Find the percentage of chlorine in NaCl.

2. Find the percentage of sodium in NaOH.

3. Find the percentage of nitrogen in KCN

4. Find the percentage of potassium in KCN.

5. Find the percentage of oxygen in CO_2.

6. Find the percentage of carbon in CO_2.

Solutions 2

1. The atomic weight of the NaCl is: 23 + 35 = 58.

   The weight of the chlorine is 35.
   The percentage of chlorine is 35/58 = .60 = 60%.

2. The atomic weight of the NaOH is: 23 + 16 + 1 = 40

   The weight of the sodium is 23.
   The percentage of sodium is 23/40 = .575 = 58%.

3. The atomic weight of the KCN is: 39 + 12 + 14 = 65

   The weight of the nitrogen is 14.
   The percentage of nitrogen is 14/65 = .215 = 22%.

4. The atomic weight of the KCN is: 39 + 12 + 14 = 65.

   The weight of the potassium is 39.
   The percentage of potassium is 39/65 = .60 = 60%.

5. The atomic weight of the CO_2 is: 12 + 16 + 16 = 44.

   The weight of the oxygen is 32.
   The percentage of oxygen is 32/44 = .727 = 73%.

6. The atomic weight of the CO_2 is: 12 + 16 + 16 = 44

   The weight of the carbon is 12.
   The percentage of carbon is 12/44 = .272 = 27%.

Isolating One Variable

Algebra involves working with variables and unknowns in equations and formulas. When see a formula such as "Distance equals Rate times Time," or D = RT, you have three variables. To solve the equation, you will need numbers for two of the variables, and then you can solve for the third. In this equation, if you were given numbers for R and T, you could just replace the R and T and multiply to find D. This is very straightforward. If the rate is 55 mph and the time is 7 hours, multiply 55 x 7 to get 385 miles, which is the distance. If the given information was 385 miles covered in 7 hours, how could you find the rate? You have two options: either plug in the numbers first, and then solve for R, or solve for R first, and then plug in the numbers. It is the second option that is the object of this lesson. The objective is to transform D = RT into R = something.

We begin with: D = RT

Divide both sides by T: $\dfrac{D}{T} = \dfrac{RT}{T}$

We now have: $\dfrac{D}{T} = R$

With this formula, we can replace D and T with the data to solve for R.

$$R = \frac{D}{T} = \frac{385 \text{ mi}}{7 \text{ h}} = 55 \text{ miles per hour or mph}$$

Some examples have more than three variables. Carefully examine the following examples. Try some of the practice problems, and compare your answers with the solutions.

Example 1

Solve for A:

Step 1: $AB = CD$

Step 2: $\dfrac{A\cancel{B}}{\cancel{B}} = \dfrac{CD}{B}$

Step 3: $A = \dfrac{CD}{B}$

The concept we employed in solving example 1 is that we can divide both sides of an equation without changing the value of the equation.

Shortcut: Notice the position of B in steps 1 and 3 above. It was on the top line (numerator) on the left hand side of the equation in step 1, but in step 3 it is on the right hand side in the bottom line (denominator). We could say it is on the opposite side (right from left), in the opposite position (denominator from numerator). By observing this and other problems, we can see that this is true.

Let's try both methods on example 2. In what I will call the long method, multiply both sides by the same variable and divide both sides by the same variable. In the short method we employ the same concept, but without as much writing.

Example 2

Solve for A: Long Method

$$\dfrac{B}{A} = \dfrac{C}{D}$$

$$\dfrac{A}{1} \times \dfrac{B}{A} = \dfrac{C}{D} \times \dfrac{A}{1} \qquad \text{Multiply both sides by A.}$$

$$B = \dfrac{CA}{D}$$

$$\dfrac{D}{C} \times B = \dfrac{CA}{D} \times \dfrac{D}{C} \qquad \text{Multiply by D/C, the reciprocal.}$$

$$\dfrac{DB}{C} = A$$

Solve for A: Short Method

$$\frac{B}{A} = \frac{C}{D}$$

$$B = \frac{CA}{D} \qquad \text{Opposite side, opposite position, for A.}$$

$$\frac{DB}{C} = A \qquad \text{Opposite side, opposite position, for D and C.}$$

Instead of opposite side, opposite position, you could say, "On the opposite side, multiply by the reciprocal," or D/C. This is another way of saying "opposite position," since D and C are both in opposite positions.

Practice Problems 1

1. Solve for A: $ABC = D$

2. Solve for B: $\frac{A}{BC} = \frac{D}{E}$

3. Solve for X: $\frac{YZ}{B} = \frac{A}{X}$

4. Solve for Y: $\frac{1}{Y} = \frac{1}{A}$

Solutions 1

1. Solve for A: $A = \frac{D}{BC}$

2. Solve for B: $\frac{EA}{DC} = B$

3. Solve for X: $X = \frac{BA}{YZ}$

4. Solve for Y: $A = Y$

There are also problems involving variables in equations that include adding and subtracting. The same shortcut can be used here. I call it opposite side, opposite sign. Example 3 is worked both ways, using the long and the short methods.

Example 3
Solve for A: Long Method

Subtract B both sides.
$$A + B = C + D$$
$$A + B - B = C + D - B$$
$$A = C + D - B$$

Solve for A: Short Method
Opposite side, opposite sign.

$$A + B = C + D$$
$$A = C + D - B$$

Comparing the last step in each method, you see that we are employing the same concept without the additional writing.

Practice Problems 2

1. Solve for C: B – A = C + D

2. Solve for X: X + Y – Z = A – B

3. Solve for P: A – P = D + E

Solutions 2

1. Solve for C: C = B – A – D

2. Solve for X: X = A – B – Y + Z

3. Solve for P: A – D – E = P

Sometimes in isolating one variable, we use a combination of several operations. Then we use both of the procedures.

Example 4
Solve for A:

$$\frac{A}{B} - C = 0$$

$$\frac{A}{B} = C \qquad \text{Add C.}$$

To add or subtract, opposite side—opposite sign

$$A = CB \qquad \text{Multiply by B.}$$

To multiply or divide, opposite side—opposite place.

Example 5
Solve for A:

$$X(A + B) = C$$

$$A + B = \frac{C}{X} \qquad \text{Divide by X.}$$

To multiply or divide, opposite side—opposite place.

$$A = \frac{C}{X} - B \qquad \text{Add -B.}$$

To add or subtract, opposite side—opposite sign.

Practice Problems 3

1. Solve for A: $A(B + C) - D = 5$

2. Solve for B: $A(B + C) - D = 5$

3. Solve for C: $A(B + C) - D = 5$

4. Solve for D: $A(B + C) - D = 5$

Solutions 3

1. Solve for A: $A = \dfrac{D + 5}{B + C}$

2. Solve for B: $B = \dfrac{D + 5}{A} - C$

3. Solve for C: $C = \dfrac{D + 5}{A} - B$

4. Solve for D: $A(B + C) - 5 = D$ or $D = A(B + C) - 5$

Ratios

You know how to do ratios. Now we are going to derive equations from ratios in order to solve for missing information. In most of the problems in this lesson, the key will be to find all the possible ratios and then choose the right one to find the correct answer. Let's do some problems to learn these two important concepts.

Example 1

In the parking lot, there are 72 motor vehicles comprised entirely of motorcycles and cars. The ratio of cars to motorcycles is three to one. How many cars are there?

You can derive three ratios from this information. They are listed below. Only one ratio was given, the cars to cycles, but we can also derive the cars to the total and the cycles to the total.

Cars to cycles $\qquad \dfrac{\text{Cars}}{\text{Cycles}} = \dfrac{3}{1}$

Cars to total vehicles $\qquad \dfrac{\text{Cars}}{\text{Total}} = \dfrac{3}{4}$

Cycles to total vehicles $\qquad \dfrac{\text{Cycles}}{\text{Total}} = \dfrac{1}{4}$

Of the three ratios, which one uses the information given (total vehicles) and the information requested (number of cars)? The second ratio, cars to total vehicles, is the one to use.

$$\frac{\text{Cars}}{\text{Total}} = \frac{3}{4} \rightarrow \frac{\text{Cars}}{72} = \frac{3}{4} \rightarrow \frac{3 \cdot 72}{4} = 54$$

We can use this information to deduce that there are 18 cycles since cars + cycles = 72.

You can also use what you know about atomic weights and ratios to find how many grams of each element are present in a given amount of that compound. Water has an atomic weight of 18 (two hydrogens at 1 each and one oxygen at 16). There are three possible ratios to derive from this compound.

In example 2, we are asked to find the mass of hydrogen if there are 1,440 grams of water.

Example 2

Hydrogen to water $\qquad \dfrac{H_2}{H_2O} = \dfrac{2}{18}$

Oxygen to water $\qquad \dfrac{O}{H_2O} = \dfrac{16}{18}$

Hydrogen to oxygen $\qquad \dfrac{H_2}{O} = \dfrac{2}{16}$

If we know there are 1,440 grams of water, we can use our ratios to find the mass of the hydrogen present and the mass of the oxygen present. To find hydrogen's mass, choose the ratio that has water and hydrogen since we have been given the amount of water, and we are looking for the amount of hydrogen.

$$\frac{H_2}{H_2O} = \frac{2}{18} \rightarrow \frac{H_2}{1440} = \frac{2}{18} \rightarrow \frac{2 \cdot 1440}{18} \rightarrow H_2 = 160 \text{ grams}$$

Example 3

Find the mass of carbon in CS_2.
There are 1,596 grams of the compound.

Carbon to compound $\qquad \dfrac{C}{CS_2} = \dfrac{12}{76}$

Carbon to sulfur $\qquad \dfrac{C}{S_2} = \dfrac{12}{64}$

Sulfur to compound $\qquad \dfrac{S_2}{CS_2} = \dfrac{64}{76}$

Of the three ratios, which one uses the information given (total grams of the compound) and the information requested (mass of the carbon)? The first ratio, carbon to compound, is the one to use.

$$\frac{C}{CS_2} = \frac{12}{76} \;\rightarrow\; \frac{C}{1596} = \frac{12}{76} \;\rightarrow\; \frac{12 \cdot 1596}{76} \;\rightarrow\; C = 252 \text{ grams}$$

Sometimes there are three elements in a compound, which increases the number of ratios you can have. Choose the best ratio to find the needed answer.

Practice Problems 1

1. The oak tree has 56 birds sitting and singing on its branches. A close look reveals only bluebirds and cardinals are present. The ratio of bluebirds to cardinals is three to five. How many cardinals are there?

2. In Atlanta, 42,000 fans came to the game. Braves' fans outnumbered Pirates' fans two to one. How many intrepid Pirates' fans were at the game?

3. Find the mass of carbon in KCN.
   There are 455 grams of the compound.

4. Find the mass of nitrogen in KCN.
   There are 455 grams of the compound.

5. Find the mass of oxygen in MgO.
   There are 1,560 grams of the compound.

6. Find the mass of magnesium in MgO.
   There are 1,560 grams of the compound.

Solutions 1

1. $\dfrac{\text{Blue}}{\text{Card}} = \dfrac{3}{5}, \dfrac{\text{Blue}}{\text{Total}} = \dfrac{3}{8}, \dfrac{\text{Card}}{\text{Total}} = \dfrac{5}{8}$

   $\dfrac{\text{Card}}{\text{Total}} = \dfrac{5}{8} \rightarrow \dfrac{\text{Card}}{56} = \dfrac{5}{8} \rightarrow \text{Card} = \dfrac{5 \cdot 56}{8} = 35$

2. $\dfrac{\text{Braves}}{\text{Pirates}} = \dfrac{2}{1}, \dfrac{\text{Braves}}{\text{Total}} = \dfrac{2}{3}, \dfrac{\text{Pirates}}{\text{Total}} = \dfrac{1}{3}$

   $\dfrac{\text{Pirates}}{\text{Total}} = \dfrac{1}{3} \rightarrow \dfrac{\text{Pirates}}{42,000} = \dfrac{1}{3} \rightarrow \text{Pirates} = \dfrac{1 \cdot 42,000}{3} = 14,000$

3. $\dfrac{K}{KCN} = \dfrac{39}{65}, \dfrac{C}{KCN} = \dfrac{12}{65}, \dfrac{N}{KCN} = \dfrac{14}{65}$

   $\dfrac{C}{KCN} = \dfrac{12}{65} \rightarrow \dfrac{C}{455} = \dfrac{12}{65} \rightarrow C = \dfrac{12 \cdot 455}{65} = 84 \text{ g}$

   With three elements present, there are other ratios between the elements themselves, such as:

   $\dfrac{K}{C} = \dfrac{39}{12} \qquad \dfrac{K}{N} = \dfrac{39}{14} \qquad \dfrac{N}{C} = \dfrac{14}{12}$

4. $\dfrac{K}{KCN} = \dfrac{39}{65}, \dfrac{C}{KCN} = \dfrac{12}{65}, \dfrac{N}{KCN} = \dfrac{14}{65}$

   $\dfrac{N}{KCN} = \dfrac{14}{65} \rightarrow \dfrac{C}{455} = \dfrac{14}{65} \rightarrow C = \dfrac{14 \cdot 455}{65} = 98 \text{ g}$

5. $\dfrac{\text{Mg}}{\text{MgO}} = \dfrac{24}{40}, \dfrac{O}{\text{MgO}} = \dfrac{16}{40}, \dfrac{O}{\text{Mg}} = \dfrac{16}{24}$

   $\dfrac{O}{\text{MgO}} = \dfrac{16}{40} \rightarrow \dfrac{O}{1560} = \dfrac{16}{40} \rightarrow O = \dfrac{16 \cdot 1560}{40} = 624 \text{ g}$

6. $\dfrac{\text{Mg}}{\text{MgO}} = \dfrac{24}{40}, \dfrac{O}{\text{MgO}} = \dfrac{16}{40}, \dfrac{O}{\text{Mg}} = \dfrac{16}{24}$

   $\dfrac{\text{Mg}}{\text{MgO}} = \dfrac{24}{40} \rightarrow \dfrac{\text{Mg}}{1560} = \dfrac{24}{40} \rightarrow \text{Mg} = \dfrac{24 \cdot 1560}{40} = 936 \text{ g}$

Unit Multipliers and Metric Conversions

To change inches to feet, inches to yards, ounces to pounds, pounds to tons, etc., you need to learn only two new skills. The first is how to make "one," and the second is how to divide so as to produce the correct unit of measure. Let's view some examples to see how these two new skills function.

Skill #1 (example 1 below) — We must make one, so that when we multiply one unit times two feet, we still have two feet. It may be more pieces and a different form, but it must still be equal to two feet. So we ask ourselves, "What has inches and feet in it that is equal to one?"

There are two possibilities: 1 ft/12 in = 1, or 12 in/ft = 1.

Both equations equal one, because the numerator and denominator are identical in value, only expressed in different ways. These fractions are called *unit multipliers*. Either of them could be multiplied times two feet without changing its value, because both are equal to one, and one times anything is still one.

Example 1
Change two feet to inches.
First we write two feet as a fraction, with feet in the numerator.

$$\frac{2 \text{ feet}}{1} \times \boxed{} = \frac{\text{inches}}{}$$

The missing term must be equal to 1. It is our unit multiplier.

$$\frac{1 \text{ foot}}{12 \text{ inches}} = 1 \text{ or } \frac{12 \text{ inches}}{1 \text{ foot}} = 1$$

Skill #2 (example 2 below) — Which one of the unit multipliers, when multiplied times two feet, will leave only inches, the desired unit of measure?

Example 2

$$\frac{2 \text{ ft}}{1} \times \frac{1 \text{ ft}}{12 \text{ in}} = \frac{2 \text{ ft}^2}{12 \text{ in}}$$

Here we still have feet in the answer, but we want just inches.

$$\frac{2 \text{ ft}}{1} \times \frac{12 \text{ in}}{1 \text{ ft}} = \frac{24 \text{ in}}{1}$$

Here we have only inches in the answer, which is what we want.

Notice that the "feet" are canceled as a result of having feet in the numerator of the first fraction and feet in the denominator of the second fraction. The key here is not the numbers, which tell how many, but the unit of measure, which tells what kind—in this example, the feet and the inches. Since we begin with feet in the numerator, skill #2 is to make sure there is a "feet" in the denominator of our unit multiplier. And since we want inches in the numerator at the end, we must have inches in the numerator of our unit multiplier. Here are two more examples:

Example 3
Change 64 ounces to pounds.

$$\frac{64 \text{ oz}}{1} \times \frac{1 \text{ pound}}{16 \text{ oz}} = \frac{4 \text{ pounds}}{1}$$

Example 4
Change one-half yard to inches.

$$\frac{1 \text{ yd}}{2} \times \frac{36 \text{ in}}{1 \text{ yd}} = \frac{18 \text{ in}}{1}$$

Practice Problems 1
Change 12 pints to quarts.

1. Select the unit multiplier to be used.

2. Which one goes in the numerator? _____

3. Which one goes in the denominator? _____

4. Solve the equation.

Change seven yards to feet.

5. Select the unit multiplier to be used.

6. Which one goes in the numerator? _____

7. Which one goes in the denominator? _____

8. Solve the equation.

Solutions 1

1. $\dfrac{1 \text{ qt}}{2 \text{ pt}}$

2. 1 qt

3. 2 pt

4. $12 \not{\text{pt}} \times \dfrac{1 \text{ qt}}{2 \not{\text{pt}}} = 6 \text{ qt}$

5. $\dfrac{3 \text{ ft}}{1 \text{ yd}}$

6. 3 ft

7. 1 yd

8. $7 \not{\text{yd}} \times \dfrac{3 \text{ ft}}{1 \not{\text{yd}}} = 21 \text{ ft}$

Square and Cubic Unit Multipliers

How many square inches in one square foot? Or how many in^2 equal 1 ft^2?
One square foot = 1 ft x 1 ft = 1 ft^2. One square inch = 1 in x 1 in = 1 in^2.

$$\frac{1 \text{ ft}}{1} \times \frac{1 \text{ ft}}{1} \times \frac{?}{?} \times \frac{?}{?} = \frac{? \text{ in}^2}{1}$$

$$\frac{1 \not{\text{ft}}}{1} \times \frac{1 \not{\text{ft}}}{1} \times \frac{12 \text{ in}}{1 \not{\text{ft}}} \times \frac{12 \text{ in}}{1 \not{\text{ft}}} = \frac{144 \text{ in}^2}{1}$$

Change 2 ft^3 (2 cubic feet) to in^3 (cubic inches).

$$\frac{2 \not{\text{ft}}}{1} \times \frac{1 \not{\text{ft}}}{1} \times \frac{1 \not{\text{ft}}}{1} \times \frac{12 \text{ in}}{1 \not{\text{ft}}} \times \frac{12 \text{ in}}{1 \not{\text{ft}}} \times \frac{12 \text{ in}}{1 \not{\text{ft}}} = \frac{3,456 \text{ in}^3}{1}$$

Sometimes you will need to use more than one unit multiplier in the same problem. In the following examples, watch as one mile is changed to inches and one gallon is changed to cups.

$$\frac{1\ \text{mi}}{1} \times \frac{5{,}280\ \text{ft}}{1\ \text{mi}} \times \frac{12\ \text{in}}{1\ \text{ft}} = \frac{63{,}360\ \text{in}}{1}$$

$$\frac{1\ \text{gal}}{1} \times \frac{4\ \text{quart}}{1\ \text{gal}} \times \frac{2\ \text{pints}}{1\ \text{quart}} \times \frac{2\ \text{cups}}{1\ \text{pint}} = \frac{16\ \text{cups}}{1}$$

Practice Problems 2

1. Change 40 ft^2 into square inches.

2. Change 8 yd^2 into square feet.

3. Change 370 cm^3 into cubic meters.

4. Change 9,500 m^2 into km^2.

5. Change 11 ft^3 into cubic inches.

6. Change 16 m^3 into cubic centimeters.

Solutions 2

1. $\dfrac{40\ \text{ft}}{1} \times \dfrac{\text{ft}}{1} \times \dfrac{12\ \text{in}}{\text{ft}} \times \dfrac{12\ \text{in}}{\text{ft}} = 5{,}760\ \text{in}^2$

2. $\dfrac{8\ \text{yd}}{1} \times \dfrac{\text{yd}}{1} \times \dfrac{3\ \text{ft}}{\text{yd}} \times \dfrac{3\ \text{ft}}{\text{yd}} = 72\ \text{ft}^2$

3. $\dfrac{370\ \text{cm}}{1} \times \dfrac{\text{cm}}{1} \times \dfrac{\text{cm}}{1} \times \dfrac{\text{m}}{100\ \text{cm}} \times \dfrac{\text{m}}{100\ \text{cm}} \times \dfrac{\text{m}}{100\ \text{cm}} = .00037\ \text{m}^3$

4. $\dfrac{9{,}500\ \text{m}}{1} \times \dfrac{\text{m}}{1} \times \dfrac{\text{km}}{1{,}000\ \text{m}} \times \dfrac{\text{km}}{1{,}000\ \text{m}} = .0095\ \text{km}^2$

5. $\dfrac{11\ \text{ft}}{1} \times \dfrac{\text{ft}}{1} \times \dfrac{\text{ft}}{1} \times \dfrac{12\ \text{in}}{\text{ft}} \times \dfrac{12\ \text{in}}{\text{ft}} \times \dfrac{12\ \text{in}}{\text{ft}} = 19{,}008\ \text{in}^3$

6. $\dfrac{16\ m}{1} \times \dfrac{m}{1} \times \dfrac{m}{1} \times \dfrac{100\ cm}{m} \times \dfrac{100\ cm}{m} \times \dfrac{100\ cm}{m} = 16{,}000{,}000\ cm^3$

Metric Conversions

There are many metric equivalents available for use when converting from metric to imperial measure and vice versa. In the following tables, I've listed the ones I feel are most important. The conversions are approximate and have been rounded for ease in calculations. Because they are rounded, you may get slightly different answers, depending on whether you choose the table on the left or the one on the right for your calculations. The easiest way to do these is to make the denominator of the unit multiplier one rather than a decimal. This way you multiply a decimal instead of dividing it. Notice examples 5 and 6 to see this illustrated.

METRIC TO IMPERIAL	IMPERIAL TO METRIC
1 centimeter ≈ .4 inches	1 inch ≈ 2.5 centimeters
1 meter ≈ 1.1 yards	1 yard ≈ .9 meters
1 kilometer ≈ .62 miles	1 mile ≈ 1.6 kilometers
1 gram ≈ .035 ounces	1 ounce ≈ 28 grams
1 kilogram ≈ 2.2 pounds	1 pound ≈ .45 kilograms
1 liter ≈ 1.06 quarts	1 quart ≈ .95 liters

Using what we've learned about unit multipliers, let's work some conversions in the following examples.

Example 5
Change five liters to quarts.

$$\dfrac{5\ liters}{1} \times \dfrac{1.06\ quarts}{1\ liter} = \dfrac{5.3\ quarts}{1}$$

It seems easier to multiply by 1.06 than to divide by .95. Notice the slightly different answers.

Example 6
Change five liters to quarts.

$$\dfrac{5\ liters}{1} \times \dfrac{1\ quart}{.95\ liter} = \dfrac{5.26\ quarts}{1}$$

Example 7
Change 34 inches to centimeters.

$$\frac{34 \text{ in}}{1} \times \frac{2.5 \text{ cm}}{1 \text{ in}} = \frac{85 \text{ cm}}{1}$$

Practice Problems 3

1. Change eight kilometers to miles.

2. Change 27 ounces to grams.

3. Change eight kilograms to pounds.

4. Change five yards to meters.

5. Change 250 grams to ounces.

6. Change 12 quarts to liters.

Solutions 3

1. $\dfrac{8 \text{ km}}{1} \times \dfrac{.62 \text{ mi}}{1 \text{ km}} = \dfrac{4.96 \text{ mi}}{1}$

2. $\dfrac{27 \text{ oz}}{1} \times \dfrac{28 \text{ g}}{1 \text{ oz}} = \dfrac{756 \text{ g}}{1}$

3. $\dfrac{8 \text{ kg}}{1} \times \dfrac{2.2 \text{ lb}}{1 \text{ kg}} = \dfrac{17.6 \text{ lb}}{1}$

4. $\dfrac{5 \text{ yd}}{1} \times \dfrac{.9 \text{ m}}{1 \text{ yd}} = \dfrac{4.5 \text{ m}}{1}$

5. $\dfrac{250 \text{ g}}{1} \times \dfrac{.035 \text{ oz}}{1 \text{ g}} = \dfrac{8.75 \text{ oz}}{1}$

6. $\dfrac{12 \text{ qt}}{1} \times \dfrac{.95 \text{ liters}}{1 \text{ qt}} = \dfrac{11.4 \text{ liters}}{1}$

Distance = Rate x Time

In this lesson, we are going to cover two variations of motion problems. I hope you don't get motion sickness easily! In our study, we are going to assume that the motion is uniform and constant. When you get to physics, you will understand what this means more fully. In other words, these are going to be relatively simple problems, but they will also be a good application of our study of rational expressions.

There are three components that we will be dealing with: D for Distance, R for Rate, and T for Time. You can remember them as DRT or D i RT. Distance is equal to Rate multiplied by Time, or D = RT. There are several ways to rearrange these variables, all of which are spinoffs of the original equation:

$$D = RT \qquad D/T = R \qquad D/R = T$$

Each of these variables will eventually be replaced by a number (which tells how many) and an accompanying value (which tells what kind). Distance could be six miles or four feet or three yards but it is always how far, as in length or distance. Time can be two hours or five seconds or seven years; it is always how long in terms of elapsed time. Rate is a combination of the distance and the time. Problems like this will remind you of similar problems done with unit multipliers. Rate is how fast a distance is covered in how long a time. The most common application is miles per hour (mph or miles/hour) or kilometers per hour (km/h or kilometers/hour). We read 60 mph as "sixty miles per hour." The equation we began with, D = RT, is convenient because all the variables are on one line. This is really a derivation of Rate = Distance/Time, or R = D/T. Rate may also be referred to as velocity.

Let's do some problems with 60 m/h as the rate, and solve for distance and time. Since there are three unknowns in this equation, when we have the information for two of the unknowns, we can solve to find the third.

$$\text{Rate} = \frac{\text{Distance}}{\text{Time}} \text{ or } 60 \text{ m/h} = \frac{60 \text{ mi}}{1 \text{ hr}}$$

Example 1
If you drive at a rate of 60 mph for 3 hours (time), how far (distance) have you driven? Since we are looking for distance, we choose D = RT as the equation. Then replace R with 60 mph and T with 3 hours, and solve for D.

$$D = R \times T = \frac{60 \text{ mi}}{1 \text{ hr}} \times \frac{3 \text{ hr}}{1} = \frac{180 \text{ mi}}{1}$$

(The values divide out as in unit multipliers.)

Example 2
How long will it take to drive 270 miles at 60 mph? Since we are looking for T, we choose T = D/R, and replace R with 60 mph and D with 270 miles.

$$T = \frac{D}{R} = \frac{270}{\frac{60 \text{ mi}}{1 \text{ hr}}} = \frac{\frac{270 \text{ mi}}{1}}{\frac{60 \text{ mi}}{1 \text{ hr}}} \times \frac{\frac{1 \text{ hr}}{60 \text{ mi}}}{\frac{1 \text{ hr}}{60 \text{ mi}}} = \frac{4.5 \text{ hr}}{1}$$

Example 3
What was your average rate of travel if you drove 420 miles in 8 hours? Since we are looking for rate, we choose R = D/T as the equation. Then replace D with 420 miles and T with 8 hours, and solve for R.

$$R = \frac{D}{T} = \frac{420 \text{ mi}}{8 \text{ hr}} = \frac{52.5 \text{ mi}}{1 \text{ hr}} = 52.5 \text{ mph}$$

Practice Problems 1

1. You ran 3.5 miles in one-half hour. What was your rate in miles per hour?

2. Walking at 4 mph, we covered 3 miles. How long did it take?

3. Jeff drove 11 hours at 85 kilometers per hour (km/h). How far did Jeff drive?

4. Sandi was driving at 80 km/h. I fell asleep at 8:30 AM. She drove 220 kilometers while I slept. What time is it now?

5. Riding the stallion for 20 minutes, Fritha covered 7 miles. How fast did she ride?

6. Gina jogged at 6 mph for 30 minutes. How far did she go?

Solutions 1

1. $R = \dfrac{D}{T} = \dfrac{3.5 \text{ mi}}{.5 \text{ hr}} = 7 \text{ mph}$

2. $T = \dfrac{D}{R} = \dfrac{3 \text{ mi}}{\frac{4 \text{ mi}}{1 \text{ hr}}} = 3/4 \text{ hours} = 45 \text{ minutes}$

3. $D = RT = \dfrac{85 \text{ km}}{1 \text{ hr}} \times \dfrac{11 \text{ hr}}{1} = 935 \text{ km}$

4. $T = \dfrac{D}{R} = \dfrac{220 \text{ km}}{\frac{80 \text{ km}}{1 \text{ hr}}} = \begin{array}{l} 2\,3/4 \text{ hours} = 2:45 \\ 8:30 + 2:45 = 11:15 \text{ AM} \end{array}$

5. $R = \dfrac{D}{T} = \dfrac{7 \text{ mi}}{\frac{1}{3} \text{ hr}} = 21 \text{ mph}$

6. $D = RT = \dfrac{6 \text{ mi}}{1 \text{ hr}} \times \dfrac{.5 \text{ hr}}{1} = 3 \text{ mi}$

Now we'll do some problems with more than one person or car. In these types of problems, drawing or sketching a picture is a great help. Now take what you learned about D = RT and apply it at another level. The best way is to consider a problem and talk our way through it.

Example 4

Isaac and Ethan were driving out to Pittsburgh. Isaac took the PA Turnpike and averaged 50 mph while Ethan took scenic Route 30 and averaged 10 mph less than Isaac. If they both left at 10:00 AM and Isaac arrived at his grandparents' home at 4:00 PM, when did Ethan arrive?

To represent Isaac's Distance, Rate, and Time, we have two different options:

$$D_{Isaac} = R_{Isaac} \, T_{Isaac} \qquad\qquad D_I = R_I \, T_I$$

Choose one that you are comfortable with. I am going to use the second option with *I* subscripts for Isaac and *E* subscripts for Ethan.

Step 1 — Identify unknowns.

$$D_I = R_I T_I \qquad\qquad D_E = R_E T_E$$

Step 2 — Draw a picture.

Phil. $\xrightarrow[\quad D_E \quad]{\quad D_I \quad}$ Pitts.

The distance for them is the same. $\qquad D_I = D_E$

Substitute RT for D in both equations. $\qquad R_I T_I = R_E T_E$

Step 3 — Figure out what we know, and then substitute.

Isaac's time is 6 hours.
    (4:00 PM – 10:00 AM)
Isaac's rate is 50 mph. $\qquad\qquad\qquad$ (50 mph)(6 hours) =
Ethan's rate is 40 mph. $\qquad\qquad\qquad\qquad$ (40 mph)(Time)
    (10 mph less than Isaac).

Step 4 — Solve for the missing information.

$$\frac{(50 \text{ mph})(6 \text{ hr})}{(40 \text{ mph})} = 7.5 \text{ hr} = (\text{Time})$$

Ethan's arrival is 7.5 hours + 10:00 AM = 5:30 PM

How far did they travel?

Choose either Isaac: D = RT, 50 mph x 6 hours = 300 miles,
or Ethan: 40 mph x 7.5 hours = 300 miles.

Example 5

Wesley and Derrick walked the long way to work. Wesley walked at 4 mph, and Derrick walked at 6 mph. Derrick arrived one hour sooner than Wesley, so how far is it to work?

Step 1 — Identify unknowns. $D_W = R_W T_W$ $D_D = R_D T_D$

Step 2 — Draw a picture.

The distance for them is the same. $D_W = D_D$
Substitute RT for D in both equations. $R_W T_W = R_D T_D$

Step 3 — Figure out what we know, and then substitute.

Wesley's time is 1 hr more than Derrick's. $T_W = T_D + 1$
Wesley's rate is 4 mph.
Derrick's rate is 6 mph. $(4 \text{ mph})(T_D + 1) = (6 \text{ mph})(T_D)$

Step 4—Solve for the missing information.

$$4T_D + 4 = 6T_D$$
$$4 = 2T_D$$
$$2 = T_D$$

If Derrick's time is 2 hours, then Wesley's is 3 hours, since Wesley's time is 1 hour more than Derrick's.

How far did they travel?
　　Choose either Derrick: D = RT, 6 mph x 2 hours = 12 miles, or Wesley: 4 mph x 3 hours = 12 miles.

Example 6

Johnny left on his bike at 9:00 AM, speeding along at 6 mph. Joseph left at 10:00 AM and caught him at the park at noon. How fast was Joseph traveling? Joseph will be F, and Johnny will be N.

Step 1 — Identify Unknowns.　　$D_N = R_N T_N$　$D_F = R_F T_F$

Step 2 — Draw a picture.

$$D_N$$
Home ———————————→ Park
$$D_F$$

The distance for them is the same.　　$D_N = D_F$
Substitute RT for D in both equations.　　$R_N T_N = R_F T_F$

Step 3 — Figure out what we know, and then substitute.

Johnny's time is 3 hours.
Joseph's time is 2 hours.　$\Big\}$ $(6 \text{ mph})(3 \text{ hrs}) = (R_F)(2 \text{ hrs})$
Johnny's rate is 6 mph.

Step 4 — Solve for the missing info.

$$18 = 2R_F$$
$$9 = R_F$$

Joseph's rate is 9 mph.

How far did they travel?
　　Choose either Joseph: D = RT, 9 mph x 2 hours = 18 miles, or Johnny: 6 mph x 3 hours = 18 miles.

Practice Problems 2

1. Claire left at 6:00 AM on a long canoe trip. She reached the bridge at 10:00 AM. Gretchen left one hour later and arrived at the bridge at the same time as Claire. Gretchen paddles 2 mph faster than Claire. How fast did they travel and how far?

2. While riding their tandem bicycle to South Park at 9 mph, Calvin and Kathie found the trip to the park was slightly downhill. On their trip home, they were able to go only 6 mph, and it took them 6 hours to get back. How long did it take to get to the park originally, and how far is it?

3. With Katie driving, the 450-mile trip to Maine took 9 hours. On the way home, Stephanie drove 5 mph less than Katie and took in the scenery. How long did it take to get home?

4. Uncle Cal had the sailboat skimming along and made it down to Jan's Marina in 2 hours. On the way back, he didn't catch the wind just right, and it took him 5 hours. His speed on the way back was 12 mph less than on the way down. What was the rate of the boat going down to the marina?

Solutions 2

1. $D_C = D_G$

               $R_C T_C = R_G T_G$

          $(R_C)(4h) = (R_C + 2)(3h)$

               $4R_C = 3R_C + 6$

                $R_C = 6$ so $R_G = 6 + 2 = 8$ mph

        Claire: 6 mph x 4 hours = 24 miles

2. $D_P = D_H$

            $R_P T_P = R_H T_H$

         $(9)(T_P) = (6)(6)$

            $9T_P = 36$

              $T_P = 4h$

            $D_P = (9)(4) = 36$ miles

3. $D_K = D_S$

            $R_K T_K = R_S T_S$ 450 mi/9 hr = 50 mph

        $(50)(9h) = (50 - 5)(T)$

            $450 = 45T$

      10 hours = T

4. $D_D = D_B$

              $R_D T_D = R_B T_B$

       $(R_D)(2 \text{ hours}) = (R_D - 12)(5 \text{ hours})$

                $2R = 5R - 60$

                $R = 20$ mph

More Motion Problems

As shown before, the key to solving a motion problem is how you draw the problem. After the sketch is made, begin by substituting known information in order to isolate the variable. Here are four examples with accompanying discussion.

Example 1

Two saleswomen left the hotel at the same time, traveling in opposite directions. Sue headed north at 60 mph while Kelly headed south at 56 mph. If they left at 7:22 AM, what time will it be when they are 290 miles apart?

Step 1 — Identify unknowns.

$$D_S = R_S T_S \qquad D_K = R_K T_K \qquad \text{S for Sue and K for Kelly}$$

Step 2 — Draw a picture.

$$\overset{D_S \qquad \text{Hotel} \qquad D_K}{\underset{\underset{N}{\longleftarrow} \qquad \underset{290 \text{ mi}}{|}}{\rule{12cm}{0.4pt}}}$$

The distance adds up to 290 miles. $\quad D_S + D_K = 290$

Substitute RT for D in both equations. $R_S T_S + R_K T_K = 290$

Step 3 — Figure out what we know, and then substitute.

$T_S = T_K$, so we can use T
   since the time is the same.

Sue's rate is 60 mph.

Kelly's rate is 56 mph.

$$(60)(T) + (56)(T) = 290$$

Step 4 — Solve for the missing information.

$(60)(T) + (56)(T) = 290$

$116\,T = 290$

$T = 2.5$ hours   The time is 7:22 + 2:30 = 9:52

Step 5 — Check the solution. $(60)(2.5) + (56)(2.5) = 290$ It works!

Example 2

On the fund-raising bike hike, Heidi rode until she was tired and then pushed the bike the rest of the way. She pushed her bike at 4 mph and rode it at 9 mph. She finished the 44-mile trek in 6 hours. How long did she push, and how long did she ride?

Step 1 — Identify unknowns.

$D_P = R_P T_P$ $D_R = R_R T_R$ P for push and R for ride.

Step 2 — Draw a picture.

The distance adds up to 44 miles. $D_P + D_R = 44$

Substitute RT for D in both equations. $R_P T_P + R_R T_R = 44$

Step 3 — Figure out what we know, and then substitute.

Pushing rate is 4 mph.
Riding rate is 9 mph. $(4)(T_P) + (9)(6 - T_P) = 44$
$T_P + T_R = 6,\ T_R = 6 - T_P$

Step 4 — Solve for the missing information.

$(4)(T_P) + (9)(6 - T_P) = 44$

$4T_P + 54 - 9T_P = 44$

$-5T_P = -10$

$T_P = 2$ Time pushing is 2, so riding is 4.

Step 5 — Check the solution. $(4)(2) + (9)(4) = 44$

Example 3

Samuel was at the park 6 miles away. Ada knew he would be leaving at 1:30 PM, so she harnessed the horse and left in the buggy at the same time. Samuel walks at 5 mph and the buggy travels at 10 mph. How soon till they meet?

Step 1 — Identify unknowns.

$$D_S = R_S T_S \qquad D_A = R_A T_A \qquad \text{S for Samuel and A for Ada}$$

Step 2 — Draw a picture.

The distance adds up to 6 miles. $D_S + D_A = 6$

Substitute RT for D in both equations. $R_S T_S + R_A T_A = 6$

Step 3 — Figure out what we know, and then substitute.

Samuel's rate is 5 mph.
Ada's rate is 10 mph. $(5)(T) + (10)(T) = 6$
The time is the same, so $T_S = T_A$.

Step 4 — Solve for the missing information.

$$15T = 6$$
$$T = 2/5 \text{ hr}$$

2/5 of an hour is 2/5 of 60 or 24 minutes,

$$1:30 \text{ PM} + 24 \text{ min} = 1:54 \text{ PM}$$

Step 5 — Check the solution. $(5)(2/5) + (10)(2/5) = 6$

Practice Problems 1

1. The CB radio had a range of 20 miles. David walked west at a rate of 4 mph and Jonathan left at the same time, jogging at a rate of 8 mph towards the east. How long did it take them to reach the maximum range of their radios?

2. The lake was 3,000 yards wide. Glenda and Jessica were on opposite shores and decided to swim toward each other. Glenda swims at a rate of 120 yards per minute while Jessica strokes along at 80 yards per minute, so Jessica was given a five-minute head start. When did they meet?

3. While rafting down a 19.5 mile stretch of the Youghiogheny River, we went through the rapids at 5 mph, and then paddled in the slower current at 3 mph. We spent twice as much time in the rapids as in the slower current. How long were we in each?

Solutions 1

1. $D_D = R_D T_D$ $D_J = R_J T_J$ D for David and J for Jonathan

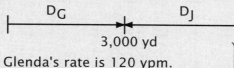

20 mi

David's rate is 4 mph.
Jonathan's rate is 8 mph.
The time is the same, so $T_D = T_J$.

$D_D + D_J = 20$

$R_D T_D + R_J T_J = 20$

$(4)(T) + (8)(T) = 20$

$12T = 20$

$T = 1\ 2/3$

$T = 1\ 2/3$ hr

2. $D_G = R_G T_G$ $D_J = R_J T_J$ G for Glenda and J for Jessica

3,000 yd

Glenda's rate is 120 ypm.
Jessica's rate is 80 ypm.

$T_J = T_G + 5$ since Jessica
is 5 minutes more than Glenda.

$D_G + D_J = 3,000$

$R_G T_G + R_J T_J = 3,000$

$(120)(T_G) + (80)(T_G + 5) = 3,000$

$200T + 400 = 3,000$

$T = 13$

13 minutes for Glenda and 18 minutes for Jessica.

3. $D_S = R_S T_S$ $D_R = R_R T_R$ R for Rapids and S for Slower

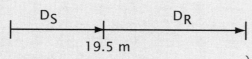

19.5 m

Rapids' rate is 5 mph.
Slower current's rate is 3 mph.

Time in rapids = 2 x time
        in slow, so $T_R = 2T_S$.

$D_S + D_R = 19.5$

$R_S T_S + R_R T_R = 19.5$

$(3)(T_S) + (5)(2T_S) = 19.5$

$13T = 19.5$

$T = 1\ 1/2$

$T = 1\ 1/2$ hr in slower current and 3 hr in rapids

Graphing Lines: Slope-Intercept Formula

You learned how to graph a line in *Algebra 1*. The line on the graph was compared to baking bread for a bake sale. The X-axis represented time and the Y-axis represented loaves of bread. If you baked three loaves of bread every hour, the line would look like the one labeled 3 in figure 1.

Two words that describe what we do in graphing lines are slope and intercept. The ***slope-intercept formula*** is $Y = mX + b$. In the formula, the ***slope*** is m, and the ***intercept*** is b. In $Y = 2X + 3$, 2 is the slope and 3 is the intercept.

Slope is the $\frac{\text{up dimension}}{\text{over dimension}}$. In our example of the bread baking, for every one hour (over), we were able to bake three loaves (up). So for every hour, we move over to the right one space and up three spaces. We continue to do this, and when we connect two or more points, we have a line that "slopes" up. We describe the slope as $\frac{3 \text{ up}}{1 \text{ over}}$ or 3. If we made five loaves each hour, it would be a steeper slope, $\frac{5 \text{ up}}{1 \text{ over}}$ or 5. The larger the slope, the steeper the line. If it takes four hours to make one loaf of bread, the slope would be $\frac{1 \text{ up}}{4 \text{ over}}$ or 1/4. It would be over four spaces and up one space. Look at figure 1 for slopes of 3, 5, and 1/4.

Figure 1   positive slopes

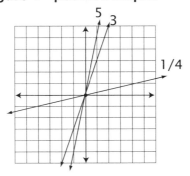

Figure 2   negative slopes

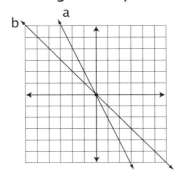

You can also have negative slopes. An example could be the businessman who loses two dollars each day. For every one day—minus two dollars. The slope is over one and down two (the opposite of up because it is minus). It will look like line *a* in figure 2 on the previous page ($Y = -2X + 0$, or $Y = -2X$). Line *b* is an example of losing one dollar per day ($Y = -1X$, or $Y = -X$).

When I think of an intercept, it brings to mind a jet. Think of the slope of the line as being the path of the jet, and the point where it "intercepts" the Y-axis as the intercept. There are an infinite number of parallel lines that have the same slope, but when you add the intercept, you narrow it down to one specific line.

Practice Problems 1
Estimate the slope and the intercept, and match it with the most probable equation given on the next page.

A.
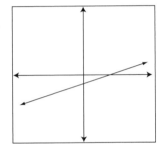

B.
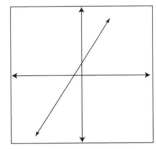

C.
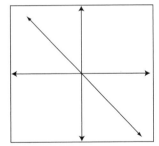

D.
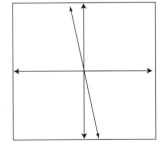

E. 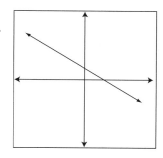   F.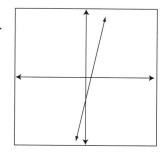

1. Y = 1/3 X – 2   2. Y = 5X – 3

3. Y = –1X + 0 or Y = –X   4. Y = 2X + 2

5. Y = –4X   6. Y = –1/3 X + 2

Solutions 1

1. A   2. F

3. C   4. B

5. D   6. E

Equation of a Line with Different Givens

The slope-intercept formula is very helpful in drawing a line on a Cartesian coordinate graph. Find the intercept and draw a point. Then move from that point to another point by counting over and up according to the slope. Once you have two points, you can connect them and draw your line. Now we will be given slightly different information, but we still need to find the slope and the intercept in order to draw an accurate line and have an exact equation.

Example 1
Given: slope = 3, through the point (1, 2) on the same line.
Drawing this:

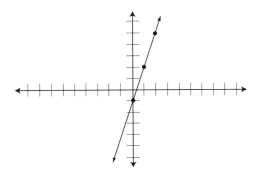

We start at the point (1, 2) and move over 1 and up 3 (slope = 3/1 or 3). Connecting these points, the line appears to intercept the Y-axis at point (0, –1). We have a pretty good idea that our intercept is (–1). To find out for sure, we'll substitute what we know from the givens into the slope-intercept formula.

Y = m X + b

Substitute 3 for the slope or *m*.
Y = 3 X + b

Substitute the point (1, 2) for X and Y.
(2) = 3(1) + b
Solve for *b*.
2 = 3 + b
–1 = b
So the intercept is –1.

Now we'll find the slope when only given two points.

Example 2

Given: point 1: (−1, 5) and point 2: (1, 1).

Plot the points and estimate the intercepts and the slope.
Then find the slope-intercept formula for the line.

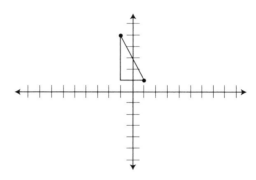

To find the slope, draw a right triangle from point 1 to point 2.
You can see that the over dimension of the triangle is 2, and the up (or
down in this case) dimension is −4. The slope is −4/2 or −2. Now we are
at the same place as in example 1. We have the slope, and we choose
one of the given points in order to find the intercept. In the following
example, I chose point 2, or (1, 1).

$Y = -2X + b$ The slope or *m*, is −2

Substituting (1, 1).

$(1) = -2 (1) + b$

$1 = -2 + b$

$3 = b$

So our intercept is 3.

The equation is $Y = -2X + 3$.

Another way of describing a line is called *the standard form of the equation of a line*. Sometimes this is just referred to as the *equation of a line*. Instead of $Y = mX + b$, the formula is written as $AX + BY = C$. In the slope-intercept formula, we want the coefficient of Y to be 1. In the standard form, the coefficient of X is A (instead of *m*), and both variables are written on the left hand side of the equation. Look at the next few examples and notice the differences and similarities.

Example 3

$2X + 3Y = 6$ equation of a line

$3Y = -2X + 6$ subtract 2X from both sides

$Y = -2/3\ X + 2$ divide by 3 to get slope-intercept formula

Example 4

Conversely

$Y = 4/5X + 2$ slope-intercept formula

$5Y = 4X + 10$ multiply both sides by 5

$-4X + 5Y = 10$ subtract 4X to get the equation of a line

Another way to find the slope without plotting the points and drawing the triangle is by finding the differences between the X- and Y-coordinates. In figures 3 and 4, I drew dotted lines from the points to the X- and Y-axes to illustrate these differences. In figure 5, notice the lengths of the sides of the triangle.

Figure 3

Change in X = 5 – 2 = 3

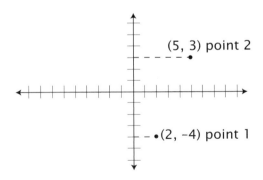

Figure 4

Change in Y = 3 – (–4) = 7

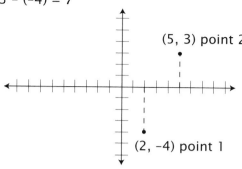

(5, 3) point 2

(2, –4) point 1

Figure 5

Slope

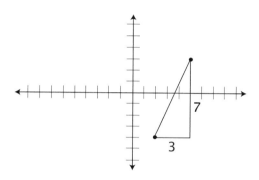

7

3

The over dimension of the triangle is 3. We can find it by taking 5 (the X-coordinate of point 2) minus 2 (the X-coordinate of point 1) to get 3. (5 – 2 = 3.) The up dimension is 7. We can find it by taking 3 (the Y-coordinate of point 2) minus negative 4 (the Y-coordinate of point 1) to get 7. More simply: 3 – (–4) = 7. To construct a formula for determining slope using this principle, we begin by labeling points 1 and 2 with subscripts. (Subscripts are little numbers below the line that help us to identify points but do not affect the value of the number.) Point 2 is written as (X_2, Y_2) and point 1 as (X_1, Y_1).

The formula for calculating slope is:

$$\text{Slope is } \frac{\text{up}}{\text{over}} = \frac{\text{Change in Y-coordinates}}{\text{Change in X-coordinates}} = \frac{Y_2 - Y_1}{X_2 - X_1} = \frac{3 - (-4)}{5 - 2} = \frac{7}{3}$$

Practice Problems 2

1. Find the slope and intercept of a line through (0, 0) and (–2, 4). Describe the line in slope-intercept form and in standard form.

2. Find the slope and intercept of a line through (4, 5) and (1, 3). Describe the line in slope-intercept form and in standard form.

3. Find the slope and intercept of a line through (–1, 1) and (3, –2). Describe the line in slope-intercept form and in standard form.

Solutions 2

1. Points:

$(0, 0)$ and $(-2, 4)$

$$m = \frac{Y_2 - Y_1}{X_2 - X_1}$$

$$m = \frac{(4) - (0)}{(-2) - (0)} = \frac{4}{-2} = -2$$

Slope-intercept form:

$$Y = mX + b$$
$$(0) = (-2)(0) + B$$
$$0 = 0 + B$$
$$0 = B$$
$$Y = -2X + 0$$
$$Y = -2X$$

Standard form:

$$Y = -2X$$
$$2X + Y = 0$$

2. Points:

$(4, 5)$ and $(1, 3)$

$$m = \frac{Y_2 - Y_1}{X_2 - X_1}$$

$$m = \frac{(3) - (5)}{(1) - (4)} = \frac{-2}{-3} = \frac{2}{3}$$

Slope-intercept form:

$$Y = mX + b$$

$$(5) = \left(\frac{2}{3}\right)(4) + b$$

$$5 = \frac{8}{3} + b$$

$$5 - \frac{8}{3} = b$$

$$\frac{15}{3} - \frac{8}{3} = b$$

$$\frac{7}{3} = b$$

$$Y = \frac{2}{3}X + \frac{7}{3}$$

Standard form:

$$Y = \frac{2}{3}X + \frac{7}{3}$$

$$3Y = 2X + 7$$

$$-2X + 3Y = 7$$

3. Points:

$(-1, 1)$ and $(3, -2)$

$$m = \frac{Y_2 - Y_1}{X_2 - X_1}$$

$$m = \frac{(-2) - (1)}{(3) - (-1)} = \frac{-3}{4} = -\frac{3}{4}$$

Slope-intercept form:

$$Y = mX + b$$

$$(1) = \left(-\frac{3}{4}\right)(-1) + b$$

$$1 = \frac{3}{4} + b$$

$$1 - \frac{3}{4} = b$$

$$\frac{4}{4} - \frac{3}{4} = b$$

$$\frac{1}{4} = b$$

$$Y = -\frac{3}{4}X + \frac{1}{4}$$

Standard form:

$$Y = -\frac{3}{4}X + \frac{1}{4}$$

$$4Y = -3X + 1$$

$$3X + 4Y = 1$$

Parallel and Perpendicular Lines; Inequalities

Parallel and Perpendicular Lines

Two or more lines that have the same slope and different Y-intercepts are *parallel*. We've talked about the fact that there are an infinite number of lines that have the same slope. The Y-intercept distinguishes one from the other. In figure 1, notice that all the lines have a slope of 2/3, and thus all are parallel. Only the Y-intercepts are different for each line.

Figure 1

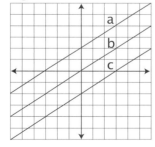

Figure 2

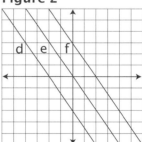

Look at figure 2. Again, all the lines are parallel and have the same slope but different intercepts. What is the slope of the lines in figure 2? Do you see a relationship between the slope in figure 1 and the slope in figure 2?

Figure 3

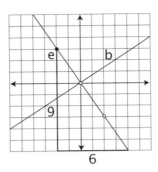

In figure 3, we've drawn line *b* from figure 1 and line *e* from figure 2. These lines are not parallel; quite the opposite — they are *perpendicular*. Notice the relationship between the slopes. (The intercept is not important at this point.) The slope of line *b* is 2/3. The reciprocal of 2/3 is 3/2. The negative reciprocal of 2/3 is –3/2, which is the slope of line *e*. Conversely, the negative reciprocal of –3/2 is –(–2/3) or +2/3, the slope of line *b*.

You can verify the slope of line *e* by making a right triangle using two points that intersect the line exactly. Try (–2, 3) and (4, –6). The right triangle shows an "up" dimension of 9 and an "over" dimension of 6, making the slope 9/6, or simplified, 3/2. We can see that the slope is negative, so it is –3/2. There are other points that could have been chosen since any two distinct points of intersection may be used to determine the slope of a line.

Verify the slope of line *b* using the same procedure. Your calculations should yield a slope of 2/3. Since that is the negative reciprocal of the slope of line *e* (–3/2), these lines are perpendicular.

Horizontal lines have a slope of zero. The slope of a *vertical line* is undefined.

Practice Problems 1
Use purchased graph paper or go to MathUSee.com and choose "Downloads" to find printable graph paper.

1. Which line is parallel to Y = 2X + 5?
   A. Y = –2X + 4
   B. 2Y = 4X + 3
   C. 3Y + 6X = 6

2. Draw a line parallel to Y = 2X + 5, passing through the point (1, –1).

3. Describe the line you drew for #2 using the slope-intercept formula.

4. Which line is perpendicular to Y = 1/2 X + 4?
   A. Y = –1/2 X + 4
   B. Y = –2X – 3
   C. 2Y = X – 4

5. Draw a line perpendicular to Y = 1/2 X + 4, passing through the point (0, 0).

6. Describe the line you drew for #5 using the slope-intercept formula.

Solutions 1

1. B, because 2Y = 4X + 3 is the same as Y = 2X + 3/2 when divided by 2.

2. on graph

3. Y = 2X – 3

4. B

5. on graph

6. Y = –2X + 0 or Y = –2X

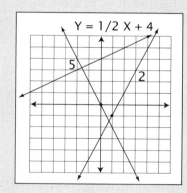

Graphing Inequalities

Up to this point, whenever we graphed a line, both sides of the equation were equal, as in $Y = 2X - 3$. But there are also other situations, known as *inequalities*, where Y may be greater than (>), greater than or equal to (≥), less than (<), or less than or equal to (≤) another term. Based on the line $Y = 2X - 3$, here are the possible inequalities:

$$Y > 2X - 3 \qquad Y \geq 2X - 3 \qquad Y < 2X - 3 \qquad Y \leq 2X - 3$$

Let's consider the line: $Y > 2X - 3$. There are three areas to consider: points to the left of the line, points on the line, and points to the right of the line. We'll examine three points, one in each of these areas, and see whether they agree or disagree with our equation. The points are: #1 (0, 0), #2 (2, 1), and #3 (4, 0). After plotting them and being sure that each of the three areas is represented, we'll substitute each set of points in the equation.

Figure 4

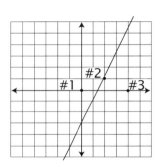

#1 $Y > 2X - 3$	#2 $Y > 2X - 3$	#3 $Y > 2X - 3$
$0 > 2(0) - 3$	$1 > 2(2) - 3$	$0 > 2(4) - 3$
$0 > -3$	$1 > 1$	$0 > 5$
This is true!	This is not true!	This is not true!

Since point #1 is **true**, then all the points in the shaded area are also true and satisfy the equation. Point #3 is **not true**, so we leave that area alone. Point #2, representing the line itself, is also **not true**, so we draw this as a dotted line to show that it is not included in the solution. The answer is the shaded area.

Figure 5

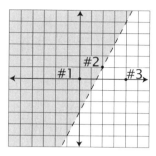

Figure 6 is the graph of Y ≥ 2X − 3. This is exactly like the previous graph, except it includes the line. This is a combination of Y > 2X − 3 (the previous graph) and Y = 2X − 3 (which is the line itself).

Figure 6

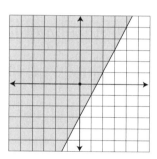

Figure 7 is the graph of $Y < 2X - 3$. This is the opposite of figure 6, and since it is "less than" and not "less than and equal to," it is a dotted line. Using the origin $(0, 0)$ as the test point, we have $(0) < 2(0) - 3$ or $0 < -3$, which is not true. So the shading is to the right of the line.

Figure 7

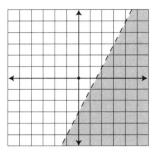

Figure 8 is the graph of $Y \leq 2X - 3$, which is exactly like the previous graph, except it includes the line. This is a combination of $Y < 2X - 3$ (the previous graph) and $Y = 2X - 3$ (which is the line itself).

Figure 8

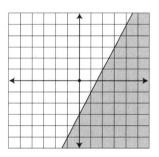

When you have an equation that has a negative Y, such as $-2Y \geq 3X + 6$, that you want to be positive and in the slope-intercept form, multiplying or dividing by a negative number changes the inequality sign. When you have an inequality, you can multiply or divide by a positive number without affecting the equation, but multiplying by a negative number is a different situation.

Adding or subtracting anything from both sides does not change the sign. Notice the following equations with real numbers to see how this works.

Example 1
8 = 8 is true.

When multiplied by +2, it is 16 = 16. This is also true.

When multiplied by –2, it is –16 = –16, which is still true.

Example 2
5 > –3 is true.

When multiplied by +3, it is 15 > –9. This is also true.

When multiplied by –3, it is –15 > 9, which is not true.

To make it true, change the inequality sign, and get –15 < 9, which is now true.

Example 3
1 < 2 is true.

When multiplied by +6, it is 6 < 12. This is also true.

When multiplied by –6, it is –6 < –12, which is not true.

To make it true, change the inequality sign, and get –6 > –12, which is now true.

So if you have $-2Y \geq 3X + 6$, divide both sides by a -2 (or multiply both sides by a -1/2) and the result is $Y \leq -3/2\ X - 3$. The equals sign is not affected by multiplying by a negative number, as we saw in example 1.

Practice Problems 2

1. $Y \leq X - 1$

2. $-Y < -2X - 1$

3. $3Y - 9 \geq X - 3$

4. $-2Y \geq 3X + 6$

Solutions 2

1. Y ≤ X − 1

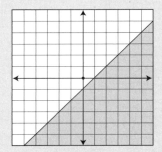

2. Y > 2X + 1

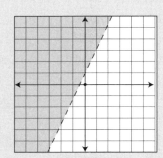

3. Y ≥ 1/3 X + 2

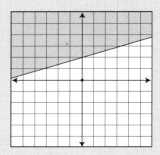

4. Y ≤ −3/2 X − 3

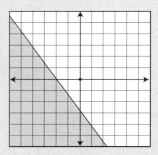

LESSON 22

Distance Formula and Midpoint Formula

To find the distance between two points in the Cartesian coordinate system, you need to know how to plot points on the X- and Y-axes and understand the Pythagorean theorem. The *distance formula* is the result of plotting points, making a right triangle, and finding the hypotenuse of the right triangle. Consider example 1.

Example 1
Find the distance between point A (1, 3) and point B (5, 6).

Step 1 — Draw a line between point A and point B. This is the distance that we are going to measure.

Step 2 — Make a right triangle with the legs parallel to the X- and Y-axes and the line in step 1 as the hypotenuse.

Step 3 — Find the length of the two legs.

Step 4 — Use the Pythagorean theorem to find the distance between the two points.

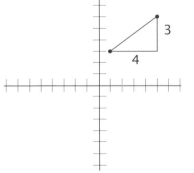

$$\left(X \text{ leg}\right)^2 + \left(Y \text{ leg}\right)^2 = \left(\text{Distance}\right)^2$$
$$4^2 + 3^2 = D^2$$
$$16 + 9 = D^2$$
$$25 = D^2$$
$$5 = D$$

The distance between point A and point B is five units.

The key to finding the distance between the two points is finding the length of the X and Y legs. Looking at the coordinates of the points, notice the change, or difference, in the X-coordinates and the Y-coordinates. These give you the length of the legs in the right triangle. The change in the X-coordinate from point A to point B is from 1 to 5, so the difference is 5 − 1 = 4.

The change in the Y-coordinate from point A to point B is from 3 to 6, so the difference is 6 − 3 = 3. The length of the legs could be written as ΔX and ΔY. The symbol Δ represents "change." ΔX means the change in X, and ΔY means the change in Y. The distance formula can be rewritten as $\Delta X^2 + \Delta Y^2 = \text{Distance}^2$.

Formula 1

$$\text{Distance}^2 = \Delta X^2 + \Delta Y^2$$

$$\text{Distance} = \sqrt{\Delta X^2 + \Delta Y^2}$$

Taking this further still, let's use variables to represent our two points. Representing point A is (X_A, Y_A) or (X_1, Y_1) since it is the first point. Thus point B is (X_B, Y_B) or (X_2, Y_2). Now we'll use these coordinates in our formula for finding the length of the legs. Subtract the X-coordinates, $X_B − X_A$, to find ΔX, and subtract the Y-coordinates, $Y_B − Y_A$, to solve for ΔY.

Substituting into the distance formula above, we have:

$$\text{Distance} = \sqrt{\Delta X^2 + \Delta Y^2} = \sqrt{\left(X_B − X_A\right)^2 + \left(Y_B − Y_A\right)^2}$$

Replacing with the actual coordinates of example 1:

$$\text{Distance} = \sqrt{\Delta X^2 + \Delta Y^2} = \sqrt{\left(5 − 1\right)^2 + \left(6 − 3\right)^2}$$

$$\text{Distance} = \sqrt{4^2 + 3^2}$$

$$\text{Distance} = \sqrt{25}$$

$$\text{Distance} = 5$$

Here is another example. Try to do it by yourself, and then check your work with the solution.

Example 2

Find the distance between point 1 (1, 5) and point 2 (–3, –2).

Step 1 — Draw a line between point A and point B.

Step 2 — Make a right triangle.

Step 3 — Find the length of the two legs.

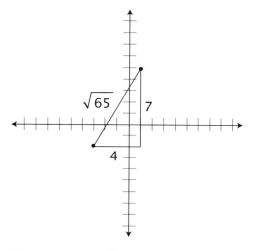

Step 4 — Use the Pythagorean theorem.

$$\left(X \text{ leg}\right)^2 + \left(Y \text{ leg}\right)^2 = \left(\text{Distance}\right)^2$$
$$4^2 + 7^2 = D^2$$
$$\sqrt{16 + 49} = D$$
$$\sqrt{65} = D$$

The distance between point 1 and point 2 is $\sqrt{65}$ units.

Or use the formula!

$$\text{Distance} = \sqrt{\Delta X^2 + \Delta Y^2} = \sqrt{\left(X_2 - X_1\right)^2 + \left(Y_2 - Y_1\right)^2}$$
from 2 to 1
$$= \sqrt{\left(-3 - 1\right)^2 + \left(-2 - 5\right)^2} = \sqrt{\left(-4\right)^2 + \left(-7\right)^2} = \sqrt{65}$$

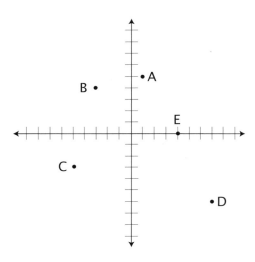

Use these five points for the following problems: A = (1, 5); B = (−3, 4); C = (−5, −3); D = (7, −6); E = (4, 0).

Practice Problems 1

Find the distance between the points:

1. A and D   2. A and C

3. B and C   4. B and E

5. C and D   6. A and B

7. B and D   8. C and E

9. D and E   10. A and E

Solutions 1

1. $\sqrt{157}$ 2. 10

3. $\sqrt{53}$ 4. $\sqrt{65}$

5. $\sqrt{153}$ 6. $\sqrt{17}$

7. $10\sqrt{2}$

8. $3\sqrt{10}$

9. $3\sqrt{5}$

10. $\sqrt{34}$

Midpoint Formula

Finding the coordinates of the midpoint, or the halfway point, between two points is much easier than finding the distance between two points. First find the midpoint of the X-coordinate, and then find the midpoint of the Y-coordinate. You now have the coordinates of one point, and that is the midpoint. Let's find the midpoint of example 2 between the points: $(-3, -2)$ and $(1, 5)$.

Adding the X-coordinates and dividing by two gives $(-3 + 1)/2 = -2/2 = -1$. Adding the Y-coordinates and dividing by two gives $(-2 + 5)/2 = 3/2$. The midpoint is $(-1, 3/2)$. Check the graph.

The *midpoint formula* is $\left(\dfrac{X_1 + X_2}{2}, \dfrac{Y_1 + Y_2}{2} \right)$.

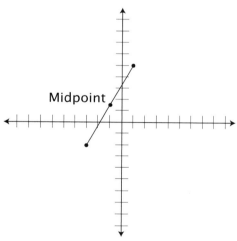

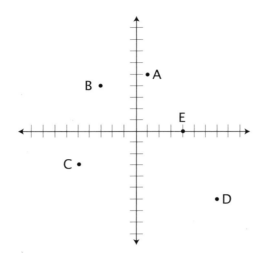

Use these five points for the following problems.

A = (1, 5)
B = (-3, 4)
C = (-5, -3)
D = (7, -6)
E = (4, 0)

Practice Problems 2

Find the midpoint between the given points.

1. A and D

2. A and C

3. B and C

4. B and E

5. C and D

6. A and B

7. B and D

8. C and E

9. D and E

10. A and E

Solutions 2

1. (4, –1/2)

2. (–2, 1)

3. (–4, 1/2)

4. (1/2, 2)

5. (1, –9/2)

6. (–1, 9/2)

7. (2, –1)

8. (–1/2, –3/2)

9. (11/2, –3)

10. (5/2, 5/2)

Conic Sections: Circle and Ellipse

The circle, ellipse, parabola, and hyperbola were introduced in *Algebra 1*. In this book, we want to cover them in more depth, both as they appear on a two-dimensional graph and their corresponding equations. These four shapes are referred to as ***conic sections***. Conic comes from cone. When a plane intersects a single or double cone, the resulting shapes are the circle, ellipse, parabola, and hyperbola. Look at the shapes below and on the DVD.

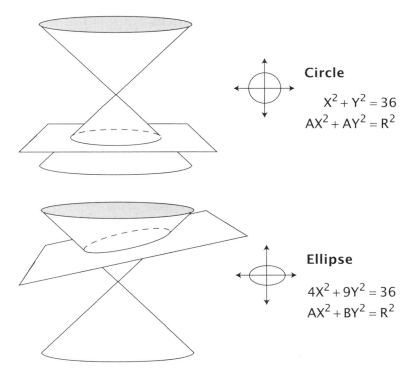

Circle

$$X^2 + Y^2 = 36$$
$$AX^2 + AY^2 = R^2$$

Ellipse

$$4X^2 + 9Y^2 = 36$$
$$AX^2 + BY^2 = R^2$$

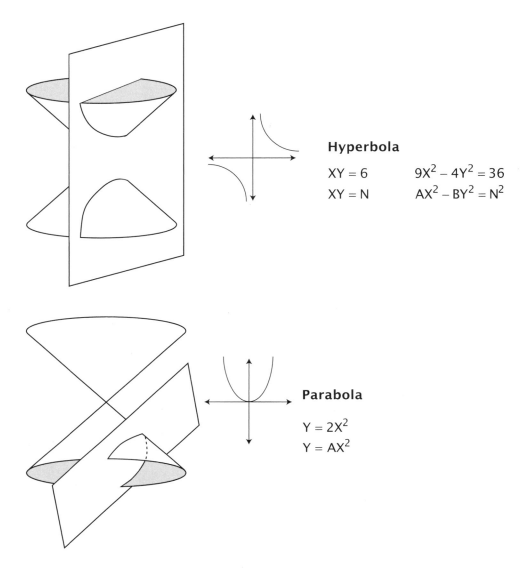

Hyperbola

$XY = 6$ $9X^2 - 4Y^2 = 36$

$XY = N$ $AX^2 - BY^2 = N^2$

Parabola

$Y = 2X^2$

$Y = AX^2$

The Circle

The ***equation of the circle*** could better be written as $(X - a)^2 + (Y - b)^2 = R^2$, with the center at (a, b), and a radius of R. The equation $X^2 + Y^2 = 9$ can be rewritten in this form as $(X - 0)^2 + (Y - 0)^2 = 3^2$, with the center at $(0, 0)$ and a radius of 3. Its graph is figure 1.

Figure 1

Example 1

Graph $X^2 + Y^2 = 25$.

The equation of the circle may be written as
$$(X - 0)^2 + (Y - 0)^2 = 5^2,$$
with the center at (0, 0) and a radius of 5. Its graph is below.

Example 2

Sometimes we get a quadratic for X and Y, such as:
$$X^2 + 2X + 1 + Y^2 - 4Y + 4 = 9.$$

Factoring $(X + 1)^2 + (Y - 2)^2 = 3^2$

The center is (–1, 2) and the radius is 3.
The graph is below.

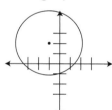

Example 3

Rewrite the equation to find the center and the radius:
$$X^2 - 6X + Y^2 + 10Y = -18.$$

To make the X and Y components perfect squares, we need to complete the squares of each.

$X^2 - 6X + \underline{\quad} + Y^2 + 10Y + \underline{\quad} = -18 + \underline{\quad}$
$X^2 - 6X + 9 + Y^2 + 10Y + 25 = -18 + 34$

Factoring $(X - 3)^2 + (Y + 5)^2 = 4^2$

The center is (3, –5) and the radius is 4.
The graph is to the right.

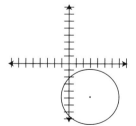

Example 4

Find the equation of the circle, given the center point and the radius. The center is (–2, 3) and the radius is 4.

Working backwards, $(X + 2)^2 + (Y - 3)^2 = 4^2$,
   which is $X^2 + 4X + 4 + Y^2 - 6Y + 9 = 16$.

Combined further, $X^2 + 4X + Y^2 - 6Y = 3$.

Practice Problems 1

Find the coordinates of the center, and the radius, and then graph the results.

1. $X^2 + Y^2 = 16$

2. $(X + 1)^2 + (Y + 1)^2 = 36$

3. $(X - 2)^2 + (Y + 3)^2 = 49$

4. $4X^2 + 4Y^2 = 9$

Given the coordinates of the center, and the radius, create the equation of each circle.

5. (1, 2) r = 4

6. (–2, –3) r = 2

7. (0, 4) r = 8

8. (3, 1/2) r = 10

By completing the square, find the center and radius of these equations, and then sketch the results.

9. $X^2 - 6X + Y^2 - 8Y = -8$.

10. $X^2 - 2X + Y^2 - 4Y = 11$

11. $X^2 + Y^2 - 8Y - 9 = 0$.

12. $X^2 - X + Y^2 + 2Y = -29/36$

Solutions 1

1. (0, 0) r = 4

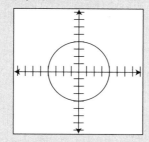

2. (−1, −1) r = 6

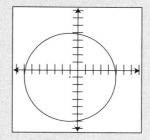

3. (2, −3) r = 7

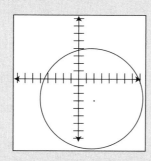

4. (0, 0) r = 3/2 (divide by 4)

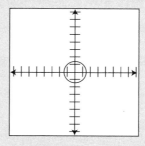

5. $(X - 1)^2 + (Y - 2)^2 = 16$

6. $(X + 2)^2 + (Y + 3)^2 = 4$

7. $(X)^2 + (Y - 4)^2 = 64$

8. $(X - 3)^2 + (Y - 1/2)^2 = 100$

9. $(X - 3)^2 + (Y - 4)^2 = 17$

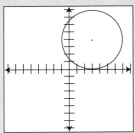

Center: (3, 4)
Radius: $\sqrt{17}$

10. $(X - 1)^2 + (Y - 2)^2 = 16$

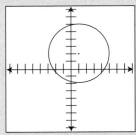

Center: (1, 2)
Radius: 4

11. $(X - 0)^2 + (Y - 4)^2 = 25$

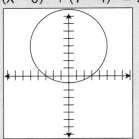

Center: (0, 4)
Radius: 5

12. $(X - 1/2)^2 + (Y + 1)^2 = 4/9$

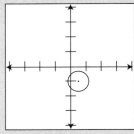

Center: (1/2, –1)
Radius: 2/3

The Ellipse

If the coefficients of X^2 and Y^2 are equal, the result is a circle. In our previous examples, both coefficients were one. If you were given an equation with coefficients of four for both X and Y, you could divide through the equation by four, and the coefficients would be one again. If the coefficients are equal, the graph will be a circle.

The equation $9X^2 + 4Y^2 = 36$ (or $X^2/4 + Y^2/9 = 1$ after dividing by 36) looks similar to an equation for a circle because you have two squares added together. In this case, observe the coefficients. If the coefficients are not equal, then the equation is for an *ellipse*.

Example 5

Plot several points and graph the ellipse $4X^2 + 9Y^2 = 36$.

The key is to find the value of each variable that makes the corresponding term equal zero. Then you know where the graph intercepts the axes.

If $X = 0$
$$4(0)^2 + 9Y^2 = 36$$
$$9Y^2 = 36$$
$$Y^2 = 4$$
$$Y = \pm 2$$

If $Y = 0$
$$4X^2 + 9(0)^2 = 36$$
$$4X^2 = 36$$
$$X^2 = 9$$
$$X = \pm 3$$

X	Y
0	±2
±3	0

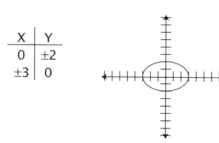

Example 6

Plot several points and graph the ellipse $9(X - 1)^2 + 16(Y - 2)^2 = 144$.

Locate the center, and then find the values of X and Y that make each term equal zero. Then you can find the extremities of the ellipse.

If $X = 1$
$$9(0) + 16(Y - 2)^2 = 144$$
$$16(Y - 2)^2 = 144$$
$$(Y - 2)^2 = 9$$
$$Y - 2 = +3, -3$$
$$Y = +5, -1$$

If $Y = 2$
$$9(X - 1)^2 + 16(0) = 144$$
$$9(X - 1)^2 = 144$$
$$(X - 1)^2 = 16$$
$$X - 1 = +4, -4$$
$$X = +5, -3$$

X	Y
1	+5
1	−1
−3	2
+5	2

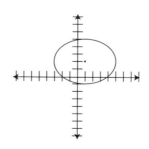

Practice Problems 2

Find the coordinates of the center and graph the result.

1. $9X^2 + 4Y^2 = 36$

2. $2(X - 1)^2 + 3(Y + 1)^2 = 48$

3. $\dfrac{(X+1)^2}{25} + \dfrac{(Y+2)^2}{20} = 1$

4. $16X^2 + 9Y^2 = 144$

Solutions 2

1.

X	Y
0	+3
0	−3
+2	0
−2	0

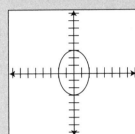

2.

X	Y
1	+3
1	−5
−4 *	−1
+6 *	−1

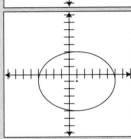

* $\sqrt{24}$ is (approximately) 5.

3.

X	Y
−1	≈ 2.5
−1	≈ −6.5
+4	−2
−6	−2

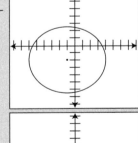

4.

X	Y
0	+4
0	−4
−3	0
+3	0

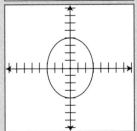

Parabola

If an equation has one variable raised to the first power (first degree) and another variable raised to the second power (second degree), then your graph will be a *parabola*. In *Algebra 1*, we dealt with parabolas of the form $Y^1 = X^2$, which look like figure 1. In $Y = X^2$, if $X = +1$ or -1, then Y is $+1$, because the numbers are squared. If $X = \pm 2$, then Y is $+4$. If there is a coefficient, as in $Y = 2X^2$, what would it do to the graph? The graph gets steeper much more quickly than without a coefficient (see figure 2).

Figure 1

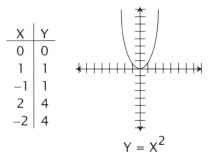

X	Y
0	0
1	1
−1	1
2	4
−2	4

$Y = X^2$

Figure 2

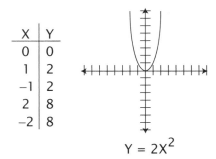

X	Y
0	0
1	2
−1	2
2	8
−2	8

$Y = 2X^2$

Figure 3

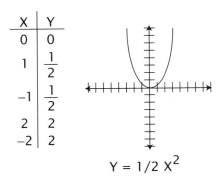

X	Y
0	0
1	$\frac{1}{2}$
−1	$\frac{1}{2}$
2	2
−2	2

$$Y = 1/2\ X^2$$

Figure 4

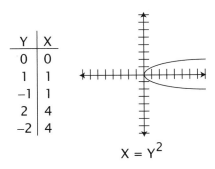

Y	X
0	0
1	1
−1	1
2	4
−2	4

$$X = Y^2$$

If the coefficient is a fraction, then the graph appears to spread out or **_dilate_** (see figure 3). You can also switch the variables and have $X^1 = Y^2$, which looks like figure 4. Look at the tables of numbers that have been plotted, as well as the graphs for each one.

If the coefficient is a negative number, as in $Y = -2X^2$, what will the graph look like? It should be the opposite of figure 2. Look at figure 5 and see whether this makes sense.

Figure 5

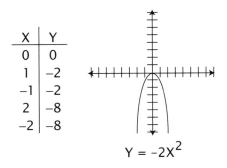

X	Y
0	0
1	−2
−1	−2
2	−8
−2	−8

$$Y = -2X^2$$

Practice Problems 1

Estimate what the graph should look like, and then plot several points to confirm your hypothesis.

1. $Y = 2X^2$

2. $Y = 1/3\ X^2$

3. $Y = -3X^2$

4. $Y = -1/4\ X^2$

5. $X = Y^2$

6. $X = -3Y^2$

Solutions 1

1.

X	Y
0	0
1	2
−1	2
2	8
−2	8

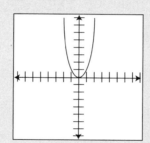

$Y = 2X^2$

2.

X	Y
0	0
1	$\frac{1}{3}$
−1	$\frac{1}{3}$
2	$\frac{4}{3}$
−2	$\frac{4}{3}$

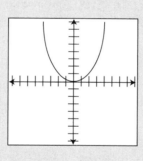

$Y = 1/3\ X^2$

3.

X	Y
0	0
1	−3
−1	−3
2	−12
−2	−12

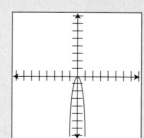

$Y = -3X^2$

4.

X	Y
0	0
1	$-\dfrac{1}{4}$
-1	$-\dfrac{1}{4}$
2	-1
-2	-1

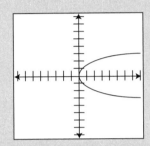

$Y = -1/4\ X^2$

5.

Y	X
0	0
1	1
-1	1
2	4
-2	4

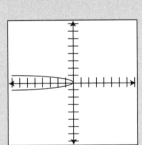

$X = Y^2$

Note: The positions of X and Y are switched in the charts for #5 and #6.

6.

Y	X
0	0
1	-3
-1	-3
2	-12
-2	-12

$X = -3Y^2$

Not only can the graph become steeper, more spread out, or inverted, but it can also move like a ***translation***. (See transformational geometry in *Geometry*.) For example: $Y = X^2 + 2$. See figure 6. It is the same parabola as figure 1, just moved, or translated, up two on the Y-axis. If the term is a negative number, the parabola would move down the Y-axis. See figure 7.

Figure 6

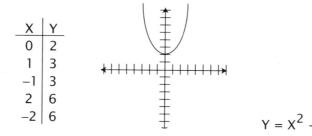

X	Y
0	2
1	3
−1	3
2	6
−2	6

$Y = X^2 + 2$

Figure 7

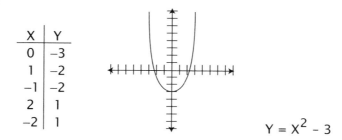

X	Y
0	−3
1	−2
−1	−2
2	1
−2	1

$Y = X^2 - 3$

Practice Problems 2

Estimate what the graph should look like, and then plot several points to confirm your hypothesis.

1. $Y = 2X^2 + 1$
2. $Y = 1/2 \, X^2 + 3$

3. $Y = -X^2 + 2$
4. $Y = -1/3 \, X^2 + 1$

5. $X = 2/3 \, Y^2 + 2$
6. $X = -2Y^2 - 1$

Solutions 2

1.

X	Y
0	1
1	3
−1	3
2	9
−2	9

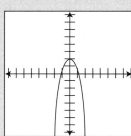

$Y = 2X^2 + 1$

2.

X	Y
0	3
1	$\frac{7}{2}$
−1	$\frac{7}{2}$
2	5
−2	5

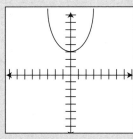

$Y = 1/2\, X^2 + 3$

3.

X	Y
0	2
1	1
−1	1
2	−2
−2	−2

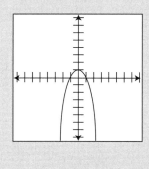

$Y = -X^2 + 2$

4.

X	Y
0	1
1	$\frac{2}{3}$
−1	$\frac{2}{3}$
2	$-\frac{1}{3}$
−2	$-\frac{1}{3}$

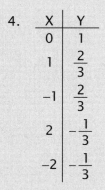

$Y = -1/3\, X^2 + 1$

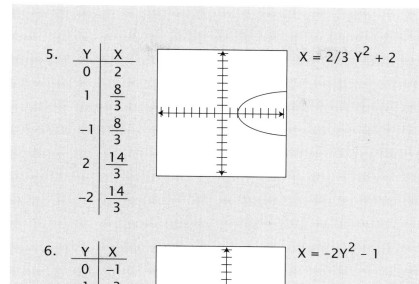

5.

Y	X
0	2
1	$\frac{8}{3}$
−1	$\frac{8}{3}$
2	$\frac{14}{3}$
−2	$\frac{14}{3}$

$X = 2/3\ Y^2 + 2$

6.

Y	X
0	−1
1	−3
−1	−3
2	−9
−2	−9

$X = -2Y^2 - 1$

In application, it is common to take a curve that has been sketched on a graph and try to find the equation that would give us that curve. If that can be done, reasonable predictions can be made about the curve at points beyond where it is shown on the graph.

If we are given a parabola, we can make observations about its translation and/or dilation to come up with the equation for the parabola. It may be helpful at this point to know that the general form for the equation of a parabola is:

$$Y = AX^2 + BX + C$$

For the parabolas shown up to this point, the value of B has been zero, which is why the second term has not been apparent.

Parabola: Maxima and Minima

So far, we have graphed parabolas that move up and down on the axes. If X has a positive coefficient, as in $Y = 2X^2$, then the lowest point, or *vertex*, is on the Y-axis. If X has a negative coefficient, as in $Y = -2X^2$, then the highest point (vertex) is also on the Y-axis. For equations of the sort $X = Y^2$ with positive or negative coefficients, the same holds true with the X-axis.

If the quadratic has a middle term, then the parabola moves off the axis. The object of this lesson is to find where the parabola moves, and to predict where the vertex will be located. The lowest point or vertex of a positive parabola is called the *minima*. The highest point or vertex of a negative parabola is called the *maxima*.

The general form for the equation of a parabola is $Y = AX^2 + BX + C$.

Example 1

Graph $Y = X^2 + 4X - 2$ by plotting several points.

$$X = -5 \quad Y = (-5)^2 + 4(-5) - 2$$
$$Y = 3$$

$$X = -2 \quad Y = (-2)^2 + 4(-2) - 2$$
$$Y = -6$$

$$X = -4 \quad Y = (-4)^2 + 4(-4) - 2$$
$$Y = -2$$

$$X = -1 \quad Y = (-1)^2 + 4(-1) - 2$$
$$Y = -5$$

$$X = -3 \quad Y = (-3)^2 + 4(-3) - 2$$
$$Y = -5$$

$$X = 0 \quad Y = (0)^2 + 4(0) - 2$$
$$Y = -2$$

$$X = 1 \quad Y = (1)^2 + 4(1) - 2$$
$$Y = 3$$

Figure 1

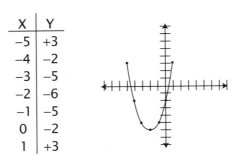

X	Y
−5	+3
−4	−2
−3	−5
−2	−6
−1	−5
0	−2
1	+3

The graph shows visually what our table of data is telling us. The pattern is that the value of Y decreases to −6 and then begins moving up again. So the vertex, or minima, is (−2, −6). How could we derive this from the original equation? We know that the −2 (the C term from the formula) moves the parabola up or down the Y-axis. So we need to focus on the middle term—the one with the B coefficient. Looking at the example, don't worry about the C term, but instead focus on $Y = X^2 + 4X$. Observe that the parabola shifted to the left instead of staying on the Y-axis. I drew a line through the vertex to split the parabola into two symmetrical parts. This line parallel to the Y-axis is called the **_axis of symmetry_**.

Figure 2

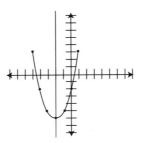

On the graph, we can see that the X-coordinate of this line is −2. Plugging −2 into the equation gives us Y = −6, which is our vertex. If we can find the value of the X-coordinate, then we can find the Y-coordinate, and we will know the location of the vertex. Let's set Y = 0 to find the value of X.

Figure 3

$$Y = X^2 + 4X$$
$$0 = X^2 + 4X$$
$$0 = X(X + 4)$$
$$X = 0 \quad \text{or} \quad X + 4 = 0$$
$$X = 0 \qquad X = -4$$

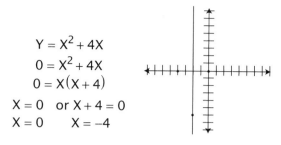

We can see that the axis of symmetry, $X = -2$, will lie halfway between these points, and we know if $X = -2$, then $Y = -6$.

Now let's run through the same process with $AX^2 + BX + C = 0$, focusing on $AX^2 + BX$.

Figure 4

$$Y = AX^2 + BX$$
$$0 = AX^2 + BX$$
$$0 = X(AX + B)$$
$$X = 0 \quad \text{or} \quad AX + B = 0$$
$$X = 0 \qquad X = -\frac{B}{A}$$

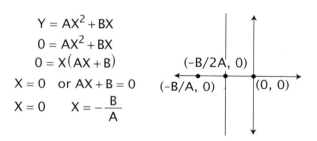

(-B/2A, 0)

(-B/A, 0) (0, 0)

If the two coordinates are 0 and $-B/A$, then according to the midpoint formula the distance halfway between them (the line of symmetry) is $-B/2A$ or $1/2$ times $-B/A$.

So when given an equation with B (a middle term), use $-B/2A$ to find the axis of symmetry (X-coordinate), use the X-coordinate in the original equation to find the value of the Y-coordinate, and you will have the vertex.

Example 2 (See figure 5 on next page.)
Graph $Y = X^2 + 5X + 2$

$A = 1, B = 5, C = 2$

$$\frac{-B}{2A} = \frac{-(5)}{2(1)} = -5/2$$

Plugging in the X-coordinate of the axis of symmetry:

$$Y = X^2 + 5X + 2$$
$$Y = (-5/2)^2 + 5(-5/2) + 2 = -17/4 \text{ or } -4\,1/4$$

Vertex = $(-2\,1/2, -4\,1/4)$

Figure 5

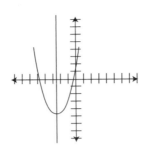

Practice Problems 1

Find the axis of symmetry and the vertex, and then sketch the graph.

1. $Y = X^2 - 6X + 1$

2. $Y = 2X^2 + 8X + 2$

3. $Y = -X^2 + 3$

4. $Y = -3X^2 + 6X$

5. $Y = 2/3\ X^2 - 4$

6. $Y = 1/2\ X^2 - 2X$

Solutions 1

1. $Y = X^2 - 6X + 1$
   $A = 1,\ B = -6$
   $\dfrac{-B}{2A} = \dfrac{-(-6)}{2(1)} = 3$
   $Y = X^2 - 6X + 1$
   $Y = (3)^2 - 6(3) + 1 = -8$
   Axis of symmetry at $X = 3$
   Vertex $= (3,\ -8)$

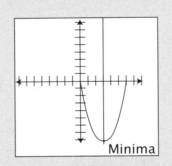

2. $Y = 2X^2 + 8X + 2$
   $A = 2,\ B = 8$
   $\dfrac{-B}{2A} = \dfrac{-(8)}{2(2)} = -2$
   $Y = 2X^2 + 8X + 2$
   $Y = 2(-2)^2 + 8(-2) + 2 = -6$
   Axis of symmetry at $X = -2$
   Vertex $= (-2,\ -6)$

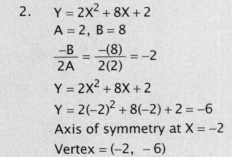

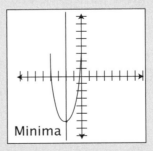

3. Graph points down with vertex at (0, 3)

4. $Y = -3X^2 + 6X$

   $A = -3, \ B = +6$

   $\dfrac{-B}{2A} = \dfrac{-(6)}{2(-3)} = 1$

   $Y = -3X^2 + 6X$

   $Y = -3(1)^2 + 6(1) = 3$

   Axis of symmetry at $X = 1$

   Vertex $= (1, \ 3)$

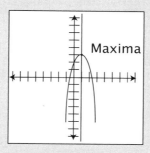

5. Graph points up with vertex at (0, –4)

6. $Y = 1/2 \ X^2 - 2X$

   $A = 1/2, \ B = -2$

   $\dfrac{-B}{2A} = \dfrac{-(-2)}{2(1/2)} = 2$

   $Y = 1/2 \ X^2 - 2X$

   $Y = 1/2(2)^2 - 2(2) = -2$

   Axis of symmetry at $X = 2$

   Vertex $= (2, \ -2)$

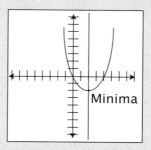

Summary of the Parabola in Algebrese

For any quadratic of the form $Y = AX^2 + BX + C$:

1. If $A > 0$, the graph of the parabola points up.
2. If $A < 0$, the graph of the parabola points down.
3. If $|A| > 1$, the graph is steeper than $Y = X^2$.
4. If $0 < |A| < 1$, the graph is flatter than $Y = X^2$.
5. C moves the graph up and down the Y-axis.
6. The axis of symmetry is $X = -B/2A$.
7. The coordinates of the vertex are:
   $$[-B/2A, \ A(-B/2A)^2 + B(-B/2A) + C].$$
8. The vertex is the maxima or the minima.

We can apply our knowledge of maxima and minima to to find the different possible areas of rectangles having the same perimeters.

Example 3

You purchased 100 feet of fence to come off the back of your house for a rectangular play area. What are the dimensions needed to give you the maximum area inside the fence?

Area is the length times the width or (X)(100−2X).

$(X)(100-2X) = 100X - 2X^2 = -2X^2 + 100X$ Parabola points down.

Using -B/2A, we have A = −2 and B = +100.

$$\frac{-(100)}{2(-2)} = 25 \text{ is the axis of symmetry.}$$

Area of play yard = $-2(25)^2 + 100(25) = 1250$.
(25, 1250) is the vertex of the parabola.

Maximum area = 25 x 50 = 1250 ft^2

Notice the different combinations and their places on the graph, and see whether the solution makes sense.

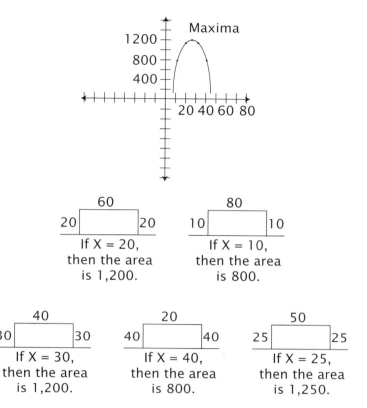

Practice Problems 2

1. Having just bought some chickens, you need to fence in a rectangular chicken yard. If you have 150 feet of fencing, what will the dimensions of the largest yard you can make be?

2. The weather man is calling for a frost. It is up to you to cover the tender shoots tonight. In the barn, there is a roll of sheet metal, 24 inches wide, that will, when folded twice, make a cover for the plants. What is the height and breadth of the rectangular dimensions that will give the most space underneath?

3. Chuck has a 20-foot piece of wood to make a sandbox. What is the most efficient use of the lumber to get maximum space with this amount of timber?

Solutions 2

1. Area = LW = (X)(150 − 2X)/2 = (X)(75 − X)
    (X)(75 − X) = 75X − X^2 = −X^2 +75X
    Parabola points down.

    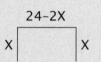

    Using −B/2A, we have A = −1, B = +75.

    $\dfrac{-(75)}{2(-1)}$ = 37.5 → This is the axis of symmetry.

    Area = −(37.5)^2 + 75(37.5) = −1406.25 + 2812.5 = 1406.25
    (37.5, 1406.25) is the vertex.
    Area = 37.5 x 37.5 = 1406.25 ft^2

2. Area = LW = (X)(24 − 2X) = 24X − 2X^2 =
    −2X^2 + 24X
    Parabola points down.

    Using −B/2A, we have A = −2, B = +24.

    $\dfrac{-(24)}{2(-2)}$ = 6 → This is the axis of symmetry.

    Area = −2(6)^2 + 24(6) = −72 + 144 = 72   (6, 72) is the vertex.
    Area = 6 x 12 = 72 ft^2

3. Area = LW = (X)(20 − 2X)/2 = 10X − X^2 =
    −X^2 + 10X
    Parabola points down.

    Using −B/2A, we have A = −1, B = +10.

    $\dfrac{-(10)}{2(-1)}$ = 5 → This is the axis of symmetry.

    Area = −(5)^2 + 10(5) = −25 + 50 = 25   (5, 25) is the vertex.
    Area = 5 x 5 = 25 ft^2

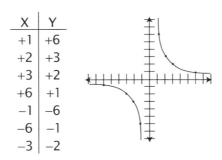

Hyperbola

In *Algebra 1*, we were introduced to the **hyperbola** with the equation XY = N, where N is some number. Let's do two examples and plot the points to get a feel for this conic section.

Example 1

Graph XY = 6 by plotting several points.

If X = 1   (1)Y = 6   If X = 3   (3)Y = 6   If X = –1   (–1)Y = 6
                  Y = 6   Y = 2   Y = –6

If X = 2   (2)Y = 6   If X = 6   (6)Y = 6   If X = –6   (–6)Y = 6
                  Y = 3   Y = 1   Y = –1

Figure 1

X	Y
+1	+6
+2	+3
+3	+2
+6	+1
–1	–6
–6	–1
–3	–2

Notice that as Y increases, X decreases, and vice versa. Looking at the original equation, do you think X or Y can ever be zero? No, because there is no number times zero that is equal to six. Both of the curves approach the axes, but they will never touch them. Just for fun, find the value of Y if X = .01. The value of Y would have to be 600. Picture that point on the graph.

The hyperbola is a visual representation of an *inverse relationship*. Another example of an inverse relationship is Distance = Rate x Time. Distance is a constant, say 100 miles. If you drive 100 miles per hour, it takes one hour (100 = 100 x 1) to go 100 miles. If you drive 50 mph, the time increases to two hours (100 = 50 x 2). If the rate decreases to 25 mph, the time increases to four hours (100 = 25 x 4). As the rate decreases, the time increases and vice versa. An example of *direct variation* is represented by the line Y=mX + b. As X increases, Y also increases.

There is another type of equation that is also graphed as a hyperbola. This type is similar to the difference of two squares. Officially, it is when you have two variables, each raised to the second power, with opposite signs. They don't have to be perfect squares, however. Examples are $A^2 - B^2 = 9$ and $3G^2 - 4H^2 = 12$.

Hyperbola Summary

$XY = +N$	The graph lies in the 1st and 3rd quadrants.
$XY = -N$	The graph lies in the 2nd and 4th quadrants.
$AX^2 - BY^2 = N^2$	The graph intersects the X-axis in two places. It looks like a C and a backwards C.
$AY^2 - BX^2 = N^2$	The graph intersects the Y-axis in two places. It looks like a U and an upside-down U.

Example 2

Graph $X^2 - Y^2 = 9$ by plotting several points.

If Y = 0 $X^2 - (0)^2 = 9$ If Y = 3 $X^2 - (3)^2 = 9$
                    $X = \pm 3$ $X = \pm 4.2$ *

If Y = 1 $X^2 - (1)^2 = 9$ If Y = 4 $X^2 - (4)^2 = 9$
                    $X = \pm 3.2$ * $X = \pm 5$

If Y = 2 $X^2 - (2)^2 = 9$ If Y = 5 $X^2 - (5)^2 = 9$
                    $X = \pm 3.6$ * $X = \pm 5.8$ *

       * approximately

Figure 2

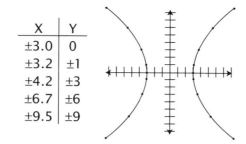

X	Y
±3.0	0
±3.2	±1
±4.2	±3
±6.7	±6
±9.5	±9

Note: If Y = 4, X = ±5, and if Y = −4, X = ±5.

That gives us four coordinates: (4,5), (4,−5), (−4,5), (−4,−5).

Practice Problems 1

1. $XY = 6$

2. $9X^2 - 4Y^2 = 36$

3. $XY = -1$

4. $2Y^2 - X^2 = 18$

5. $XY = -8$

6. $X^2 - 4Y^2 = 16$

7. $XY = 12$

8. $X^2 + Y^2 = 4$

Solutions 1

1-2.

X	Y
+1	+6
+2	+3
+3	+2
+6	+1
−1	−6
−6	−1

X	Y
±2.0	0
±2.8	±3
±4.5	±6
±6.3	±9

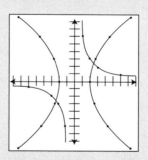

3-4.

X	Y
+.25	−4
1	−1
+4	−.25
−.25	+4
−1	+1
−4	+.25

X	Y
0	±3.0
±3	±3.7
±6	±5.2
±9	±7.0

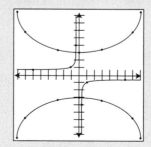

5-6.

X	Y
+1	−8
+2	−4
+4	−2
+8	−1
−1	+8
−8	+1

X	Y
±4.0	0
±4.5	±1
±5.7	±2
±7.2	±3
±8.9	±4

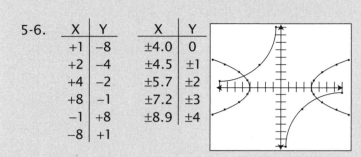

7-8.

X	Y
+2	+6
+3	+4
+4	+3
+6	+2
−2	−6
−6	−2

X	Y
±2.0	0
0	±2

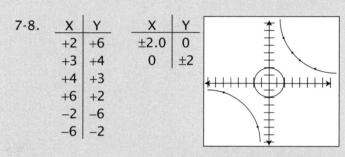

Either the X or the Y value may be chosen first. In some cases the computed value is approximate.

Solving Systems of Equations
Lines and Conic Sections

In *Algebra 1*, we learned how to find the solution of two lines that intersect at a single point. We did this first by graphing, and then by using substitution and elimination. In this lesson, we will use the same techniques to find the intersection of a line (linear equation with first-degree variables) and either a circle, ellipse, parabola, or hyperbola (non-linear equations with second-degree variables). When this is mastered, we will move to the solutions of two non-linear equations.

The best way to start is to do some examples and talk our way through them. If you get stuck on the first example, you can do examples 2 and 3 first, and then come back to this one.

Example 1
Find solution of $\begin{cases} X^2 + Y^2 = 16 \\ Y = 2X - 1 \end{cases}$

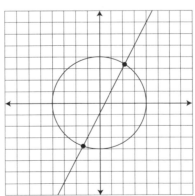

1. Identify the nature of the equations. Are they linear or non-linear, and if non-linear, what conic section is represented? The first equation represents a circle and the second equation a line.

2. Sketch a graph of each to give an estimate of the solutions. Use graph paper and a compass for more accuracy.

3. We have two options, either to square the second equation and then use elimination, or substitute the second equation into the first. I choose the second option.

$$X^2 + (2X - 1)^2 = 16$$
$$X^2 + 4X^2 - 4X + 1 = 16$$
$$5X^2 - 4X - 15 = 0$$

4. First we look to see whether it is possible to find the solutions by factoring. It isn't, so we use the quadratic formula.

$$X = \frac{-B \pm \sqrt{B^2 - 4AC}}{2A} = \frac{-(-4) \pm \sqrt{(-4)^2 - 4(5)(-15)}}{2(5)}$$
$$X = \frac{4 \pm \sqrt{16 + 300}}{10} = \frac{4 \pm 2\sqrt{79}}{10}$$
$$X = \frac{2 + \sqrt{79}}{5}, \frac{2 - \sqrt{79}}{5}$$

5. Next we put these answers into $Y = 2X - 1$ to find the solutions for Y.

$$Y = 2\left(\frac{2 + \sqrt{79}}{5}\right) - 1 = \frac{4}{5} + \frac{2\sqrt{79}}{5} - \frac{5}{5} = \frac{2\sqrt{79}}{5} - \frac{1}{5} = \frac{2\sqrt{79} - 1}{5}$$
$$Y = 2\left(\frac{2 - \sqrt{79}}{5}\right) - 1 = \frac{4}{5} - \frac{2\sqrt{79}}{5} - \frac{5}{5} = -\frac{1}{5} - \frac{2\sqrt{79}}{5} = \frac{-1 - 2\sqrt{79}}{5}$$

6. Here are the final solutions. For the second line, approximate $\sqrt{79}$ to be 9.

$$\left(\frac{2+\sqrt{79}}{5}, \frac{2\sqrt{79}-1}{5}\right) \text{ and } \left(\frac{2-\sqrt{79}}{5}, \frac{-1-2\sqrt{79}}{5}\right)$$

$$\left(\frac{2+9}{5}, \frac{(2\cdot9)-1}{5}\right) = \left(\frac{11}{5}, \frac{17}{5}\right) \text{ and } \left(\frac{2-9}{5}, \frac{-1-(2\cdot9)}{5}\right) = \left(-\frac{7}{5}, -\frac{19}{5}\right)$$

Plot these points on the graph, and they fit!

Note that in some cases, the graphs of two equations will not intersect at all. In these cases, there are no real solutions that satisfy both equations. For example, a line and a circle may intersect in two places, in only one place, or not at all. Other combinations of linear equations and/or conic sections have other possibilities.

Let's back up to gain perspective on this problem. First we substituted $Y = 2X - 1$ into the first equation so that we had only one variable to solve for, in this case X. Since we couldn't factor the equation, we used the quadratic formula to find the two values of X. We knew by our graph that there were two points where the graphs intersected, so we expected two solutions.

After we found the two values for X, we had to replace them in either of the equations to find the corresponding values of Y. We chose the easier of the two equations, and as a result, we had the final solution. Finally, to make sure that our solution was the same as the coordinates on the graph, we substituted 9 for the approximate value of the square root of 79. And sure enough, the points are where we expected to find them.

If this seems hard, relax, the problems are not usually as difficult as this one. Let's do another example—an easier one.

Example 2

Find solution of $\begin{cases} XY = 6 \\ 3X + 2Y = 12 \end{cases}$

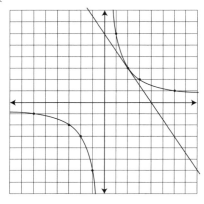

$$2Y = -3X + 12$$
$$Y = \frac{-3X}{2} + 6$$

1. Identify the nature of the equations. The first equation represents a hyperbola, and the second equation represents a line.

2. Sketch a graph of each equation to give an estimate of the solutions. Use graph paper and a compass for more accuracy.

3. Substitute the second equation into the first.

$$X\left(-\frac{3}{2}X + 6\right) = 6$$
$$-\frac{3}{2}X^2 + 6X = 6$$
$$-\frac{3}{2}X^2 + 6X - 6 = 0$$
$$-3X^2 + 12X - 12 = 0$$

4. Simplify by dividing each side by -3. Then look to see whether it is possible to factor to find the solutions. It is, and the solution is X = 2.

$$X^2 - 4X + 4 = 0$$
$$(X - 2)(X - 2) = 0$$

5. Next we put this answer into XY = 6 to find the solution for Y.

$$(2)Y = 6$$
$$Y = 3$$

6. The final solution is (2, 3), which agrees with the graph.

Example 3

Find solution of $\begin{cases} XY = 4 \\ X^2 + Y^2 = 8 \end{cases}$

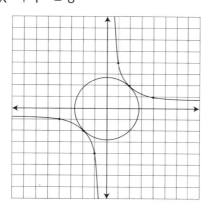

1. The first equation is a hyperbola and the second, a circle.

2. Sketch a graph of each to give an estimate of the solutions.

3. Substitute the first equation into the second to solve for one variable.

$$\left(\frac{4}{Y}\right)^2 + Y^2 = 8$$
$$\frac{16}{Y^2} + Y^2 = 8$$
$$16 + Y^4 = 8Y^2$$
$$Y^4 - 8Y^2 + 16 = 0$$

4. Factor to find the solutions.

$$\left(Y^2 - 4\right)^2 = 0$$
$$\left(Y^2 - 4\right)\left(Y^2 - 4\right) = 0$$
$$(Y - 2)(Y + 2)(Y - 2)(Y + 2) = 0$$

5. Put $Y = \pm 2$ into $XY = 4$ to find the solution for X.

$$(2)X = 4 \quad (-2)X = 4$$
$$X = 2 \quad\quad X = -2$$

6. The final solutions are $(2, 2)$ and $(-2, -2)$, which agrees with the graph.

Practice Problems 1

1. $\begin{cases} Y = X + 3 \\ X^2 + Y^2 = 9 \end{cases}$

2. $\begin{cases} XY = 8 \\ -6X + 3Y = 18 \end{cases}$

3. $\begin{cases} Y = X^2 \\ Y = 4 \end{cases}$

4. $\begin{cases} 4X^2 + Y^2 = 36 \\ Y = X + 1 \end{cases}$

5. $\begin{cases} X^2 - Y^2 = 24 \\ X^2 + Y^2 = 36 \end{cases}$

6. $\begin{cases} Y = X^2 + 2 \\ X^2 + Y^2 = 4 \end{cases}$

7. $\begin{cases} XY = 6 \\ X^2 + 9Y^2 = 36 \end{cases}$

8. $\begin{cases} Y = 2X \\ XY = 10 \end{cases}$

Solutions 1

1. Find the solutions of $\begin{cases} Y = X + 3 \\ X^2 + Y^2 = 9 \end{cases}$

    A. The first equation is a line
       and the second is a circle.

    B. Sketch a graph of each to give
       an estimate of the solutions.

    C. Substitute the first equation into
       the second to solve for one variable.

$$X^2 + (X + 3)^2 = 9$$
$$X^2 + X^2 + 6X + 9 = 9$$
$$2X^2 + 6X = 0$$

    D. Factor to find the solutions. $\quad 2\left(X^2 + 3X\right) = 0(2)$

$$X(X + 3) = 0$$
$$X = 0 \quad X = -3$$

    E. Put X = 0 and X = –3 $\qquad Y = 0 + 3 \quad Y = -3 + 3$
       into Y = X + 3 to find Y. $\qquad Y = 3 \qquad\quad Y = 0$

    F. The final solutions are (0, 3) and (–3, 0).

2. Find the solutions of $\begin{cases} XY = 8 \\ -6X + 3Y = 18 \end{cases}$

    $-6X + 3Y = 18 \rightarrow Y = 2X + 6$

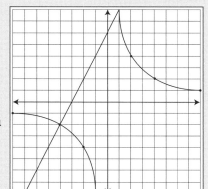

    A. The first equation is a hyperbola
       and the second is a line.

    B. Sketch a graph of each to give
       an estimate of the solutions.

C. Substitute the first into the second
   to solve for one variable.

$$X(2X + 6) = 8$$

$$2X^2 + 6X - 8 = 0$$

$$X^2 + 3X - 4 = 0$$

D. Factor to find the solutions.

$$(X - 1)(X + 4) = 0$$
$$X = 1 \quad X = -4$$

E. Put $X = 1$ and $X = -4$
   into $XY = 8$ to find Y.

$$(1)Y = 8 \quad (-4)Y = 8$$
$$Y = 8 \qquad Y = -2$$

F. The final solutions are (1, 8) and (–4, –2).

3. Find the solutions of $\begin{cases} Y = X^2 \\ Y = 4 \end{cases}$

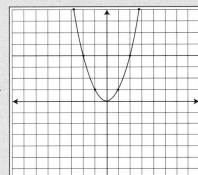

A. The first equation is a parabola
   and the second is a line.

B. Sketch a graph of each to give
   an estimate of the solutions.

C. Solve for one variable. $4 = X^2$

D. Factor to find the solutions. $\pm 2 = X$

E. $X = 2$, $X = -2$ and $Y = 4$, so solutions are (2, 4) and (–2, 4).

4. Find the solutions of $\begin{cases} 4X^2 + Y^2 = 36 \\ Y = X + 1 \end{cases}$

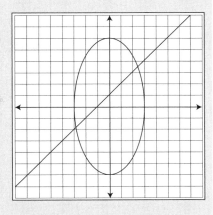

A. The first equation is an ellipse and the second is a line.

B. Sketch a graph of each to give an estimate of the solutions.

C. Solve for one variable.

$$4X^2 + (X+1)^2 = 36$$

$$4X^2 + (X^2 + 2X + 1) = 36$$

$$5X^2 + 2X - 35 = 0$$

D. Use the quadratic formula. $\quad X = \dfrac{-1 + 4\sqrt{11}}{5} \text{ or } \dfrac{-1 - 4\sqrt{11}}{5}$

E. $Y = X + 1$, so: $\left(\dfrac{-1 + 4\sqrt{11}}{5}, \dfrac{4 + 4\sqrt{11}}{5} \right) \left(\dfrac{-1 - 4\sqrt{11}}{5}, \dfrac{4 - 4\sqrt{11}}{5} \right)$

5. Find the solutions of $\begin{cases} X^2 - Y^2 = 24 \\ X^2 + Y^2 = 36 \end{cases}$

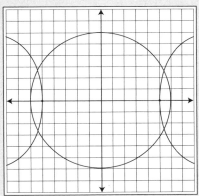

A. The first equation is a hyperbola and the second is a circle.

B. Sketch a graph of each to give an estimate of the solutions.

C. Eliminate to solve for one variable.

$$2X^2 = 60$$

$$X^2 = 30$$

D. Take the square root of each side. $X = \pm\sqrt{30}$

E. Put $X = \sqrt{30}$, and $X = -\sqrt{30}$ into $X^2 + Y^2 = 36$.

F. The final solutions are:

$$\left(\sqrt{30},\sqrt{6}\right)\left(\sqrt{30},-\sqrt{6}\right)\left(-\sqrt{30},\sqrt{6}\right)\left(-\sqrt{30},-\sqrt{6}\right)$$

6.   Find the solutions of $\begin{cases} Y = X^2 + 2 \\ X^2 + Y^2 = 4 \end{cases}$

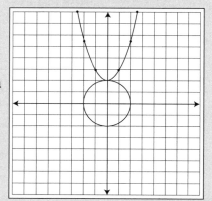

A. The first equation is a parabola
   and the second is a circle.

B. Sketch a graph of each to give
   an estimate of the solutions.

C. Substitute the first into the second to solve for one variable.

$$(Y - 2) + Y^2 = 4$$

$$Y^2 + Y - 6 = 0$$

D. Factor to find the solutions. $(Y + 3)(Y - 2) = 0$

$$Y = -3 * \quad Y = 2$$

E. The final solution is (0, 2).

*Sometimes when working with quadratics, all the roots do not
work. It is good idea to check the roots to see if they work.

In this case Y = -3 does not work in either equation. So we have
only one root, which is 2.

7. Find the solutions of $\begin{cases} XY = 6 \\ X^2 + 9Y^2 = 36 \end{cases}$

A. The first equation is a hyperbola
   and the second is an ellipse.

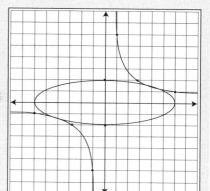

B. Sketch a graph of each to give
   an estimate of the solutions.

C. Substitute the first into the second to solve for one variable.

$$\left(\frac{6}{Y}\right)^2 + 9Y^2 = 36$$

$$\frac{36}{Y^2} + 9Y^2 = 36$$

$$36 + 9Y^4 = 36Y^2$$

D. Factor to find one variable.

$$9Y^4 - 36Y^2 + 36 = 0$$

$$Y^4 - 4Y^2 + 4 = 0$$

$$(Y^2 - 2)^2 = 0$$

$$Y = \pm\sqrt{2}$$

E. Put $Y = \pm\sqrt{2}$ into $XY = 6$ to find X.

$$X\left(+\sqrt{2}\right) = 6, X = 3\sqrt{2}$$
$$X\left(-\sqrt{2}\right) = 6, X = -3\sqrt{2}$$

F. The solutions are: $\left(+3\sqrt{2}, +\sqrt{2}\right)$ and $\left(-3\sqrt{2}, -\sqrt{2}\right)$.

8. Find the solutions of $\begin{cases} Y = 2X \\ XY = 10 \end{cases}$

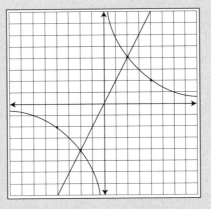

A. The first equation is a line and the second is a hyperbola.

B. Sketch a graph of each to give an estimate of the solutions.

C. Substitute the first into the second to solve for one variable.

$$X(2X) = 10$$
$$2X^2 = 10$$
$$X^2 = 5$$

D. Take the square root of each side. $X = \pm\sqrt{5}$

E. Put $X = \sqrt{5}$ and $X = -\sqrt{5}$ into $Y = 2X$ to find Y.

F. The solutions are: $\left(\sqrt{5}, 2\sqrt{5}\right)$ and $\left(-\sqrt{5}, -2\sqrt{5}\right)$.

LESSON 28

Coins, Consecutive Integers, and Mixtures

The first part of this lesson is a review from *Algebra 1*. If you feel comfortable with the material after doing the practice problems, continue on to the next part. If this is new, spend some time until you have learned this material, and then move on to the next section.

Coin Problems

We can use what we've learned about solving simultaneous equations and apply it to some interesting coin problems. Did you ever wonder how to find out how many of each kind of coin there are in someone's pocket, given the amount of money and the number of coins? Here is how you do it.

Example 1

I have seven coins in my pocket. They are all either dimes or nickels. The value of the coins is $.55. How many of each kind do I have? There are two equations present: the number of coins (how many) and the dollar amount of the coins (how much). Each may be represented by an equation.

How many: nickels plus dimes equals 7
or N + D = 7

How much: nickels (.05) plus dimes (.10) equals .55
or .05N + .10D = .55

Using the LCM, we can multiply the second equation by 100, making it 5N + 10D = 55. Putting our two equations together yields:

$$\begin{array}{ll}
N + D = 7 \text{ times } (-5) = & -5N - 5D = -35 \\
5N + 10D = 55 & \underline{5N + 10D = 55} \\
& 5D = 20 \quad \text{If } D = 4 \text{ and } N + D = 7, \\
& D = 4 \qquad \text{then } N = 3.
\end{array}$$

Checking it to make sure: 4 dimes is $.40 and 3 nickels is $.15, which adds up to $.55.

The key to remembering the two equations is count and amount. Count describes how many of each kind, and amount describes how much each kind is worth.

Practice Problems 1

1. I have 11 coins in my pocket. They are all either dimes or nickels. The value of the coins is $.70. How many of each coin do I have?

2. I have 12 coins in my pocket. They are all either pennies or nickels. The value of the coins is $.32. How many of each coin do I have?

Solutions 1

1. How many: nickels plus dimes equals 11
   or N + D = 11

   How much: nickels (.05) plus dimes (.10) equals .70
   or .05N + .10D = .70

   $$\begin{array}{ll}
   N + D = 11 \text{ times } (-5) = & -5N - 5D = -55 \\
   5N + 10D = 70 & \underline{5N + 10D = 70} \\
   & 5D = 15 \quad \text{If } D = 3 \text{ and } N + D = 11 \\
   & D = 3 \qquad \text{then } N = 8.
   \end{array}$$

---

2. How many: pennies plus dimes equals 12
    or P + N = 12

    How much: pennies (.01) plus nickels (.05) equals .32
    or .01P + .05N = .32

$$P + N = 12 \text{ times } (-1) = \quad -P + -N = -12$$
$$1P + 5N = 32 \qquad\qquad \underline{P + 5N = 32}$$
$$4N = 20 \quad \text{If } N = 5 \text{ and } P + N = 12,$$
$$N = 5 \quad \text{then } P = 7.$$

Consecutive Integers

Integers include whole numbers, their negative counterparts, and zero. Examples of *consecutive integers* are 2, 3, 4, or 10, 11, 12. They begin with the smallest number and increase by one. Consecutive **even** integers begin with the smallest number, which is even, and increase by two, such as 6, 8, 10 or 22, 24, 26. Consecutive **odd** integers begin with the smallest number, which is odd, and increase by two, such as 7, 9, 11, or 33, 35, 37. Consecutive integers may also include negative numbers. Three consecutive even integers could be −14, −12, and −10, the smallest being −14 and the largest −10.

Representing these relationships with algebra would look like this:

Consecutive Integers:
    N, N +1, N + 2
    If N = 5, then N + 1 = 6 and N + 2 = 7, so 5, 6, 7.

Consecutive Even Integers:
    N, N + 2, N + 4
    If N = 12, then N + 2 = 14 and N + 4 = 16, so 12, 14, 16.

Consecutive Odd Integers:
    N, N + 2, N + 4
    If N = 23, then N + 2 = 25 and N + 4 = 27, so 23, 25, 27.

Example 2

Find three consecutive integers where three times the first integer, plus two times the third integer, is equal to 29.

3(first) + 2(third) = 29

N = first, N +1 = second , N + 2 = third

$3N + 2(N + 2) = 29$ Since N = 5, then N + 1 = 6 and N + 2 = 7

$3N + 2N + 4 = 29$ The solution (3 consecutive integers) is 5, 6, 7.

$5N + 4 = 29$

$N = 5$

To check: 3(5) + 2(7) = 15 + 14 = 29. It checks!

Practice Problems 2

1. Find three consecutive integers such that five times the first integer, plus two times the second, is equal to four times the third.

2. Find three consecutive even integers such that two times the first integer, plus two times the second, is equal to six times the third.

Solutions 2

1. Find three consecutive integers such that five times the first integer, plus two times the second integer, is equal to four times the third.
   5 (first) + 2 (second) = 4 (third)

   N = first, N + 1 = second, N + 2 = third

   $5N + 2(N + 1) = 4(N + 2)$ Since N = 2, then N + 1 = 3 and N + 2 = 4

   $5N + 2N + 2 = 4N + 8$ The 3 consecutive integers are 2, 3, 4.

   $7N + 2 = 4N + 8$

   $3N = 6$

   $N = 2$

   To check 5(2) + 2(3) = 4(4)
   10 + 6 = 16   It checks!

2. Find three consecutive even integers such that two times the first integer, plus two times the second integer, is equal to six times the third.   2(first) + 2(second) = 6(third)

   N = first, N + 2 = second, N + 4 = third

   $2N + 2(N + 2) = 6(N + 4)$ Since N = −10, then N + 2 = −8
   and N + 4 = −6.

   $2N + 2N + 4 = 6N + 24$ The 3 consecutive even integers

   $4N + 4 = 6N + 24$ are −10, −8, −6.

   $-2N = 20$

   $N = -10$

   To check 2(−10) + 2(−8) = 6(−6)
   −20 − 16 = −36   It checks!

Mixtures

These problems are very similar to those above. What sounds like one problem may be separated into two. The basis for separating the equations is "how much," and "what kind," which in these problems will be "what percent." As before, let's do a real problem and figure out the patterns as we go.

Example 3

As a painter, you are setting out to create a unique watercolor for the color of the sea. On hand, you have a bluish mixture that is 50% blue colorant and 50% water. This is to be mixed with a solution that is 20% blue colorant and 80% water. The goal is to have 30 ml of a mixture that is 40% blue colorant and 60% water. How much of each of the original liquids do we add together to form this new mixture?

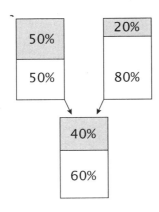

We can do this by using the percentage of water in the mixture or by the percentage of colorant. In this problem, I choose to look at the percentage of colorant for the "what kind." I'll refer to the two original solutions as B_F for blue 50% and B_T for blue 20% (F for Fifty, T for Twenty).

How much? $B_F + B_T = 30 \text{ ml}$ → $B_F + B_T = 30$

What kind? $50\%B_F + 20\%B_T = 40\%(30)$ → $.5B_F + .2B_T = .4(30)$

$$
\begin{aligned}
-2B_F - 2B_T &= -60 \\
\underline{5B_F + 2B_T} &= \underline{120} \\
3B_F &= 60 \\
B_F &= 20 \text{ ml}
\end{aligned}
$$

→   first line multiplied by −2
→   second line multiplied by 10

Add to eliminate a variable.

If B_F is 20 ml, then B_T is 10 ml, because they add up to 30.

It might help our thinking to use the blocks to represent the 20 ml of 50% and the 10 ml of 20% that equal the 30 ml of 40% solution. Use averages to make the number of colored blocks the same in each 10 ml container.

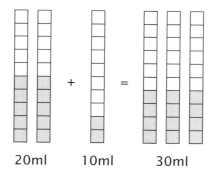

20ml   10ml   30ml

Let's do example 1 again focusing on the percentage of water instead of the colorant. W_F for water 50% and W_E for water 80%.

How much? $\qquad W_F + W_E = 30 \text{ ml} \qquad \rightarrow \qquad W_F + W_E = 30$

What kind? $\qquad 50\%W_F + 80\%W_E = 60\%(30) \qquad \rightarrow \qquad .5W_F + .8W_E = .6(30)$

$$\begin{array}{l} -5W_F - 5W_E = -150 \\ \underline{5W_F + 8W_E = 180} \\ 3W_E = 30 \\ W_E = 10 \text{ ml} \end{array}$$

$\rightarrow$ first line multiplied by -5

$\rightarrow$ second line multiplied by 10

Then add to eliminate a variable.

If W_E is 10 ml, then W_F is 20, because they add up to 30.

Before we begin doing practice problems, notice that when combining the solutions the percentage of the resultant mixture is always between the percentage of the two initial solutions. If you start with a 30% and a 60%, how can you possibly get to a 70% or a 10% by just using these two solutions? You can't.

Practice Problems 3

1. In your seaside laboratory, you have a mixture of saltwater that is 70% salt and 30% water. This is to be mixed with a solution that is 30% salt and 70% water. The goal is to have 40 ml of a mixture that is 60% salt and 40% water. How much of each of the original liquids do you add together to form this new mixture?

2. Some changes in your saltwater are in order. Now you have one mixture that is 25% salt and 75% water and another that is 60% salt and 40% water. The new goal is to have 14 liters of a mixture that is 30% salt and 70% water. How much of each of the original liquids do you add together to form this new mixture?

3. Do number 1 from the perspective of the water and compare your answers.

4. As the swimming pool chemist, you have two beakers of a chlorine/water solution. The first is 5% chlorine and 95% water, and the second is 1% chlorine and 99% water. The goal is to have 60 liters of a mixture that is 2% chlorine. How much from each of the beakers do you need to form the desired solution?

5. At the beginning of the work day, you found two containers of a chlorine/water solution. The first is 6% chlorine and the second is 2% chlorine. There is an order for 32 liters of a mixture that is 4.5% chlorine. How much of each of the beakers do you need to form the desired solution?

6. Do number 5 from the opposite perspective of how you did it the first time and compare your answers.

7. As a budding artist, you find a mixture of red watercolor that is 80% red colorant and another bottle that is 30% red. For your masterpiece, you need 60 ml of 55% red solution. How much of the original liquids do you add to form this watercolor?

8. Do number 7 from the opposite perspective of how you did it the first time and compare your answers.

9. Now you want to use oil-based paint. The present two mixtures are 35% yellow with 65% mineral spirits and 60% yellow with 40% mineral spirits. Today you need 35 ml of 45% yellow solution. How much of the solutions is needed for the yellow?

10. Do number 9 from the opposite perspective of how you did it the first time, and compare your answers.

Solutions 3

1. How much? $S_1 + S_2 = 40$ ml $\rightarrow$ $S_1 + S_2 = 40$
   What kind? $70\%S_1 + 30\%S_2 = 60\%(40)$ $\rightarrow$ $.70S_1 + .30S_2 = .60(40)$

   $\begin{array}{r} -3S_1 - 3S_2 = -120 \\ 7S_1 + 3S_2 = \ 240 \\ \hline 4S_1 = \ 120 \end{array}$ $\begin{array}{l} \rightarrow \quad \text{multiplied by} -3 \\ \rightarrow \quad \text{multiplied by 10} \end{array}$

   $S_1 = 30$ ml $S_1 + S_2 = 40$
   $(30) + S_2 = 40$
   $S_2 = 10$ ml

2. How much? $S_1 + S_2 = 14$ litres $\rightarrow$ $S_1 + S_2 = 14$
   What kind? $25\%S_1 + 60\%S_2 = 30\%(14)$ $\rightarrow$ $.25S_1 + .6S_2 = .3(14)$

   $\begin{array}{r} -25S_1 - 25S_2 = -350 \\ 25S_1 + 60S_2 = \ 420 \\ \hline 35S_2 = \ 70 \end{array}$ $\begin{array}{l} \rightarrow \quad \text{multiplied by} -25 \\ \rightarrow \quad \text{multiplied by 100} \end{array}$

   $S_2 = 2$ litres $S_1 + S_2 = 14$
   $S_1 + (2) = 14$
   $S_1 = 12$ litres

3. How much? $W_1 + W_2 = 40$ ml $\rightarrow$ $W_1 + W_2 = 40$

   What kind? $30\%W_1 + 70\%W_2 = 40\%(40)$ $\rightarrow$ $.3W_1 + .7W_2 = .4(40)$

   $\begin{array}{l} -3W_1 - 3W_2 = -120 \\ \underline{3W_1 + 7W_2 = 160} \\ 4W_2 = 40 \end{array}$ $\quad \begin{array}{l} \rightarrow \text{ multiplied by } -3 \\ \rightarrow \text{ multiplied by } 10 \end{array}$

   $W_2 = 10$ ml $\quad W_1 + W_2 = 40$

   $ W_1 + (10) = 40$

   $ W_1 = 30$ ml

4. How much? $C_1 + C_2 = 60$ L $\rightarrow$ $C_1 + C_2 = 60$

   What kind? $5\%C_1 + 1\%C_2 = 2\%(60)$ $\rightarrow$ $.05C_1 + .01C_2 = .02(60)$

   $\begin{array}{l} -1C_1 - 1C_2 = -60 \\ \underline{5C_1 + 1C_2 = 120} \\ 4C_1 = 60 \end{array}$ $\quad \begin{array}{l} \rightarrow \text{ multiplied by } -1 \\ \rightarrow \text{ multiplied by } 100 \end{array}$

   $C_1 = 15$ L $\quad C_1 + C_2 = 60$

   $ (15) + C_2 = 60$

   $ C_2 = 45$ L

5. How much? $C_1 + C_2 = 32$ L $\rightarrow$ $C_1 + C_2 = 32$

   What kind? $6\%C_1 + 2\%C_2 = 4.5\%(32)$ $\rightarrow$ $.06C_1 + .02C_2 = .045(32)$

   $\begin{array}{l} -20C_1 - 20C_2 = -640 \\ \underline{60C_1 + 20C_2 = 1440} \\ 40C_1 = 800 \end{array}$ $\quad \begin{array}{l} \rightarrow \text{ multiplied by } -20 \\ \rightarrow \text{ multiplied by } 1000 \end{array}$

   $C_1 = 20$ L $\quad C_1 + C_2 = 32$

   $ (20) + C_2 = 32$

   $ C_2 = 12$ L

6. How much? $W_1 + W_2 = 32$ L $\rightarrow$ $W_1 + W_2 = 32$

What kind? $94\%W_1 + 98\%W_2 = 95.5\%(32)$ $\rightarrow$ $.94W_1 + .98W_2 = .955(32)$

$$-94W_1 - 94W_2 = -3008 \quad \rightarrow \quad \text{multiplied by} -94$$

$$\underline{94W_1 + 98W_2 = 3056} \quad \rightarrow \quad \text{multiplied by } 100$$

$$4W_2 = 48$$

$$W_2 = 12 \text{ L} \qquad W_1 + W_2 = 32$$

$$W_1 + (12) = 32$$

$$W_1 = 20 \text{ L}$$

7. How much? $R_1 + R_2 = 60$ ml $\rightarrow$ $R_1 + R_2 = 60$

What kind? $80\%R_1 + 30\%R_2 = 55\%(60)$ $\rightarrow$ $.8R_1 + .3R_2 = .55(60)$

$$-30R_1 - 30R_2 = -1800 \quad \rightarrow \quad \text{multiplied by} -30$$

$$\underline{80R_1 + 30R_2 = 3300} \quad \rightarrow \quad \text{multiplied by } 100$$

$$50R_1 = 1500$$

$$R_1 = 30 \text{ ml} \qquad R_1 + R_2 = 60$$

$$(30) + R_2 = 30$$

$$R_2 = 30 \text{ ml}$$

8. How much? $W_1 + W_2 = 60$ ml $\rightarrow$ $W_1 + W_2 = 60$

What kind? $20\%W_1 + 70\%W_2 = 45\%(60)$ $\rightarrow$ $.2W_1 + .7W_2 = .45(60)$

$$-20W_1 - 20W_2 = -1200 \quad \rightarrow \quad \text{multiplied by} -20$$

$$\underline{20W_1 + 70W_2 = 2700} \quad \rightarrow \quad \text{multiplied by } 100$$

$$50W_2 = 1500$$

$$W_2 = 30 \text{ ml} \qquad W_1 + W_2 = 60$$

$$W_1 + (30) = 60$$

$$W_1 = 30 \text{ ml}$$

9. How much? $Y_1 + Y_2 = 35$ ml $\rightarrow$ $Y_1 + Y_2 = 35$

What kind? $35\%Y_1 + 60\%Y_2 = 45\%(35)$ $\rightarrow$ $.35Y_1 + .6Y_2 = .45(35)$

$$-35Y_1 - 35Y_2 = -1225 \quad \rightarrow \quad \text{multiplied by} - 35$$
$$\underline{35Y_1 + 60Y_2 = \;\;1575} \quad \rightarrow \quad \text{multiplied by 100}$$
$$25Y_2 = \;\;350$$
$$Y_2 = 14 \text{ ml} \qquad Y_1 + Y_2 = 35$$
$$Y_1 + (14) = 35$$
$$Y_1 = 21 \text{ ml}$$

10. How much? $W_1 + W_2 = 35$ ml $\rightarrow$ $W_1 + W_2 = 35$

What kind? $65\%W_1 + 40\%W_2 = 55\%(35)$ $\rightarrow$ $.65W_1 + .4W_2 = .55(35)$

$$-40W_1 - 40W_2 = -1400 \quad \rightarrow \quad \text{multiplied by} - 40$$
$$\underline{65W_1 + 40W_2 = 1925} \quad \rightarrow \quad \text{multiplied by 100}$$
$$25W_1 = 525$$
$$W_1 = 21 \text{ ml} \qquad W_1 + W_2 = 35$$
$$(21) + W_2 = 35$$
$$W_2 = 14 \text{ ml}$$

Age and Boat in the Current Problems

As in previous problems, we'll be using substitution and elimination to solve the equations that we generate. With most word problems, the key is setting them up properly, and then choosing the appropriate variables.

Age Problems

The most effective way to learn problems of this sort is to study an example.

Example 1
In nine years Steve will be twice as old as John. Four years ago, John was one-third the age of Steve. How old are they now?

Two pieces of information were given, so we make two equations from them. S will represent the age of Steve now, and J will represent John's age. The first statement speaks of nine years from now, so that is $S + 9$ and $J + 9$, and the equation is $(S + 9) = (J + 9) \times 2$. Read the information again as you look at the equation. The second statement is four years ago, so $S - 4$ and $J - 4$ represent four years ago. The equation is: $1/3 (S - 4) = (J - 4)$.

Equation 1

$$S + 9 = 2(J + 9)$$
$$S + 9 = 2J + 18$$
$$S = 2J + 9$$

Equation 2

$$J - 4 = \frac{1}{3}(S - 4)$$
$$3(J - 4) = S - 4$$
$$3J - 12 = S - 4$$

Substitute S from equation 1 into equation 2.

$$3J - 12 = (2J + 9) - 4$$
$$3J - 12 = 2J + 5$$
$$J = 17$$

If J = 17, you can put this in either equation to find S.

$$S + 9 = 2(17 + 9)$$
$$S = 43$$

Practice Problems 1

1. In seven years, Joseph will be twice as old as Emmitt. Last year Emmitt was one-sixth the age of Joseph. How old are they now?

2. In 10 years, Sandi Beth will be twice as old as Ethan. Five years ago, Sandi Beth was 3 1/2 times the age of Ethan. How old are they now?

Solutions 1

1. J represents the age of Joseph now, and E represents Emmitt's age.

   The first statement speaks of seven years from now,
   so that is J + 7 and E + 7, and the equation is: (J + 7) = (E + 7) x 2.

   The second statement is about last year.
   J – 1 and E – 1 are last year.

   The equation is 1/6 (J – 1) = (E – 1).

Equation 1

$$(J+7) = (E+7) \times 2$$
$$J+7 = 2E+14$$
$$J = 2E+7$$

$$J = 2(3)+7$$
$$J = 13$$

Equation 2

$$1/6(J-1) = E-1$$
$$J-1 = 6(E-1)$$
$$J-1 = 6E-6$$
$$J = 6E-5$$
$$2E+7 = 6E-5$$
$$12 = 4E$$
$$3 = E$$

2. S represents the age of Sandi Beth now, and E years represents Ethan's age.

The first statement speaks of 10 years from now, so that is S + 10 and E + 10, and the equation is (S + 10) = (E + 10) x 2.

The second statement is about five years ago: S–5 and E–5. The equation is (S–5) = 3.5(E–5).

Equation 1

$$(S+10) = (E+10) \times 2$$
$$S+10 = 2E+20$$
$$S = 2E+10$$

$$S = 2E+10$$
$$S = 2(15)+10$$
$$S = 40$$

Equation 2

$$(S-5) = 3.5(E-5)$$
$$S-5 = 3.5E-17.5$$
$$S = 3.5E-12.5$$
$$2E+10 = 3.5E-12.5$$
$$22.5 = 1.5E$$
$$15 = E$$

Boat in the Current

If you have ever canoed, you know that it is a pleasurable experience to move with the current, but a chore to paddle against the current. For the sake of clarity, whenever we say downstream, we mean with the current, while upstream denotes against the current. If you happen to be in a stream with a 3 mph current, and you just lie back and look at the sky, you will move along at 3 mph as well. If you decide to paddle at 5 mph, then you will be going 8 mph, 3 mph from the current, and 5 mph from paddling.

$$\text{Rate}_{\text{downstream}} = \text{Rate}_{\text{boat}} + \text{Rate}_{\text{water}}$$
$$R_D = R_B + R_W$$
$$8 \text{ mph} = 5 \text{ mph} + 3 \text{ mph}$$

If you are traveling upstream at a rate of 5 mph in the same current, this is what the equation would look like.

$$\text{Rate}_{\text{upstream}} = \text{Rate}_{\text{boat}} - \text{Rate}_{\text{water}}$$
$$R_U = R_B - R_W$$
$$2 \text{ mph} = 5 \text{ mph} - 3 \text{ mph}$$

Another way to write these two equations is: $R_D = B + W$ and $R_U = B - W$.

Remember from the motion problems that $D = RT$. In the example, if the rate downstream is 8 mph, then $B + W = 5 + 3$. If we were paddling for 3 hours, then distance downstream equals rate downstream multiplied by time downstream, or:

$$D_D = R_D \times T_D$$
$$D_D = 8 \text{ mph} \times 3 \text{ hours}$$
$$24 \text{ miles} = 8 \text{ mph} \times 3 \text{ hours}$$

Using this equation from motion and the one we just learned, we can put them together and form two other equations.

$$R_D = B + W \qquad\qquad R_U = B - W$$
$$D_D = R_D T_D \qquad\qquad D_U = R_U T_U$$
$$D_D = (B + W)T_D \qquad\qquad D_U = (B - W) T_U$$

Example 2

Jeff's boat travels 64 miles downstream in 4 hours. The same boat travels 20 miles upstream in 5 hours. What is the speed of the boat and the current?

$$D_D = R_D \times T_D$$
$$(64) = R_D \times (4)$$
$$16 = R_D \qquad \rightarrow \quad R_D = B + W$$
$$(16) = B + W$$

$$D_U = R_U \times T_U$$
$$(20) = R_U \times (5)$$
$$4 = R_U \qquad \rightarrow \quad R_U = B - W$$
$$(4) = B - W$$

$$16 = B + W$$
$$\underline{4 = B - W}$$
$$20 = 2B$$
$$10 = B \qquad \rightarrow \ (4) = (10) - W$$
$$6 = W$$

The rate of the boat is 10 mph and the rate of the current is 6 mph.

Example 3

The Gateway Clipper can go 12 miles upstream in the same time that it takes to go 24 miles downstream. The Allegheny river flows at 3 mph. What is the rate of the boat? In this problem, time is the same, so there is no need for subscripts on the T.

$$D_D = R_D T_D$$
$$(24) = (B + W)T$$
$$24 = BT + 3T$$

$$D_U = R_U T_U$$
$$(12) = (B - W)T$$
$$12 = BT - 3T$$
$$-12 = -BT + 3T$$

$$
\begin{aligned}
24 &= BT + 3T \\
-12 &= -BT + 3T \\
\hline
12 &= 6T \\
2 &= T
\end{aligned}
$$

$\rightarrow$
$$24 = (B + 3)T$$
$$24 = (B + 3)(2)$$
$$12 = B + 3$$
$$9 = B$$

The rate of the boat is 9 mph, the rate of the current is 3 mph, and the time is 2 hours.

Practice Problems 2

1. The Delta Queen traveled at a steady rate of 15 mph without the current. It took her 8 hours to travel downstream to the port and 16 hours to travel upstream to her home. How far did she travel to the port, and what is the speed of the water?

2. The Deerslayer and the Mohican chief were paddling furiously to escape the Iroquois. They paddled the canoe at a steady rate for 60 miles in water at a rate of 5 mph. When it was safe, they paddled back upstream at the same rate and went 6 miles in 3 hours. What is the rate of the canoe? How long did they travel downstream?

Solutions 2

1. $D_D = R_D T_D$
$D_D = (B + W)T$
$D_D = ((15) + W)(8)$

$D_U = R_U T_U$
$D_U = (B - W)T$
$D_U = ((15) - W)(16)$

$(15 + W)8 = (15 - W)16$
$120 + 8W = 240 - 16W$
$24W = 120$
$W = 5$ mph

$D_D = (15 + (5))8$
$D_D = 160$ miles

2. $D_U = R_U T_U$
$(6) = R_U(3)$
$2 = R_U$

$R_U = (C - W)$
$(2) = (C - (5))$
$7 = C$
$C = 7$ mph

$D_D = R_D T_D$
$(60) = ((7) + (5))T$
$60 = 12T$
$5 = T$
$T = 5$ hours

Solving Equations with Three Variables

We know two methods for solving equations with two variables: substitution and elimination. To solve equations with three or more variables, the first step is to simplify them until we have two equations with two variables in each. Label the equations A, B, and C. Then transform any two of them at a time (A and B, A and C, or B and C) until you've eliminated the same variable in each of them. Label these two new equations D and E.

Using substitution or elimination, find the value of one of the variables in the new equations. Take this answer and use it to find the other variable in D or E. Now substitute the two known variables to find a solution for the third variable. What we are looking for is a solution that satisfies all three equations. Finally, check your values for X, Y, and Z in the equations to make sure they are correct.

Example 1
Find the solution that will satisfy all three equations.

$$\boxed{A} \quad 2X + 3Y - Z = 11$$
$$\boxed{B} \quad 3X - Y + 2Z = 4$$
$$\boxed{C} \quad 4X + 2Y + 3Z = 8$$

Use A + B = D and A + C = E to eliminate Z.

$$\boxed{A} \quad 2(2X + 3Y - Z = 11) \Rightarrow 4X + 6Y - 2Z = 22$$
$$\boxed{B} \quad (3X - Y + 2Z = 4) \Rightarrow \underline{3X - Y + 2Z = 4}$$
$$7X + 5Y = 26 \quad \rightarrow \boxed{D} \ 7X + 5Y = 26$$

$$\boxed{A} \quad 3(2X + 3Y - Z = 11) \Rightarrow 6X + 9Y - 3Z = 33$$
$$\boxed{C} \quad (4X + 2Y + 3Z = 8) \Rightarrow \underline{4X + 2Y + 3Z = 8}$$
$$10X + 11Y = 41 \quad \rightarrow \boxed{E} \ 10X + 11Y = 41$$

Eliminate Y.

$$\boxed{D} \quad -11(7X+5Y=26) \Rightarrow \quad -77X-55Y=-286$$
$$\boxed{E} \quad 5(10X+11Y=41) \Rightarrow \quad \underline{50X+55Y=\;\;205}$$
$$-27X\qquad\quad=\;-81 \;\to\; X=3$$

Put X = 3 in D.

$$\boxed{D} \quad 7(3)+5Y=26$$
$$21+5Y=26$$
$$5Y=5$$
$$Y=1$$

Put X = 3 and Y = 1 in A.

$$\boxed{A} \quad 2(3)+3(1)-Z=11$$
$$9-Z=11$$
$$-Z=2$$
$$Z=-2$$

Check X = 3, Y = 1, Z = -2.

$$\boxed{A}\; 2(3)+3(1)-(-2)=11 \quad \boxed{B}\; 3(3)-(1)+2(-2)=4 \quad \boxed{C}\; 4(3)+2(1)+3(-2)=8$$
$$9+2=11 \qquad\qquad 9-1-4=4 \qquad\qquad 12+2-6=8$$
$$11=11 \qquad\qquad\quad 4=4 \qquad\qquad\qquad 8=8$$

Example 2

Find the solution that will satisfy all three equations.

$$\boxed{A} \quad 3X+2Y+4Z=9$$
$$\boxed{B} \quad 4X+3Y-2Z=6$$
$$\boxed{C} \quad 5X+4Y-3Z=8$$

Use A + B = D and A + C = E to eliminate Y.

$$\boxed{A} \quad -3(3X+2Y+4Z=9) \Rightarrow \quad -9X-6Y-12Z=-27$$
$$\boxed{B} \quad 2(4X+3Y-2Z=6) \Rightarrow \quad \underline{8X+6Y-\;4Z=\;12}$$
$$-X\qquad\;-16Z=-15 \;\to\; \boxed{D}\; -X-16Z=-15$$

$$\boxed{A} \quad -2(3X+2Y+4Z=9) \Rightarrow \quad -6X-4Y-8Z=-18$$
$$\boxed{C} \quad (5X+4Y-3Z=8) \Rightarrow \quad \underline{5X+4Y-3Z=\;\;8}$$
$$-X\qquad-11Z=-10 \;\to\; \boxed{E}\; -X-11Z=-10$$

Eliminate X.

$$\boxed{D} \quad -1(-X - 16Z = -15) \Rightarrow \quad X + 16Z = 15$$
$$\boxed{E} \quad (-X - 11Z = -10) \Rightarrow \quad \underline{-X - 11Z = -10}$$
$$5Z = 5 \rightarrow Z = 1$$

Put Z = 1 in D.

$$\boxed{D} \quad -X - 16(1) = -15$$
$$-X - 16 = -15$$
$$-X = 1$$
$$X = -1$$

Put Z = 1 and X = −1 in A.

$$\boxed{A} \quad 3(-1) + 2Y + 4(1) = 9$$
$$-3 + 2Y + 4 = 9$$
$$2Y = 8$$
$$Y = 4$$

Check X = −1, Y = 4, Z = 1.

$$\boxed{A} \; 3(-1) + 2(4) + 4(1) = 9 \quad \boxed{B} \; 4(-1) + 3(4) - 2(1) = 6 \quad \boxed{C} \; 5(-1) + 4(4) - 3(1) = 8$$
$$-3 + 12 = 9 \qquad\qquad 8 - 2 = 6 \qquad\qquad -5 + 13 = 8$$
$$9 = 9 \qquad\qquad\qquad 6 = 6 \qquad\qquad\qquad 8 = 8$$

Practice Problems 1

Find the solution that will satisfy all three equations.

1. $\boxed{A}$ X + 2Y − Z = 2
   $\boxed{B}$ 2X − Y + 3Z = −1
   $\boxed{C}$ −2X + 3Y − 4Z = 1

2. $\boxed{A}$ X − 2Y + 4Z = −4
   $\boxed{B}$ 3X + 4Y − 5Z = 25
   $\boxed{C}$ 5X − 3Y + 2Z = 12

Solutions 1

1. Use A + B = D and B + C = E to eliminate Y.

$\boxed{A}$ $(X + 2Y - Z = 2) \Rightarrow$ $X + 2Y - Z = 2$
$\boxed{B}$ $2(2X - Y + 3Z = -1) \Rightarrow$ $\dfrac{4X - 2Y + 6Z = -2}{5X \qquad + 5Z = 0} \rightarrow \boxed{D}\ 5X + 5Z = 0$

$\boxed{B}$ $3(2X - Y + 3Z = -1) \Rightarrow$ $6X - 3Y + 9Z = -3$
$\boxed{C}$ $(-2X + 3Y - 4Z = 1) \Rightarrow$ $\dfrac{-2X + 3Y - 4Z = 1}{4X \qquad + 5Z = -2} \rightarrow \boxed{E}\ 4X + 5Z = -2$

Eliminate Z.

$\boxed{D}$ $-1(5X + 5Z = 0) \Rightarrow$ $-5X - 5Z = 0$
$\boxed{E}$ $(4X + 5Z = -2) \Rightarrow$ $\dfrac{4X + 5Z = -2}{-X \qquad = -2} \rightarrow X = 2$

Put X = 2 in D.

$\boxed{D}$ $5(2) + 5Z = 0$
$10 + 5Z = 0$
$5Z = -10$
$Z = -2$

Put X = 2 and Z = -2 in A.

$\boxed{A}$ $(2) + 2Y - (-2) = 2$
$2Y + 4 = 2$
$2Y = -2$
$Y = -1$

Check X = 2, Z = -2, Y = -1.

$\boxed{A}$ $(2) + 2(-1) - (-2) = 2$ $\quad$ $\boxed{B}$ $2(2) - (-1) + 3(-2) = -1$ $\quad$ $\boxed{C}$ $-2(2) + 3(-1) - 4(-2) = 1$
$\qquad\quad 2 - 2 + 2 = 2$ $\qquad\qquad\qquad\quad 4 + 1 - 6 = -1$ $\qquad\qquad\qquad\quad -4 - 3 + 8 = 1$
$\qquad\qquad\quad 2 = 2$ $\qquad\qquad\qquad\qquad\quad -1 = -1$ $\qquad\qquad\qquad\qquad\quad 1 = 1$

2. Use A + B = D and A + C = E to eliminate X.

$\boxed{A}$ $-3(X - 2Y + 4Z = -4) \Rightarrow -3X + 6Y - 12Z = 12$
$\boxed{B}$ $(3X + 4Y - 5Z = 25) \Rightarrow \underline{3X + 4Y - 5Z = 25}$
$ 10Y - 17Z = 37 \rightarrow \boxed{D}\ 10Y - 17Z = 37$

$\boxed{A}$ $-5(X - 2Y + 4Z = -4) \Rightarrow -5X + 10Y - 20Z = 20$
$\boxed{C}$ $(5X - 3Y + 2Z = 12) \Rightarrow \underline{5X - 3Y + 2Z = 12}$
$ 7Y - 18Z = 32 \rightarrow \boxed{E}\ 7Y - 18Z = 32$

Eliminate Y.

$\boxed{D}$ $-7(10Y - 17Z = 37) \Rightarrow -70Y + 119Z = -259$
$\boxed{E}$ $10(7Y - 18Z = 32) \Rightarrow \underline{70Y - 180Z = 320}$
$ -61Z = 61 \quad \rightarrow Z = -1$

Put Z = –1 in D.

$\boxed{D}$ $\quad 10Y - 17(-1) = 37$
$ 10Y + 17 = 37$
$ 10Y = 20$
$ Y = 2$

Put Z = –1 and Y = 2 in A.

$\boxed{A}$ $\quad X - 2(2) + 4(-1) = -4$
$ X - 4 - 4 = -4$
$ X - 8 = -4$
$ X = 4$

Check X = 4, Y = 2, Z = –1.

$\boxed{A}\ (4) - 2(2) + 4(-1) = -4$ $\quad$ $\boxed{B}\ 3(4) + 4(2) - 5(-1) = 25$ $\quad$ $\boxed{C}\ 5(4) - 3(2) + 2(-1) = 12$
$ 4 - 4 - 4 = -4$ $ 12 + 8 + 5 = 25$ $ 20 - 6 - 2 = 12$
$ -4 = -4$ $ 25 = 25$ $ 12 = 12$

Vectors

Vectors have two components: ***direction*** and ***magnitude***. They are shown graphically as arrows. Motions in one dimension (along a line) give their direction in terms of a sign (+ or −). We have been studying a simple form of one-dimensional vectors, without using the term, since we began studying positive and negative numbers in *Pre-Algebra*. Consider the following examples:

+4 ⟶ + +4 ⟶ = +8 ⟶   (+4)+(+4) = (+8)

+3 ⟶ + −3 ⟵ = 0   (+3)+(−3) = (0)

+4 ↑ + −1 ↓ = +3 ↑

(+4)+(−1) = (+3)

Vectors move in two dimensions, over and up, instead of in one direction as in a number line. This movement or motion takes place in a two-dimensional plane.

Example 1

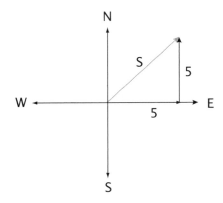

You leave your hotel to solve a national emergency and head five blocks east, and then turn at a right angle and head five blocks north. When you arrive there, you call Superman. Since he doesn't need to go on foot, he wants to know how far away you are and in what direction. The length of the distance between the hotel and where you are now is called the magnitude of the vector. On the picture, S stands for Superman. The direction is called simply that. Every vector has these two components, magnitude and direction.

Since we have a right triangle, we can figure out the length of S by using the Pythagorean theorem. $S = \sqrt{50}$ or $5\sqrt{2}$, which our calculator shows us is 7.07 blocks.

We have computed the distance, and the direction is north and east, or northeast. So we tell the man in red and blue to head northeast for 7.07 blocks.

$$5^2 + 5^2 = 50 = S^2$$

$$\sqrt{50} = S$$

7.07 blocks = S, which is the magnitude of the resultant vector

Most of our vector problems will not be drawn as a grid of downtown but with another familiar grid—the Cartesian coordinate system, which consists of two number lines. Instead of north and south, we have the Y-axis. We refer to the X-axis in place of east and west.

In our first example, we knew that the direction of the resultant vector was 45° because of what we learned in geometry about special right triangles. If the two legs of a right triangle are the same length, then the triangle is a 45°-45°-90° right triangle. We could have told the Man of Steel to travel 45° for 7.07 blocks. The direction is referred to as angle theta, using the Greek letter θ.

Example 2

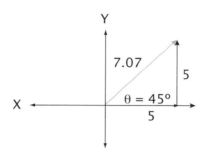

Suppose you begin at your home and travel three miles due east and then travel an additional four miles due north. In order to compute the magnitude of the journey from start to finish, we use the Pythagorean theorem. We are adding two one-dimensional vectors that are at right angles to each other. The answer is called the magnitude of the resultant vector, or the *resultant*.

Example 3

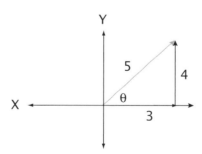

$$3^2 + 4^2 = 25 = \text{resultant}^2 \text{ or magnitude}^2$$
$$\text{Magnitude} = 5 \text{ miles}$$

Theta appears to be a little larger than 45°, so we estimate that it is about 50° to 55°. We'll find out how to determine the measure of angle theta later on. This process of discovering the angle using the length of the sides is taught in detail in trigonometry, which you will study more fully in *PreCalculus*.

Find the magnitude of a vector that is the result of two vectors—one moving along the X-axis six blocks, and added to the first one, another vector moving three blocks in the direction of Y.

Example 4

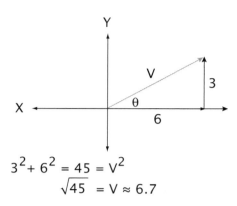

$$3^2 + 6^2 = 45 = V^2$$
$$\sqrt{45} = V \approx 6.7$$

If the magnitude of the vector is 10 and the short leg is 5, what is the length of the long leg and what is the measure of theta?

Example 5

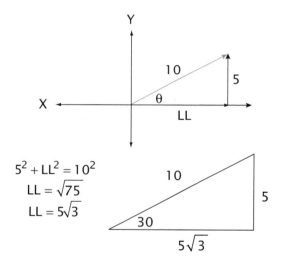

$$5^2 + LL^2 = 10^2$$
$$LL = \sqrt{75}$$
$$LL = 5\sqrt{3}$$

Using geometry skills, we can see that the length of the X vector is $5\sqrt{3}$. Using what we know of special right triangles, we can see this is a 30°- 60°- 90° triangle because the short side, or short leg, is one-half the length of the hypotenuse (our resultant vector). This ratio of one to two (1/2 or .5) is the door to our understanding trigonometry, which helps us find theta, or direction.

Trigonometric Ratios

Now for a quick review of basic trigonometry as presented in the Math-U-See *Geometry* course.

Let's begin by describing all the angles and all the sides of a right triangle. We already know that one angle is 90°, and that the side opposite the right angle is the **hypotenuse**. This leaves two angles and two sides, or legs, to name. I've decided to call the angles θ (theta) and α (alpha), both letters in the Greek alphabet. The sides are described in reference to the angles. If I am standing in angle θ, then the leg furthest away from me will be the **opposite** side. The side or leg that touches me (where my feet are standing in the illustration) is the **adjacent** side. If I move to stand in angle α, then the side furthest away from me becomes the "opposite" side and the side touching me is the "adjacent" side. The names for the sides depend on what angle you are referring to.

I'll illustrate this with our old friend, the 3 - 4 - 5 right triangle.

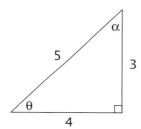

Standing at θ, the opposite side is three units long and the adjacent side is four units long. If I move to α, then four is opposite and three is adjacent. In both instances, the hypotenuse is five.

Now we come to the three main trigonometric ratios, using the terminology of opposite, adjacent, and hypotenuse. The ratios are *sine, cosine,* and *tangent.* The sine of either angle in a right triangle (in our example either θ or α) is described as the ratio of the opposite side over the hypotenuse. See the chart on the next page.

A fun way to remember these three trigonometric ratios is the result of dropping a brick on your big toe. What would you do? Probably get a pan of water and "soak your toe," or SOH-CAH-TOA.

SOH stands for sin $= \dfrac{\text{opposite}}{\text{hypotenuse}} \Rightarrow S = \dfrac{O}{H}$

CAH stands for cos $= \dfrac{\text{adjacent}}{\text{hypotenuse}} \Rightarrow C = \dfrac{A}{H}$

TOA stands for tan $= \dfrac{\text{opposite}}{\text{adjacent}} \Rightarrow T = \dfrac{O}{A}$

Notice the commonly used abbreviations for sine, cosine and tangent.

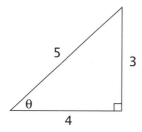

Use the 3-4-5 right triangle, the sine of angle θ is $\dfrac{\text{opposite}}{\text{hypotenuse}} = \dfrac{3}{5}$ or $\sin \theta = \dfrac{3}{5}$.

The cosine of θ is the adjacent over the hypotenuse, or $\cos \theta = \dfrac{4}{5}$.

And the tangent of θ is the opposite over the adjacent, or $\tan \theta = \dfrac{3}{4}$.

Another relationship found in these trigonometric expressions is that sine and cosine are complementary angles. In a right triangle (with one right angle), the other two angles always add up to 90°, so they are *complementary*.

Using the 3 - 4 - 5 triangles, let's look at the ratios once again. We learned that the ratios are constant for a 45°- 45°- 90° triangle or a 30°- 60°- 90° triangle. In the latter, the short side is always one-half the hypotenuse. The length of the sides of different 30°- 60°- 90° triangles may vary, but the short side will always be one-half of the hypotenuse.

$$\sin 30° = \frac{\text{opposite}}{\text{hypotenuse}} = \frac{\text{opposite}}{\text{hypotenuse}} = \frac{1}{2}$$

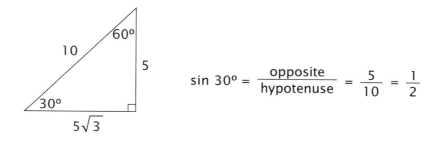

What if the sides of the triangle had lengths such as shown below?

$$\sin 30° = \frac{\text{opposite}}{\text{hypotenuse}} = \frac{5}{10} = \frac{1}{2}$$

What if the side lengths varied and you had this triangle?

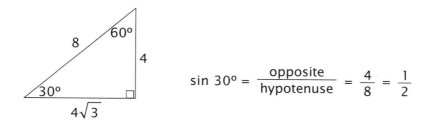

$$\sin 30° = \frac{\text{opposite}}{\text{hypotenuse}} = \frac{4}{8} = \frac{1}{2}$$

Observe that the ratio of the small side to the hypotenuse remains constant.

If you type in 30° on your calculator and then hit the key that says SIN, it will give you the ratio as .5 because the ratio is always one-half or .5. Trig ratios are usually expressed as a decimal with four digits after the decimal point, so .5 would be written as .5000. If you know the angle, you can always find the ratio. The inverse is also true. In our examples, we know the lengths of the sides but not the angle. We are hoping that if we have the ratio, we can go in reverse and find the angle, and this is true. Calculators differ, so you may have to get out the manual to find out how to compute trig functions on your model. Use sin 30° = .5 as your standard to figure out how to use your calculator.

To find the measure of the angle, first make sure that your calculator gives the answer in degrees and not radians or some other kind of measure. The inverse of X is $1/X$ or X^{-1} power. So also in trigonometry, the *inverse* of sin is $\sin^{-1}$. There is usually a second line or inverse key on your calculator. It is similar to a shift key and is used to get the inverse of the sin, cos, and tan.

If you type in .5000 on your calculator and then hit the key that says SIN^{-1}, it will give you the answer as 30°. Given the ratio, you can always find the angle. Try the exercise below to find the angles using the ratios of the 3 - 4 - 5 right triangle.

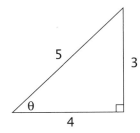

$$\sin \theta = \frac{3}{5} = .6000 \qquad \text{enter .6000, hit SIN}^{-1} \qquad \theta = 36.9°$$

$$\cos \theta = \frac{4}{5} = .8000 \qquad \text{enter .8000, hit COS}^{-1} \qquad \theta = 36.9°$$

$$\tan \theta = \frac{3}{4} = .7500 \qquad \text{enter .7500, hit TAN}^{-1} \qquad \theta = 36.9°$$

Example 6
Find the magnitude and direction of the resultant vector.

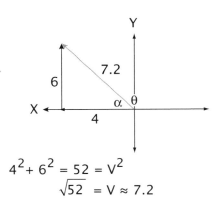

$$4^2 + 6^2 = 52 = V^2$$
$$\sqrt{52} = V \approx 7.2$$

We can see that the direction is more than 90°. If we find the angle inside the triangle, which we'll call α, we can subtract from 180° to find the measure of θ, since $\alpha + \theta = 180°$. Using the tangent, we see the side opposite over the side adjacent is 6/4 or 1.5. Entering 1.5 and pushing the key for TAN^{-1}, we see the angle is close to 56° inside the triangle, so angle θ is 180° − 56° or 124°. The magnitude is 7.2 and the direction is 124°.

If you are adding multiple vectors, make sure you draw them with the tail of the second vector placed over the tip or head of the first vector. See example 7. The first vector is over three and up two. The second vector, added to it, is over one and up four. The sum of these two vectors is shown in the second figure.

Example 7

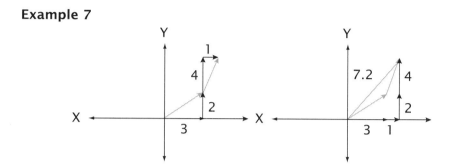

You can find this vector's over and up coordinates by adding the X, or over, dimensions (3 + 1 = 4), and then adding the Y dimensions (2 + 4 = 6). The final vector is over four and up six, which is what we had in the previous example, so we know the magnitude is 7.2 and the direction is 56°.

Student Solutions

Solutions are shown in detail. The student may use canceling and other shortcuts as long as the answers match.

Lesson Practice 1A

1. $3^{-2} = \dfrac{1}{3^2} = \dfrac{1}{9}$

2. $X^{-1} = \dfrac{1}{X^1} = \dfrac{1}{X}$

3. $\left(-\dfrac{2}{3}\right)^2 = -\dfrac{2^2}{3^2} = \dfrac{4}{9}$

4. $\dfrac{1}{2^{-3}} = \dfrac{2^3}{1} = 2^3 = 8$

5. $\dfrac{1}{X^{-5}} = \dfrac{X^5}{1} = X^5$

6. $\left(\dfrac{1}{2}\right)^3 = \dfrac{1^3}{2^3} = \dfrac{1}{8}$

7. $2^2 \cdot 2^6 \cdot 2^3 = 2^{2+6+3} = 2^{11}$ or $2{,}048$

8. $R \cdot R^4 \cdot R^2 = R^{1+4+2} = R^7$

9. $X^{-1} \cdot X^3 \cdot X^{-4} =$
   $X^{(-1)+3+(-4)} = X^{-2}$ or $\dfrac{1}{X^2}$

10. $3^A \cdot 3^B = 3^{A+B}$

11. $4^0 \cdot 4^{-2} \cdot 4^2 = 4^{0+(-2)+2} = 4^0 = 1$

12. $X^A \cdot X^{2A} \cdot X^B = X^{A+2A+B} = X^{3A+B}$

13. $5^2 \div 5^6 = 5^{2-6} = 5^{-4} = \dfrac{1}{5^4}$ or $\dfrac{1}{625}$

14. $Y^3 \div Y^0 = Y^{3-0} = Y^3$

15. $\dfrac{2^{-4}}{2^{-2}} = 2^{-4-(-2)} = 2^{-2} = \dfrac{1}{2^2}$ or $\dfrac{1}{4}$

16. $X^{16} \div X^7 = X^{16-7} = X^9$

17. $B^{-2} \div B^4 = B^{-2-4} = B^{-6}$ or $\dfrac{1}{B^6}$

18. $\dfrac{Y^4}{Y^8} = Y^{4-8} = Y^{-4}$ or $\dfrac{1}{Y^4}$

19. $\left(X^3\right)^4 = X^{3\cdot4} = X^{12}$

20. $\left(5^0\right)^3 = 5^{0\cdot3} = 5^0 = 1$

21. $\left(\left(A^2\right)^2\right)^3 = A^{2\cdot2\cdot3} = A^{12}$

22. $\dfrac{A^3B^3}{A^{-2}B^4} = A^3A^2B^3B^{-4} =$
    $A^{3+2}B^{3+(-4)} = A^5B^{-1}$ or $\dfrac{A^5}{B}$

23. $\dfrac{H^{-4}N^6}{HN^2} = H^{-4}H^{-1}N^6N^{-2} =$
    $H^{-4+(-1)}N^{6+(-2)} = H^{-5}N^4$ or $\dfrac{N^4}{H^5}$

24. $\dfrac{P^{-2}Q^0P^3}{Q^{-1}Q^6P^4} = P^{-2}P^3P^{-4}Q^0Q^1Q^{-6} =$
    $P^{-2+3+(-4)}Q^{0+1+(-6)} = P^{-3}Q^{-5}$ or $\dfrac{1}{P^3Q^5}$

Lesson Practice 1B

1. $4^{-3} = \dfrac{1}{4^3}$ or $\dfrac{1}{64}$

2. $7^3 = 343$

3. $\left(\dfrac{-1}{-3}\right)^3 = \dfrac{-1^3}{-3^3} = \dfrac{-1}{-27} = \dfrac{1}{27}$

4. $\dfrac{1}{A^{-1}} = \dfrac{A^1}{1} = A$

5. $\dfrac{1}{3^{-2}} = \dfrac{3^2}{1} = \dfrac{9}{1} = 9$

6. $\left(\dfrac{1}{3}\right)^2 = \dfrac{1^2}{3^2} = \dfrac{1}{9}$

7. $3^1 \cdot 3^0 \cdot 3^4 = 3^{1+0+4} = 3^5$ or 243

8. $X^2 \cdot X^{-3} \cdot X^{10} = X^{2+(-3)+10} = X^9$

9. $Y^{-6} \cdot Y^0 = Y^{-6+0} = Y^{-6}$ or $\dfrac{1}{Y^6}$

10. $2^A \cdot 2^{3B} = 2^{A+3B}$

11. $6^4 \cdot 6^2 \cdot 6^{-5} \cdot 6^1 = 6^{4+2+(-5)+1} = 6^2 = 36$

12. $A^9 \cdot A^{-9} = A^{9+(-9)} = A^0 = 1$

13. $B^2 \div B^3 = B^{2-3} = B^{-1}$ or $\dfrac{1}{B}$

14. $X^0 \div X^{-8} = X^{0-(-8)} = X^8$

15. $\dfrac{3^2}{3^3} = 3^{2-3} = 3^{-1} = \dfrac{1}{3}$

16. $6^{A+B} \div 6^B = 6^{(A+B)-B} = 6^A$

17. $Y^{16} \div Y^{12} = Y^{16-12} = Y^4$

18. $\dfrac{A^{-4}}{A^5} = A^{-4-5} = A^{-9}$ or $\dfrac{1}{A^9}$

19. $\left(Y^2\right)^{-10} = Y^{(2)(-10)} = Y^{-20}$ or $\dfrac{1}{Y^{20}}$

20. $\left(8^2\right)^0 = 8^{(2)(0)} = 8^0 = 1$

21. $\left(\left(Z^3\right)^{-2}\right)^5 = Z^{(3)(-2)(5)} = Z^{-30}$ or $\dfrac{1}{Z^{30}}$

22. $\dfrac{X^2Y^{-2}}{X^{-4}Y^{-5}} = X^2X^4Y^{-2}Y^5 =$
$X^{2+4}Y^{-2+5} = X^6Y^3$

23. $\dfrac{3^A3^{2A}}{3^B3^{2B}} = \dfrac{3^{A+2A}}{3^{B+2B}} = \dfrac{3^{3A}}{3^{3B}} = 3^{3A}3^{-3B} =$
$3^{3A+(-3B)} = 3^{3A-3B}$

24. $\dfrac{R^5S^{-3}S^2}{SR^6R^3} = R^5R^{-6}R^{-3}S^{-3}S^2S^{-1} =$
$R^{5+(-6)+(-3)}S^{-3+2+(-1)} = R^{-4}S^{-2}$ or $\dfrac{1}{R^4S^2}$

Systematic Review 1C

1. $5^{-2} = \dfrac{1}{5^2}$ or $\dfrac{1}{25}$

2. $4^5 \cdot 4^2 \cdot 4^{-4} = 4^{5+2+(-4)} = 4^3$ or 64

3. $X^A \div X^B = X^{A-B}$

4. $(31-5)^0 = (26)^0 = 1$

5. $\left(Y^D\right)^G = Y^{DG}$

6. $\left(\left(2A\right)^3\right)^2 = (2A)^{(3)(2)} = (2A)^6 =$
$2^6A^6 = 64A^6$

7. $\dfrac{X^4Y^{-3}}{X^0X^{-5}} = X^4X^0X^5Y^{-3} = X^{4+0+5}Y^{-3} =$
X^9Y^{-3} or $\dfrac{X^9}{Y^3}$

8. $\dfrac{X^{-3}Y^{-4}Y^2X}{X^{-5}Y^2} = X^{-3}X^1X^5Y^{-4}Y^2Y^{-2} =$
$X^{-3+1+5}Y^{-4+2+(-2)} = X^3Y^{-4}$ or $\dfrac{X^3}{Y^4}$

9. $-14^2 = -(14)(14) = -196$

10. $(-15)^2 = (-15)(-15) = 225$

11. given

12. $1.1 - .086 = 2.4B$
$(1,000)1.1 - (1,000).086 = (1,000)2.4B$
$1,100 - 86 = 2,400B$
$1,014 = 2,400B$
$\dfrac{1,014}{2,400} = B$
$\dfrac{169}{400} = B$

13. $3 + 2\dfrac{1}{5} = \dfrac{1}{2}M - \dfrac{3}{10}$
$5\dfrac{1}{5} = \dfrac{1}{2}M - \dfrac{3}{10}$
$\dfrac{26}{5} = \dfrac{1}{2}M - \dfrac{3}{10}$
$\left(\dfrac{10}{1}\right)\left(\dfrac{26}{5}\right) = \left(\dfrac{10}{1}\right)\left(\dfrac{1}{2}\right)M - \left(\dfrac{10}{1}\right)\left(\dfrac{3}{10}\right)$
$\dfrac{260}{5} = \dfrac{10}{2}M - \dfrac{30}{10}$
$52 = 5M - 3$
$55 = 5M$
$\dfrac{55}{5} = M = 11$

14. $.388 = 1.3 + .3Q$
$(1,000).388 = (1,000)1.3 + (1,000).3Q$
$388 = 1,300 + 300Q$
$-912 = 300Q$
$\dfrac{-912}{300} = Q = -3\dfrac{1}{25}$

15. $12A^2 - 6AB = 6A(2A - B)$

16. $15AB^3 + 18BA - 21B^2 =$
$3B(5AB^2 + 6A - 7B)$

17. $-13(X^2 - 2X + 4) = -13X^2 + 26X - 52$

18. $X^2(XY + X^3) = X^3Y + X^5$

19. $(15 - 6 + 8^2 \div 2 \div 4) \times |9^2 - 10^2| =$
$(15 - 6 + 64 \div 2 \div 4) \times |81 - 100| =$
$(15 - 6 + 32 \div 4) \times |-19| =$
$(15 - 6 + 8) \times 19 =$
$(9 + 8) \times 19 =$
$(17) \times 19 = 323$

20. $(4 \times 8 - 5 + 2)^2 - |3 - 6 - 7^2 \times 9| =$
$(32 - 5 + 2)^2 - |3 - 6 - 49 \times 9| =$
$(27 + 2)^2 - |3 - 6 - 441| =$
$(29)^2 - |-3 - 441| =$
$841 - |-444| =$
$841 - 444 = 397$

Systematic Review 1D

1. $X^0 = 1$

2. $2^{-4} = \dfrac{1}{2^4} = \dfrac{1}{16}$

3. $5^{-2} \cdot 5^{-6} = 5^{-2+(-6)} = 5^{-8}$ or $\dfrac{1}{5^8}$

4. $9^9 \div 9^3 = 9^{9-3} = 9^6$

5. $\left(8^7\right)^5 = 8^{(7)(5)} = 8^{35}$

6. $\left(9^2\right)^0 = 9^{(2)(0)} = 9^0 = 1$

7. $3^2 M^{-2} N^4 M^{-3} M^{-2} N^1 = 3^2 M^{-2+(-3)+(-2)} N^{4+1}$
$= 9M^{-7} N^5$ or $\dfrac{9N^5}{M^7}$

8. $\dfrac{R^{-6} R P^{-4}}{P^9 R^{-4}} = R^{-6} R R^4 P^{-4} P^{-9}$
$= R^{-6+1+4} P^{-4+(-9)}$
$= R^{-1} P^{-13}$ or $\dfrac{1}{RP^{13}}$

9. $\left(-1\dfrac{1}{5}\right)^2 = \left(-\dfrac{6}{5}\right)^2 = \dfrac{(-6)^2}{(-5)^2}$
$= \dfrac{36}{25} = 1\dfrac{11}{25}$

10. $-2^3 = -(2)(2)(2) = -8$

11. $\dfrac{3}{4} - \dfrac{5}{6} R = \dfrac{7}{10}$
$(60)\dfrac{3}{4} - (60)\dfrac{5}{6} R = (60)\dfrac{7}{10}$
$\dfrac{180}{4} - \dfrac{300}{6} R = \dfrac{420}{10}$
$45 - 50R = 42$
$-50R = -3$
$R = \dfrac{-3}{-50} = \dfrac{3}{50}$

12. $.5Y + .3 = .002$
$(1,000).5Y + (1,000).3 = (1,000).002$
$500Y + 300 = 2$
$500Y = -298$
$Y = \dfrac{-298}{500} = -\dfrac{149}{250}$

13. $3\dfrac{2}{3} - \dfrac{5}{12} K = 1\dfrac{1}{4}$
$\dfrac{11}{3} - \dfrac{5}{12} K = \dfrac{5}{4}$
$(12)\dfrac{11}{3} - (12)\dfrac{5}{12} K = (12)\dfrac{5}{4}$
$\dfrac{132}{3} - \dfrac{60}{12} K = \dfrac{60}{4}$
$44 - 5K = 15$
$-5K = -29$
$K = \dfrac{-29}{-5} = 5\dfrac{4}{5}$

14. $1.203H + .9 = -.6$
$(1,000)1.203H + (1,000).9 =$
$(1,000)(-.6)$
$1,203H + 900 = -600$
$1,203H = -1500$
$H = \dfrac{-1500}{1203} = -1\dfrac{297}{1203} - 1\dfrac{99}{401}$

15. $56X - 49XA - 28X^2 =$
$7X(8 - 7A - 4X)$

16. $4X - 16X^3 =$
$4X(1 - 4X^2)$

17. $(2X)^3 (X - 5 + 3X^2) = 8X^4 - 40X^3 + 24X^5$
or $24X^5 + 8X^4 - 40X^3$
It is customary to put equations in
reverse order by value of the exponents.

18. $5X^2Y(3X + 4YX - X^2Y^3) =$

$15X^3Y + 20X^3Y^2 - 5X^4Y^4$

or $-5X^4Y^4 + 20X^3Y^2 + 15X^3Y$

19. $((10+3)^2 - 9) \div 20 =$

$((13)^2 - 9) \div 20 =$

$(169 - 9) \div 20 =$

$(160) \div 20 = 8$

20. $[42 \div 6 - 2] \times 11 - 13^2 =$

$[7 - 2] \times 11 - 169 =$

$[5] \times 11 - 169 =$

$55 - 169 = -114$

Systematic Review 1E

1. $\left(\dfrac{1}{3}\right)^2 = \dfrac{1^2}{3^2} = \dfrac{1}{9}$

2. $\left(10^2\right)^{-4} = 10^{(2)(-4)} = 10^{-8}$ or $\dfrac{1}{10^8}$

3. $4^A \cdot 4^B = 4^{A+B}$

4. $11 \div 11^0 = 11^1 \div 11^0 = 11^{1-0} = 11^1 = 11$

5. $\left(3^D\right)^4 = 3^{4D}$

6. $\left(5^3\right)^2 = 5^{(3)(2)} = 5^6$

7. $\dfrac{B^5B^2C^{-5}}{B^{-4}C^{-3}} = B^5B^2B^4C^{-5}C^3 =$

$B^{5+2+4}C^{-5+3} = B^1C^{-2}$ or $\dfrac{B^{11}}{C^2}$

8. $\dfrac{D^6C^{-4}D^2}{D^{-4}C^0C^2} = D^6D^2D^4C^{-4}C^0C^{-2} =$

$C^{-4+0+(-2)}D^{6+2+4} = C^{-6}D^{12}$ or $\dfrac{D^{12}}{C^6}$

9. $\left(-2\dfrac{3}{4}\right)^2 = \left(-\dfrac{11}{4}\right)^2 = \dfrac{11^2}{4^2} = \dfrac{121}{16} = 7\dfrac{9}{16}$

10. $(-10)^4 = 10,000$

11. $-5\dfrac{1}{2}Y - \dfrac{2}{9} = \dfrac{5}{18}$

$-\dfrac{11}{2}Y - \dfrac{2}{9} = \dfrac{5}{18}$

$(18)\left(-\dfrac{11}{2}\right)Y + (18)\left(-\dfrac{2}{9}\right) = (18)\left(\dfrac{5}{18}\right)$

$-\dfrac{198}{2}Y - \dfrac{36}{9} = \dfrac{90}{18}$

$-99Y - 4 = 5$

$-99Y = 9$

$Y = \dfrac{9}{-99} = -\dfrac{1}{11}$

12. $-.7A + .8A = 1.2$

$(10)(-.7A) + (10)(.8A) = (10)(1.2)$

$-7A + 8A = 12$

$A = 12$

13. $1\dfrac{2}{3} = -2\dfrac{1}{4} + 1\dfrac{3}{5}A$

$\dfrac{5}{3} = -\dfrac{9}{4} + \dfrac{8}{5}A$

$(60)\left(\dfrac{5}{3}\right) = (60)\left(-\dfrac{9}{4}\right) + (60)\left(\dfrac{8}{5}\right)A$

$\dfrac{300}{3} = -\dfrac{540}{4} + \dfrac{480}{5}A$

$100 = -135 + 96A$

$235 = 96A$

$\dfrac{235}{96} = A = 2\dfrac{43}{96}$

14. $3X - 1.6 = .34$

$(100)(3X) - (100)(1.6) = (100)(.34)$

$300X - 160 = 34$

$300X = 194$

$X = \dfrac{194}{300} = \dfrac{97}{150}$

15. $9M - 10M^3 + 19M^2 = (M)(9 - 10M^2 + 19M)$

16. $-36M - 72M^2 + 45M^2 = (9M)(-4 - 8M + 5M) =$

$(9M)(-4 - 3M)$ or $(-9M)(4 + 3M)$

17. $A^3(XA^1 + 2X^2A^1 - A^2) =$

$XA^4 + 2X^2A^4 - A^5$

18. $AB^1(A^2 - 4AB^1 + 2B^1) =$

$A^3B - 4A^2B^2 + 2AB^2$

19. $-19 - |(7)(-2)| + 6^2 = -19 - |-14| + 6^2 =$

$-19 - 14 + 6^2 = -19 - 14 + 36 =$

$-33 + 36 = 3$

20. $5 \times 3 + 7^2 - 7 + |-8 \div 4| =$
$5 \times 3 + 7^2 - 7 + |-2| =$
$5 \times 3 + 7^2 - 7 + 2 = 5 \times 3 + 49 - 7 + 2 =$
$15 + 49 - 7 + 2 = 64 - 7 + 2 = 57 + 2 = 59$

Lesson Practice 2A

1. correct

2. correct

3. incorrect: $\dfrac{X^2 + 3X + 6}{X^2 + 2X + 7} =$

$\dfrac{X^2}{X^2 + 2X + 7} + \dfrac{3X}{X^2 + 2X + 7} + \dfrac{6}{X^2 + 2X + 7}$

4. incorrect: $\dfrac{B}{B} + B^0 = 1 + 1 = 2$

5. $\dfrac{4X^2 + X}{X} = \dfrac{(X)(4X + 1)}{X} = \dfrac{4X + 1}{1} = 4X + 1$

6. $\dfrac{Y^2 + 2Y}{Y} = \dfrac{(Y)(Y + 2)}{Y} = \dfrac{Y + 2}{1} = Y + 2$

7. $\dfrac{4X + 4Y}{2} = \dfrac{(4)(X + Y)}{2} = \dfrac{2(X + Y)}{1} =$
$2(X + Y)$ or $2X + 2Y$

8. $\dfrac{12AB + 16A^2}{4A} = \dfrac{(4A)(3B + 4A)}{4A} = 3B + 4A$

9. $\dfrac{5XY + 20XYZ}{5YZ} = \dfrac{(5Y)(X + 4XZ)}{(5Y)(Z)} = \dfrac{X + 4XZ}{Z}$

This fraction cannot be reduced further because Z cannot be factored out of the numerator.

10. $\dfrac{2X^2Y - XY^2}{XY} = \dfrac{(XY)(2X - Y)}{XY} =$

$\dfrac{2X - Y}{1} = 2X - Y$

11. $\dfrac{6}{X + 2} + \dfrac{4X}{X + 2} = \dfrac{6 + 4X}{X + 2}$

12. $\dfrac{3}{4} + \dfrac{3}{X} = \dfrac{3(X)}{4(X)} + \dfrac{3(4)}{X(4)} =$

$\dfrac{3X}{4X} + \dfrac{12}{4X} = \dfrac{3X + 12}{4X}$

13. $\dfrac{7}{4X} - \dfrac{3}{4Y} = \dfrac{7(Y)}{4X(Y)} - \dfrac{3(X)}{4Y(X)} = \dfrac{7Y - 3X}{4XY}$

14. $\dfrac{A}{B} - \dfrac{B}{A} = \dfrac{A(A)}{B(A)} - \dfrac{B(B)}{A(B)} = \dfrac{A^2 - B^2}{AB}$

15. $\dfrac{3X}{Y - 1} + \dfrac{2X}{Y + 1} =$

$\dfrac{3X(Y + 1)}{(Y - 1)(Y + 1)} + \dfrac{2X(Y - 1)}{(Y + 1)(Y - 1)} =$

$\dfrac{3X(Y + 1) + 2X(Y - 1)}{(Y + 1)(Y - 1)} =$

$\dfrac{3XY + 3X + 2XY - 2X}{(Y + 1)(Y - 1)} = \dfrac{5XY + X}{(Y + 1)(Y - 1)}$

or: $\dfrac{5XY + X}{Y^2 + Y - Y - 1} = \dfrac{5XY + X}{Y^2 - 1}$

16. $\dfrac{R}{T} + \dfrac{RS}{RT} = \dfrac{R}{T} + \dfrac{S}{T} = \dfrac{R + S}{T}$

Lesson Practice 2B

1. incorrect:

$\dfrac{3}{X + 1} + \dfrac{6}{X - 1} =$

$\dfrac{3(X - 1)}{(X + 1)(X - 1)} + \dfrac{6(X + 1)}{(X - 1)(X + 1)} =$

$\dfrac{(3X - 3) + (6X + 6)}{(X - 1)(X + 1)} =$

$\dfrac{9X + 3}{(X - 1)(X + 1)}$ or $\dfrac{9X + 3}{X^2 - 1}$

2. incorrect: $\dfrac{1}{4} + \dfrac{2}{4} = \dfrac{1 + 2}{4} = \dfrac{3}{4}$

3. correct

4. correct

5. $\dfrac{Y^4 + Y^2}{Y^2} = \dfrac{(Y^2)(Y^2 + 1)}{Y^2} = \dfrac{Y^2 + 1}{1} = Y^2 + 1$

6. $\dfrac{6X + 3A + 3X}{3} = \dfrac{(3)(2X + A + X)}{3} =$

$\dfrac{2X + A + X}{1} = 2X + A + X = 3X + A$

7. $\dfrac{16X^3Y + 8X^2Y}{2XY} = \dfrac{(2XY)(8X^2 + 4X)}{2XY} =$

$\dfrac{8X^2 + 4X}{1} = 8X^2 + 4X$ or $4X(2X + 1)$

8. $\dfrac{A^2B^3C^4 - ABC^2}{AB} = \dfrac{(ABC^2)(AB^2C^2 - 1)}{AB} =$

$C^2(AB^2C^2 - 1)$ or $AB^2C^4 - C^2$

9. $\dfrac{2R + 2}{(R + 1)} = \dfrac{2(R + 1)}{(R + 1)} = \dfrac{2}{1} = 2$

10. $\dfrac{5X^2 - 5A^2}{5A} = \dfrac{(5)\left(X^2 - A^2\right)}{(5)A} = \dfrac{X^2 - A^2}{A}$

11. $\dfrac{8}{X+2} + \dfrac{-1}{X-2} =$

$\dfrac{8(X-2)}{(X+2)(X-2)} + \dfrac{-1(X+2)}{(X-2)(X+2)} =$

$\dfrac{8(X-2) - 1(X+2)}{(X+2)(X-2)} = \dfrac{8X - 16 - X - 2}{(X+2)(X-2)} =$

$\dfrac{7X - 18}{(X+2)(X-2)}$ or $\dfrac{7X - 18}{X^2 - 4}$

12. $\dfrac{1}{2} + \dfrac{6}{AB} = \dfrac{1(AB)}{2(AB)} + \dfrac{6(2)}{AB(2)} = \dfrac{AB + 12}{2AB}$

13. $\dfrac{8}{2A} - \dfrac{3}{2B} = \dfrac{8(B)}{2A(B)} - \dfrac{3(A)}{2B(A)} = \dfrac{8B - 3A}{2AB}$

14. $\dfrac{XY}{Z} + \dfrac{ZY}{XY} = \dfrac{XY}{Z} + \dfrac{Z}{X} =$

$\dfrac{XY(X)}{Z(X)} + \dfrac{Z(Z)}{X(Z)} = \dfrac{X^2Y + Z^2}{XZ}$

15. $\dfrac{X^2}{3} - \dfrac{X}{2} + \dfrac{2}{3} = \dfrac{X^2(2)}{3(2)} - \dfrac{X(3)}{2(3)} + \dfrac{2(2)}{3(2)} =$

$\dfrac{2X^2}{6} - \dfrac{3X}{6} + \dfrac{4}{6} = \dfrac{2X^2 - 3X + 4}{6}$

16. $\dfrac{X}{Y} + \dfrac{3Y}{X+1} = \dfrac{X(X+1)}{Y(X+1)} + \dfrac{3Y(Y)}{(X+1)(Y)} =$

$\dfrac{X^2 + X + 3Y^2}{Y(X+1)}$ or $\dfrac{X^2 + X + 3Y^2}{XY + Y}$

Systematic Review 2C

1. $\dfrac{X}{X} - 2^0 = 1 - 1 = 0$: correct

2. incorrect: $\dfrac{2X+3}{X+6} = \dfrac{2X}{X+6} + \dfrac{3}{X+6}$

3. $\dfrac{BX + BC}{B} = \dfrac{B(X+C)}{B} = \dfrac{X+C}{1} = X + C$

4. $\dfrac{27Y^2 - 54}{9} = \dfrac{9\left(3Y^2 - 6\right)}{9} =$

$\dfrac{3Y^2 - 6}{1} = 3Y^2 - 6 = 3(Y^2 - 2)$

5. $\dfrac{8X^2Y^4 + 4XY^3}{2Y^3} = \dfrac{2Y^3\left(4X^2Y + 2X\right)}{2Y^3} =$

$\dfrac{4X^2Y + 2X}{1} = 4X^2Y + 2X = 2X(2XY + 1)$

6. $\dfrac{A}{B} + \dfrac{C}{2B} = \dfrac{A(2)}{B(2)} + \dfrac{C}{2B} = \dfrac{2A + C}{2B}$

7. $\dfrac{X}{4} + \dfrac{Y}{7} = \dfrac{X(7)}{4(7)} + \dfrac{Y(4)}{7(4)} = \dfrac{7X + 4Y}{28}$

8. $\dfrac{5}{X} + \dfrac{X}{5} = \dfrac{5(5)}{X(5)} + \dfrac{X(X)}{5(X)} =$

$\dfrac{5^2 + X^2}{5X} = \dfrac{25 + X^2}{5X}$

9. $3X^4 \cdot 2X^5 = (3)(2)X^4X^5 = 6X^{4+5} = 6X^9$

10. $X^2X^{-2} = X^{2+(-2)} = X^0 = 1$

11. $\left(10^2\right)^{-4} = 10^{(2)(-4)} = 10^{-8}$ or

$\dfrac{1}{10^8}$ or $\dfrac{1}{100,000,000}$

12. $2^{15} = \left(2^3\right)^5$ so $X = 3$

13. $\dfrac{X^{-3}Y^{-2}Y^{-1}}{Y^{-3}X^{-5}} = X^{-3}Y^{-2}Y^{-1}Y^3X^5 =$

$X^{-3+5}Y^{-2+(-1)+3} = X^2Y^0 = X^2(1) = X^2$

14. $\dfrac{A^3A^{-2}B^2}{B^{-2}A^4} = A^3A^{-2}B^2B^2A^{-4} =$

$A^{3+(-2)+(-4)}B^{2+2} = A^{-3}B^4$ or $\dfrac{B^4}{A^3}$

15. $2\dfrac{1}{2} = \dfrac{2}{5}P - 1\dfrac{3}{7}$

$\dfrac{5}{2} = \dfrac{2}{5}P - \dfrac{10}{7}$

$(70)\dfrac{5}{2} = (70)\dfrac{2}{5}P - (70)\dfrac{10}{7}$

$\dfrac{350}{2} = \dfrac{140}{5}P - \dfrac{700}{7}$

$175 = 28P - 100$

$275 = 28P$

$\dfrac{275}{28} = P = 9\dfrac{23}{28}$

16. $.2X + .03X = .69$

$(100)(.2X) + (100)(.03X) = (100)(.69)$

$20X + 3X = 69$

$23X = 69$

$X = \dfrac{69}{23} = 3$

17. $7\left(-B + 2^2 + 3B\right) = 7(-B + 4 + 3B) =$

$-7B + 28 + 21B$ or $28 + 14B$

18. $XYZ(X + Y + Z) = X^2YZ + XY^2Z + XYZ^2$

19. $-4 \cdot 3^2 + 2 - 5 - |6 - 22| =$

$-4 \cdot 3^2 + 2 - 5 - |-16| = -4 \cdot 3^2 + 2 - 5 - 16 =$

$-4 \cdot 9 + 2 - 5 - 16 = -36 + 2 - 5 - 16 = -55$

20. $(6 + 3)^2 - (4 - 8)^2 + 3 \div \dfrac{1}{3} =$

$9^2 - (-4)^2 + 3 \div \dfrac{1}{3} = 81 - 16 + 3 \div \dfrac{1}{3} =$

$81 - 16 + 3 \times \dfrac{3}{1} = 81 - 16 + 9 = 74$

Systematic Review 2D

1. incorrect: $\dfrac{1}{3} + \dfrac{1}{2} = \dfrac{1(2)}{3(2)} + \dfrac{1(3)}{2(3)} =$

$\dfrac{2 + 3}{6} = \dfrac{5}{6}$

2. correct

3. $\dfrac{4X^2X - X^2}{X} = \dfrac{X(4X^2 - X)}{X} =$

$\dfrac{4X^2 - X}{1} = 4X^2 - X$

4. $\dfrac{(X + 1)(X - 1)}{X - 1} = \dfrac{X + 1}{1} = X + 1$

5. $\dfrac{28 - 14A^2}{7} = \dfrac{7(4 - 2A^2)}{7} =$

$\dfrac{4 - 2A^2}{1} = 4 - 2A^2 = 2(2 - A^2)$

6. $\dfrac{8}{X + 1} + \dfrac{10}{X + 2} = \dfrac{8(X + 2)}{(X + 1)(X + 2)} +$

$\dfrac{10(X + 1)}{(X + 2)(X + 1)} = \dfrac{8X + 16 + 10X + 10}{(X + 1)(X + 2)} =$

$\dfrac{18X + 26}{(X + 1)(X + 2)}$ or $\dfrac{18X + 26}{X^2 + 3X + 2}$

7. $\dfrac{7}{4Y} - \dfrac{9}{2Y} = \dfrac{7}{4Y} - \dfrac{9(2)}{2Y(2)} =$

$\dfrac{7 - 18}{4Y} = \dfrac{-11}{4Y}$

8. $\dfrac{A}{B} + \dfrac{B}{C} = \dfrac{A(C)}{B(C)} + \dfrac{B(B)}{C(B)} = \dfrac{AC + B^2}{BC}$

9. $8^{-5} \div 8^2 = 8^{-5-2} = 8^{-7}$ or $\dfrac{1}{8^7}$

10. $\left(\dfrac{1}{2}\right)^{-3} = \dfrac{1^{-3}}{2^{-3}} = \dfrac{2^3}{1^3} = \dfrac{8}{1} = 8$

11. $\left(7^2\right)^{-4} = 7^{(2)(-4)} = 7^{-8}$ or $\dfrac{1}{7^8}$

12. $1{,}000{,}000 = 10^6 = \left(10^2\right)^3$ so $X = 3$

13. $\dfrac{6A^3B^3A^{-2}B^{-4}}{2A^{-1}B^0} = \dfrac{6A^{3 + (-2)}B^{3 + (-4)}}{2A^{-1}(1)} =$

$\dfrac{\overset{3}{6}A^1B^{-1}}{\underset{}{2}A^{-1}} = 3A^1B^{-1}A^1 = 3A^{1+1}B^{-1} =$

$3A^2B^{-1}$ or $\dfrac{3A^2}{B}$

14. $\dfrac{3A^{-2}B^4A^2}{18B^1A^1A^3} = \dfrac{3A^{-2+2}B^4}{18B^1A^{1+3}} =$

$\dfrac{\overset{}{3}A^0B^4}{\underset{6}{18}B^1A^4} = \dfrac{B^4B^{-1}}{6A^4} =$

$\dfrac{B^{4 + (-1)}}{6A^4} = \dfrac{B^3}{6A^4}$ or $\dfrac{1}{6}B^3A^{-4}$

15. $\dfrac{1}{2} - \dfrac{2}{5}P = \dfrac{3}{7}$

$(70)\dfrac{1}{2} - (70)\dfrac{2}{5}P = (70)\dfrac{3}{7}$

$\dfrac{70}{2} - \dfrac{140}{5}P = \dfrac{210}{7}$

$35 - 28P = 30$

$-28P = -5$

$P = \dfrac{-5}{-28} = \dfrac{5}{28}$

16. $7.2 - 3 = .07X$

$(100)(7.2) - (100)(3) = (100)(.07X)$

$720 - 300 = 7X$

$420 = 7X$

$\dfrac{420}{7} = X = 60$

17. $-72XY^2 + 45X^2Y = (9XY)(-8Y + 5X)$

18. $18A^2 - 24AB^3 = (6A)(3A - 4B^3)$

19. $(1 - 7)^2 - 8N + 11 = -3$

$(-6)^2 - 8N + 11 = -3$

$36 - 8N + 11 = -3$

$\qquad -8N = -3 - 36 - 11$

$\qquad -8N = -50$

$\qquad N = \dfrac{-50}{-8} = \dfrac{25}{4} = 6\dfrac{1}{4}$

20.
$$B(6+6)^2 + |100 - 1^2| - 14 = 5 \cdot 9 + 4$$
$$B(12)^2 + |100 - 1| - 14 = 5 \cdot 9 + 4$$
$$144B + |99| - 14 = 45 + 4$$
$$144B + 99 - 14 = 49$$
$$144B + 85 = 49$$
$$144B = 49 - 85$$
$$144B = -36$$
$$B = \frac{-36}{144} = -\frac{1}{4}$$

Systematic Review 2E

1. correct: $\dfrac{X+3}{X} = \dfrac{X}{X} + \dfrac{3}{X} = 1\dfrac{3}{X}$

2. incorrect: $\dfrac{2}{X+1} + \dfrac{3}{X} =$

$$\frac{2(X)}{(X+1)(X)} + \frac{3(X+1)}{X(X+1)} =$$
$$\frac{2X + 3X + 3}{X(X+1)} = \frac{5X+3}{X^2 + X}$$

3. $\dfrac{AX - 6Y + 6X}{2} =$

$$\frac{AX}{2} - \frac{6Y}{2} + \frac{6X}{2} = \frac{AX}{2} - 3Y + 3X$$

4. $\dfrac{B^4 - B^2}{B^2} = \dfrac{(B^2)(B^2 - 1)}{B^2} = \dfrac{B^2 - 1}{1} = B^2 - 1$

5. $\dfrac{6A^2 + 6A}{12A} = \dfrac{(6A)(A+1)}{(6A)(2)} =$

$$\frac{A+1}{2} \text{ or } \frac{A}{2} + \frac{1}{2}$$

6. $\dfrac{4}{X} + \dfrac{1}{3} = \dfrac{4(3)}{X(3)} + \dfrac{1(X)}{3(X)} = \dfrac{12 + X}{3X}$

7. $\dfrac{X}{Y} + \dfrac{4Y}{X+2} = \dfrac{X(X+2)}{Y(X+2)} + \dfrac{4Y(Y)}{(X+2)(Y)} =$

$$\frac{X^2 + 2X + 4Y^2}{(X+2)(Y)} \text{ or } \frac{X^2 + 2X + 4Y^2}{XY + 2Y}$$

8. $\dfrac{3}{Q+1} + \dfrac{2}{Q} = \dfrac{3(Q)}{Q(Q+1)} + \dfrac{2(Q+1)}{Q(Q+1)} =$

$$\frac{3Q + 2Q + 2}{Q(Q+1)} = \frac{5Q+2}{Q(Q+1)} \text{ or } \frac{5Q+2}{Q^2 + Q}$$

9. $2^2 X^3 \cdot 2^3 X^{-1} = 2^{2+3} X^{3+(-1)} = 2^5 X^2 \text{ or } 32X^2$

10. $\dfrac{Y^3}{Y^3} = Y^{3-3} = Y^0 = 1$

11. $\left((5^2)^4 \right)^{-3} = 5^{(2)(4)(-3)} = 5^{-24} \text{ or } \dfrac{1}{5^{24}}$

12. $(49^3) = (7^2)^3 = 7^{(2)(3)} = 7^6$

13. $(X^2)^3 (X^{-4})^2 = X^{(2)(3)} X^{(-4)(2)} =$

$$X^6 X^{-8} = X^{6 + (-8)} = X^{-2} \text{ or } \frac{1}{X^2}$$

14. $(P^{-4})^{-2} P^3 P^{-1} = P^{(-4)(-2)} P^3 P^{-1} =$

$$P^8 P^3 P^{-1} = P^{8+3+(-1)} = P^{10}$$

15. $.024F + F = .56$

$$(1,000)(.024F) + (1,000)(F) = (1,000)(.56)$$
$$24F + 1,000F = 560$$
$$1,024F = 560$$
$$F = \frac{560}{1,024} = \frac{35}{64}$$

16. $10\dfrac{2}{3} B + 3\dfrac{1}{6} = 1\dfrac{7}{8}$

$$\frac{32}{3} B + \frac{19}{6} = \frac{15}{8}$$
$$(24)\frac{32}{3} B + (24)\frac{19}{6} = (24)\frac{15}{8}$$
$$\frac{768}{3} B + \frac{456}{6} = \frac{360}{8}$$
$$256B + 76 = 45$$
$$256B = -31$$
$$B = -\frac{31}{256}$$

17. $100(2.3X - .07Y) = 230X - 7Y$

18. $1,000(.009A + .02 + 3) =$

$$9A + 20 + 3,000 = 9A + 3,020$$

19. $(6 \div 9) \cdot 2 - 9Y = 8(Y - 4 + 7)$

$$\left(\frac{2}{3}\right) \cdot 2 - 9Y = 8(Y + 3)$$
$$\frac{4}{3} - 9Y = 8Y + 24$$
$$\frac{4}{3} - 24 = 8Y + 9Y$$
$$\frac{4}{3} - \frac{72}{3} = 17Y$$
$$\frac{-68}{3} = 17Y$$
$$\frac{1}{17} \times \frac{-68}{3} = 17Y \times \frac{1}{17}$$
$$\frac{-68}{51} = Y = -\frac{4}{3} \text{ or } -1\frac{1}{3}$$

20. $(11-4)^2 \div 7 - |3-9| = 14(R-2R)$

$(7)^2 \div 7 - |-6| = 14(-R)$

$49 \div 7 - 6 = -14R$

$7 - 6 = -14R$

$1 = -14R$

$\dfrac{1}{-14} = R = -\dfrac{1}{14}$

Lesson Practice 3A

1. $85,000,000 = 8.5 \times 10^7$

2. $.341 = 3.41 \times 10^{-1}$

3. $.00038 = 3.8 \times 10^{-4}$

4. $9,700,000,000 = 9.7 \times 10^9$

5. $.000073 \times .0054 =$
$(7.3 \times 10^{-5})(5.4 \times 10^{-3}) =$
$(7.3 \times 5.4)(10^{-5} \times 10^{-3}) =$
$39.42 \times 10^{-8} = 3.942 \times 10^{-7}$

6. $(58,000,000)(650) =$
$(5.8 \times 10^7)(6.5 \times 10^2) =$
$(5.8 \times 6.5)(10^7 \times 10^2) =$
$37.7 \times 10^9 = 3.77 \times 10^{10}$

7. $(.098)(.006) =$
$(9.8 \times 10^{-2})(6.0 \times 10^{-3}) =$
$(9.8 \times 6.0)(10^{-2} \times 10^{-3}) =$
$58.8 \times 10^{-5} = 5.88 \times 10^{-4}$

8. $1,800,000,000 \times 2,400,000 =$
$(1.8 \times 10^9)(2.4 \times 10^6) =$
$(1.8 \times 2.4)(10^9 \times 10^6) = 4.32 \times 10^{15}$

9. $27,000 \div .009 =$
$(2.7 \times 10^4) \div (9.0 \times 10^{-3}) =$
$(2.7 \div 9.0)(10^4 \div 10^{-3}) =$
$.3 \times 10^7 = 3 \times 10^6$

10. $.0021 \div 340,000 =$
$(2.1 \times 10^{-3}) \div (3.4 \times 10^5) =$
$(2.1 \div 3.4)(10^{-3} \div 10^5) =$
$.62 \times 10^{-8} \approx 6.2 \times 10^{-9}$

11. $\dfrac{.00042}{.00006} = (4.2 \times 10^{-4}) \div (6.0 \times 10^{-5}) =$
$(4.2 \div 6.0)(10^{-4} \div 10^{-5}) =$
$.7 \times 10^1 = 7.0 \times 10^0 = 7 \times 1 = 7$

12. $\dfrac{6,800,000,000}{430,000} =$
$(6.8 \times 10^9) \div (4.3 \times 10^5) =$
$(6.8 \div 4.3)(10^9 \div 10^5) \approx 1.58 \times 10^4$

13. $\dfrac{2X^3}{X^3Y^{-2}} - \dfrac{8XY^3}{XY} + \dfrac{7X^3Y^0}{X^4Y^2} =$
$\dfrac{2}{Y^{-2}} - \dfrac{8Y^3}{Y^1} + \dfrac{7X^3(1)}{X^4Y^2} =$
$2Y^2 - 8Y^3Y^{-1} + 7X^3X^{-4}Y^{-2} =$
$2Y^2 - 8Y^{3+(-1)} + 7X^{3+(-4)}Y^{-2} =$
$2Y^2 - 8Y^2 + 7X^{-1}Y^{-2} = -6Y^2 + 7X^{-1}Y^{-2}$

14. $\dfrac{AX^3}{X^{-3}} - \dfrac{BX^3}{B^{-1}} + \dfrac{ABX}{X^2B^2} =$
$AX^3X^3 - B^1X^3B^1 + AB^1X^1X^{-2}B^{-2} =$
$AX^{3+3} - B^{1+1}X^3 + AB^{1+(-2)}X^{1+(-2)} =$
$AX^6 - B^2X^3 + AB^{-1}X^{-1}$

15. $3AB^{-1} + \dfrac{13AB}{A^{-2}B^1} + \dfrac{3}{A^{-1}B} =$
$3AB^{-1} + 13A^1B^1A^2B^{-1} + 3A^1B^{-1} =$
$3AB^{-1} + 13A^{1+2}B^{1+(-1)} + 3AB^{-1} =$
$6AB^{-1} + 13A^3B^0 = 6AB^{-1} + 13A^3$

16. $3X^2Y^2X^{-3}Y^0 + 2X^1Y^{-3}6X^1Y^1 =$
$3X^{2+(-3)}Y^{2+0} + (2)(6)X^{1+1}Y^{-3+1} =$
$3X^{-1}Y^2 + 12X^2Y^{-2}$

Lesson Practice 3B

1. $360,000,000 = 3.6 \times 10^8$

2. $.0000001 = 1 \times 10^{-7}$

3. $.0059 = 5.9 \times 10^{-3}$

4. $425,000 = 4.25 \times 10^5$

5. $.0063 \times 1,200,000 =$
$(6.3 \times 10^{-3})(1.2 \times 10^6) =$
$(6.3 \times 1.2)(10^{-3} \times 10^6) = 7.56 \times 10^3$

6. $(16,000)(.007) =$
$(1.6 \times 10^4)(7 \times 10^{-3}) =$
$(1.6 \times 7)(10^4 \times 10^{-3}) =$
$11.2 \times 10^1 = 1.12 \times 10^2$

7. $(.92)(.0009) =$
$(9.2 \times 10^{-1})(9 \times 10^{-4}) =$
$(9.2 \times 9)(10^{-1} \times 10^{-4}) =$
$82.8 \times 10^{-5} = 8.28 \times 10^{-4}$

8. $2,300,000 \times 4,000,000,000,000 =$
$(2.3 \times 10^6)(4.0 \times 10^{12}) =$
$(2.3 \times 4.0)(10^6 \times 10^{12}) = 9.2 \times 10^{18}$

9. $.0024 \div .000003 =$
$(2.4 \times 10^{-3}) \div (3 \times 10^{-6}) =$
$(2.4 \div 3)(10^{-3} \div 10^{-6}) =$
$.8 \times 10^3 = 8 \times 10^2$

10. $29,000,000 \div 15,000 =$
$(2.9 \times 10^7) \div (1.5 \times 10^4) =$
$(2.9 \div 1.5)(10^7 \div 10^4) \approx 1.93 \times 10^3$

11. $\dfrac{9,600,000,000,000}{.02} =$
$(9.6 \times 10^{12}) \div (2 \times 10^{-2}) =$
$(9.6 \div 2)(10^{12} \div 10^{-2}) = 4.8 \times 10^{14}$

12. $\dfrac{.00016}{.000008} = (1.6 \times 10^{-4}) \div (8 \times 10^{-6}) =$
$(1.6 \div 8.0)(10^{-4} \div 10^{-6}) =$
$.2 \times 10^2 = 2 \times 10^1$

13. $\dfrac{R^{12}R^{-3}R^6}{R^1S^{-3}} + \dfrac{R^3R^{-2}R^0}{R^1} =$
$R^{12}R^{-3}R^6R^{-1}S^3 + R^3R^{-2}R^0R^{-1} =$
$R^{12+(-3)+6+(-1)}S^3 + R^{3+(-2)+0+(-1)} =$
$R^{14}S^3 + R^0 = R^{14}S^3 + 1$

14. $\dfrac{2X^3Y^3}{X^{-1}Y^{-1}} + \dfrac{8X^5Y^5}{XY} - \dfrac{3^{-1}X^4Y^5}{X^2Y^2} =$
$2X^3Y^3X^1Y^1 + 8X^5Y^5X^{-1}Y^{-1} - \dfrac{X^4Y^5X^{-2}Y^{-2}}{3^1} =$
$2X^{3+1}Y^{3+1} + 8X^{5+(-1)}Y^{5+(-1)} - \dfrac{X^{4+(-2)}Y^{5+(-2)}}{3} =$
$2X^4Y^4 + 8X^4Y^4 - \dfrac{X^2Y^3}{3} = 10X^4Y^4 - \dfrac{X^2Y^3}{3}$

15. $16A^0B^4 + \dfrac{4B^4}{B^2B^5} - 4A =$
$16(1)B^4 + 4B^4B^{-2}B^{-5} - 4A =$
$16B^4 + 4B^{4+(-2)+(-5)} - 4A =$
$16B^4 + 4B^{-3} - 4A = 16B^4 + \dfrac{4}{B^3} - 4A$

16. $2A^6B^{-3}C^3A^{-2} + C^1C^0C^2A^4B^{-3} =$
$2A^{6+(-2)}B^{-3}C^3 + C^{1+0+2}A^4B^{-3} =$
$2A^4B^{-3}C^3 + A^4B^{-3}C^3 =$
$\dfrac{2A^4C^3}{B^3} + \dfrac{A^4C^3}{B^3} = \dfrac{3A^4C^3}{B^3}$

Systematic Review 3C

1. $62,000 = 6.2 \times 10^4$

2. $.75 = 7.5 \times 10^{-1}$

3. $.0048 = 4.8 \times 10^{-3}$

4. $3,080,000 = 3.08 \times 10^6$

5. $(62,000)(.75) =$
$(6.2 \times 10^4)(7.5 \times 10^{-1}) =$
$(6.2 \times 7.5)(10^4 \times 10^{-1}) =$
$46.5 \times 10^3 = 4.65 \times 10^4$

6. $(3,080,000)(.0048) =$
$(3.08 \times 10^6)(4.8 \times 10^{-3}) =$
$(3.08 \times 4.8)(10^6 \times 10^{-3}) =$
$14.784 \times 10^3 = 1.4784 \times 10^4$

7. $(3,080,000) \div (.75) =$
$(3.08 \times 10^6) \div (7.5 \times 10^{-1}) =$
$(3.08 \div 7.5)(10^6 \div 10^{-1}) =$
$.4107 \times 10^7 \approx 4.107 \times 10^6$

8. $(62,000) \div (.0048) =$
$(6.2 \times 10^4) \div (4.8 \times 10^{-3}) =$
$(6.2 \div 4.8)(10^4 \div 10^{-3}) \approx 1.29 \times 10^7$

9. $8XXY - YX^2Y + \dfrac{2XY}{X^{-1}} =$
$8X^2Y - X^2Y^2 + 2XYX^1 =$
$8X^2Y - X^2Y^2 + 2X^2Y = 10X^2Y - X^2Y^2$

10. $4X^2X^{-1} + \dfrac{12X^2}{X^{-3}} + \dfrac{8XX^3}{\left(X^2\right)^2} =$

$4X^{2+(-1)} + 12X^2X^3 + \dfrac{8X^4}{X^{(2)(2)}} =$

$4X^1 + 12X^{2+3} + \dfrac{8X^4}{X^4} =$

$4X + 12X^5 + \dfrac{8}{1} = \quad 4X + 12X^5 + 8$ or

$12X^5 + 4X + 8$

11. $4A + \dfrac{9AB^2B^{-1}}{B^1} + 8AB =$

$4A + 9AB^2B^{-1}B^{-1} + 8AB =$

$4A + 9AB^{2+(-1)+(-1)} + 8AB =$

$4A + 9A + 8AB = 13A + 8AB$

12. $2A BA^{-1} + 3AB - 5B =$

$2A^{1+(-1)}B + 3AB - 5B =$

$2A^0B + 3AB - 5B = 2(1)B + 3AB - 5B =$

$2B + 3AB - 5B = 3AB - 3B$ or $3B(A - 1)$

13. $\dfrac{75A^2Y - 50AX}{5AX} = \dfrac{(5A)(15AY - 10X)}{5AX} =$

$\dfrac{15AY - 10X}{X}$ or $\dfrac{5(3AY - 2X)}{X}$

14. $\dfrac{12XY + 30Y}{3Y} = \dfrac{(3Y)(4X + 10)}{3Y} =$

$\dfrac{4X + 10}{1} = 4X + 10$ or $2(2X + 5)$

15. $\dfrac{Y}{X+1} + \dfrac{3}{2Y} = \dfrac{Y(2Y)}{(X+1)(2Y)} + \dfrac{3(X+1)}{2Y(X+1)} =$

$\dfrac{Y(2Y) + 3(X+1)}{2Y(X+1)} = \dfrac{2Y^2 + 3X + 3}{2XY + 2Y}$

16. $\dfrac{3}{7} + \dfrac{A}{B} = \dfrac{3(B)}{7(B)} + \dfrac{A(7)}{B(7)} = \dfrac{3B + 7A}{7B}$

17. $\left(10^3\right)^4 = 10^{(3)(4)} = 10^{12}$

18. $\dfrac{X^{-3}X^2Y^4}{Y^{-2}Y^{-1}X^{-1}} = \dfrac{X^{-3+2}Y^4}{Y^{-2+(-1)}X^{-1}} = \dfrac{X^{-1}Y^4}{Y^{-3}X^{-1}} =$

$X^{-1}Y^4Y^3X^1 = X^{-1+1}Y^{4+3} = X^0Y^7 = Y^7$

19. $XY(3X - 4Y) = 3X^2Y - 4XY^2$

20. $\dfrac{Y^2}{X}\left(\dfrac{X^3}{Y} + \dfrac{Y^2}{Y}\right) =$

$\dfrac{Y^2X^3}{XY} + \dfrac{Y^4}{XY} = Y^2Y^{-1}X^3X^{-1} + \dfrac{Y^4Y^{-1}}{X} = YX^2 + \dfrac{Y^3}{X}$

Systematic Review 3D

1. $25{,}000{,}000 = 2.5 \times 10^7$

2. $.000039 = 3.9 \times 10^{-5}$

3. $.000000014 = 1.4 \times 10^{-8}$

4. $760 = 7.6 \times 10^2$

5. $(.000000014)(.000039) =$

$\left(1.4 \times 10^{-8}\right)\left(3.9 \times 10^{-5}\right) =$

$(1.4 \times 3.9)\left(10^{-8} \times 10^{-5}\right) = 5.46 \times 10^{-13}$

6. $(25{,}000{,}000)(760) =$

$\left(2.5 \times 10^7\right)\left(7.6 \times 10^2\right) =$

$(2.5 \times 7.6)\left(10^7 \times 10^2\right) =$

$19 \times 10^9 = 1.9 \times 10^{10}$

7. $(.000000014) \div (.000039) =$

$\left(1.4 \times 10^{-8}\right) \div \left(3.9 \times 10^{-5}\right) =$

$(1.4 \div 3.9)\left(10^{-8} \div 10^{-5}\right) =$

$.359 \times 10^{-3} = 3.59 \times 10^{-4}$

8. $(25{,}000{,}000) \div (760) =$

$\left(2.5 \times 10^7\right) \div \left(7.6 \times 10^2\right) =$

$(2.5 \div 7.6)\left(10^7 \div 10^2\right) =$

$.329 \times 10^5 = 3.29 \times 10^4$

9. $\dfrac{X^3Y^2X^{-1}}{Y} - \dfrac{4Y^2X^2Y^{-2}}{Y} - \dfrac{3X^2Y}{X^2X^1} =$

$\dfrac{X^{3+(-1)}Y^2}{Y} - \dfrac{4Y^{2+(-2)}X^2}{Y} - \dfrac{3X^2Y}{X^{2+1}} =$

$\dfrac{X^2Y^2}{Y^1} - \dfrac{4Y^0X^2}{Y^1} - \dfrac{3X^2Y}{X^3} =$

$X^2Y^2Y^{-1} - 4(1)X^2Y^{-1} - 3X^2YX^{-3} =$

$X^2Y^{2+(-1)} - 4X^2Y^{-1} - 3X^{2+(-3)}Y =$

$X^2Y - 4X^2Y^{-1} - 3X^{-1}Y$ or $X^2Y - \dfrac{4X^2}{Y} - \dfrac{3Y}{X}$

10. $\dfrac{30X^2Y^{-1}}{Y^1X^3} - \dfrac{11X^1Y^{-1}}{Y^3X^2} - \dfrac{18^0X^0X^{-1}}{Y^4X^2Y^{-2}} =$

$\dfrac{30X^2Y^{-1}}{Y^1X^3} - \dfrac{11X^1Y^{-1}}{Y^3X^2} - (1)(1)X^{-1}Y^{-4}X^{-2}Y^2 =$

$\dfrac{30X^2Y^{-1}}{Y^1X^3} - \dfrac{11X^1Y^{-1}}{Y^3X^2} - X^{-3}Y^{-2} =$

$30X^2Y^{-1}Y^{-1}X^{-3} - 11X^1Y^{-1}Y^{-3}X^{-2} - X^{-3}Y^{-2} =$

$30X^{-1}Y^{-2} - 11X^{-1}Y^{-4} - X^{-3}Y^{-2}$ or

$\dfrac{30}{XY^2} - \dfrac{11}{XY^4} - \dfrac{1}{X^3Y^2}$

11. $\dfrac{13A}{B^0} - \dfrac{2A^3B^1}{B^1AA} - \dfrac{4B^2B^{-2}}{24B^2} =$

$\dfrac{13A}{1} - \dfrac{2A^3}{A^2} - \dfrac{4B^0}{24B^2} =$

$13A - 2A^3A^{-2} - \dfrac{4(1)}{4(6B^2)} =$

$13A - 2A - \dfrac{1}{6B^2} = 11A - \dfrac{1}{6B^2}$

12. $3ABB^3B^{-2} - 17BA^{-1}A^2 - 42B^{-2}A =$

$3AB^{1+3+(-2)} - 17A^{-1+2}B - 42AB^{-2} =$

$3AB^2 - 17AB - 42AB^{-2}$ or

$3AB^2 - 17AB - \dfrac{42A}{B^2}$

13. $\dfrac{48B^2Y + 72B^2Y}{24B^2} = \dfrac{24B^2(2Y+3Y)}{24B^2} =$

$\dfrac{2Y+3Y}{1} = 2Y + 3Y = 5Y$

14. $\dfrac{21X^2 - 35XY}{14X} = \dfrac{7X(3X-5Y)}{7X(2)} = \dfrac{3X-5Y}{2}$

15. $\dfrac{2X}{Y+1} + \dfrac{7}{X-2} =$

$\dfrac{2X(X-2)}{(Y+1)(X-2)} + \dfrac{7(Y+1)}{(X-2)(Y+1)} =$

$\dfrac{2X(X-2) + 7(Y+1)}{(Y+1)(X-2)} = \dfrac{2X^2 - 4X + 7Y + 7}{(Y+1)(X-2)}$ or

$\dfrac{2X^2 - 4X + 7Y + 7}{XY + X - 2Y - 2}$

16. $\dfrac{A}{7} + \dfrac{7}{A^2} = \dfrac{A(A^2)}{7(A^2)} + \dfrac{7(7)}{A^2(7)} = \dfrac{A^3 + 49}{7A^2}$

17. $-\left(2^{-3}\right)^2 = -\left(2^{(-3)(2)}\right) = -\left(2^{-6}\right) =$

$-\dfrac{1}{2^6} = -\dfrac{1}{64}$

18. $\dfrac{A^{-6}B^3A^{-2}A^7}{B^{-2}B^{-1}} = \dfrac{A^{-6+(-2)+7}B^3}{B^{-2+(-1)}} = \dfrac{A^{-1}B^3}{B^{-3}} =$

$A^{-1}B^3B^3 = A^{-1}B^{3+3} = A^{-1}B^6$ or $\dfrac{B^6}{A}$

19. $2A^{-2}B^5\left(3AB^1 - 5B^{-3}\right) =$

$2A^{-2}B^5(3)AB^1 - 2A^{-2}B^5(5)B^{-3} =$

$6A^{-2+1}B^{5+1} - 10A^{-2}B^{5+(-3)} =$

$6A^{-1}B^6 - 10A^{-2}B^2$ or $\dfrac{6B^6}{A} - \dfrac{10B^2}{A^2}$

20. $\dfrac{X}{Y^{-2}}\left(\dfrac{X^2Y}{X} + \dfrac{Y}{X^3}\right) = \dfrac{X^3Y}{XY^{-2}} + \dfrac{XY}{Y^{-2}X^3} =$

$X^{-1}X^3Y^1Y^2 + X^1Y^1Y^2X^{-3} = X^2Y^3 + X^{-2}Y^3$ or

$X^2Y^3 + \dfrac{Y^3}{X^2}$

Systematic Review 3E

1. $9{,}400 = 9.4 \times 10^3$

2. $.00000053 = 5.3 \times 10^{-7}$

3. $.012 = 1.2 \times 10^{-2}$

4. $160{,}000 = 1.6 \times 10^5$

5. $(9{,}400)(.012) =$

$\left(9.4 \times 10^3\right)\left(1.2 \times 10^{-2}\right) =$

$(9.4 \times 1.2)\left(10^3 \times 10^{-2}\right) =$

$11.28 \times 10^1 = 1.128 \times 10^2$

6. $(160{,}000)(.00000053) =$

$\left(1.6 \times 10^5\right)\left(5.3 \times 10^{-7}\right) =$

$(1.6 \times 5.3)\left(10^5 \times 10^{-7}\right) =$

8.48×10^{-2}

7. $(9{,}400) \div (.012) =$

$\left(9.4 \times 10^3\right) \div \left(1.2 \times 10^{-2}\right) =$

$(9.4 \div 1.2)\left(10^3 \div 10^{-2}\right) \approx 7.83 \times 10^5$

8. $(160{,}000) \div (.00000053) =$

$\left(1.6 \times 10^5\right) \div \left(5.3 \times 10^{-7}\right) =$

$(1.6 \div 5.3)\left(10^5 \div 10^{-7}\right) =$

$.302 \times 10^{12} \approx 3.02 \times 10^{11}$

9. $\dfrac{25X^{-1}Y^2}{YX} - \dfrac{11Y^0X^0}{XXY^{-1}} - \dfrac{2(x)}{(x)Y^{-2}} =$

$25X^{-1}Y^2Y^{-1}X^{-1} - \dfrac{11(1)(1)}{X^2Y^{-1}} - \dfrac{2}{Y^{-2}} =$

$25X^{-2}Y - 11X^{-2}Y^1 - 2Y^2 =$

$14X^{-2}Y - 2Y^2$ or $\dfrac{14Y}{X^2} - 2Y^2$

10. $\dfrac{13X^2Y}{XY^2} - \dfrac{7X^2Y}{Y^{-1}X^{-1}} + \dfrac{5X^0Y^{-1}X}{Y^3} =$

$13X^2YX^{-1}Y^{-2} - 7X^2Y^1X^1Y^1 + 5(1)Y^{-1}XY^{-3} =$

$13XY^{-1} - 7X^3Y^2 + 5XY^{-4}$ or

$\dfrac{13X}{Y} - 7X^3Y^2 + \dfrac{5X}{Y^4}$

11. $\dfrac{3A^2}{B} - \dfrac{2^1 2BB}{A^{-2}A} - 4A^{-2}B =$

$3A^2B^{-1} - 4B^2A^2A^{-1} - 4A^{-2}B =$

$3A^2B^{-1} - 4AB^2 - 4A^{-2}B$

or $\dfrac{3A^2}{B} - 4AB^2 - \dfrac{4B}{A^2}$

12. $4A^1BA^{-1}B + 3AB^{-2}B^3 - 5B^3A^{-1}B^{-1}A^1 =$

$4B^2 + 3AB - 5B^2 = 3AB - B^2$

13. $\dfrac{6X^2 + 4X^3}{12X} = \dfrac{(2X)(3X + 2X^2)}{(2X)(6)} =$

$\dfrac{3X + 2X^2}{6} = \dfrac{3X}{6} + \dfrac{2X^2}{6} = \dfrac{X}{2} + \dfrac{X^2}{3}$

14. $\dfrac{32X^2 + 4X + 16}{8X} = \dfrac{(4)(8X^2 + X + 4)}{(4)2X} =$

$\dfrac{8X^2 + X + 4}{2X} = \dfrac{8X^2}{2X} + \dfrac{X}{2X} + \dfrac{4}{2X} =$

$\dfrac{4X}{1} + \dfrac{1}{2} + \dfrac{2}{X} = 4X + \dfrac{1}{2} + \dfrac{2}{X}$

15. $\dfrac{4A}{A+5} + \dfrac{6}{A} = \dfrac{4A(A)}{(A+5)(A)} + \dfrac{6(A+5)}{A(A+5)} =$

$\dfrac{4A^2 + 6A + 30}{A(A+5)}$ or $\dfrac{4A^2 + 6A + 30}{A^2 + 5A}$

16. $\dfrac{X}{X^2Y} + \dfrac{2Y}{XY} = \dfrac{1}{XY} + \dfrac{2Y}{XY} = \dfrac{1+2Y}{XY}$

17. $\left(-4^2\right)^{-1} = -4^{(2)(-1)} = -4^{-2} =$

$-\dfrac{1}{4^2} = -\dfrac{1}{(4)(4)} = -\dfrac{1}{16}$

18. $\dfrac{3^3A^{-3}BC^{-2}}{3^{-1}C^2A^{-4}B^3} = 3^3 3^1 A^{-3}A^4 BB^{-3}C^{-2}C^{-2} =$

$3^4AB^{-2}C^{-4} = 81AB^{-2}C^{-4}$ or $\dfrac{81A}{B^2C^4}$

19. $5^2A^2B^{-3}\left(A^{-4} + AB^2\right) =$

$5^2A^2B^{-3}A^{-4} + 5^2A^2B^{-3}AB^2 =$

$25A^{-2}B^{-3} + 25A^3B^{-1}$ or $\dfrac{25}{A^2B^3} + \dfrac{25A^3}{B}$

20. $\dfrac{X}{Y}\left(\dfrac{X^3}{Y^{-1}} + \dfrac{Y^3X}{X^3}\right) = \dfrac{XX^3}{Y^1Y^{-1}} + \dfrac{X^1Y^3X^1}{Y^1X^3} =$

$\dfrac{X^4}{1} + X^1Y^3X^1Y^{-1}X^{-3} =$

$X^4 + X^{-1}Y^2$ or $X^4 + \dfrac{Y^2}{X}$

Lesson Practice 4A

1. $3\sqrt{2} + 6\sqrt{2} = (3+6)\sqrt{2} = 9\sqrt{2}$

2. $5\sqrt{7} - 2\sqrt{5} = 5\sqrt{7} - 2\sqrt{5}$

cannot be simplified

3. $6\sqrt{X} - 8\sqrt{X} = (6-8)\sqrt{X} = -2\sqrt{X}$

4. $4\sqrt{3} + 16\sqrt{3} = (4+16)\sqrt{3} = 20\sqrt{3}$

5. $\left(2\sqrt{5}\right)\left(3\sqrt{6}\right) = (2)(3)\sqrt{5}\sqrt{6} = 6\sqrt{30}$

6. $\dfrac{10\sqrt{5}}{2\sqrt{5}} = \dfrac{10}{2} = 5$

7. $\left(9\sqrt{X}\right)\left(2\sqrt{Y}\right) =$

$(9)(2)\sqrt{X}\sqrt{Y} = 18\sqrt{XY}$

8. $\dfrac{16\sqrt{20}}{8\sqrt{10}} = \dfrac{16\sqrt{2}}{8} = \dfrac{2\sqrt{2}}{1} = 2\sqrt{2}$

9. $\dfrac{5}{\sqrt{2}} = \dfrac{5\sqrt{2}}{\sqrt{2}\sqrt{2}} = \dfrac{5\sqrt{2}}{\sqrt{4}} = \dfrac{5\sqrt{2}}{2}$

10. $\dfrac{4\sqrt{6}}{\sqrt{3}} = \dfrac{4\sqrt{2}}{1} = 4\sqrt{2}$

11. $\dfrac{\sqrt{12}}{\sqrt{6}} = \dfrac{\sqrt{2}}{1} = \sqrt{2}$

12. $\dfrac{9\sqrt{27}}{\sqrt{2}} = \dfrac{9\sqrt{27}\sqrt{2}}{\sqrt{2}\sqrt{2}} = \dfrac{9\sqrt{54}}{\sqrt{4}} =$

$\dfrac{9\sqrt{9}\sqrt{6}}{2} = \dfrac{9(3)\sqrt{6}}{2} = \dfrac{27\sqrt{6}}{2}$

13. $\dfrac{2}{\sqrt{5}} + \dfrac{4}{\sqrt{6}} = \dfrac{2\sqrt{5}}{\sqrt{5}\sqrt{5}} + \dfrac{4\sqrt{6}}{\sqrt{6}\sqrt{6}} =$

$\dfrac{2\sqrt{5}}{\sqrt{25}} + \dfrac{4\sqrt{6}}{\sqrt{36}} = \dfrac{2\sqrt{5}}{5} + \dfrac{4\sqrt{6}}{6} =$

$\dfrac{2\sqrt{5}}{5} + \dfrac{2\sqrt{6}}{3} = \dfrac{2\sqrt{5}(3)}{5(3)} + \dfrac{2\sqrt{6}(5)}{3(5)} =$

$\dfrac{6\sqrt{5}}{15} + \dfrac{10\sqrt{6}}{15} = \dfrac{6\sqrt{5} + 10\sqrt{6}}{15}$

14. $-\sqrt{2}\left(3\sqrt{12}+2\sqrt{18}\right)=$
$-3\sqrt{12}\sqrt{2}-2\sqrt{18}\sqrt{2}=$
$-3\sqrt{24}-2\sqrt{36}=-3\sqrt{4}\sqrt{6}-2(6)=$
$-3(2)\sqrt{6}-12=-6\sqrt{6}-12$

15. $\dfrac{X}{\sqrt{2}}+\dfrac{X}{\sqrt{7}}=\dfrac{X\sqrt{2}}{\sqrt{2}\sqrt{2}}+\dfrac{X\sqrt{7}}{\sqrt{7}\sqrt{7}}=$
$\dfrac{X\sqrt{2}}{\sqrt{4}}+\dfrac{X\sqrt{7}}{\sqrt{49}}=\dfrac{X\sqrt{2}}{2}+\dfrac{X\sqrt{7}}{7}=$
$\dfrac{X\sqrt{2}\,(7)}{2(7)}+\dfrac{X\sqrt{7}\,(2)}{7(2)}=\dfrac{7X\sqrt{2}+2X\sqrt{7}}{14}$

16. $4\left(2\sqrt{10}-\sqrt{20}\right)=4(2)\sqrt{10}-4\sqrt{20}=$
$8\sqrt{10}-4\sqrt{4}\sqrt{5}=8\sqrt{10}-4(2)\sqrt{5}=$
$8\sqrt{10}-8\sqrt{5}$

Lesson Practice 4B

1. $7\sqrt{A}-5\sqrt{A}=(7-5)\sqrt{A}=2\sqrt{A}$

2. $2\sqrt{10}+4\sqrt{10}=$
$(2+4)\sqrt{10}=6\sqrt{10}$

3. $6\sqrt{3}+6\sqrt{5}=6\sqrt{3}+6\sqrt{5}$
cannot be simplified

4. $3\sqrt{11}-5\sqrt{11}=(3-5)\sqrt{11}=-2\sqrt{11}$

5. $\left(3\sqrt{6}\right)\left(3\sqrt{5}\right)=(3)(3)\sqrt{6}\sqrt{5}=9\sqrt{30}$

6. $\dfrac{24\sqrt{14}}{6\sqrt{2}}=\dfrac{4\sqrt{14}}{\sqrt{2}}=\dfrac{4\sqrt{7}}{1}=4\sqrt{7}$

7. $\left(7\sqrt{A}\right)\left(2\sqrt{A}\right)=(7)(2)\sqrt{A}\sqrt{A}=$
$14\sqrt{A^2}=14A$

8. $\dfrac{14\sqrt{X}}{7\sqrt{X}}=\dfrac{14}{7}=2$

9. $\dfrac{X}{\sqrt{Y}}=\dfrac{X\sqrt{Y}}{\sqrt{Y}\sqrt{Y}}=\dfrac{X\sqrt{Y}}{\sqrt{Y^2}}=\dfrac{X\sqrt{Y}}{Y}$

10. $\dfrac{3\sqrt{7}}{\sqrt{2}}=\dfrac{3\sqrt{7}\sqrt{2}}{\sqrt{2}\sqrt{2}}=\dfrac{3\sqrt{14}}{\sqrt{4}}=\dfrac{3\sqrt{14}}{2}$

11. $\dfrac{2\sqrt{3}}{\sqrt{6}}=\dfrac{2}{\sqrt{2}}=\dfrac{2\sqrt{2}}{\sqrt{2}\sqrt{2}}=\dfrac{2\sqrt{2}}{\sqrt{4}}=$
$\dfrac{2\sqrt{2}}{2}=\dfrac{\sqrt{2}}{1}=\sqrt{2}$

12. $\dfrac{5\sqrt{5}}{\sqrt{10}}=\dfrac{5}{\sqrt{2}}=\dfrac{5\sqrt{2}}{\sqrt{2}\sqrt{2}}=\dfrac{5\sqrt{2}}{\sqrt{4}}=\dfrac{5\sqrt{2}}{2}$

13. $\dfrac{6}{\sqrt{2}}+\dfrac{10}{\sqrt{7}}=\dfrac{6\sqrt{2}}{\sqrt{2}\sqrt{2}}+\dfrac{10\sqrt{7}}{\sqrt{7}\sqrt{7}}=$
$\dfrac{6\sqrt{2}}{\sqrt{4}}+\dfrac{10\sqrt{7}}{\sqrt{49}}=\dfrac{6\sqrt{2}}{2}+\dfrac{10\sqrt{7}}{7}=$
$\dfrac{3\sqrt{2}}{1}+\dfrac{10\sqrt{7}}{7}=\dfrac{3\sqrt{2}\,(7)}{1(7)}+\dfrac{10\sqrt{7}}{7}=$
$\dfrac{21\sqrt{2}+10\sqrt{7}}{7}$

14. $\sqrt{6}\left(2\sqrt{7}+8\sqrt{5}\right)=$
$2\sqrt{7}\sqrt{6}+8\sqrt{5}\sqrt{6}=2\sqrt{42}+8\sqrt{30}$

15. $\dfrac{-B}{\sqrt{A}}+\dfrac{-A}{\sqrt{B}}=\dfrac{-B\sqrt{A}}{\sqrt{A}\sqrt{A}}+\dfrac{-A\sqrt{B}}{\sqrt{B}\sqrt{B}}=$
$\dfrac{-B\sqrt{A}}{A}+\dfrac{-A\sqrt{B}}{B}=\dfrac{-B\sqrt{A}\,(B)}{A(B)}+\dfrac{-A\sqrt{B}\,(A)}{B(A)}=$
$\dfrac{-B^2\sqrt{A}-A^2\sqrt{B}}{AB}$ or $-\dfrac{B^2\sqrt{A}+A^2\sqrt{B}}{AB}$

16. $11\left(3\sqrt{24}+2\sqrt{60}\right)=$
$33\sqrt{24}+22\sqrt{60}=33\sqrt{4}\sqrt{6}+22\sqrt{4}\sqrt{15}=$
$33(2)\sqrt{6}+22(2)\sqrt{15}=66\sqrt{6}+44\sqrt{15}$

Systematic Review 4C

1. $3\sqrt{169A^4}=3\sqrt{169}\sqrt{A^4}=3(13)A^2=39A^2$

2. $3\sqrt{X}+4\sqrt{X}=(3+4)\sqrt{X}=7\sqrt{X}$

3. $\left(3\sqrt{2}\right)\left(4\sqrt{8}\right)=(3)(4)\sqrt{2}\sqrt{8}=$
$12\sqrt{16}=12(4)=48$

4. $\sqrt{2}\left(6\sqrt{2}+5\sqrt{8}\right)=6\sqrt{2}\sqrt{2}+5\sqrt{8}\sqrt{2}=$
$6\sqrt{4}+5\sqrt{16}=6(2)+5(4)=12+20=32$

5. $\dfrac{5\sqrt{30}}{\sqrt{6}}=\dfrac{5\sqrt{5}}{1}=5\sqrt{5}$

6. $\dfrac{\sqrt{56}}{\sqrt{8}}=\dfrac{\sqrt{7}}{1}=\sqrt{7}$

7. $2\sqrt{27}=2\sqrt{9}\sqrt{3}=2(3)\sqrt{3}=6\sqrt{3}$

8. $4\sqrt{75}=4\sqrt{25}\sqrt{3}=4(5)\sqrt{3}=20\sqrt{3}$

9. $\dfrac{5}{\sqrt{3}}=\dfrac{5\sqrt{3}}{\sqrt{3}\sqrt{3}}=\dfrac{5\sqrt{3}}{\sqrt{9}}=\dfrac{5\sqrt{3}}{3}$

10. $\dfrac{3}{\sqrt{5}} + \dfrac{4}{\sqrt{6}} = \dfrac{3\sqrt{5}}{\sqrt{5}\,\sqrt{5}} + \dfrac{4\sqrt{6}}{\sqrt{6}\,\sqrt{6}} =$

$\dfrac{3\sqrt{5}}{\sqrt{25}} + \dfrac{4\sqrt{6}}{\sqrt{36}} = \dfrac{3\sqrt{5}}{5} + \dfrac{4\sqrt{6}}{6} =$

$\dfrac{3\sqrt{5}}{5} + \dfrac{2\sqrt{6}}{3} = \dfrac{3\sqrt{5}\,(3)}{5(3)} + \dfrac{2\sqrt{6}\,(5)}{3(5)} =$

$\dfrac{9\sqrt{5} + 10\sqrt{6}}{15}$

11. $(6,100)(.000045) =$
$\left(6.1\times10^3\right)\left(4.5\times10^{-5}\right) =$
$(6.1\times4.5)\left(10^3\times10^{-5}\right) =$
$27.45\times10^{-2} = 2.745\times10^{-1}$

12. $(.0000098)(140) =$
$\left(9.8\times10^{-6}\right)\left(1.4\times10^2\right) =$
$(9.8\times1.4)\left(10^{-6}\times10^2\right) =$
$13.72\times10^{-4} = 1.372\times10^{-3}$

13. $(630,000) \div (9,000,000,000) =$
$\left(6.3\times10^5\right) \div \left(9\times10^9\right) =$
$(6.3\div9)\left(10^5\div10^9\right) =$
$.7\times10^{-4} = 7\times10^{-5}$

14. $\dfrac{(.000093)(.00000006)}{(300)} =$

$\dfrac{\left(9.3\times10^{-5}\right)\left(6\times10^{-8}\right)}{3\times10^2} =$

$\dfrac{(9.3\times6)\left(10^{-5}\times10^{-8}\right)}{3\times10^2} = \dfrac{55.8\times10^{-13}}{3\times10^2} =$

$(55.8\div3)\left(10^{-13}\div10^2\right) =$
$18.6\times10^{-15} = 1.86\times10^{-14}$

15. $\dfrac{3A^{-2}B^{-4}C^{-2}}{AB^{-2}} + \dfrac{6ABC^{-3}}{C^{-1}B^{-1}} - \dfrac{4AB^4C}{C^3B^2} =$

$3A^{-2}B^{-4}C^{-2}A^{-1}B^2 + \dfrac{6ABC^{-3}}{C^{-1}B^{-1}} - \dfrac{4AB^4C}{C^3B^2} =$

$3A^{-3}B^{-2}C^{-2} + 6ABC^{-3}C^1B^1 - \dfrac{4AB^4C^1}{C^3B^2} =$

$3A^{-3}B^{-2}C^{-2} + 6AB^2C^{-2} - 4AB^4C^1C^{-3}B^{-2} =$
$3A^{-3}B^{-2}C^{-2} + 6AB^2C^{-2} - 4AB^2C^{-2} =$

$3A^{-3}B^{-2}C^{-2} + 2AB^2C^{-2}$ or $\dfrac{3}{A^3B^2C^2} + \dfrac{2AB^2}{C^2}$

16. $\dfrac{2X}{5} - \dfrac{3X}{10} = 25$

$(10)\dfrac{2X}{5} - (10)\dfrac{3X}{10} = (10)25$

$\dfrac{20X}{5} - \dfrac{30X}{10} = 250$

$4X - 3X = 250$
$X = 250$

17. $\left(-2X^2Y\right)\left(3XY^5\right) =$
$(-2)(3)X^2Y^1X^1Y^5 = -6X^3Y^6$

18. $36X^5Y^2Z^{-2} \div 9X^0Y^4Z^{-3} =$
$(36\div9)\left(X^{5-0}Y^{2-4}Z^{-2-(-3)}\right) =$
$4X^5Y^{-2}Z$ or $\dfrac{4X^5Z}{Y^2}$

19. $\dfrac{5X}{3} = 30$

$(3)\dfrac{5X}{3} = (3)30$

$\dfrac{15X}{3} = 90$

$5X = 90$
$X = \dfrac{90}{5} = 18$

20. $.3X + 20 = 10 + .5X$
$(10)(.3X) + (10)(20) = (10)(10) + (10)(.5X)$
$3X + 200 = 100 + 5X$
$200 - 100 = 5X - 3X$
$100 = 2X$
$\dfrac{100}{2} = X = 50$

Systematic Review 4D

1. $4\sqrt{\dfrac{25}{64}X^2} = 4\dfrac{\sqrt{25}}{\sqrt{64}}\sqrt{X^2} =$

$4\left(\dfrac{5}{8}\right)X = \dfrac{20}{8}X = \dfrac{5}{2}X$

2. $2\sqrt{10} - 3\sqrt{5} = 2\sqrt{10} - 3\sqrt{5}$
can't be simplified

3. $\left(8\sqrt{11}\right)\left(2\sqrt{11}\right) = (8)(2)\sqrt{11}\sqrt{11} =$
$16\sqrt{121} = 16(11) = 176$

4. $\sqrt{3}\left(8\sqrt{10} - 9\sqrt{5}\right) =$
$8\sqrt{10}\sqrt{3} - 9\sqrt{5}\sqrt{3} = 8\sqrt{30} - 9\sqrt{15}$

5. $\dfrac{4\sqrt{48}}{\sqrt{8}} = \dfrac{4\sqrt{6}}{1} = 4\sqrt{6}$

6. $\dfrac{\sqrt{84}}{\sqrt{2}} = \dfrac{\sqrt{42}}{1} = \sqrt{42}$

7. $5\sqrt{80} = 5\sqrt{16}\sqrt{5} = 5(4)\sqrt{5} = 20\sqrt{5}$

8. $6\sqrt{125} = 6\sqrt{25}\sqrt{5} = 6(5)\sqrt{5} = 30\sqrt{5}$

9. $\dfrac{7}{\sqrt{2}} = \dfrac{7\sqrt{2}}{\sqrt{2}\sqrt{2}} = \dfrac{7\sqrt{2}}{\sqrt{4}} = \dfrac{7\sqrt{2}}{2}$

10. $\dfrac{9}{\sqrt{3}} + \dfrac{6}{\sqrt{2}} = \dfrac{9\sqrt{3}}{\sqrt{3}\sqrt{3}} + \dfrac{6\sqrt{2}}{\sqrt{2}\sqrt{2}} =$

$\dfrac{9\sqrt{3}}{\sqrt{9}} + \dfrac{6\sqrt{2}}{\sqrt{4}} = \dfrac{9\sqrt{3}}{3} + \dfrac{6\sqrt{2}}{2} =$

$\dfrac{3\sqrt{3}}{1} + \dfrac{3\sqrt{2}}{1} = 3\sqrt{3} + 3\sqrt{2}$

11. $(58)(.000000037) =$

$\left(5.8 \times 10^1\right)\left(3.7 \times 10^{-8}\right) =$

$(5.8 \times 3.7)\left(10^1 \times 10^{-8}\right) =$

$21.46 \times 10^{-7} = 2.146 \times 10^{-6}$

12. $(.0000000046)(82,000,000) =$

$\left(4.6 \times 10^{-9}\right)\left(8.2 \times 10^7\right) =$

$(4.6 \times 8.2)\left(10^{-9} \times 10^7\right) =$

$37.72 \times 10^{-2} = 3.772 \times 10^{-1}$

13. $(.0000096) \div (32) =$

$\left(9.6 \times 10^{-6}\right) \div \left(3.2 \times 10^1\right) =$

$(9.6 \div 3.2)\left(10^{-6} \div 10^1\right) = 3 \times 10^{-7}$

14. $\dfrac{(.0000012)(18)}{.00000054} =$

$\dfrac{\left(1.2 \times 10^{-6}\right)\left(1.8 \times 10^1\right)}{5.4 \times 10^{-7}} =$

$\dfrac{(1.2 \times 1.8)\left(10^{-6} \times 10^1\right)}{5.4 \times 10^{-7}} = \dfrac{2.16 \times 10^{-5}}{5.4 \times 10^{-7}} =$

$\left(2.16 \times 10^{-5}\right) \div \left(5.4 \times 10^{-7}\right) =$

$(2.16 \div 5.4)\left(10^{-5} \times 10^{-7}\right) =$

$.4 \times 10^2 = 4 \times 10^1$

15. $\dfrac{9X^2YZ}{YX} + \dfrac{2XY^{-1}Z^2}{Z^3Y^{-2}} - \dfrac{6X^{-2}YZ^3}{X^{-3}Z^4} =$

$9X^2Y^1ZY^{-1}X^{-1} + 2XY^{-1}Z^2Z^{-3}Y^2 - \dfrac{6X^{-2}YZ^3}{X^{-3}Z^4} =$

$9XZ + 2XYZ^{-1} - \dfrac{6X^{-2}YZ^3}{X^{-3}Z^4} =$

$9XZ + 2XYZ^{-1} - 6X^{-2}YZ^3X^3Z^{-4} =$

$9XZ + 2XYZ^{-1} - 6XYZ^{-1} =$

$9XZ - 4XYZ^{-1}$ or $9XZ - \dfrac{4XY}{Z}$

16. $\dfrac{1}{8X} - \dfrac{1}{7X} = 2$

$(56X)\dfrac{1}{8X} - (56X)\dfrac{1}{7X} = (56X)2$

$\dfrac{56X}{8X} - \dfrac{56X}{7X} = 112X$

$7 - 8 = 112X$

$-1 = 112X$

$-\dfrac{1}{112} = X$

17. $\left(4X^3Y\right)\left(3X^{-4}\right)\left(2X^2\right) = (4)(3)(2)X^3YX^{-4}X^2 =$

$24X^{3+(-4)+2}Y = 24XY$

18. $48D^{-7}E^5F^4 \div 16E^{-6}F^3 = \dfrac{48D^{-7}E^5F^4}{16E^{-6}F^3} =$

$\dfrac{3D^{-7}E^5F^4}{E^{-6}F^3} = 3D^{-7}E^5F^4E^6F^{-3} =$

$3D^{-7}E^{11}F$ or $\dfrac{3E^{11}F}{D^7}$

19. $X + 6 = \dfrac{4X}{7}$

$(7)X + (7)6 = (7)\dfrac{4X}{7}$

$7X + 42 = \dfrac{28X}{7}$

$7X + 42 = 4X$

$7X - 4X = -42$

$3X = -42$

$X = \dfrac{-42}{3} = -14$

20.

$$\frac{3X+2}{4} - 5 = \frac{X+8}{2} - 8$$

$$(4)\frac{3X+2}{4} - (4)5 = (4)\frac{X+8}{2} - (4)8$$

$$\frac{(4)(3X+2)}{4} - 20 = \frac{(4)(X+8)}{2} - 32$$

$$3X+2-20 = \frac{(2)(X+8)}{1} - 32$$

$$3X-18 = 2X+16-32$$

$$3X-2X = 16-32+18$$

$$X = 2$$

Systematic Review 4E

1. $X\sqrt{49X^2Y^2} = X\sqrt{49}\sqrt{X^2}\sqrt{Y^2} =$
$X(7)XY = 7X^2Y$

2. $4\sqrt{6} + 11\sqrt{6} = (4+11)\sqrt{6} = 15\sqrt{6}$

3. $(5\sqrt{X})(6\sqrt{Y}) = (5)(6)\sqrt{X}\sqrt{Y} = 30\sqrt{XY}$

4. $\sqrt{6}(\sqrt{7} + 4\sqrt{6}) = \sqrt{6}\sqrt{7} + 4\sqrt{6}\sqrt{6} =$
$\sqrt{42} + 4\sqrt{36} = \sqrt{42} + 4(6) = \sqrt{42} + 24$

5. $\frac{10\sqrt{63}}{\sqrt{7}} = \frac{10\sqrt{9}}{1} = 10(3) = 30$

6. $\frac{\sqrt{128}}{\sqrt{8}} = \frac{\sqrt{16}}{1} = 4$

7. $\sqrt{200} = \sqrt{100}\sqrt{2} = 10\sqrt{2}$

8. $\left(\frac{1}{3}\right)\sqrt{72} = \left(\frac{1}{3}\right)\sqrt{36}\sqrt{2} =$
$\left(\frac{1}{3}\right)(6)\sqrt{2} = 2\sqrt{2}$

9. $\frac{8}{\sqrt{10}} = \frac{8\sqrt{10}}{\sqrt{10}\sqrt{10}} = \frac{8\sqrt{10}}{\sqrt{100}} =$
$\frac{8\sqrt{10}}{10} = \frac{4\sqrt{10}}{5}$

10. $\frac{10}{\sqrt{7}} + \frac{16}{\sqrt{11}} = \frac{10\sqrt{7}}{\sqrt{7}\sqrt{7}} + \frac{16\sqrt{11}}{\sqrt{11}\sqrt{11}} =$
$\frac{10\sqrt{7}}{\sqrt{49}} + \frac{16\sqrt{11}}{\sqrt{121}} = \frac{10\sqrt{7}}{7} + \frac{16\sqrt{11}}{11} =$
$\frac{10\sqrt{7}(11)}{7(11)} = \frac{16\sqrt{11}(7)}{11(7)} = \frac{110\sqrt{7} + 112\sqrt{11}}{77}$

11. $(.00034)(.00000026) =$
$(3.4\times10^{-4})(2.6\times10^{-7}) =$
$(3.4\times2.6)(10^{-4}\times10^{-7}) = 8.84\times10^{-11}$

12. $(77,000)(740,000,000) =$
$(7.7\times10^4)(7.4\times10^8) =$
$(7.7\times7.4)(10^4\times10^8) =$
$56.98\times10^{12} = 5.698\times10^{13}$

13. $(490,000) \div (.007) =$
$(4.9\times10^5) \div (7 \div 10^{-3}) =$
$(4.9 \div 7)(10^5 \div 10^{-3}) =$
$.7\times10^8 = 7\times10^7$

14. $\frac{(28,000,000)(210,000,000)}{.98} =$
$\frac{(2.8\times10^7)(2.1\times10^8)}{(9.8\times10^{-1})} =$
$\frac{(2.8\times2.1)(10^7\times10^8)}{(9.8\times10^{-1})} = \frac{5.88\times10^{15}}{9.8\times10^{-1}} =$
$(5.88\times10^{15}) \div (9.8\times10^{-1}) =$
$(5.88 \div 9.8)(10^{15} \div 10^{-1}) =$
$.6\times10^{16} = 6\times10^{15}$

15. $\frac{2Q^{-1}R^0T^2}{T^3R^{-1}Q^2} - \frac{5Q^2R^{-3}T^4}{T^1R^{-2}Q^3} + \frac{Q^5R^3T^4}{RRQ^2} =$
$2Q^{-1}R^0T^2T^{-3}R^1Q^{-2} - \frac{5Q^2R^{-3}T^4}{T^1R^{-2}Q^3} + \frac{Q^5R^3T^4}{RRQ^2} =$
$2Q^{-3}R^1T^{-1} - 5Q^2R^{-3}T^4T^{-1}R^2Q^{-3} + \frac{Q^5R^3T^4}{RRQ^2} =$
$2Q^{-3}R^1T^{-1} - 5Q^{-1}R^{-1}T^3 + Q^5R^3T^4R^{-1}R^{-1}Q^{-2} =$
$2Q^{-3}RT^{-1} - 5Q^{-1}R^{-1}T^3 + Q^3RT^4$ or
$\frac{2R}{Q^3T} - \frac{5T^3}{QR} + Q^3RT^4$

16. $\frac{X}{3} - \frac{X}{5} = \frac{2}{15}$
$(15)\frac{X}{3} - (15)\frac{X}{5} = (15)\frac{2}{15}$
$\frac{15X}{3} - \frac{15X}{5} = \frac{30}{15}$
$5X - 3X = 2$
$2X = 2$
$X = \frac{2}{2} = 1$

17. $\left(7AX^3Y^2\right)\left(-X^2Y^{-2}\right) =$

$(7)(-1)AX^3Y^2X^2Y^{-2} = -7AX^5$

18. $135A^2B^5C^{-3} \div 15A^{-2}B^{-3} =$

$\dfrac{135A^2B^5C^{-3}}{15A^{-2}B^{-3}} = \dfrac{9A^2B^5C^{-3}}{A^{-2}B^{-3}} =$

$9A^2B^5C^{-3}A^2B^3 = 9A^4B^8C^{-3}$ or $\dfrac{9A^4B^8}{C^3}$

19. $\dfrac{3X}{8} = 11 - 2$

$(8)\dfrac{3X}{8} = (8)11 - (8)2$

$\dfrac{24X}{8} = 88 - 16$

$3X = 72$

$X = \dfrac{72}{3} = 24$

20.

$30 - .15X = .6X - 15$

$100(30) - 100(.15X) = 100(.6X) - 100(15)$

$3{,}000 - 15X = 60X - 1{,}500$

$4{,}500 = 75X$

$\dfrac{4{,}500}{75} = X = 60$

Lesson Practice 5A-1

Remember that finding the solution for a factoring problem often involves trial and error. Review the teaching material for tips if necessary. If you have the blocks, you may find them helpful.

Some problems may be factored more than one way. If you have a different answer, check it by multiplying the factors back to what you started with.

1. $X^2 + 7X + 12 = (X + 4)(X + 3)$

2. $X^2 + 3X + 2 = (X + 1)(X + 2)$

3. $X^2 + X - 6 = (X - 2)(X + 3)$

4. $X^2 - 11X + 30 = (X - 6)(X - 5)$

5. $X^2 - Y^2 = (X - Y)(X + Y)$

6. $A^2 - 81 = (A - 9)(A + 9)$

7. $2X^2 + X - 3 = (2X + 3)(X - 1)$

8. $3X^2 + 17X + 10 = (3X + 2)(X + 5)$

9. $5X^2 + 14X - 3 = (5X - 1)(X + 3)$

10. $4X^2 + 21X + 5 = (4X + 1)(X + 5)$

11. $2X^2 + 12X + 16 =$

$(2)\left(X^2 + 6X + 8\right) =$

$(2)(X + 2)(X + 4)$

12. $X^3 + 6X^2 + 9X =$

$(X)\left(X^2 + 6X + 9\right) =$

$(X)(X + 3)(X + 3)$

13. $A^4 - 81 = \left(A^2 - 9\right)\left(A^2 + 9\right) =$

$(A - 3)(A + 3)\left(A^2 + 9\right)$

14. $X^4 - 17X^2 + 16 = \left(X^2 - 16\right)\left(X^2 - 1\right) =$

$(X - 4)(X + 4)(X - 1)(X + 1)$

For this problem, it may help to express X^2 as Y in the first step. Since $X^4 = X^{2^2}$, X^4 would be equal to Y^2, so $Y^2 - 17Y + 16 = (Y - 16)(Y - 1)$. After this step, replace the Y with X^2 and solve.

15. $3X^2 + 10X + 12 = 4$

$3X^2 + 10X + 8 = 0$

$(3X + 4)(X + 2) = 0$

$3X + 4 = 0 \qquad 3\left(-\dfrac{4}{3}\right)^2 + 10\left(-\dfrac{4}{3}\right) + 12 = 4$

$\quad 3X = -4 \qquad 3\left(\dfrac{16}{9}\right) - \dfrac{40}{3} + 12 = 4$

$\quad X = -\dfrac{4}{3} \qquad \dfrac{48}{9} - \dfrac{120}{9} + \dfrac{108}{9} = 4$

$\qquad\qquad\qquad\qquad \dfrac{36}{9} = 4$

$\qquad\qquad\qquad\qquad\qquad 4 = 4 \text{ ok}$

$X + 2 = 0 \qquad 3(-2)^2 + 10(-2) + 12 = 4$

$\quad X = -2 \qquad\quad 3(4) - 20 + 12 = 4$

$\qquad\qquad\qquad\qquad 12 - 20 + 12 = 4$

$\qquad\qquad\qquad\qquad\qquad 4 = 4 \text{ ok}$

16.

$X^2 - 49 = 0$

$(X - 7)(X + 7) = 0$

$X - 7 = 0 \qquad (7)^2 - 49 = 0$

$\quad X = 7 \qquad\qquad 0 = 0 \text{ ok}$

$X + 7 = 0 \qquad (-7)^2 - 49 = 0$

$\quad X = -7 \qquad\qquad 0 = 0 \text{ ok}$

17.
$$2X^4 = 72X^2$$
$$2X^4 - 72X^2 = 0$$
$$2X^2(X-6)(X+6) = 0$$

$$2X^2 = 0 \qquad 2(0)^4 = 72(0)^2$$
$$X^2 = 0 \qquad 2(0) = 72(0)$$
$$X = 0 \qquad\quad 0 = 0 \text{ ok}$$

$$X - 6 = 0 \qquad 2(6)^4 = 72(6)^2$$
$$X = 6 \qquad 2(1296) = 72(36)$$
$$2592 = 2592 \text{ ok}$$

$$X + 6 = 0 \qquad 2(-6)^4 = 72(-6)^2$$
$$X = -6 \qquad 2(1296) = 72(36)$$
$$2592 = 2592 \text{ ok}$$

18.
$$X^4 - 26X^2 + 27 = 2$$
$$X^4 - 26X^2 + 25 = 0$$
$$(X^2 - 1)(X^2 - 25) = 0$$
$$(X-1)(X+1)(X-5)(X+5) = 0$$

$$X = 1 \qquad (1)^4 - 26(1)^2 + 27 = 2$$
$$1 - 26 + 27 = 2$$
$$2 = 2 \text{ ok}$$

$$X = -1 \qquad (-1)^4 - 26(-1)^2 + 27 = 2$$
$$1 - 26 + 27 = 2$$
$$2 = 2 \text{ ok}$$

$$X = 5 \qquad (5)^4 - 26(5)^2 + 27 = 2$$
$$625 - 650 + 27 = 2$$
$$2 = 2 \text{ ok}$$

$$X = -5 \qquad (-5)^4 - 26(-5)^2 + 27 = 2$$
$$625 - 650 + 27 = 2$$
$$2 = 2 \text{ ok}$$

Lesson Practice 5A-2

1. $X^2 + X - 12 = (X+4)(X-3)$

2. $X^2 - 10X + 24 = (X-4)(X-6)$

3. $X^2 - 8X - 9 = (X-9)(X+1)$

4. $X^2 + 7X + 10 = (X+5)(X+2)$

5. $25A^2 - 25B^2 = (5A - 5B)(5A + 5B)$
or $25(A - B)(A + B)$

6. $4X^2 - 64 = (2X - 8)(2X + 8)$ or
$(4)(X^2 - 16) = (4)(X - 4)(X + 4)$

7. $3A^2 + 4AB + B^2 = (3A + B)(A + B)$

8. $6X^2 - 2X - 4 = (2)(3X^2 - X - 2) =$
$(2)(3X + 2)(X - 1)$

9. $2X^2 + X - 15 = (2X - 5)(X + 3)$

10. $3X^2 + 20X - 32 = (3X - 4)(X + 8)$

11. $2X^3 + 12X^2 + 16X = (2X)(X^2 + 6X + 8) =$
$(2X)(X + 2)(X + 4)$

12. $3A^2 - 21A + 18 = (3)(A^2 - 7A + 6) =$
$(3)(A - 1)(A - 6)$

13. $X^8 - 1 = (X^4 - 1)(X^4 + 1) =$
$(X^2 - 1)(X^2 + 1)(X^4 + 1) =$
$(X - 1)(X + 1)(X^2 + 1)(X^4 + 1)$

14. $X^4 - 109X^2 + 900 = (X^2 - 9)(X^2 - 100) =$
$(X - 3)(X + 3)(X - 10)(X + 10)$

15.
$$6X^3 + 10X^2 = -4X$$
$$6X^3 + 10X^2 + 4X = 0$$
$$3X^3 + 5X^2 + 2X = 0$$
$$X(3X^2 + 5X + 2) = 0$$
$$X(3X + 2)(X + 1) = 0$$

$$X = 0$$

$$3X + 2 = 0 \qquad 6\left(-\frac{2}{3}\right)^3 + 10\left(-\frac{2}{3}\right)^2 = -4\left(-\frac{2}{3}\right)$$
$$3X = -2 \qquad 6\left(-\frac{8}{27}\right) + 10\left(\frac{4}{9}\right) = -4\left(-\frac{2}{3}\right)$$
$$X = -\frac{2}{3} \qquad -\frac{48}{27} + \frac{40}{9} = \frac{8}{3}$$
$$-\frac{16}{9} + \frac{40}{9} = \frac{24}{9}$$
$$\frac{24}{9} = \frac{24}{9} \text{ ok}$$

$$X + 1 = 0 \qquad 6(-1)^3 + 10(-1)^2 = -4(-1)$$
$$X = -1 \qquad 6(-1) + 10(1) = 4$$
$$-6 + 10 = 4$$
$$4 = 4 \text{ ok}$$

16. $2X^2 - 8X - 14 = 10$

$X^2 - 4X - 7 = 5$

$X^2 - 4X - 12 = 0$

$(X - 6)(X + 2) = 0$

$X - 6 = 0$ $\qquad$ $2(6)^2 - 8(6) - 14 = 10$

$\quad X = 6$ $\qquad\quad$ $2(36) - 48 - 14 = 10$

$\qquad\qquad\qquad\quad$ $72 - 48 - 14 = 10$

$\qquad\qquad\qquad\qquad\qquad$ $10 = 10$ ok

$X + 2 = 0$ $\qquad$ $2(-2)^2 - 8(-2) - 14 = 10$

$\quad X = -2$ $\qquad\quad$ $2(4) + 16 - 14 = 10$

$\qquad\qquad\qquad\quad$ $8 + 16 - 14 = 10$

$\qquad\qquad\qquad\qquad\quad$ $10 = 10$ ok

17. $\qquad$ $X^3 - 50X = 50X$

$\qquad\qquad$ $X^3 - 100X = 0$

$\qquad\qquad$ $X(X^2 - 100) = 0$

$\qquad$ $X(X - 10)(X + 10) = 0$

$\quad X = 0$

$\qquad\qquad\qquad\qquad$ $(10)^3 - 50(10) = 50(10)$

$X - 10 = 0$ $\qquad$ $1000 - 500 = 500$

$\quad X = 10$ $\qquad\qquad$ $500 = 500$ ok

$X + 10 = 0$ $\qquad$ $(-10)^3 - 50(-10) = 50(-10)$

$\quad X = -10$ $\qquad$ $-1000 + 500 = -500$

$\qquad\qquad\qquad\qquad$ $-500 = -500$ ok

18. $\qquad$ $-8 = A^2 - 16A + 20$

$\qquad\qquad$ $0 = A^2 - 16A + 28$

$\qquad\qquad$ $0 = (A - 2)(A - 14)$

$A - 2 = 0$ $\qquad$ $-8 = (2)^2 - 16(2) + 20$

$\quad A = 2$ $\qquad\qquad$ $-8 = 4 - 32 + 20$

$\qquad\qquad\qquad\qquad$ $-8 = -8$ ok

$A - 14 = 0$ $\qquad$ $-8 = (14)^2 - 16(14) + 20$

$\quad A = 14$ $\qquad\quad$ $-8 = 196 - 224 + 20$

$\qquad\qquad\qquad\qquad$ $-8 = -8$ ok

Lesson Practice 5B-1

1. $\dfrac{2}{X-1} + \dfrac{6}{X+2} + \dfrac{3}{X^2+X-2} =$

$\dfrac{2(X+2)}{(X-1)(X+2)} + \dfrac{6(X-1)}{(X-1)(X+2)} + \dfrac{3}{X^2+X-2} =$

$\dfrac{2X+4+6X-6}{X^2+X-2} + \dfrac{3}{X^2+X-2} = \dfrac{8X+1}{X^2+X-2}$

2. $\dfrac{X+2}{X-2} - \dfrac{X+2}{X+2} = \dfrac{X+2}{X-2} - \dfrac{1}{1} =$

$\dfrac{X+2}{X-2} - \dfrac{1(X-2)}{1(X-2)} = \dfrac{X+2-(X-2)}{X-2} =$

$\dfrac{X+2-X+2}{X-2} = \dfrac{4}{X-2}$

3. $\dfrac{3}{A} + \dfrac{5}{A+1} = \dfrac{3(A+1)}{A(A+1)} + \dfrac{5(A)}{(A+1)(A)} =$

$\dfrac{3A+3+5A}{A^2+A} = \dfrac{8A+3}{A^2+A}$

4. $\dfrac{3X}{X+3} - \dfrac{2X}{X+2} =$

$\dfrac{3X(X+2)}{(X+3)(X+2)} - \dfrac{2X(X+3)}{(X+2)(X+3)} =$

$\dfrac{(3X^2+6X)-(2X^2+6X)}{X^2+5X+6} = \dfrac{X^2}{X^2+5X+6}$

5. $\dfrac{7}{X+2} + \dfrac{4}{3-X} - \dfrac{2X+1}{X^2-X-6} =$

$\dfrac{7}{X+2} + \dfrac{4(-1)}{(3-X)(-1)} - \dfrac{2X+1}{X^2-X-6} =$

$\dfrac{7}{X+2} + \dfrac{-4}{X-3} - \dfrac{2X+1}{X^2-X-6} =$

$\dfrac{7(X-3)}{(X+2)(X-3)} + \dfrac{-4(X+2)}{(X-3)(X+2)} - \dfrac{2X+1}{X^2-X-6} =$

$\dfrac{(7X-21)+(-4X-8)-(2X+1)}{X^2-X-6} = \dfrac{X-30}{X^2-X-6}$

6. $\dfrac{2X}{X^2-4} + \dfrac{8X}{X+2} - \dfrac{4}{X-2} =$

$\dfrac{2X}{X^2-4} + \dfrac{8X(X-2)}{(X+2)(X-2)} - \dfrac{4(X+2)}{(X-2)(X+2)} =$

$\dfrac{2X}{X^2-4} + \dfrac{(8X^2-16X)}{X^2-4} - \dfrac{(4X+8)}{X^2-4} =$

$\dfrac{2X+8X^2-16X-4X-8}{X^2-4} = \dfrac{8X^2-18X-8}{X^2-4}$

7. $\dfrac{\dfrac{2}{X}}{\dfrac{X+3}{4X}} = \dfrac{\dfrac{2}{X} \times \dfrac{4X}{X+3}}{\dfrac{X+3}{4X} \times \dfrac{4X}{X+3}} = \dfrac{2}{X} \times \dfrac{4X}{X+3} =$

$\dfrac{2(4X)}{X(X+3)} = \dfrac{2(4)}{X+3} = \dfrac{8}{X+3}$

8. $\dfrac{2+\dfrac{1}{2}}{6-\dfrac{2}{3}} = \dfrac{\dfrac{4}{2}+\dfrac{1}{2}}{\dfrac{18}{3}-\dfrac{2}{3}} = \dfrac{\dfrac{5}{2}}{\dfrac{16}{3}} =$

$\dfrac{\dfrac{5}{2} \times \dfrac{3}{16}}{\dfrac{16}{3} \times \dfrac{3}{16}} = \dfrac{15}{32}$

9. $\dfrac{2-\dfrac{3}{A}}{4+\dfrac{1}{A-1}} = \dfrac{\dfrac{2A}{A}-\dfrac{3}{A}}{\dfrac{4(A-1)}{(A-1)}+\dfrac{1}{A-1}} =$

$\dfrac{\dfrac{2A-3}{A}}{\dfrac{4A-4}{A-1}+\dfrac{1}{A-1}} = \dfrac{\dfrac{2A-3}{A} \times \dfrac{A-1}{4A-3}}{\dfrac{4A-3}{A-1} \times \dfrac{A-1}{4A-3}} =$

$\dfrac{(2A-3)(A-1)}{(A)(4A-3)} = \dfrac{2A^2-5A+3}{4A^2-3A}$

10. $\dfrac{\dfrac{X^2+7X+12}{X^2+X-12}}{\dfrac{X^2+3X+2}{X^2-9}} =$

$\dfrac{\dfrac{X^2+7X+12}{X^2+X-12} \times \dfrac{X^2-9}{X^2+3X+2}}{\dfrac{X^2+3X+2}{X^2-9} \times \dfrac{X^2-9}{X^2+3X+2}} =$

$\dfrac{X^2+7X+12}{X^2+X-12} \times \dfrac{X^2-9}{X^2+3X+2} =$

$\dfrac{(X+3)\cancel{(X+4)}}{\cancel{(X+4)}\cancel{(X-3)}} \times \dfrac{\cancel{(X-3)}(X+3)}{(X+2)(X+1)} =$

$\dfrac{(X+3)(X+3)}{(X+2)(X+1)} = \dfrac{X^2+6X+9}{X^2+3X+2}$

11. $\dfrac{X-\dfrac{5}{Y}}{X+\dfrac{4}{Y}} = \dfrac{\dfrac{XY}{Y}-\dfrac{5}{Y}}{\dfrac{XY}{Y}+\dfrac{4}{Y}} = \dfrac{\dfrac{XY-5}{Y}}{\dfrac{XY+4}{Y}} =$

$\dfrac{\dfrac{XY-5}{Y} \times \dfrac{Y}{XY+4}}{\dfrac{XY+4}{Y} \times \dfrac{Y}{XY+4}} = \dfrac{XY-5}{\cancel{Y}} \times \dfrac{\cancel{Y}}{XY+4} =$

$\dfrac{XY-5}{XY+4}$

12. $\dfrac{\dfrac{X^2+X-6}{X^2-11X+30}}{\dfrac{X^2-7X+10}{X^2-10X+24}} =$

$\dfrac{\dfrac{X^2+X-6}{X^2-11X+30} \times \dfrac{X^2-10X+24}{X^2-7X+10}}{\dfrac{X^2-7X+10}{X^2-10X+24} \times \dfrac{X^2-10X+24}{X^2-7X+10}} =$

$\dfrac{X^2+X-6}{X^2-11X+30} \times \dfrac{X^2-10X+24}{X^2-7X+10} =$

$\dfrac{\cancel{(X-2)}(X+3)}{(X-5)\cancel{(X-6)}} \times \dfrac{\cancel{(X-6)}(X-4)}{\cancel{(X-2)}(X-5)} =$

$\dfrac{(X+3)(X-4)}{(X-5)(X-5)} = \dfrac{X^2-X-12}{X^2-10X+25}$

Lesson Practice 5B-2

1. $\dfrac{10}{X+4} + \dfrac{3}{X-4} - \dfrac{2}{X^2-16} =$

$\dfrac{10(X-4)}{(X+4)(X-4)} + \dfrac{3(X+4)}{(X-4)(X+4)} - \dfrac{2}{X^2-16} =$

$\dfrac{(10X-40)+(3X+12)}{X^2-16} - \dfrac{2}{X^2-16} =$

$\dfrac{13X-30}{X^2-16}$

2. $\dfrac{A+B}{A-B} + \dfrac{2A}{B} = \dfrac{(A+B)(B)}{(A-B)(B)} + \dfrac{(2A)(A-B)}{(B)(A-B)} =$

$\dfrac{AB+B^2+2A^2-2AB}{AB-B^2} = \dfrac{2A^2-AB+B^2}{AB-B^2}$

3. $\dfrac{15}{X} + \dfrac{20}{X-1} = \dfrac{(15)(X-1)}{(X)(X-1)} + \dfrac{(20)(X)}{(X-1)(X)} =$

$\dfrac{15X-15+20X}{X^2-X} = \dfrac{35X-15}{X^2-X}$

4. $\dfrac{4X}{X+1} - \dfrac{3Y}{X+1} = \dfrac{4X-3Y}{X+1}$

5. $\dfrac{4}{B-4} + \dfrac{5}{B-5} + \dfrac{B-5}{B^2-9B+20} =$

$\dfrac{(4)(B-5)}{(B-4)(B-5)} + \dfrac{(5)(B-4)}{(B-5)(B-4)} + \dfrac{B-5}{B^2-9B+20} =$

$\dfrac{4B-20+5B-20+B-5}{B^2-9B+20} = \dfrac{10B-45}{B^2-9B+20}$

6. $\dfrac{2X+3}{4X^2+6X} + \dfrac{2X}{2X+3} + \dfrac{3}{2X} =$

$\dfrac{2X+3}{4X^2+6X} + \dfrac{2X(2X)}{(2X+3)(2X)} + \dfrac{3(2X+3)}{2X(2X+3)} =$

$\dfrac{(2X+3)+(4X^2)+(6X+9)}{4X^2+6X} = \dfrac{4X^2+8X+12}{4X^2+6X} =$

$\dfrac{2(2X^2+4X+6)}{2(2X^2+3X)} = \dfrac{2X^2+4X+6}{2X^2+3X}$

7. $\dfrac{\frac{A}{B}}{\frac{A+B}{AB}} = \dfrac{\frac{A}{B} \times \frac{AB}{A+B}}{\frac{A+B}{AB} \times \frac{AB}{A+B}} =$

$\dfrac{A}{B} \times \dfrac{AB}{A+B} =$

$\dfrac{A^2B}{AB+B^2} = \dfrac{(B)(A^2)}{(B)(A+B)} = \dfrac{A^2}{A+B}$

8. $\dfrac{3-\frac{1}{3}}{5+\frac{3}{5}} = \dfrac{\frac{9}{3}-\frac{1}{3}}{\frac{25}{5}+\frac{3}{5}} = \dfrac{\frac{8}{3}}{\frac{28}{5}} =$

$\dfrac{\frac{8}{3} \times \frac{5}{28}}{\frac{28}{5} \times \frac{5}{28}} = \dfrac{8}{3} \times \dfrac{5}{28} = \dfrac{40}{84} = \dfrac{10}{21}$

9. $\dfrac{4+\frac{1}{X}}{5+\frac{X}{X+1}} = \dfrac{\frac{4X}{X}+\frac{1}{X}}{\frac{5(X+1)}{X+1}+\frac{X}{X+1}} =$

$\dfrac{\frac{4X+1}{X}}{\frac{5X+5+X}{X+1}} = \dfrac{\frac{4X+1}{X} \times \frac{X+1}{6X+5}}{\frac{6X+5}{X+1} \times \frac{X+1}{6X+5}} =$

$\dfrac{4X+1}{X} \times \dfrac{X+1}{6X+5} = \dfrac{4X^2+5X+1}{6X^2+5X}$

10. $\dfrac{\frac{X^2+4X-5}{X^2-3X-18}}{\frac{X^2+6X+5}{X^2-8X+12}} =$

$\dfrac{\frac{X^2+4X-5}{X^2-3X-18} \times \frac{X^2-8X+12}{X^2+6X+5}}{\frac{X^2+6X+5}{X^2-8X+12} \times \frac{X^2-8X+12}{X^2+6X+5}} =$

$\dfrac{X^2+4X-5}{X^2-3X-18} \times \dfrac{X^2-8X+12}{X^2+6X+5} =$

$\dfrac{(X-1)\cancel{(X+5)}}{(X+3)\cancel{(X-6)}} \times \dfrac{(X-2)\cancel{(X-6)}}{\cancel{(X+5)}(X+1)} =$

$\dfrac{(X-1)(X-2)}{(X+3)(X+1)} = \dfrac{X^2-3X+2}{X^2+4X+3}$

11. $\dfrac{Y-\frac{2}{3}}{Y-\frac{1}{4}} = \dfrac{\frac{3Y}{3}-\frac{2}{3}}{\frac{4Y}{4}-\frac{1}{4}} = \dfrac{\frac{3Y-2}{3}}{\frac{4Y-1}{4}} =$

$\dfrac{\frac{3Y-2}{3} \times \frac{4}{4Y-1}}{\frac{4Y-1}{4} \times \frac{4}{4Y-1}} = \dfrac{3Y-2}{3} \times \dfrac{4}{4Y-1} =$

$\dfrac{4(3Y-2)}{3(4Y-1)} = \dfrac{12Y-8}{12Y-3}$

12. $\dfrac{\frac{X^4-16}{X^2-5X+4}}{\frac{X^2-4}{X^2+3X-28}} =$

$\dfrac{\frac{X^4-16}{X^2-5X+4} \times \frac{X^2+3X-28}{X^2-4}}{\frac{X^2-4}{X^2+3X-28} \times \frac{X^2+3X-28}{X^2-4}} =$

$\dfrac{X^4-16}{X^2-5X+4} \times \dfrac{X^2+3X-28}{X^2-4} =$

$\dfrac{\cancel{(X^2-4)}(X^2+4)}{\cancel{(X-4)}(X-1)} \times \dfrac{\cancel{(X-4)}(X+7)}{\cancel{(X^2-4)}} =$

$\dfrac{(X^2+4)(X+7)}{X-1} = \dfrac{X^3+7X^2+4X+28}{X-1}$

Systematic Review 5C

1. $X^2 + 9X + 20 = (X+4)(X+5)$

2. $X^2 - 9X + 20 = (X-4)(X-5)$

3. $X^2 - 36 = (X-6)(X+6)$

4. $4X^2 + 8X + 3 = (2X+1)(2X+3)$

5. $6X^2 + X - 2 = (3X+2)(2X-1)$

6. $X^2 - X - 20 = (X-5)(X+4)$

7. $20X^4 + 10X^3 - 30X^2 =$
$(10X^2)(2X^2 + X - 3) = (10X^2)(2X+3)(X-1)$

8. $X^4 - 16 = (X^2 - 4)(X^2 + 4) =$
$(X-2)(X+2)(X^2+4)$

9. $\qquad X^2 - 2X = -X + 6$
$\qquad X^2 - X - 6 = 0$
$\qquad (X-3)(X+2) = 0$

$X - 3 = 0 \qquad\quad (3)^2 - 2(3) = -(3) + 6$
$\qquad X = 3 \qquad\qquad\quad 9 - 6 = -3 + 6$
$\qquad\qquad\qquad\qquad\qquad\quad 3 = 3 \text{ ok}$

$X + 2 = 0 \qquad\quad (-2)^2 - 2(-2) = -(-2) + 6$
$\qquad X = -2 \qquad\qquad\quad 4 + 4 = 2 + 6$
$\qquad\qquad\qquad\qquad\qquad\quad 8 = 8 \text{ ok}$

10. $\qquad 7 = 4X + X^2 - 5$
$\qquad 0 = X^2 + 4X - 12$
$\qquad 0 = (X+6)(X-2)$

$X + 6 = 0 \qquad\quad 7 = 4(-6) + (-6)^2 - 5$
$\qquad X = -6 \qquad\qquad 7 = -24 + 36 - 5$
$\qquad\qquad\qquad\qquad\qquad 7 = 7 \text{ ok}$

$X - 2 = 0 \qquad\quad 7 = 4(2) + (2)^2 - 5$
$\qquad X = 2 \qquad\qquad\quad 7 = 8 + 4 - 5$
$\qquad\qquad\qquad\qquad\qquad 7 = 7 \text{ ok}$

11. $\dfrac{5}{X} - \dfrac{4}{X-1} = \dfrac{5(X-1)}{X(X-1)} - \dfrac{4(X)}{(X-1)(X)} =$

$\dfrac{5X - 5 - 4X}{X^2 - X} = \dfrac{X-5}{X^2 - X}$

12. $\dfrac{3}{X+2} - \dfrac{6}{X-3} + \dfrac{4X}{X^2 - X - 6} =$

$\dfrac{3(X-3)}{(X+2)(X-3)} - \dfrac{6(X+2)}{(X-3)(X+2)} + \dfrac{4X}{X^2 - X - 6} =$

$\dfrac{(3X-9) - (6X+12) + 4X}{X^2 - X - 6} = \dfrac{X - 21}{X^2 - X - 6}$

13. $\dfrac{1 + \dfrac{1}{3}}{1 - \dfrac{1}{3}} = \dfrac{\dfrac{3}{3} + \dfrac{1}{3}}{\dfrac{3}{3} - \dfrac{1}{3}} = \dfrac{\dfrac{4}{3}}{\dfrac{2}{3}} = \dfrac{\dfrac{4}{3} \times \dfrac{3}{2}}{\dfrac{2}{3} \times \dfrac{3}{2}} =$

$\dfrac{4}{3} \times \dfrac{3}{2} = \dfrac{12}{6} = 2$

14. $\dfrac{\dfrac{4}{3X}}{\dfrac{X-5}{X}} = \dfrac{\dfrac{4}{3X} \times \dfrac{X}{X-5}}{\dfrac{X-5}{X} \times \dfrac{X}{X-5}} = \dfrac{4}{3X} \times \dfrac{X}{X-5} =$

$\dfrac{4}{3} \times \dfrac{1}{X-5} = \dfrac{4}{3(X-5)} = \dfrac{4}{3X - 15}$

15. $\dfrac{3\sqrt{32}}{\sqrt{2}} = \dfrac{3\sqrt{2}\sqrt{16}}{\sqrt{2}} = \dfrac{3\sqrt{16}}{1} = 3\sqrt{16} = 3(4) = 12$

16. $6\sqrt{300} = 6\sqrt{100}\sqrt{3} = 6(10)\sqrt{3} = 60\sqrt{3}$

17. $\dfrac{7}{\sqrt{5}} = \dfrac{7\sqrt{5}}{\sqrt{5}\sqrt{5}} = \dfrac{7\sqrt{5}}{\sqrt{25}} = \dfrac{7\sqrt{5}}{5}$

18. $\dfrac{2}{\sqrt{7}} + \dfrac{8}{\sqrt{11}} = \dfrac{2\sqrt{7}}{\sqrt{7}\sqrt{7}} + \dfrac{8\sqrt{11}}{\sqrt{11}\sqrt{11}} =$

$\dfrac{2\sqrt{7}}{\sqrt{49}} + \dfrac{8\sqrt{11}}{\sqrt{121}} = \dfrac{2\sqrt{7}}{7} + \dfrac{8\sqrt{11}}{11} =$

$\dfrac{2\sqrt{7}(11)}{7(11)} + \dfrac{8\sqrt{11}(7)}{11(7)} = \dfrac{22\sqrt{7}}{77} + \dfrac{56\sqrt{11}}{77} =$

$\dfrac{22\sqrt{7} + 56\sqrt{11}}{77}$

19. $\dfrac{(51,000)(600)}{(1,700)(.012)} = \dfrac{(5.1 \times 10^4)(6 \times 10^2)}{(1.7 \times 10^3)(1.2 \times 10^{-2})} =$

$\dfrac{30.6 \times 10^6}{2.04 \times 10^1} = (30.6 \times 10^6) \div (2.04 \times 10^1) =$

$(30.6 \div 2.04)(10^6 \div 10^1) = 15 \times 10^5 = 1.5 \times 10^6$

20. $\dfrac{X+4}{X^2 + 6X + 8} = \dfrac{(X+4)}{(X+2)(X+4)} = \dfrac{1}{X+2}$

Systematic Review 5D

1. $X^2 + 3X - 4 = (X+4)(X-1)$

2. $X^2 - 2X - 15 = (X-5)(X+3)$

3. $X^2 + 7X + 10 = (X+5)(X+2)$

4. $X^2 + 17X + 70 = (X+10)(X+7)$

5. $X^2 - 13X + 42 = (X-6)(X-7)$

6. $5X^2 + X - 4 = (5X-4)(X+1)$

7. $3X^2 + 13X - 10 = (3X-2)(X+5)$

8. $9X^2 - 1 = (3X - 1)(3X + 1)$

9. $X^2 + 13X + 20 = -22$

$X^2 + 13X + 42 = 0$

$(X + 6)(X + 7) = 0$

$X + 6 = 0 \qquad (-6)^2 + 13(-6) + 20 = -22$

$\qquad X = -6 \qquad 36 - 78 + 20 = -22$

$\qquad\qquad\qquad\qquad -22 = -22 \ \ \text{ok}$

$X + 7 = 0 \qquad (-7)^2 + 13(-7) + 20 = -22$

$\qquad X = -7 \qquad 49 - 91 + 20 = -22$

$\qquad\qquad\qquad\qquad -22 = -22 \ \ \text{ok}$

10. $X^2 + X + 7 = 16 + X$

$X^2 - 9 = 0$

$(X - 3)(X + 3) = 0$

$X - 3 = 0 \qquad (3)^2 + (3) + 7 = 16 + (3)$

$\qquad X = 3 \qquad\quad 9 + 3 + 7 = 19$

$\qquad\qquad\qquad\qquad\quad 19 = 19 \ \ \text{ok}$

$X + 3 = 0 \qquad (-3)^2 + (-3) + 7 = 16 + (-3)$

$\qquad X = -3 \qquad\quad 9 - 3 + 7 = 13$

$\qquad\qquad\qquad\qquad\quad 13 = 13 \ \ \text{ok}$

11. $\dfrac{2X}{Y} - \dfrac{3}{X} + \dfrac{4}{Y} =$

$\dfrac{2X(X)}{Y(X)} - \dfrac{3(Y)}{X(Y)} + \dfrac{4(X)}{Y(X)} = \dfrac{2X^2 - 3Y + 4X}{XY}$

12. $\dfrac{5}{X - 4} - \dfrac{9}{X} + \dfrac{8X}{X^2 - 4X} =$

$\dfrac{5(X)}{(X - 4)(X)} - \dfrac{9(X - 4)}{X(X - 4)} + \dfrac{8X}{X^2 - 4X} =$

$\dfrac{5X - (9X - 36) + 8X}{X^2 - 4X} = \dfrac{4X + 36}{X^2 - 4X}$

13. $\dfrac{2 + \frac{1}{2}}{5 - \frac{1}{8}} = \dfrac{\frac{4}{2} + \frac{1}{2}}{\frac{40}{8} - \frac{1}{8}} = \dfrac{\frac{5}{2}}{\frac{39}{8}} =$

$\dfrac{\frac{5}{2} \times \frac{8}{39}}{\frac{39}{8} \times \frac{8}{39}} = \dfrac{5}{2} \times \dfrac{8}{39} = \dfrac{40}{78} = \dfrac{20}{39}$

14. $\dfrac{X - \frac{1}{7}}{\frac{1}{7} - X} = \dfrac{\frac{7X}{7} - \frac{1}{7}}{\frac{1}{7} - \frac{7X}{7}} = \dfrac{\frac{7X - 1}{7}}{\frac{1 - 7X}{7}} =$

$\dfrac{\frac{7X - 1}{7} \times \frac{7}{1 - 7X}}{\frac{1 - 7X}{7} \times \frac{7}{1 - 7X}} = \dfrac{7X - 1}{7} \times \dfrac{7}{1 - 7X} =$

$\dfrac{7X - 1}{1 - 7X} = \dfrac{(7X - 1)(-1)}{(1 - 7X)(-1)} = \dfrac{-(7X - 1)}{(7X - 1)} = -1$

15. $\dfrac{\sqrt{12}}{3\sqrt{2}} = \dfrac{\sqrt{2}\sqrt{6}}{3\sqrt{2}} = \dfrac{\sqrt{6}}{3}$

16. $\dfrac{1}{5\sqrt{6}} = \dfrac{\sqrt{6}}{5\sqrt{6}\sqrt{6}} = \dfrac{\sqrt{6}}{5\sqrt{36}} = \dfrac{\sqrt{6}}{5(6)} = \dfrac{\sqrt{6}}{30}$

17. $7\sqrt{80} = 7\sqrt{16}\sqrt{5} = 7(4)\sqrt{5} = 28\sqrt{5}$

18. $\dfrac{5}{\sqrt{10}} + \dfrac{4}{\sqrt{13}} = \dfrac{5\sqrt{10}}{\sqrt{10}\sqrt{10}} + \dfrac{4\sqrt{13}}{\sqrt{13}\sqrt{13}} =$

$\dfrac{5\sqrt{10}}{\sqrt{100}} + \dfrac{4\sqrt{13}}{\sqrt{169}} = \dfrac{5\sqrt{10}}{10} + \dfrac{4\sqrt{13}}{13} =$

$\dfrac{\sqrt{10}}{2} + \dfrac{4\sqrt{13}}{13} = \dfrac{\sqrt{10}(13)}{2(13)} + \dfrac{4\sqrt{13}(2)}{13(2)} =$

$\dfrac{13\sqrt{10} + 8\sqrt{13}}{26}$

19. $\dfrac{(140,000)(27,000)}{420} =$

$\dfrac{(1.4 \times 10^5)(2.7 \times 10^4)}{4.2 \times 10^2} =$

$\dfrac{(1.4 \times 2.7)(10^5 \times 10^4)}{4.2 \times 10^2} = \dfrac{3.78 \times 10^9}{4.2 \times 10^2} =$

$(3.78 \times 10^9) \div (4.2 \times 10^2) =$

$(3.78 \div 4.2)(10^9 \div 10^2) = .9 \times 10^7 = 9 \times 10^6$

20. $\dfrac{X^2 - 9}{X^2 + 6X + 9} = \dfrac{(X - 3)(X + 3)}{(X + 3)(X + 3)} = \dfrac{X - 3}{X + 3}$

Systematic Review 5E

1. $X^2 + 2X - 24 = (X + 6)(X - 4)$

2. $X^2 + 10X + 9 = (X + 9)(X + 1)$

3. $X^2 - 7X + 10 = (X - 5)(X - 2)$

4. $64 - X^2 = (8 - X)(8 + X)$

5. $2X^2 - 17X + 30 = (2X - 5)(X - 6)$

6. $3X^2 + 8X - 3 = (3X - 1)(X + 3)$

7. $4X^2 - 19X + 12 = (4X - 3)(X - 4)$

8. $X^2 - X - 6 = (X - 3)(X + 2)$

9. $2X^2 - 20X = -36 - 2X$

   $X^2 - 10X = -18 - X$

   $X^2 - 9X + 18 = 0$

   $(X - 6)(X - 3) = 0$

   $X - 6 = 0$ $2(6)^2 - 20(6) = -36 - 2(6)$

       $X = 6$ $2(36) - 120 = -36 - 12$

                      $72 - 120 = -48$

                             $-48 = -48$ ok

   $X - 3 = 0$ $2(3)^2 - 20(3) = -36 - 2(3)$

       $X = 3$ $2(9) - 60 = -36 - 6$

                      $18 - 60 = -42$

                          $-42 = -42$ ok

10. $9X^2 - 20X = -16 + 4X$

    $9X^2 - 24X + 16 = 0$

    $(3X - 4)(3X - 4) = 0$

    $3X - 4 = 0$

    $3X = 4$

    $X = \dfrac{4}{3}$

    $9\left(\dfrac{4}{3}\right)^2 - 20\left(\dfrac{4}{3}\right) = -16 + 4\left(\dfrac{4}{3}\right)$

    $9\left(\dfrac{16}{9}\right) - \dfrac{80}{3} = -16 + \dfrac{16}{3}$

    $\dfrac{144}{9} - \dfrac{240}{9} = -\dfrac{144}{9} + \dfrac{48}{9}$

    $-\dfrac{96}{9} = -\dfrac{96}{9}$ ok

11. $\dfrac{X - 3}{2X} - \dfrac{X - 2}{2Y} = \dfrac{(X - 3)(Y)}{2X(Y)} - \dfrac{(X - 2)(X)}{2Y(X)} =$

    $\dfrac{(XY - 3Y) - (X^2 - 2X)}{2XY} = \dfrac{XY - 3Y - X^2 + 2X}{2XY}$

12. $\dfrac{8X - 2}{X^2 + 5X + 6} - \dfrac{X + 2}{X + 3} =$

    $\dfrac{8X - 2}{(X + 3)(X + 2)} - \dfrac{X + 2}{X + 3} =$

    $\dfrac{8X - 2}{(X + 3)(X + 2)} - \dfrac{(X + 2)(X + 2)}{(X + 3)(X + 2)} =$

    $\dfrac{8X - 2 - (X^2 + 4X + 4)}{X^2 + 5X + 6} = \dfrac{-X^2 + 4X - 6}{X^2 + 5X + 6}$

13. $\dfrac{4 + \dfrac{1}{4}}{6 - 1\dfrac{2}{3}} = \dfrac{\dfrac{16}{4} + \dfrac{1}{4}}{\dfrac{18}{3} - \dfrac{5}{3}} = \dfrac{\dfrac{17}{4}}{\dfrac{13}{3}} =$

    $\dfrac{\dfrac{17}{4} \times \dfrac{3}{13}}{\dfrac{13}{3} \times \dfrac{3}{13}} = \dfrac{17}{4} \times \dfrac{3}{13} = \dfrac{51}{52}$

14. $\dfrac{\dfrac{5X}{2} + 1}{2X - \dfrac{4}{3X}} = \dfrac{\dfrac{5X}{2} + \dfrac{2}{2}}{\dfrac{2X(3X)}{3X} - \dfrac{4}{3X}} =$

    $\dfrac{\dfrac{5X + 2}{2}}{\dfrac{6X^2 - 4}{3X}} = \dfrac{\dfrac{5X + 2}{2} \times \dfrac{3X}{6X^2 - 4}}{\dfrac{6X^2 - 4}{3X} \times \dfrac{3X}{6X^2 - 4}} =$

    $\dfrac{5X + 2}{2} \times \dfrac{3X}{6X^2 - 4} =$

    $\dfrac{3X(5X + 2)}{2(6X^2 - 4)} = \dfrac{15X^2 + 6X}{12X^2 - 8}$

15. $\dfrac{20\sqrt{15}}{5\sqrt{3}} = \dfrac{20\sqrt{3}\sqrt{5}}{5\sqrt{3}} = \dfrac{20\sqrt{5}}{5} = \dfrac{4\sqrt{5}}{1} = 4\sqrt{5}$

16. $\dfrac{2}{\sqrt{10}} = \dfrac{2\sqrt{10}}{\sqrt{10}\sqrt{10}} = \dfrac{2\sqrt{10}}{\sqrt{100}} =$

    $\dfrac{2\sqrt{10}}{10} = \dfrac{\sqrt{10}}{5}$

17. $9\sqrt{40} = 9\sqrt{4}\sqrt{10} = 9(2)\sqrt{10} = 18\sqrt{10}$

18. $\dfrac{6}{\sqrt{7}} + \dfrac{9}{\sqrt{5}} = \dfrac{6\sqrt{7}}{\sqrt{7}\sqrt{7}} + \dfrac{9\sqrt{5}}{\sqrt{5}\sqrt{5}} =$

    $\dfrac{6\sqrt{7}}{\sqrt{49}} + \dfrac{9\sqrt{5}}{\sqrt{25}} = \dfrac{6\sqrt{7}}{7} + \dfrac{9\sqrt{5}}{5} =$

    $\dfrac{6\sqrt{7}\,(5)}{7(5)} + \dfrac{9\sqrt{5}\,(7)}{5(7)} = \dfrac{30\sqrt{7} + 63\sqrt{5}}{35}$

19. $\dfrac{(26,000)(.00004)}{(1,300,000)(200,000,000)} =$

    $\dfrac{(2.6 \times 10^4)(4 \times 10^{-5})}{(1.3 \times 10^6)(2 \times 10^8)} =$

    $\dfrac{(2.6 \times 4)(10^4 \times 10^{-5})}{(1.3 \times 2)(10^6 \times 10^8)} =$

    $\dfrac{10.4 \times 10^{-1}}{2.6 \times 10^{14}} =$

    $(10.4 \times 10^{-1}) \div (2.6 \times 10^{14}) =$

    $(10.4 \div 2.6)(10^{-1} \div 10^{14}) = 4 \times 10^{-15}$

20. $\dfrac{X^2+7X+10}{X^2+4X+4} = \dfrac{(X+2)(X+5)}{(X+2)(X+2)} = \dfrac{X+5}{X+2}$

18. $\sqrt{\sqrt[4]{81}} = \left(81^{\frac{1}{4}}\right)^{\frac{1}{2}} = 3^{\frac{1}{2}}$ or $\sqrt{3}$

19. $\sqrt{\sqrt{A^{16}}} = \left(\left(A^{16}\right)^{\frac{1}{2}}\right)^{\frac{1}{2}}$

$= A^{(16)\left(\frac{1}{2}\right)\left(\frac{1}{2}\right)} = A^4$

20. $\left(\sqrt[3]{8}\right)^5 = \left(8^{\frac{1}{3}}\right)^5 = 2^5 = 32$

Lesson Practice 6A

1. $\left(16^{\frac{1}{2}}\right)^3 = 4^3 = 64$

2. $\left(X^{\frac{3}{4}}\right)^{\frac{8}{3}} = X^{\left(\frac{3}{4}\right)\left(\frac{8}{3}\right)} = X^{\frac{24}{12}} = X^2$

3. $\left(2^6\right)^{\frac{1}{3}} = 64^{\frac{1}{3}} = 4$

4. $\left((-4)^2\right)^{\frac{3}{4}} = (16)^{\frac{3}{4}} = 2^3 = 8$

5. $\left(3^{-4}\right)^{\frac{1}{2}} = \left(\dfrac{1}{3^4}\right)^{\frac{1}{2}} = \left(\dfrac{1}{81}\right)^{\frac{1}{2}} = \dfrac{1}{9}$

6. $\left[\left(\dfrac{4}{9}\right)^{\frac{1}{2}}\right]^3 = \left[\dfrac{2}{3}\right]^3 = \dfrac{8}{27}$

7. $\left(\dfrac{1}{2}\right)^{-3} = \left(\dfrac{2}{1}\right)^3 = 2^3 = 8$

8. $\left(X^{AB}\right)^{\frac{1}{A}} = X^{(AB)\left(\frac{1}{A}\right)} = X^{\frac{AB}{A}} = X^B$

9. $\left((-6)^2\right)^{\frac{1}{2}} = (36)^{\frac{1}{2}} = 6$

10. $\left(27^{\frac{2}{3}}\right)^2 = \left(3^2\right)^2 = 9^2 = 81$

11. $\sqrt{\sqrt{X}} = \left(X^{\frac{1}{2}}\right)^{\frac{1}{2}} = X^{\frac{1}{4}}$

12. $\left(\sqrt[3]{125}\right)^2 = \left(125^{\frac{1}{3}}\right)^2 = 5^2 = 25$

13. $\sqrt[3]{B}^5 = \left(B^{\frac{1}{3}}\right)^5 = B^{\frac{5}{3}}$

14. $\sqrt{\sqrt[3]{64}} = \left(64^{\frac{1}{3}}\right)^{\frac{1}{2}} = 4^{\frac{1}{2}} = 2$

15. $\left(\sqrt{36}\right)^3 = \left(36^{\frac{1}{2}}\right)^3 = 6^3 = 216$

16. $\sqrt{\sqrt{25}} = \left(25^{\frac{1}{2}}\right)^{\frac{1}{2}} = 5^{\frac{1}{2}}$ or $\sqrt{5}$

17. $\left(\sqrt[6]{64}\right)^{-3} = \left(64^{\frac{1}{6}}\right)^{-3} = 2^{-3} = \dfrac{1}{2^3} = \dfrac{1}{8}$

Lesson Practice 6B

1. $\left(32^{\frac{2}{5}}\right)^2 = \left(2^2\right)^2 = 4^2 = 16$

2. $\left(2^{\frac{2}{3}}\right)^{\frac{1}{4}} = 2^{\left(\frac{2}{3}\right)\left(\frac{1}{4}\right)} = 2^{\frac{2}{12}} = 2^{\frac{1}{6}}$

3. $\left(X^3\right)^{\frac{1}{4}} = X^{(3)\left(\frac{1}{4}\right)} = X^{\frac{3}{4}}$

4. $\left((-3)^3\right)^{\frac{2}{9}} = (-27)^{\frac{2}{9}} = (-27)^{\left(\frac{1}{3}\right)\left(\frac{2}{3}\right)}$

$= \left((-27)^{\frac{1}{3}}\right)^{\frac{2}{3}} = (-3)^{\frac{2}{3}} = (-3)^{(2)\left(\frac{1}{3}\right)}$

$= \left((-3)^2\right)^{\frac{1}{3}} = 9^{\frac{1}{3}}$

5. $\left(2^{-3}\right)^{\frac{1}{3}} = \left(\dfrac{1}{2^3}\right)^{\frac{1}{3}} = \left(\dfrac{1}{8}\right)^{\frac{1}{3}} = \dfrac{1}{2}$

6. $\left[\left(\dfrac{16}{81}\right)^{\frac{1}{8}}\right]^2 = \left(\dfrac{16}{81}\right)^{\frac{1}{4}} = \dfrac{2}{3}$

7. $\left(\dfrac{1}{3}\right)^{-4} = \left(\dfrac{3}{1}\right)^4 = 3^4 = 81$

8. $\left(B^{\frac{Y}{X}}\right)^{\frac{2X}{Y}} = B^{\left(\frac{Y}{X}\right)\left(\frac{2X}{Y}\right)} = B^{\frac{2XY}{XY}} = B^2$

9. $\left((5)^2\right)^{-\frac{1}{2}} = 25^{-\frac{1}{2}} = \left(\dfrac{1}{25}\right)^{\frac{1}{2}} = \dfrac{1}{5}$

10. $\left(9^{\frac{1}{4}}\right)^2 = 9^{\left(\frac{1}{4}\right)(2)} = 9^{\frac{1}{2}} = 3$

11. $\sqrt{\sqrt{X^4}} = \left[\left(X^4\right)^{\frac{1}{2}}\right]^{\frac{1}{2}} = X^{(4)\left(\frac{1}{2}\right)\left(\frac{1}{2}\right)} = X^1 = X$

12. $\left(\sqrt[3]{64}\right)^4 = \left(64^{\frac{1}{3}}\right)^4 = 4^4 = 256$

13. $\sqrt[3]{8^5} = \left(8^5\right)^{\frac{1}{3}} = 8^{\frac{5}{3}} = \left(8^{\frac{1}{3}}\right)^5 = 2^5 = 32$

14. $\sqrt{\sqrt[4]{16}} = \left(16^{\frac{1}{4}}\right)^{\frac{1}{2}} = 2^{\frac{1}{2}}$ or $\sqrt{2}$

15. $\left(\sqrt{49}\right)^2 = \left(49^{\frac{1}{2}}\right)^2 = 7^2 = 49$

16. $\sqrt[4]{A^8} = \left(A^{\frac{1}{4}}\right)^8 = A^{\left(\frac{1}{4}\right)(8)} = A^{\frac{8}{4}} = A^2$

17. $\left(\sqrt[3]{216}\right)^{-2} = \left(216^{\frac{1}{3}}\right)^{-2} = 6^{-2} = \frac{1}{6^2} = \frac{1}{36}$

18. $\sqrt{\sqrt{100}} = \left(100^{\frac{1}{2}}\right)^{\frac{1}{2}} = 10^{\frac{1}{2}}$ or $\sqrt{10}$

19. $\sqrt{\sqrt{81}} = \left(81^{\frac{1}{2}}\right)^{\frac{1}{2}} = 9^{\frac{1}{2}} = 3$

20. $\left(\sqrt[5]{32}\right)^4 = \left(32^{\frac{1}{5}}\right)^4 = 2^4 = 16$

Systematic Review 6C

1. $\left(16^{\frac{1}{4}}\right)^3 = 2^3 = 8$

2. $\left(5^6\right)^{\frac{1}{3}} = 5^{(6)\left(\frac{1}{3}\right)} = 5^{\frac{6}{3}} = 5^2 = 25$

3. $\left(X^{\frac{2}{3}}\right)^3 = X^{\left(\frac{2}{3}\right)(3)} = X^{\frac{6}{3}} = X^2$

4. $\left(100^{\frac{1}{2}}\right)^5 = 10^5 = 100,000$

5. $\sqrt[3]{\sqrt{X}} = \left(X^{\frac{1}{2}}\right)^{\frac{1}{3}} = X^{\left(\frac{1}{2}\right)\left(\frac{1}{3}\right)} = X^{\frac{1}{6}}$

6. $\left(\sqrt[3]{27}\right)^2 = \left(27^{\frac{1}{3}}\right)^2 = 3^2 = 9$

7. $\sqrt{\sqrt[4]{16}} = \left(16^{\frac{1}{4}}\right)^{\frac{1}{2}} = 2^{\frac{1}{2}}$ or $\sqrt{2}$

8. $\left(\sqrt{25}\right)^4 = \left(25^{\frac{1}{2}}\right)^4 = 5^4 = 625$

9. $X^2 - 5X - 14 = (X - 7)(X + 2)$

10. $25X^2 - 1 = (5X - 1)(5X + 1)$

11. $2X^2 - 5X - 3 = (2X + 1)(X - 3)$

12. $\frac{1}{9}X^2 - \frac{36}{25} = \left(\frac{1}{3}X - \frac{6}{5}\right)\left(\frac{1}{3}X + \frac{6}{5}\right)$

13. $5X^2 - 20X - 10 = 5X - 40$

$\quad X^2 - 4X - 2 = X - 8$

$\quad X^2 - 5X + 6 = 0$

$\quad (X - 3)(X - 2) = 0$

$X - 3 = 0 \qquad 5(3)^2 - 20(3) - 10 = 5(3) - 40$

$\quad X = 3 \qquad\quad 5(9) - 60 - 10 = 15 - 40$

$\qquad\qquad\qquad\quad 45 - 60 - 10 = -25$

$\qquad\qquad\qquad\qquad\quad -25 = -25 \text{ ok}$

$X - 2 = 0 \qquad 5(2)^2 - 20(2) - 10 = 5(2) - 40$

$\quad X = 2 \qquad\quad 5(4) - 40 - 10 = 10 - 40$

$\qquad\qquad\qquad\quad 20 - 40 - 10 = -30$

$\qquad\qquad\qquad\qquad\quad -30 = -30 \text{ ok}$

14. $\qquad\qquad X^2 + 25 = -10X$

$\quad X^2 + 10X + 25 = 0$

$\quad (X + 5)(X + 5) = 0$

$X + 5 = 0 \qquad (-5)^2 + 25 = -10(-5)$

$\quad X = -5 \qquad\quad 25 + 25 = 50$

$\qquad\qquad\qquad\qquad 50 = 50 \text{ ok}$

15. $\frac{7X}{X + 2} - \frac{2X}{X + 4} =$

$\frac{7X(X + 4)}{(X + 2)(X + 4)} - \frac{2X(X + 2)}{(X + 4)(X + 2)} =$

$\frac{\left(7X^2 + 28X\right) - \left(2X^2 + 4X\right)}{X^2 + 6X + 8} = \frac{5X^2 + 24X}{X^2 + 6X + 8}$

16. $\frac{3X}{X + 5} - \frac{5X}{X^2 - 25} + \frac{8}{X - 5} =$

$\frac{3X(X - 5)}{(X + 5)(X - 5)} - \frac{5X}{X^2 - 25} + \frac{8(X + 5)}{(X - 5)(X + 5)} =$

$\frac{\left(3X^2 - 15X\right) + (8X + 40)}{X^2 - 25} - \frac{5X}{X^2 - 25} =$

$\frac{3X^2 - 12X + 40}{X^2 - 25}$

17. $\dfrac{4 - \dfrac{1}{X}}{X + \dfrac{1}{2X}} = \dfrac{\dfrac{4X}{X} - \dfrac{1}{X}}{\dfrac{X(2X)}{2X} + \dfrac{1}{2X}} = \dfrac{\dfrac{4X-1}{X}}{\dfrac{2X^2+1}{2X}} =$

$\dfrac{\dfrac{4X-1}{X} \times \dfrac{2X}{2X^2+1}}{\dfrac{2X^2+1}{2X} \times \dfrac{2X}{2X^2+1}} = \dfrac{4X-1}{X} \times \dfrac{2X}{2X^2+1} =$

$\dfrac{(4X-1)(2X)}{(X)(2X^2+1)} = \dfrac{(4X-1)(2)}{2X^2+1} = \dfrac{8X-2}{2X^2+1}$

18. $\dfrac{X^3 - X}{2X^2 + 12X + 18} \div \dfrac{X^2 + 2X + 1}{X^3 - 9X} =$

$\dfrac{(X)(X^2-1)}{(2)(X^2+6X+9)} \times \dfrac{(X)(X^2-9)}{(X+1)(X+1)} =$

$\dfrac{(X)(X-1)\cancel{(X+1)}}{(2)(X+3)\cancel{(X+3)}} \times \dfrac{(X)(X-3)\cancel{(X+3)}}{\cancel{(X+1)}(X+1)} =$

$\dfrac{(X)(X)(X-1)(X-3)}{(2)(X+3)(X+1)} = \dfrac{X^2(X^2-4X+3)}{2(X^2+4X+3)} =$

$\dfrac{X^4 - 4X^3 + 3X^2}{2X^2 + 8X + 6}$

19. $\dfrac{4\sqrt{2}}{\sqrt{5}} = \dfrac{4\sqrt{2}\sqrt{5}}{\sqrt{5}\sqrt{5}} = \dfrac{4\sqrt{10}}{\sqrt{25}} = \dfrac{4\sqrt{10}}{5}$

20. $\dfrac{7}{\sqrt{8}} - \dfrac{8}{\sqrt{9}} = \dfrac{7\sqrt{2}}{\sqrt{8}\sqrt{2}} - \dfrac{8}{3} =$

$\dfrac{7\sqrt{2}}{\sqrt{16}} - \dfrac{8}{3} = \dfrac{7\sqrt{2}}{4} - \dfrac{8}{3} =$

$\dfrac{7\sqrt{2}\,(3)}{4(3)} - \dfrac{8(4)}{3(4)} = \dfrac{21\sqrt{2} - 32}{12}$

Systematic Review 6D

1. $\left(81^{\frac{3}{4}}\right)^2 = \left(3^3\right)^2 = 27^2 = 729$

2. $\left(27^{\frac{2}{3}}\right)^2 = \left(3^2\right)^2 = 9^2 = 81$

3. $\left(16^{\frac{5}{4}}\right)^2 = \left(2^5\right)^2 = 32^2 = 1{,}024$

4. $\left(32^{\frac{2}{5}}\right)^3 = \left(2^2\right)^3 = 4^3 = 64$

5. $\sqrt[3]{\sqrt{64}} = \left(64^{\frac{1}{2}}\right)^{\frac{1}{3}} = 8^{\frac{1}{3}} = 2$

6. $\left(\sqrt[3]{8}\right)^5 = \left(8^{\frac{1}{3}}\right)^5 = 2^5 = 32$

7. $\sqrt{\sqrt[3]{125}} = \left(125^{\frac{1}{3}}\right)^{\frac{1}{2}} = 5^{\frac{1}{2}}$ or $\sqrt{5}$

8. $\left(\sqrt{100}\right)^4 = \left(100^{\frac{1}{2}}\right)^4 = 10^4 = 10{,}000$

9. $X^2 + 6X + 9 = (X+3)(X+3)$

10. $X^2 - 25 = (X-5)(X+5)$

11. $X^2 - 2X - 99 = (X-11)(X+9)$

12. $4X^2 - 12X + 9 = (2X-3)(2X-3)$

13. $X^2 - 42 + 3X = 2X$

$X^2 + X - 42 = 0$

$(X+7)(X-6) = 0$

$X + 7 = 0$ $\qquad (-7)^2 - 42 + 3(-7) = 2(-7)$

$\quad X = -7$ $\qquad\quad 49 - 42 - 21 = -14$

$\qquad\qquad\qquad\qquad\qquad -14 = -14$ ok

$X - 6 = 0$ $\qquad (6)^2 - 42 + 3(6) = 2(6)$

$\quad X = 6$ $\qquad\quad 36 - 42 + 18 = 12$

$\qquad\qquad\qquad\qquad\qquad 12 = 12$ ok

14. $2X + 15 = X^2$

$0 = X^2 - 2X - 15$

$0 = (X-5)(X+3)$

$X - 5 = 0$ $\qquad 2(5) + 15 = (5)^2$

$\quad X = 5$ $\qquad\quad 10 + 15 = 25$

$\qquad\qquad\qquad\qquad 25 = 25$ ok

$X + 3 = 0$ $\qquad 2(-3) + 15 = (-3)^2$

$\quad X = -3$ $\qquad\quad -6 + 15 = 9$

$\qquad\qquad\qquad\qquad 9 = 9$ ok

15. $\dfrac{X+4}{X-8} = 7$

$X + 4 = 7(X-8)$

$X + 4 = 7(X-8)$

$X + 4 = 7X - 56$

$4 + 56 = 7X - X$

$60 = 6X$

$10 = X$

16. $\dfrac{X-2}{X+3} - \dfrac{X-4}{X^2-9} + \dfrac{X+4}{X-3} =$

$\dfrac{(X-2)(X-3)}{(X+3)(X-3)} - \dfrac{X-4}{X^2-9} + \dfrac{(X+4)(X+3)}{(X-3)(X+3)} =$

$\dfrac{\left(X^2-5X+6\right) - (X-4) + \left(X^2+7X+12\right)}{X^2-9} =$

$\dfrac{2X^2+X+22}{X^2-9}$

17. $\dfrac{\dfrac{2}{X}+\dfrac{3}{Y}}{\dfrac{2}{XY}} = \dfrac{\dfrac{2(Y)}{X(Y)}+\dfrac{3(X)}{Y(X)}}{\dfrac{2}{XY}} =$

$\dfrac{\dfrac{2Y+3X}{XY}}{\dfrac{2}{XY}} = \dfrac{\dfrac{2Y+3X}{XY} \times \dfrac{XY}{2}}{\dfrac{2}{XY} \times \dfrac{XY}{2}} =$

$\dfrac{2Y+3X}{\cancel{XY}} \times \dfrac{\cancel{XY}}{2} = \dfrac{2Y+3X}{2}$

18. $\dfrac{X^2-25}{2X^2-18} \div \dfrac{5X-25}{2X^2-6X} =$

$\dfrac{X^2-25}{2X^2-18} \times \dfrac{2X^2-6X}{5X-25} =$

$\dfrac{\cancel{(X-5)}(X+5)}{\cancel{(2)}(X^2-9)} \times \dfrac{\cancel{(2X)}(X-3)}{(5)\cancel{(X-5)}} =$

$\dfrac{X+5}{\cancel{(X-3)}(X+3)} \times \dfrac{(X)\cancel{(X-3)}}{5} =$

$\dfrac{X(X+5)}{(X+3)5} = \dfrac{X^2+5X}{5X+15}$

19. $\dfrac{8}{\sqrt{10}} = \dfrac{8\sqrt{10}}{\sqrt{10}\sqrt{10}} = \dfrac{8\sqrt{10}}{\sqrt{100}} =$

$\dfrac{8\sqrt{10}}{10} = \dfrac{4\sqrt{10}}{5}$

20. $\dfrac{6}{\sqrt{2}} - \dfrac{3}{\sqrt{6}} = \dfrac{6\sqrt{2}}{\sqrt{2}\sqrt{2}} - \dfrac{3\sqrt{6}}{\sqrt{6}\sqrt{6}} =$

$\dfrac{6\sqrt{2}}{\sqrt{4}} - \dfrac{3\sqrt{6}}{\sqrt{36}} = \dfrac{6\sqrt{2}}{2} - \dfrac{3\sqrt{6}}{6} =$

$3\sqrt{2} - \dfrac{\sqrt{6}}{2}$ or $\dfrac{6\sqrt{2}}{2} - \dfrac{\sqrt{6}}{2} = \dfrac{6\sqrt{2}-\sqrt{6}}{2}$

Systematic Review 6E

1. $\left(49^{\frac{1}{2}}\right)^3 = 7^3 = 343$

2. $(125)^{\frac{4}{3}} = 5^4 = 625$

3. $\left(1,000^{\frac{5}{3}}\right) = 10^5 = 100,000$

4. $\left(-32^{\frac{3}{5}}\right)^2 = \left(-2^3\right)^2 = (-8)^2 = 64$

5. $\sqrt{\sqrt{81}} = \left(81^{\frac{1}{2}}\right)^{\frac{1}{2}} = 9^{\frac{1}{2}} = 3$

6. $\left(\sqrt{36}\right)^3 = \left(36^{\frac{1}{2}}\right)^3 = 6^3 = 216$

7. $\sqrt{\sqrt[4]{X^8}} = \left[\left(X^8\right)^{\frac{1}{4}}\right]^{\frac{1}{2}} = X^{(8)\left(\frac{1}{4}\right)\left(\frac{1}{2}\right)}$

$= X^{\frac{8}{8}} = X^1 = X$

8. $\left(\sqrt[3]{1,000}\right)^{-5} = \left(1,000^{\frac{1}{3}}\right)^{-5}$

$= 10^{-5} = \dfrac{1}{10^5} = \dfrac{1}{100,000}$

9. $X^2-4X+4 = (X-2)(X-2)$

10. $X^2+10X+25 = (X+5)(X+5)$

11. $X^2-12X+36 = (X-6)(X-6)$

12. $3X^2+14X-5 = (3X-1)(X+5)$

13. $42-3X^2 = 15X$

$14-X^2 = 5X$

$0 = X^2+5X-14$

$0 = (X+7)(X-2)$

$X+7 = 0 \qquad 42-3(-7)^2 = 15(-7)$
$X = -7 \qquad 42-3(49) = -105$
$\qquad\qquad 42-147 = -105$
$\qquad\qquad -105 = -105 \text{ ok}$

$X-2 = 0 \qquad 42-3(2)^2 = 15(2)$
$X = 2 \qquad 42-3(4) = 30$
$\qquad\qquad 42-12 = 30$
$\qquad\qquad 30 = 30 \text{ ok}$

14.
$$X^2 - 25 = X - 5$$
$$X^2 - X - 20 = 0$$
$$(X - 5)(X + 4) = 0$$

$$X - 5 = 0 \qquad (5)^2 - 25 = (5) - 5$$
$$X = 5 \qquad 25 - 25 = 0$$
$$0 = 0 \text{ ok}$$

$$X + 4 = 0 \qquad (-4)^2 - 25 = (-4) - 5$$
$$X = -4 \qquad 16 - 25 = -9$$
$$-9 = -9 \text{ ok}$$

15.
$$\frac{8}{3X} - \frac{2}{2X} - \frac{5}{6X^2} =$$

$$\frac{8(2X)}{3X(2X)} - \frac{2(3X)}{2X(3X)} - \frac{5}{6X^2} =$$

$$\frac{16X - 6X - 5}{6X^2} = \frac{10X - 5}{6X^2}$$

16.
$$\frac{X - 3}{X - 2} - \frac{4X + 3}{X^2 - 4} - \frac{X + 3}{X + 2} =$$

$$\frac{(X - 3)(X + 2)}{(X - 2)(X + 2)} - \frac{4X + 3}{X^2 - 4} - \frac{(X + 3)(X - 2)}{(X + 2)(X - 2)} =$$

$$\frac{(X^2 - X - 6) - (4X + 3) - (X^2 + X - 6)}{X^2 - 4} =$$

$$\frac{-6X - 3}{X^2 - 4}$$

17.
$$\frac{\dfrac{1}{9} - \dfrac{X}{3}}{\dfrac{X}{12} + \dfrac{5}{8}} = \frac{\dfrac{1}{9} - \dfrac{3X}{9}}{\dfrac{2X}{24} + \dfrac{15}{24}} = \frac{\dfrac{1 - 3X}{9}}{\dfrac{2X + 15}{24}} =$$

$$\frac{\dfrac{1 - 3X}{9} \times \dfrac{24}{2X - 15}}{\dfrac{2X + 15}{24} \times \dfrac{24}{2X + 15}} =$$

$$\frac{1 - 3X}{9_3} \times \frac{\overset{8}{\cancel{24}}}{2X + 15} =$$

$$\frac{8(1 - 3X)}{3(2X + 15)} = \frac{8 - 24X}{6X + 45}$$

18.
$$\frac{X^2 - 6X - 16}{X + 2} \div \frac{X^2 - 8X + 16}{X - 4} =$$

$$\frac{X^2 - 6X - 16}{X + 2} \times \frac{X - 4}{X^2 - 8X + 16} =$$

$$\frac{(X - 8)\cancel{(X + 2)}}{\cancel{(X + 2)}} \times \frac{\cancel{(X - 4)}}{(X - 4)\cancel{(X - 4)}} =$$

$$\frac{X - 8}{X - 4}$$

19.
$$\frac{\sqrt{5}}{\sqrt{3}} = \frac{\sqrt{5}\sqrt{3}}{\sqrt{3}\sqrt{3}} = \frac{\sqrt{15}}{\sqrt{9}} = \frac{\sqrt{15}}{3}$$

20.
$$\frac{1}{\sqrt{7}} - \frac{2}{\sqrt{8}} = \frac{\sqrt{7}}{\sqrt{7}\sqrt{7}} - \frac{2\sqrt{2}}{\sqrt{8}\sqrt{2}} =$$

$$\frac{\sqrt{7}}{\sqrt{49}} - \frac{2\sqrt{2}}{\sqrt{16}} = \frac{\sqrt{7}}{7} - \frac{2\sqrt{2}}{4} =$$

$$\frac{\sqrt{7}}{7} - \frac{\sqrt{2}}{2} = \frac{\sqrt{7}(2)}{7(2)} - \frac{\sqrt{2}(7)}{2(7)} =$$

$$\frac{2\sqrt{7}}{14} - \frac{7\sqrt{2}}{14} = \frac{2\sqrt{7} - 7\sqrt{2}}{14}$$

Lesson Practice 7A

1. $\sqrt{-1} = i$

2. $\sqrt{-49} = \sqrt{49}\sqrt{-1} = 7i$

3. $\sqrt{-64X^6} = \sqrt{64}\sqrt{X^3X^3}\sqrt{-1} = 8X^3i$

4. $\sqrt{\dfrac{-121}{144}} = \sqrt{\dfrac{121}{144}}\sqrt{-1} = \dfrac{11}{12}i$

5. $\sqrt{-4} + \sqrt{-100} = 2i + 10i = 12i$

6. $2\sqrt{-9} + \sqrt{36} = 2(3)i + 6 = 6i + 6$

7. $\sqrt{-20X^2} = \sqrt{4}\sqrt{X^2}\sqrt{-1}\sqrt{5} = 2Xi\sqrt{5}$

8. $\sqrt{-A} + \sqrt{-B} = i\sqrt{A} + i\sqrt{B}$

9. $3\sqrt{-12} + 4\sqrt{-162} =$
$3\sqrt{4}\sqrt{-1}\sqrt{3} + 4\sqrt{81}\sqrt{-1}\sqrt{2} =$
$3(2)i\sqrt{3} + 4(9)i\sqrt{2} = 6i\sqrt{3} + 36i\sqrt{2}$

10. $13\sqrt{-1} - 2\sqrt{-81} = 13i - 2\sqrt{81}\sqrt{-1} =$
$13i - 2(9)i = 13i - 18i = -5i$

11. $2\sqrt{-25} + \sqrt{16} = 2\sqrt{25}i + 4 =$
$2(5)i + 4 = 10i + 4$

12. $\sqrt{3X^2} + \sqrt{4i^2} =$
$\sqrt{X^2}\sqrt{3} + \sqrt{4}\sqrt{i^2} = X\sqrt{3} + 2i$

13. $i \cdot i \cdot i \cdot i = (i^2)(i^2) = (-1)(-1) = 1$

14. $i \cdot i \cdot i \cdot i \cdot i \cdot i^3 = i^8 =$
$(i^2)(i^2)(i^2)(i^2) = (-1)^4 = 1$

15. $i^5 = i \cdot i \cdot i \cdot i \cdot i = (i^2)(i^2) \cdot i = (-1)^2 i = 1i = i$

16. $(i^3)^3 = i^{(3)(3)} = i^9 =$
$(i^2)(i^2)(i^2)(i^2)i = (-1)^4 i = 1i = i$

17. $(15i)(-8i) = (15)(-8)(i)(i) =$
$(-120)(-1) = 120$

18. $3i\sqrt{-169} = 3i\sqrt{169}\sqrt{-1} = 3i(13)(i) =$
$(3)(13)(i)(i) = 39(-1) = -39$

19. $\sqrt{-6}\sqrt{-6} = \sqrt{6}\sqrt{6}\sqrt{-1}\sqrt{-1} =$
$\sqrt{36}(i)(i) = 6(-1) = -6$

20. $\left(2\sqrt{225}\right)\left(6\sqrt{-4}\right) = 2(15)(6)\sqrt{4}\sqrt{-1} =$
$180(2)i = 360i$

17. $(-10i)(-5i) = (-10)(-5)(i)(i) = 50(-1) = -50$

18. $14i\sqrt{-1} = 14(i)(i) = 14(-1) = -14$

19. $\sqrt{-75}\sqrt{-75} = \sqrt{75}\sqrt{75}\sqrt{-1}\sqrt{-1} =$
$(75)(-1) = -75$

20. $\left(6\sqrt{-169}\right)\left(2\sqrt{-81}\right) =$
$(6)(2)\sqrt{169}\sqrt{81}\sqrt{-1}\sqrt{-1} =$
$12(13)(9)(i)(i) = 1,404(-1) = -1,404$

Lesson Practice 7B

1. $\sqrt{-225} = \sqrt{225}\sqrt{-1} = 15i$

2. $\sqrt{-121} = \sqrt{121}\sqrt{-1} = 11i$

3. $\sqrt{-49A^4} = \sqrt{49}\sqrt{A^4}\sqrt{-1} = 7A^2i$

4. $\sqrt{\dfrac{-100}{25}} = \sqrt{\dfrac{100}{25}}\sqrt{-1} = \dfrac{10}{5}i = 2i$

5. $\sqrt{-64} - \sqrt{-16} =$
$\sqrt{64}\sqrt{-1} - \sqrt{16}\sqrt{-1} = 8i - 4i = 4i$

6. $3\sqrt{36} - 2\sqrt{-4} = 3(6) - 2\sqrt{4}\sqrt{-1} =$
$18 - 2(2)i = 18 - 4i$

7. $\sqrt{-45X^9} = \sqrt{9}\sqrt{5}\sqrt{-1}\sqrt{X^4X^4X^1} = 3X^4i\sqrt{5X}$

8. $\sqrt{-X^2Y^4} + \sqrt{X^2Y^4} =$
$\sqrt{X^2}\sqrt{Y^4}\sqrt{-1} + \sqrt{X^2}\sqrt{Y^4} = XY^2i + XY^2$

9. $6\sqrt{-200} - 5\sqrt{25} = 6\sqrt{100}\sqrt{-1}\sqrt{2} - 5(5) =$
$6(10)i\sqrt{2} - 25 = 60i\sqrt{2} - 25$

10. $4\sqrt{-2} + 2\sqrt{-50} = 4\sqrt{-1}\sqrt{2} + 2\sqrt{25}\sqrt{-1}\sqrt{2} =$
$4i\sqrt{2} + 2(5)i\sqrt{2} = 4i\sqrt{2} + 10i\sqrt{2} = 14i\sqrt{2}$

11. $A\sqrt{-9} + A\sqrt{-81} = A\sqrt{9}\sqrt{-1} + A\sqrt{81}\sqrt{-1} =$
$A(3)i + A(9)i = 3Ai + 9Ai = 12Ai$

12. $\sqrt{-X^4} + \sqrt{16X^4i^2} = \sqrt{X^4}\sqrt{-1} + \sqrt{16}\sqrt{X^4}\sqrt{i^2} =$
$X^2i + 4X^2i = 5X^2i$

13. $2i^2 \cdot 3i^2 = 2(-1) \cdot 3(-1) = (-2)(-3) = 6$

14. $i^2 \cdot i^3 \cdot i^5 = i^{10} = (i^2)^5 = (i^2)(i^2)(i^2)(i^2)(i^2) =$
$(-1)^4(-1)^1 = 1^2(-1) = 1(-1) = -1$

15. $i^7 = i^6i = -1^3i = (-1)^2(-1)i = 1(-1)i = -1i = -i$

16. $\left(i^4\right)^2 = i^8 = (-1)^4 = 1^2 = 1$

Systematic Review 7C

1. $\sqrt{-81} = \sqrt{81}\sqrt{-1} = 9i$

2. $\sqrt{-169} = \sqrt{169}\sqrt{-1} = 13i$

3. $\sqrt{-64X^2} = \sqrt{64}\sqrt{X^2}\sqrt{-1} = 8Xi$

4. $\sqrt{\dfrac{-16}{25}} = \sqrt{\dfrac{16}{25}}\sqrt{-1} = \dfrac{4}{5}i$

5. $\sqrt{-4} + \sqrt{-36} = \sqrt{4}\sqrt{-1} + \sqrt{36}\sqrt{-1} =$
$2i + 6i = 8i$

6. $\sqrt{-9} + \sqrt{100} = \sqrt{9}\sqrt{-1} + 10 = 3i + 10$

7. $5\sqrt{-8} + 7\sqrt{-242} =$
$5\sqrt{4}\sqrt{-1}\sqrt{2} + 7\sqrt{121}\sqrt{-1}\sqrt{2} =$
$5(2)i\sqrt{2} + 7(11)i\sqrt{2} =$
$10i\sqrt{2} + 77i\sqrt{2} = 87i\sqrt{2}$

8. $(18i)(-7i) = (18)(-7)(i)(i) = -126(-1) = 126$

9. $(i \cdot i \cdot i) = i^2 \cdot i = -1i = -i$

10. $\left(4\sqrt{-196}\right)\left(3\sqrt{49}\right) = (4)(3)\sqrt{196}\sqrt{49}\sqrt{-1} =$
$12(14)(7)(i) = 1,176i$

11. $\left(125^{\frac{1}{3}}\right)^2\left(25^{\frac{1}{2}}\right)^3 = 5^2 5^3 = 5^5 \text{ or } 3,125$

12. $\left(X^4\right)^{\frac{1}{3}}\left(X^2\right)^{\frac{2}{3}} = X^{(4)\left(\frac{1}{3}\right)}X^{(2)\left(\frac{2}{3}\right)} =$
$X^{\frac{4}{3}}X^{\frac{4}{3}} = X^{\frac{4}{3}+\frac{4}{3}} = X^{\frac{8}{3}}$

13. $\left(\sqrt[4]{10,000}\right)^3 = \left(10,000^{\frac{1}{4}}\right)^3 = 10^3 = 1,000$

14. $\sqrt[3]{\sqrt{X^4}} = \left[\left(X^4\right)^{\frac{1}{2}}\right]^{\frac{1}{3}} = X^{(4)\left(\frac{1}{2}\right)\left(\frac{1}{3}\right)} = X^{\frac{4}{6}} = X^{\frac{2}{3}}$

15.

$$9X - 3 = -2X^2 + 8$$
$$2X^2 + 9X - 11 = 0$$
$$(2X + 11)(X - 1) = 0$$

$$9\left(-\frac{11}{2}\right) - 3 = -2\left(-\frac{11}{2}\right)^2 + 8$$

$$2X + 11 = 0$$
$$2X = -11$$
$$X = -\frac{11}{2}$$

$$-\frac{99}{2} - \frac{6}{2} = -2\left(\frac{121}{4}\right) + 8$$

$$-\frac{99}{2} - \frac{6}{2} = \frac{-242}{4} + \frac{32}{4}$$

$$-\frac{105}{2} = -\frac{210}{4}$$

$$-\frac{105}{2} = -\frac{105}{2} \text{ ok}$$

$$X - 1 = 0 \qquad 9(1) - 3 = -2(1)^2 + 8$$
$$X = 1 \qquad 9 - 3 = -2(1) + 8$$
$$6 = -2 + 8$$
$$6 = 6 \text{ ok}$$

16.

$$\frac{1}{4}X^2 = 9$$

$$\frac{1}{4}X^2 - 9 = 0$$

$$\left(\frac{1}{2}X + 3\right)\left(\frac{1}{2}X - 3\right) = 0$$

$$\frac{1}{2}X + 3 = 0 \qquad \frac{1}{4}(-6)^2 = 9$$
$$\frac{1}{2}X = -3 \qquad \frac{1}{4}(36) = 9$$
$$X = -3(2) \qquad\qquad 9 = 9 \text{ ok}$$
$$X = -6$$

$$\frac{1}{2}X - 3 = 0 \qquad \frac{1}{4}(6)^2 = 9$$
$$\frac{1}{2}X = 3 \qquad \frac{1}{4}(36) = 9$$
$$X = 3(2) \qquad\qquad 9 = 9 \text{ ok}$$
$$X = 6$$

17.

$$\frac{2X^2}{X^2 - 16} \div \frac{X}{4 - X} = \frac{2X^2}{X^2 - 16} \times \frac{4 - X}{X} =$$

$$\frac{2X^2}{(X - 4)(X + 4)} \times \frac{(-1)(-4 + X)}{X} =$$

$$\frac{2X^2}{(X - 4)(X + 4)} \times \frac{(-1)(X - 4)}{X} =$$

$$\frac{(-1)2X^2}{(X + 4)(X)} = \frac{-2X}{X + 4}$$

18.

$$\sqrt{\frac{4}{5}} - \sqrt{\frac{1}{2}} = \frac{\sqrt{4}}{\sqrt{5}} - \frac{\sqrt{1}}{\sqrt{2}} = \frac{2}{\sqrt{5}} - \frac{1}{\sqrt{2}} =$$

$$\frac{2\sqrt{5}}{\sqrt{5}\sqrt{5}} - \frac{1\sqrt{2}}{\sqrt{2}\sqrt{2}} = \frac{2\sqrt{5}}{\sqrt{25}} - \frac{\sqrt{2}}{\sqrt{4}} =$$

$$\frac{2\sqrt{5}}{5} - \frac{\sqrt{2}}{2} = \frac{2\sqrt{5}(2)}{5(2)} - \frac{\sqrt{2}(5)}{2(5)} =$$

$$\frac{4\sqrt{5}}{10} - \frac{5\sqrt{2}}{10} = \frac{4\sqrt{5} - 5\sqrt{2}}{10}$$

19.

$$(1,400)(.00021) \div (.49) =$$
$$(1.4 \times 10^3)(2.1 \times 10^{-4}) \div (4.9 \times 10^{-1}) =$$
$$(1.4 \times 2.1 \div 4.9)(10^3 \times 10^{-4} \div 10^{-1}) =$$
$$.6 \times 10^0 = 6 \times 10^{-1}$$

20.

$$\frac{4^{-1}X^2Y^{-3}}{X^{-1}Y} - \frac{3^2YXY^0}{X^2Y^2} + \frac{2^2Y^{-2}}{XX^{-2}} =$$

$$4^{-1}X^2Y^{-3}X^1Y^{-1} - \frac{3^2YXY^0}{X^2Y^2} + \frac{2^2Y^{-2}}{XX^{-2}} =$$

$$4^{-1}X^3Y^{-4} - 3^2Y^1X^1Y^0X^{-2}Y^{-2} + 2^2Y^{-2}X^{-1}X^2 =$$

$$4^{-1}X^3Y^{-4} - 3^2X^{-1}Y^{-1} + 2^2X^1Y^{-2} =$$

$$4^{-1}X^3Y^{-4} - 9X^{-1}Y^{-1} + 4X^1Y^{-2} \text{ or}$$

$$\frac{X^3}{4Y^4} - \frac{9}{XY} + \frac{4X}{Y^2}$$

Systematic Review 7D

1. $\sqrt{-16} = \sqrt{16}\sqrt{-1} = 4i$

2. $\sqrt{144} = 12$

3. $\sqrt{25X^4} = \sqrt{25}\sqrt{X^4} = 5X^2$

4. $\sqrt{\frac{-16}{25}} = \sqrt{\frac{16}{25}}\sqrt{-1} = \frac{4}{5}i$

5. $\sqrt{-9} + \sqrt{-81} = \sqrt{9}\sqrt{-1} + \sqrt{81}\sqrt{-1} = 3i + 9i = 12i$

6. $\sqrt{16} + \sqrt{-36} = 4 + \sqrt{36}\sqrt{-1} = 4 + 6i$

7. $\sqrt{20} + 2\sqrt{45} = \sqrt{4}\sqrt{5} + 2\sqrt{9}\sqrt{5} =$
$2\sqrt{5} + 2(3)\sqrt{5} = 2\sqrt{5} + 6\sqrt{5} = 8\sqrt{5}$

8. $(9i)(-8i) = (9)(-8)(i)(i) = -72(-1) = 72$

9. $(2i \cdot 2i) = (2)(2)(i)(i) = 4(-1) = -4$

10. $\left(7\sqrt{-64}\right)\left(2\sqrt{-81}\right) =$
$(7)(2)\sqrt{64}\sqrt{81}\sqrt{-1}\sqrt{-1} =$
$14(8)(9)(i)(i) = 1{,}008(-1) = -1{,}008$

11. $(343)^{\frac{2}{3}}(8)^{\frac{2}{3}} = 7^2 2^2 = (49)(4) = 196$

12. $(1{,}000)^{\frac{1}{3}}(10{,}000)^{\frac{2}{4}} =$
$10 \cdot 10^2 = 10^3 = 1{,}000$

13. $\left(\sqrt{8{,}100}\right)^{-1} = \left(8{,}100^{\frac{1}{2}}\right)^{-1} = 90^{-1} = \frac{1}{90}$

14. $\sqrt{\sqrt[5]{32}} = \left(32^{\frac{1}{5}}\right)^{\frac{1}{2}} = 2^{\frac{1}{2}}$ or $\sqrt{2}$

15. $\frac{1}{9}X^2 + \frac{25}{9} = \frac{10}{9}X$
$X^2 + 25 = 10X$
$X^2 - 10X + 25 = 0$
$(X - 5)(X - 5) = 0$

$\begin{array}{ll} X - 5 = 0 & \frac{1}{9}(5)^2 + \frac{25}{9} = \frac{10}{9}(5) \\ X = 5 & \frac{1}{9}(25) + \frac{25}{9} = \frac{50}{9} \\ & \frac{25}{9} + \frac{25}{9} = \frac{50}{9} \\ & \frac{50}{9} = \frac{50}{9} \text{ ok} \end{array}$

16. $8X^2 - 40X = -50$
$4X^2 - 20X = -25$
$4X^2 - 20X + 25 = 0$
$(2X - 5)(2X - 5) = 0$

$\begin{array}{ll} 2X - 5 = 0 & 8\left(\frac{5}{2}\right)^2 - 40\left(\frac{5}{2}\right) = -50 \\ 2X = 5 & \\ X = \frac{5}{2} & 8\left(\frac{25}{4}\right) - \frac{200}{2} = -50 \\ & \frac{200}{4} - 100 = -50 \\ & 50 - 100 = -50 \\ & -50 = -50 \text{ ok} \end{array}$

17. $\frac{X - 5}{X^2 - 10X + 25} \div \frac{X + 6}{X^2 - 3X - 10} =$
$\frac{(X - 5)}{(X - 5)(X - 5)} \div \frac{(X + 6)}{(X - 5)(X + 2)} =$
$\frac{\cancel{(X - 5)}}{\cancel{(X - 5)}(X - 5)} \times \frac{\cancel{(X - 5)}(X + 2)}{(X + 6)} = \frac{X + 2}{X + 6}$

18. $\sqrt{\frac{2}{3}} - \sqrt{\frac{3}{5}} = \frac{\sqrt{2}}{\sqrt{3}} - \frac{\sqrt{3}}{\sqrt{5}}$
$= \frac{\sqrt{2}\sqrt{3}}{\sqrt{3}\sqrt{3}} - \frac{\sqrt{3}\sqrt{5}}{\sqrt{5}\sqrt{5}} = \frac{\sqrt{6}}{\sqrt{9}} - \frac{\sqrt{15}}{\sqrt{25}}$
$= \frac{\sqrt{6}}{3} - \frac{\sqrt{15}}{5} = \frac{\sqrt{6}(5)}{3(5)} - \frac{\sqrt{15}(3)}{5(3)}$
$= \frac{5\sqrt{6} - 3\sqrt{15}}{15}$

19. $(.03)(60{,}000{,}000)(400) =$
$\left(3 \times 10^{-2}\right)\left(6 \times 10^7\right)\left(4 \times 10^2\right) =$
$(3 \times 6 \times 4)\left(10^{-2} \times 10^7 \times 10^2\right) =$
$\qquad 72 \times 10^7 = 7.2 \times 10^8$

20. $\frac{3X^2A}{X} - \frac{7X^{-2}A}{X^{-3}} - 5XA =$
$\frac{3XA}{1} - 7X^{-2}X^3A - 5XA =$
$\qquad 3XA - 7XA - 5XA = -9XA$

Systematic Review 7E

1. $\sqrt{-4} = \sqrt{4}\sqrt{-1} = 2i$

2. $\sqrt{-121} = \sqrt{121}\sqrt{-1} = 11i$

3. $\sqrt{-X^2} = \sqrt{X^2}\sqrt{-1} = Xi$

4. $\sqrt{\frac{-81}{4}} = \sqrt{\frac{81}{4}}\sqrt{-1} = \frac{9}{2}i$ or $4.5i$

5. $\sqrt{-16} + \sqrt{25} = \sqrt{16}\sqrt{-1} + 5 = 4i + 5$

6. $\sqrt{-81} + \sqrt{-1} = \sqrt{81}\sqrt{-1} + i = 9i + i = 10i$

7. $5\sqrt{-12} + 7\sqrt{-75} =$
$5\sqrt{4}\sqrt{-1}\sqrt{3} + 7\sqrt{25}\sqrt{-1}\sqrt{3} =$
$5(2)i\sqrt{3} + 7(5)i\sqrt{3} = 10i\sqrt{3} + 35i\sqrt{3} = 45i\sqrt{3}$

8. $(10i)(10i)(2i) = (10)(10)(2)(i)(i)(i) =$
$200i^2 i = 200(-1)i = -200i$

9. $(i \cdot i \cdot i \cdot 3i) = (i)(i)(i)(i)(3) =$
$(-1)(-1)(3) = 1(3) = 3$

10. $\left(6\sqrt{25}\right)\left(5\sqrt{-16}\right) = (6)(5)\sqrt{25}\sqrt{16}\sqrt{-1}$

$\qquad = 30(5)(4)i = 600i$

11. $\left(x^3\right)^{\frac{2}{3}}\left(x^5\right)^{\frac{4}{5}} = x^{(3)\left(\frac{2}{3}\right)}x^{(5)\left(\frac{4}{5}\right)}$

$\qquad = x^{\frac{6}{3}}x^{\frac{20}{5}} = x^2x^4 = x^6$

12. $\left(x^0\right)^2\left(x^{\frac{3}{3}}\right)^{\frac{1}{3}} = x^{(0)(2)}x^{\left(\frac{3}{3}\right)\left(\frac{1}{3}\right)}$

$\qquad = x^0x^{\frac{3}{9}} = x^0x^{\frac{1}{3}}$

$\qquad = x^{0+\frac{1}{3}} = x^{\frac{1}{3}}$ or $\sqrt[3]{x}$

13. $\left(\sqrt[3]{8}\right)^{-2} = \left(8^{\frac{1}{3}}\right)^{-2} = 2^{-2} = \frac{1}{2^2} = \frac{1}{4}$

14. $\sqrt[3]{\sqrt{64}} = \left(64^{\frac{1}{2}}\right)^{\frac{1}{3}} = 8^{\frac{1}{3}} = 2$

15. $\left(\frac{4}{25}\right)x^2 = 1$

$\qquad 4x^2 = 25$

$\qquad 4x^2 - 25 = 0$

$\qquad (2x-5)(2x+5) = 0$

$2x - 5 = 0 \qquad \frac{4}{25}\left(\frac{5}{2}\right)^2 = 1$

$\quad 2x = 5 \qquad \frac{4}{25}\left(\frac{25}{4}\right) = 1$

$\qquad x = \frac{5}{2} \qquad \frac{100}{100} = 1$

$\qquad\qquad\qquad\qquad 1 = 1 \;\; ok$

$2x + 5 = 0 \qquad \frac{4}{25}\left(-\frac{5}{2}\right)^2 = 1$

$\quad 2x = -5 \qquad \frac{4}{25}\left(\frac{25}{4}\right) = 1$

$\qquad x = -\frac{5}{2} \qquad \frac{100}{100} = 1$

$\qquad\qquad\qquad\qquad 1 = 1 \;\; ok$

16. $\frac{9}{4}x^2 - 4 = 0$

$\qquad 9x^2 - 16 = 0$

$(3x-4)(3x+4) = 0$

$3x - 4 = 0 \qquad \frac{9}{4}\left(\frac{4}{3}\right)^2 - 4 = 0$

$\quad 3x = 4 \qquad\quad \frac{9}{4}\left(\frac{16}{9}\right) - 4 = 0$

$\qquad x = \frac{4}{3} \qquad\qquad \frac{144}{36} - 4 = 0$

$\qquad\qquad\qquad\qquad\quad 4 - 4 = 0$

$\qquad\qquad\qquad\qquad\qquad 0 = 0 \;\; ok$

$3x + 4 = 0 \qquad \frac{9}{4}\left(-\frac{4}{3}\right)^2 - 4 = 0$

$\quad 3x = -4 \qquad \frac{9}{4}\left(\frac{16}{9}\right) - 4 = 0$

$\qquad x = -\frac{4}{3} \qquad\qquad \frac{144}{36} - 4 = 0$

$\qquad\qquad\qquad\qquad\quad 4 - 4 = 0$

$\qquad\qquad\qquad\qquad\qquad 0 = 0 \;\; ok$

17. $\frac{2x^2 + 2x - 4}{5x - 5} \div \frac{6x^2 - 6x - 36}{3x + 15} =$

$\qquad \frac{2(x^2 + x - 2)}{5(x - 1)} \div \frac{6(x^2 - x - 6)}{3(x + 5)} =$

$\qquad \frac{2\cancel{(x+2)}\cancel{(x-1)}}{5\cancel{(x-1)}} \times \frac{3(x+5)}{6(x-3)\cancel{(x+2)}} =$

$\qquad \frac{\cancel{2}\times\cancel{3}(x+5)}{5\times\cancel{6}(x-3)} = \frac{(x+5)}{5(x-3)} = \frac{x+5}{5x-15}$

18. $\sqrt{\frac{4}{7}} - \sqrt{\frac{1}{4}} = \frac{\sqrt{4}}{\sqrt{7}} - \frac{\sqrt{1}}{\sqrt{4}} = \frac{2}{\sqrt{7}} - \frac{1}{2} =$

$\qquad \frac{2\sqrt{7}}{\sqrt{7}\sqrt{7}} - \frac{1}{2} = \frac{2\sqrt{7}}{\sqrt{49}} - \frac{1}{2} = \frac{2\sqrt{7}}{7} - \frac{1}{2} =$

$\qquad \frac{2\sqrt{7}(2)}{7(2)} - \frac{1(7)}{2(7)} = \frac{4\sqrt{7} - 7}{14}$

19. $(.0000007)(.0018) \div (3,000) =$

$\qquad \left(7 \times 10^{-7}\right)\left(1.8 \times 10^{-3}\right) \div \left(3 \times 10^3\right) =$

$\qquad \left(7 \times 1.8 \div 3\right)\left(10^{-7} \times 10^{-3} \div 10^3\right) =$

$\qquad 4.2 \times 10^{-13}$

20. $-\dfrac{4X}{A} - \dfrac{A^1A^0}{A^2X^{-1}} + \dfrac{5A^{-2}}{X^1} =$

$-\dfrac{4X}{A} - A^1A^0A^{-2}X^1 + \dfrac{5}{A^2X} =$

$-\dfrac{4X}{A} - A^{-1}X + \dfrac{5}{A^2X} = -\dfrac{4X}{A} - \dfrac{X}{A} + \dfrac{5}{A^2X} =$

$\dfrac{-4X - X}{A} + \dfrac{5}{A^2X} = \dfrac{-5X}{A} + \dfrac{5}{A^2X}$

Lesson Practice 8A

1. $A - B$
2. $3X + 8$
3. $6 - \sqrt{2}$
4. $1 + 5i$
5. $(2B + 4)(2B - 4) = 4B^2 - 16$
6. $(3 + 2i)(3 - 2i) = 9 - 4i^2 =$
   $9 - 4(-1) = 9 + 4 = 13$
7. $(2 + 7i)(2 - 7i) = 4 - 49i^2 =$
   $4 - 49(-1) = 4 + 49 = 53$
8. $\left(4 + \sqrt{7}\right)\left(4 - \sqrt{7}\right) = 16 - 7 = 9$
9. $\dfrac{X}{3 + 4i} = \dfrac{X(3 - 4i)}{(3 + 4i)(3 - 4i)} = \dfrac{3X - 4Xi}{9 - 16i^2} =$
   $\dfrac{3X - 4Xi}{9 - 16(-1)} = \dfrac{3X - 4Xi}{9 + 16} = \dfrac{3X - 4Xi}{25}$
10. $\dfrac{11}{2 + i} = \dfrac{11(2 - i)}{(2 + i)(2 - i)} = \dfrac{22 - 11i}{4 - (-1)} =$
    $\dfrac{22 - 11i}{4 + 1} = \dfrac{22 - 11i}{5}$
11. $\dfrac{4i}{6 - 3i} = \dfrac{4i(6 + 3i)}{(6 - 3i)(6 + 3i)} = \dfrac{24i + 12i^2}{36 - (-9)} =$
    $\dfrac{24i + (-12)}{36 + 9} = \dfrac{24i - 12}{45} = \dfrac{8i - 4}{15}$
12. $\dfrac{i^2}{4 + 5i} = \dfrac{-1(4 - 5i)}{(4 + 5i)(4 - 5i)} =$
    $\dfrac{-4 + 5i}{16 - (-25)} = \dfrac{-4 + 5i}{41}$
13. $\dfrac{Z}{Z + \sqrt{5}} = \dfrac{Z(Z - \sqrt{5})}{(Z + \sqrt{5})(Z - \sqrt{5})} = \dfrac{Z^2 - Z\sqrt{5}}{Z^2 - 5}$

14. $\dfrac{8}{8 - \sqrt{8}} = \dfrac{8(8 + \sqrt{8})}{(8 - \sqrt{8})(8 + \sqrt{8})} =$

$\dfrac{64 + 8\sqrt{4}\sqrt{2}}{64 - 8} = \dfrac{64 + 16\sqrt{2}}{56} =$

$\dfrac{8(8 + 2\sqrt{2})}{8(7)} = \dfrac{8 + 2\sqrt{2}}{7}$

15. $\dfrac{7X}{2 - 2\sqrt{X}} = \dfrac{7X(2 + 2\sqrt{X})}{(2 - 2\sqrt{X})(2 + 2\sqrt{X})} =$

$\dfrac{14X + 14X\sqrt{X}}{4 - 4X} = \dfrac{2(7X + 7X\sqrt{X})}{2(2 - 2X)} =$

$\dfrac{7X + 7X\sqrt{X}}{2 - 2X}$

16. $\dfrac{3}{8i + i\sqrt{2}} = \dfrac{3(8i - i\sqrt{2})}{(8i + i\sqrt{2})(8i - i\sqrt{2})} =$

$\dfrac{24i - 3i\sqrt{2}}{-64 - (-2)} = \dfrac{24i - 3i\sqrt{2}}{-62}$

Lesson Practice 8B

1. $X^2 - Y^2$
2. $7X - 4$
3. $-3 + \sqrt{3}$
4. $5 - 3i^2$
5. $(3 + 8X)(3 - 8X) = 9 - 64X^2$
6. $(6 - 3i^5)(6 + 3i^5) = 36 - 9i^{10} = 36 - 9(i^2)^5 =$
   $36 - 9(-1)^5 = 36 - (-9) = 36 + 9 = 45$
7. $(2 + i)(2 - i) = 4 - (-1) = 4 + 1 = 5$
8. $\left(4 - 5\sqrt{2}\right)\left(4 + 5\sqrt{2}\right) =$
   $16 - 25(2) = 16 - 50 = -34$
9. $\dfrac{A}{4A + i} = \dfrac{A(4A - i)}{(4A + i)(4A - i)} =$
   $\dfrac{4A^2 - Ai}{16A^2 - (-1)} = \dfrac{4A^2 - Ai}{16A^2 + 1}$
10. $\dfrac{9}{3 - i} = \dfrac{9(3 + i)}{(3 - i)(3 + i)} = \dfrac{27 + 9i}{9 - (-1)} = \dfrac{27 + 9i}{10}$
11. $\dfrac{7i^2}{5 - 6i} = \dfrac{7i^2(5 + 6i)}{(5 - 6i)(5 + 6i)} =$
    $\dfrac{7(-1)(5 + 6i)}{25 - (-36)} = \dfrac{-7(5 + 6i)}{25 + 36} = \dfrac{-35 - 42i}{61}$

12. $\dfrac{3i}{2+8i} = \dfrac{3i(2-8i)}{(2+8i)(2-8i)} =$

$\dfrac{6i-24i^2}{4-(-64)} = \dfrac{6i-24(-1)}{4+64} =$

$\dfrac{6i+24}{68} = \dfrac{2(3i+12)}{2(34)} = \dfrac{3i+12}{34}$

13. $\dfrac{X^2}{X-\sqrt{4X}} = \dfrac{X^2\left(X+\sqrt{4X}\right)}{\left(X-\sqrt{4X}\right)\left(X+\sqrt{4X}\right)} =$

$\dfrac{X^3+X^2\sqrt{4X}}{X^2-4X} = \dfrac{X\left(X^2+X\sqrt{4X}\right)}{X(X-4)} =$

$\dfrac{X^2+X\sqrt{4X}}{X-4} = \dfrac{X^2+X\sqrt{4}\sqrt{X}}{X-4} = \dfrac{X^2+2X\sqrt{X}}{X-4}$

14. $\dfrac{6-2}{4+\sqrt{-4}} = \dfrac{4\left(4-\sqrt{-4}\right)}{\left(4+\sqrt{-4}\right)\left(4-\sqrt{-4}\right)} =$

$\dfrac{4\left(4-\sqrt{4}\sqrt{-1}\right)}{16-(-4)} = \dfrac{4(4-2i)}{20} = \dfrac{4-2i}{5}$

15. $\dfrac{3X+\sqrt{2}}{3X-\sqrt{2}} = \dfrac{\left(3X+\sqrt{2}\right)\left(3X+\sqrt{2}\right)}{\left(3X-\sqrt{2}\right)\left(3X+\sqrt{2}\right)} =$

$\dfrac{9X^2+3X\sqrt{2}+3X\sqrt{2}+2}{9X^2-2} = \dfrac{9X^2+6X\sqrt{2}+2}{9X^2-2}$

16. $\dfrac{5i}{2i+i\sqrt{3}} = \dfrac{5i\left(2i-i\sqrt{3}\right)}{\left(2i+i\sqrt{3}\right)\left(2i-i\sqrt{3}\right)} =$

$\dfrac{10i^2-5i^2\sqrt{3}}{4i^2-3i^2} = \dfrac{-10-\left(-5\sqrt{3}\right)}{4(-1)-3(-1)} =$

$\dfrac{-10+5\sqrt{3}}{-4-(-3)} = \dfrac{-10+5\sqrt{3}}{-4+3} =$

$\dfrac{-10+5\sqrt{3}}{-1} = 10-5\sqrt{3}$

Systematic Review 8C

1. $3X+i$

2. $10-2\sqrt{7}$

3. $(5+4i)(5-4i) = 25-16i^2 =$
$25-16(-1) = 25+16 = 41$

4. $\left(3X+\sqrt{11}\right)\left(3X-\sqrt{11}\right) = 9X^2-11$

5. $4X^2-3 = 0$
$\left(2X-\sqrt{3}\right)\left(2X+\sqrt{3}\right) = 0$

6. $2X-\sqrt{3}=0 \qquad 2X+\sqrt{3}=0$

$\qquad 2X=\sqrt{3} \qquad\qquad 2X=-\sqrt{3}$

$\qquad X=\dfrac{\sqrt{3}}{2} \qquad\qquad X=-\dfrac{\sqrt{3}}{2}$

$X=\pm\dfrac{\sqrt{3}}{2}$

7. $3Y^2-\dfrac{1}{9}=0$

$\left(Y\sqrt{3}-\dfrac{1}{3}\right)\left(Y\sqrt{3}+\dfrac{1}{3}\right)=0$

8. $Y\sqrt{3}-\dfrac{1}{3}=0 \qquad\qquad Y\sqrt{3}+\dfrac{1}{3}=0$

$\qquad Y\sqrt{3}=\dfrac{1}{3} \qquad\qquad\qquad Y\sqrt{3}=-\dfrac{1}{3}$

$\qquad Y=\dfrac{1}{3\sqrt{3}} \qquad\qquad\qquad Y=-\dfrac{1}{3\sqrt{3}}$

$\qquad Y=\dfrac{\sqrt{3}}{9} \qquad\qquad\qquad Y=-\dfrac{\sqrt{3}}{9}$

$Y=\pm\dfrac{\sqrt{3}}{9}$

9. $1\sqrt{-12}+6\sqrt{-3} = 1\sqrt{4}\sqrt{-1}\sqrt{3}+6\sqrt{-1}\sqrt{3} =$
$11(2)i\sqrt{3}+6i\sqrt{3} = 22i\sqrt{3}+6i\sqrt{3} = 28i\sqrt{3}$

10. $(7i)\left(\sqrt{-64}\right) = 7i\sqrt{64}\sqrt{-1} = 7i8i = 56i^2 = -56$

11. $\sqrt{-144}\div 4i = \sqrt{144}\sqrt{-1}\div 4i = 12i\div 4i = 3$

12. $i^2\cdot i^2 = (-1)(-1) = 1$

13. $\left(9^{\frac{3}{2}}\right)^{-2} = \left(3^3\right)^{-2} = 3^{(3)(-2)} = 3^{-6} = \dfrac{1}{3^6} = \dfrac{1}{729}$

14. $\left(\sqrt[4]{X^8}\right)^{\frac{1}{2}} = \left[\left(X^8\right)^{\frac{1}{4}}\right]^{\frac{1}{2}} =$
$X^{(8)\left(\frac{1}{4}\right)\left(\frac{1}{2}\right)} = X^{\frac{8}{8}} = X^1 = X$

15. $6X^2+3X+3 = -4X+1$
$\qquad 6X^2+7X+2 = 0$
$\qquad (3X+2)(2X+1) = 0$

$3X+2=0 \qquad\qquad 2X+1=0$
$\qquad 3X=-2 \qquad\qquad\qquad 2X=-1$
$\qquad X=-\dfrac{2}{3} \qquad\qquad\qquad X=-\dfrac{1}{2}$

$$6\left(-\frac{2}{3}\right)^2 + 3\left(-\frac{2}{3}\right) + 3 = -4\left(-\frac{2}{3}\right) + 1$$

$$6\left(\frac{4}{9}\right) - \frac{6}{3} + 3 = \frac{8}{3} + \frac{3}{3}$$

$$\frac{24}{9} - \frac{18}{9} + \frac{27}{9} = \frac{11}{3}$$

$$\frac{33}{9} = \frac{11}{3}$$

$$\frac{11}{3} = \frac{11}{3} \text{ ok}$$

$$6\left(-\frac{1}{2}\right)^2 + 3\left(-\frac{1}{2}\right) + 3 = -4\left(-\frac{1}{2}\right) + 1$$

$$6\left(\frac{1}{4}\right) - \frac{3}{2} + 3 = \frac{4}{2} + \frac{2}{2}$$

$$\frac{6}{4} - \frac{6}{4} + \frac{12}{4} = \frac{6}{2}$$

$$\frac{12}{4} = 3$$

$$3 = 3 \text{ ok}$$

16.
$$9\left(X^2 + X\right) = 25 + 9X$$

$$9X^2 + 9X = 25 + 9X$$

$$9X^2 - 25 = 0$$

$$(3X - 5)(3X + 5) = 0$$

$$3X - 5 = 0 \qquad 3X + 5 = 0$$

$$3X = 5 \qquad 3X = -5$$

$$X = \frac{5}{3} \qquad X = -\frac{5}{3}$$

$$9\left[\left(\frac{5}{3}\right)^2 + \left(\frac{5}{3}\right)\right] = 25 + 9\left(\frac{5}{3}\right)$$

$$9\left[\frac{25}{9} + \frac{5}{3}\right] = 25 + \frac{45}{3}$$

$$9\left[\frac{25}{9} + \frac{15}{9}\right] = 25 + 15$$

$$9\left[\frac{40}{9}\right] = 40$$

$$40 = 40 \text{ ok}$$

$$9\left[\left(-\frac{5}{3}\right)^2 + \left(-\frac{5}{3}\right)\right] = 25 + 9\left(-\frac{5}{3}\right)$$

$$9\left[\frac{25}{9} - \frac{5}{3}\right] = 25 - \frac{45}{3}$$

$$9\left[\frac{25}{9} - \frac{15}{9}\right] = 25 - 15$$

$$9\left[\frac{10}{9}\right] = 10$$

$$10 = 10 \text{ ok}$$

17.
$$\frac{2X^2}{X^2 - 16} \div \frac{X}{4 - X} = \frac{2X^2}{(X-4)(X+4)} \times \frac{4-X}{X}$$

$$= \frac{2X^2}{(X-4)(X+4)} \times \frac{(-1)(X-4)}{X}$$

$$= \frac{2X^2(-1)\cancel{(X-4)}}{\cancel{(X-4)}(X+4)(X)}$$

$$= \frac{-2X^2}{(X+4)(X)} = \frac{-2X}{X+4}$$

18.
$$12\sqrt{\frac{1}{3}} - 9\sqrt{\frac{2}{5}} = 12\left(\frac{\sqrt{1}}{\sqrt{3}}\right) - 9\left(\frac{\sqrt{2}}{\sqrt{5}}\right)$$

$$= \frac{12}{\sqrt{3}} - \frac{9\sqrt{2}}{\sqrt{5}}$$

$$= \frac{12\sqrt{3}}{\sqrt{3}\sqrt{3}} - \frac{9\sqrt{2}\sqrt{5}}{\sqrt{5}\sqrt{5}}$$

$$= \frac{12\sqrt{3}}{\sqrt{9}} - \frac{9\sqrt{10}}{\sqrt{25}}$$

$$= \frac{12\sqrt{3}}{3} - \frac{9\sqrt{10}}{5}$$

$$= \frac{4\sqrt{3}}{1} - \frac{9\sqrt{10}}{5}$$

$$= \frac{4\sqrt{3}\,(5)}{(5)} - \frac{9\sqrt{10}}{5}$$

$$= \frac{20\sqrt{3}}{5} - \frac{9\sqrt{10}}{5}$$

$$= \frac{20\sqrt{3} - 9\sqrt{10}}{5}$$

19.
$$\frac{60X^{-2}YZ^4}{24X^{-1}Z^3Y^{-1}} = \frac{5X^{-2}YZ^4X^1Z^{-3}Y^1}{2}$$

$$= \frac{5X^{-1}Y^2Z}{2} = \frac{5Y^2Z}{2X}$$

20.
$$\frac{4 + \dfrac{X}{Y}}{2 - \dfrac{2X}{Y}} = \frac{\dfrac{4Y}{Y} + \dfrac{X}{Y}}{\dfrac{2Y}{Y} - \dfrac{2X}{Y}} = \frac{\dfrac{4Y + X}{Y}}{\dfrac{2Y - 2X}{Y}}$$

$$= \frac{\dfrac{4Y + X}{Y} \times \dfrac{Y}{2Y - 2X}}{\dfrac{2Y - 2X}{Y} \times \dfrac{Y}{2Y - 2X}}$$

$$= \frac{4Y + X}{\cancel{Y}} \times \frac{\cancel{Y}}{2Y - 2X} = \frac{4Y + X}{2Y - 2X}$$

Systematic Review 8D

1. $4 - 8i$

2. $2 - 3\sqrt{91}$

3. $(12 + 3i)(12 - 3i) = 144 - 9i^2 =$
   $144 - 9(-1) = 144 - (-9) = 144 + 9 = 153$

4. $\left(x + \sqrt{2}\right)\left(x - \sqrt{2}\right) = x^2 - 2$

5. $81X^2 - 3 = 0$
   $\left(9X - \sqrt{3}\right)\left(9X + \sqrt{3}\right) = 0$

6. $9X - \sqrt{3} = 0 \qquad 9X + \sqrt{3} = 0$
   $\quad 9X = \sqrt{3} \qquad\qquad 9X = -\sqrt{3}$
   $\quad\ X = \dfrac{\sqrt{3}}{9} \qquad\qquad X = -\dfrac{\sqrt{3}}{9}$

7. $7Y^2 - 9 = 0$
   $\left(Y\sqrt{7} - 3\right)\left(Y\sqrt{7} + 3\right) = 0$

8. $Y\sqrt{7} - 3 = 0 \qquad Y\sqrt{7} + 3 = 0$
   $\quad Y\sqrt{7} = 3 \qquad\qquad Y\sqrt{7} = -3$
   $\qquad Y = \dfrac{3}{\sqrt{7}} \qquad\qquad Y = -\dfrac{3}{\sqrt{7}}$
   $\qquad Y = \dfrac{3\sqrt{7}}{7} \qquad\qquad Y = -\dfrac{3\sqrt{7}}{7}$

9. $6\sqrt{-50} - 5\sqrt{-18} =$
   $6\sqrt{25}\sqrt{-1}\sqrt{2} - 5\sqrt{9}\sqrt{-1}\sqrt{2} =$
   $6(5)i\sqrt{2} - 5(3)i\sqrt{2} = 30i\sqrt{2} - 15i\sqrt{2} = 15i\sqrt{2}$

10. $(5i)\left(2\sqrt{-49}\right) = (5)(2)i\sqrt{-1}\sqrt{49} =$
    $10i^2(7) = (10)(-1)(7) = -70$

11. $\sqrt{-225} \div 3 = \sqrt{225}\sqrt{-1} \div 3 = 15i \div 3 = 5i$

12. $i^3 \cdot i^3 = i^6 = \left(i^2\right)^3 = (-1)^3 = -1$

13. $\left(9^{\frac{2}{3}}\right)^3 = 9^{\left(\frac{2}{3}\right)(3)} = 9^{\frac{6}{3}} = 9^2 = 81$

14. $\left(\sqrt{400}\right)^{-1} = 20^{-1} = \dfrac{1}{20}$

15. $7X^2 + 11X = 2X^2 + 3X + 4$
    $5X^2 + 8X - 4 = 0$
    $(5X - 2)(X + 2) = 0$

    $7\left(\dfrac{2}{5}\right)^2 + 11\left(\dfrac{2}{5}\right) = 2\left(\dfrac{2}{5}\right)^2 + 3\left(\dfrac{2}{5}\right) + 4$

    $5X - 2 = 0 \qquad 7\left(\dfrac{4}{25}\right) + \dfrac{22}{5} = 2\left(\dfrac{4}{25}\right) + \dfrac{6}{5} + 4$
    $\quad 5X = 2$
    $\quad\ X = \dfrac{2}{5} \qquad\qquad \dfrac{28}{25} + \dfrac{110}{25} = \dfrac{8}{25} + \dfrac{30}{25} + \dfrac{100}{25}$

    $\qquad\qquad\qquad\qquad \dfrac{138}{25} = \dfrac{138}{25} \text{ ok}$

    $\qquad\qquad\qquad 7(-2)^2 + 11(-2) = 2(-2)^2 + 3(-2) + 4$
    $X + 2 = 0 \qquad 7(4) - 22 = 2(4) - 6 + 4$
    $\quad X = -2 \qquad\quad 28 - 22 = 8 - 6 + 4$
    $\qquad\qquad\qquad\qquad\quad 6 = 6 \text{ ok}$

16. $12X^2 - 6X - 15 = X - 3$
    $12X^2 - 7X - 12 = 0$
    $(3X - 4)(4X + 3) = 0$

    $12\left(\dfrac{4}{3}\right)^2 - 6\left(\dfrac{4}{3}\right) - 15 = \left(\dfrac{4}{3}\right) - 3$

    $3X - 4 = 0 \qquad 12\left(\dfrac{16}{9}\right) - \dfrac{24}{3} - 15 = \dfrac{4}{3} - \dfrac{9}{3}$
    $\quad 3X = 4$
    $\quad\ X = \dfrac{4}{3} \qquad\qquad \dfrac{192}{9} - \dfrac{72}{9} - \dfrac{135}{9} = -\dfrac{5}{3}$

    $\qquad\qquad\qquad\qquad\qquad\quad -\dfrac{15}{9} = -\dfrac{5}{3}$

    $\qquad\qquad\qquad\qquad\qquad\quad -\dfrac{5}{3} = -\dfrac{5}{3} \text{ ok}$

    $12\left(-\dfrac{3}{4}\right)^2 - 6\left(-\dfrac{3}{4}\right) - 15 = \left(-\dfrac{3}{4}\right) - 3$

    $4X + 3 = 0 \qquad 12\left(\dfrac{9}{16}\right) + \dfrac{18}{4} - 15 = -\dfrac{3}{4} - \dfrac{12}{4}$
    $\quad 4X = -3$
    $\quad\ X = -\dfrac{3}{4} \qquad\qquad \dfrac{108}{16} + \dfrac{72}{16} - \dfrac{240}{16} = -\dfrac{15}{4}$

    $\qquad\qquad\qquad\qquad\qquad\quad -\dfrac{60}{16} = -\dfrac{15}{4}$

    $\qquad\qquad\qquad\qquad\qquad\quad -\dfrac{15}{4} = -\dfrac{15}{4} \text{ ok}$

17.

$$\frac{6X^2+3X}{4X^2-1} \div \frac{3X+12}{2X^2+X-1} =$$

$$\frac{6X^2+3X}{4X^2-1} \times \frac{2X^2+X-1}{3X+12} =$$

$$\frac{(3X)(2X+1)}{(2X-1)(2X+1)} \times \frac{(2X-1)(X+1)}{(3)(X+4)} = \frac{(3X)(X+1)}{(3)(X+4)}$$

$$= \frac{X(X+1)}{X+4}$$

$$= \frac{X^2+X}{X+4}$$

18.

$$\sqrt{\frac{2}{X}} + \sqrt{\frac{3}{X}} = \frac{\sqrt{2}}{\sqrt{X}} + \frac{\sqrt{3}}{\sqrt{X}}$$

$$= \frac{\sqrt{2}\sqrt{X}}{\sqrt{X}\sqrt{X}} + \frac{\sqrt{3}\sqrt{X}}{\sqrt{X}\sqrt{X}}$$

$$= \frac{\sqrt{2X}}{X} + \frac{\sqrt{3X}}{X} = \frac{\sqrt{2X}+\sqrt{3X}}{X}$$

19.

$$\frac{5^1X^{-3}Y^2Z^{-1}}{5^{-2}X^1Y^{-2}Z^1} = 5^1X^{-3}Y^2Z^{-1}5^2X^{-1}Y^2Z^{-1} = 5^3X^{-4}Y^4Z^{-2}$$

$$= 125X^{-4}Y^4Z^{-2} \text{ or } \frac{125Y^4}{X^4Z^2}$$

20.

$$\frac{5-\frac{X-2}{X}}{X+\frac{5}{2X}} = \frac{\frac{5X}{X}-\frac{X-2}{X}}{\frac{X(2X)}{(2X)}+\frac{5}{2X}}$$

$$= \frac{\frac{5X-(X-2)}{X}}{\frac{2X^2+5}{2X}}$$

$$= \frac{\frac{5X-(X-2)}{X} \times \frac{2X}{2X^2+5}}{\frac{2X^2+5}{2X} \times \frac{2X}{2X^2+5}}$$

$$= \frac{5X-(X-2)}{X} \times \frac{2X}{2X^2+5}$$

$$= \frac{5X-X+2}{X} \times \frac{2X}{2X^2+5}$$

$$= \frac{(4X+2)(2X)}{(X)(2X^2+5)}$$

$$= \frac{(4X+2)(2)}{2X^2+5} = \frac{8X+4}{2X^2+5}$$

Systematic Review 8E

1. $3+2i$

2. $4+5\sqrt{8}$

3. $(A+14i)(A-14i) = A^2-196i^2 = A^2-(-196) = A^2+196$

4. $(2X+\sqrt{3})(2X-\sqrt{3}) = 4X^2-3$

5. $16X^2-15 = 0$
$(4X-\sqrt{15})(4X+\sqrt{15}) = 0$

6. $4X-\sqrt{15} = 0 \qquad 4X+\sqrt{15} = 0$
$4X = \sqrt{15} \qquad 4X = -\sqrt{15}$
$X = \frac{\sqrt{15}}{4} \qquad X = -\frac{\sqrt{15}}{4}$

7. $5X^2-10 = 0$
$(\sqrt{5}X-\sqrt{10})(\sqrt{5}X+\sqrt{10}) = 0$

8. $\sqrt{5}X-\sqrt{10} = 0 \qquad \sqrt{5}X+\sqrt{10} = 0$
$\sqrt{5}X = \sqrt{10} \qquad \sqrt{5}X = -\sqrt{10}$
$X = \frac{\sqrt{10}}{\sqrt{5}} \qquad X = -\frac{\sqrt{10}}{\sqrt{5}}$
$X = \sqrt{2} \qquad X = -\sqrt{2}$

9. $11\sqrt{-12}+6\sqrt{-3} = 11\sqrt{4}\sqrt{-1}\sqrt{3}+6\sqrt{-1}\sqrt{3} = 11(2)i\sqrt{3}+6i\sqrt{3} = 22i\sqrt{3}+6i\sqrt{3} = 28i\sqrt{3}$

10. $(-3i)(5\sqrt{-81}) = (-3i)(5)\sqrt{81}\sqrt{-1} = -15i(9)i = -135i^2 = -135(-1) = 135$

11. $\sqrt{-196} \div 2i = \sqrt{196}\sqrt{-1} \div 2i = 14i \div 2i = 7$

12. $i^4 \cdot i^4 = (i^2)^2 \cdot (i^2)^2 = (-1)^2 \cdot (-1)^2 = 1 \cdot 1 = 1$

13. $(8^2+15^2)^{\frac{3}{2}} = (64+225)^{\frac{3}{2}} = 289^{\frac{3}{2}} = 17^3 = 4,913$

14. $(\sqrt[3]{64})^{\frac{3}{2}} = 4^{\frac{3}{2}} = 2^3 = 8$

15. $\frac{4}{3}X^2 = \frac{4}{3}X+5$
$4X^2 = 4X+15$
$4X^2-4X-15 = 0$
$(2X-5)(2X+3) = 0$

$$2X - 5 = 0$$
$$2X = 5$$
$$X = \frac{5}{2}$$

$$\frac{4}{3}\left(\frac{5}{2}\right)^2 = \frac{4}{3}\left(\frac{5}{2}\right) + 5$$
$$\frac{4}{3}\left(\frac{25}{4}\right) = \frac{20}{6} + \frac{30}{6}$$
$$\frac{100}{12} = \frac{50}{6}$$
$$\frac{25}{3} = \frac{25}{3} \quad ok$$

18. $\sqrt{\dfrac{1}{2A}} - \sqrt{\dfrac{3}{2}} = \dfrac{\sqrt{1}}{\sqrt{2A}} - \dfrac{\sqrt{3}}{\sqrt{2}}$

$$= \frac{\sqrt{2A}}{\sqrt{2A}\sqrt{2A}} - \frac{\sqrt{3}\sqrt{2}}{\sqrt{2}\sqrt{2}}$$
$$= \frac{\sqrt{2A}}{2A} - \frac{\sqrt{6}}{2}$$
$$= \frac{\sqrt{2A}}{2A} - \frac{\sqrt{6}\,(A)}{2(A)}$$
$$= \frac{\sqrt{2A} - A\sqrt{6}}{2A}$$

$$2X + 3 = 0$$
$$2X = -3$$
$$X = -\frac{3}{2}$$

$$\frac{4}{3}\left(-\frac{3}{2}\right)^2 = \frac{4}{3}\left(-\frac{3}{2}\right) + 5$$
$$\frac{4}{3}\left(\frac{9}{4}\right) = -\frac{12}{6} + \frac{30}{6}$$
$$\frac{36}{12} = \frac{18}{6}$$
$$3 = 3 \quad ok$$

19. $\dfrac{3^{-1}X^{-4}Y^3X^2}{3^0Y^{-1}Y^2} = 3^{-1}X^{-4}Y^3X^23^0Y^1Y^{-2}$

$$= 3^{-1}X^{-2}Y^2 \text{ or } \frac{Y^2}{3X^2}$$

16. $8 - 6X = 3X^2 + X + 10$
$$0 = 3X^2 + 7X + 2$$
$$0 = (3X + 1)(X + 2)$$

$$8 - 6\left(-\frac{1}{3}\right) = 3\left(-\frac{1}{3}\right)^2 + \left(-\frac{1}{3}\right) + 10$$

$$3X + 1 = 0 \qquad 8 + \frac{6}{3} = 3\left(\frac{1}{9}\right) - \frac{1}{3} + 10$$
$$3X = -1$$
$$X = -\frac{1}{3} \qquad \frac{24}{3} + \frac{6}{3} = \frac{3}{9} - \frac{3}{9} + \frac{90}{9}$$
$$\frac{30}{3} = \frac{90}{9}$$
$$10 = 10 \quad ok$$

20. $\dfrac{3X + \dfrac{5X+8}{3X}}{2 - \dfrac{4}{X^2}} = \dfrac{\dfrac{3X(3X)}{3X} + \dfrac{5X+8}{3X}}{\dfrac{2(X^2)}{(X^2)} - \dfrac{4}{X^2}}$

$$= \frac{\dfrac{9X^2 + 5X + 8}{3X}}{\dfrac{2X^2 - 4}{X^2}}$$

$$X + 2 = 0 \qquad 8 - 6(-2) = 3(-2)^2 + (-2) + 10$$
$$X = -2 \qquad 8 + 12 = 3(4) - 2 + 10$$
$$20 = 12 - 2 + 10$$
$$20 = 20 \quad ok$$

$$= \frac{\dfrac{9X^2 + 5X + 8}{3X} \times \dfrac{X^2}{2X^2 - 4}}{\dfrac{2X^2 - 4}{X^2} \times \dfrac{X^2}{2X^2 - 4}}$$

$$= \frac{9X^2 + 5X + 8}{3X} \times \frac{X^2}{2X^2 - 4}$$

$$= \frac{9X^2 + 5X + 8}{3} \times \frac{X}{2X^2 - 4}$$

$$= \frac{9X^3 + 5X^2 + 8X}{6X^2 - 12}$$

17. $\dfrac{X^2 + 2X - 24}{X^2 + 8X + 12} \div \dfrac{X^2 - 8X + 16}{X^2 + 4X + 4} =$

$$\frac{X^2 + 2X - 24}{X^2 + 8X + 12} \times \frac{X^2 + 4X + 4}{X^2 - 8X + 16} =$$

$$\frac{\cancel{(X+6)}\cancel{(X-4)}}{\cancel{(X+6)}(X+2)} \times \frac{(X+2)\cancel{(X+2)}}{\cancel{(X-4)}(X-4)} = \frac{X+2}{X-4}$$

Lesson Practice 9A

1. $(X + 3)^2 = X^2 + 6X + 9$

2. $(A - 3)^2 = A^2 - 6A + 9$

3. $(3X + 4)^2 = 9X^2 + 24X + 16$

4. $(2X + 1)^2 = 4X^2 + 4X + 1$

5. $X^2 + 8X + 16 = (X + 4)^2$

6. $X^2 - 18X + 81 = (X - 9)^2$

7. $A^2 + 16A + 64 = (A + 8)^2$

8. $A^2 - 16A + 64 = (A - 8)^2$

9. $(X + 5)^3 = (X)^3 + 3(X)^2(5) + 3(X)(5)^2 + (5)^3$
$= X^3 + 15X^2 + 75X + 125$

10. $(3X - 2)^3 = (3X)^3 - 3(3X)^2(2) + 3(3X)(2)^2 - (2)^3$
$= 27X^3 - 54X^2 + 36X - 8$

11. $(2A + 2B)^3$
$= (2A)^3 + 3(2A)^2(2B) + 3(2A)(2B)^2 + (2B)^3$
$= 8A^3 + 24A^2B + 24AB^2 + 8B^3$

12. $(X - 1)^3 = (X)^3 - 3(X)^2(1) + 3(X)(1)^2 - (1)^3$
$= X^3 - 3X^2 + 3X - 1$

13. $\left(Y - \dfrac{1}{4}\right)^3$
$= (Y)^3 - 3(Y)^2\left(\dfrac{1}{4}\right) + 3(Y)\left(\dfrac{1}{4}\right)^2 - \left(\dfrac{1}{4}\right)^3$
$= Y^3 - \dfrac{3}{4}Y^2 + \dfrac{3}{16}Y - \dfrac{1}{64}$

14. $(2R + 3)^3 = (2R)^3 + 3(2R)^2(3) + 3(2R)(3)^2 + (3)^3$
$= 8R^3 + 36R^2 + 54R + 27$

15.
```
          1
        1   1
      1   2   1
    1   3   3   1
  1   4   6   4   1
1   5  10  10   5   1
```

7. $X^2 - 4X + 4 = (X - 2)^2$

8. $R^2 + 24R + 144 = (R + 12)^2$

9. $(X - 10)^3$
$= (X)^3 + 3(X)^2(-10) + 3(X)(-10)^2 + (-10)^3$
$= X^3 - 30X^2 + 300X - 1,000$

10. $(2X + 1)^3 = (2X)^3 + 3(2X)^2(1) + 3(2X)(1)^2 + 1^3$
$= 8X^3 + 12X^2 + 6X + 1$

11. $(A + B)^3 = (A)^3 + 3(A)^2(B) + 3(A)(B)^2 + (B)^3$
$= A^3 + 3A^2B + 3AB^2 + B^3$

12. $(X - 5)^3 = (X)^3 + 3(X)^2(-5) + 3(X)(-5)^2 + (-5)^3$
$= X^3 - 15X^2 + 75X - 125$

13. $\left(P - \dfrac{1}{3}\right)^3$
$= (P)^3 + 3(P)^2\left(-\dfrac{1}{3}\right) + 3(P)\left(-\dfrac{1}{3}\right)^2 + \left(-\dfrac{1}{3}\right)^3$
$= P^3 - P^2 + \dfrac{1}{3}P - \dfrac{1}{27}$

14. $(4F + 2)^3 =$
$= (4F)^3 + 3(4F)^2(2) + 3(4F)(2)^2 + (2)^3$
$= 64F^3 + 96F^2 + 48F + 8$

15.
```
    1   5  10  10   5   1
   1  6  15  20  15   6   1
  1  7  21  35  35  21   7   1
 1  8  28  56  70  56  28   8   1
```

Lesson Practice 9B

1. $(A - 4)^2 = A^2 - 8A + 16$

2. $(X - Y)^2 = X^2 - 2XY + Y^2$

3. $(2X + 5)^2 = 4X^2 + 20X + 25$

4. $(3X - 2)^2 = 9X^2 - 12X + 4$

5. $B^2 + 20B + 100 = (B + 10)^2$

6. $X^2 - 2X + 1 = (X - 1)^2$

Systematic Review 9C

1. $(X + 5)^2 = X^2 + 10X + 25$

2. $(2X - 3)^2 = 4X^2 - 12X + 9$

3. $4X^2 + 20X + 25 = (2X + 5)^2$

4. $X^2 - 14X + 49 = (X - 7)^2$

5. $(X + 4)^3 = (X)^3 + 3(X)^2(4) + 3(X)(4)^2 + (4)^3$
$= X^3 + 12X^2 + 48X + 64$

6. $(X-4)^3 = (X)^3 + 3(X)^2(-4) + 3(X)(-4)^2 + (-4)^3$
$= X^3 - 12X^2 + 48X - 64$

7. $(2X+1)^3 = (2X)^3 + 3(2X)^2(1) + 3(2X)(1)^2 + (1)^3$
$= 8X^3 + 12X^2 + 6X + 1$

8. $(3X-2)^3$
$= (3X)^3 + 3(3X)^2(-2) + 3(3X)(-2)^2 + (-2)^3$
$= 27X^3 - 54X^2 + 36X - 8$

9. $13 - 2i\sqrt{5}$

10. $9X^2 - 5 = (3X - \sqrt{5})(3X + \sqrt{5})$

11. $\dfrac{4\sqrt{6}}{2\sqrt{8}+1} = \dfrac{4\sqrt{6}(2\sqrt{8}-1)}{(2\sqrt{8}+1)(2\sqrt{8}-1)} =$
$\dfrac{8\sqrt{48} - 4\sqrt{6}}{4(8)-1} = \dfrac{8\sqrt{16}\sqrt{3} - 4\sqrt{6}}{32-1} =$
$\dfrac{8(4)\sqrt{3} - 4\sqrt{6}}{31} = \dfrac{32\sqrt{3} - 4\sqrt{6}}{31}$

12. $\dfrac{7}{4-3i} = \dfrac{7(4+3i)}{(4-3i)(4+3i)} = \dfrac{28+21i}{16-9i^2} =$
$\dfrac{28+21i}{16-(-9)} = \dfrac{28+21i}{16+9} = \dfrac{28+21i}{25}$

13. $(8i)(2\sqrt{-2}) = 16i\sqrt{-1}\sqrt{2} = 16i^2\sqrt{2} = -16\sqrt{2}$

14. $(i^3)^2 = i^{(3)(2)} = i^6 = (-1)^3 = -1$

15. $\left(100^{\frac{3}{2}}\right)^{\frac{1}{2}} = (10^3)^{\frac{1}{2}} = 1,000^{\frac{1}{2}} = \sqrt{1,000}$
$= \sqrt{100}\sqrt{10} = 10\sqrt{10}$

16. $\left(\sqrt[3]{X^6}\right)^4 = \left[(X^6)^{\frac{1}{3}}\right]^4 = X^{(6)\left(\frac{1}{3}\right)(4)} = X^{\frac{24}{3}} = X^8$

17. $6X - 12 = -2X^2 + 4X$
$3X - 6 = -X^2 + 2X$
$X^2 + X - 6 = 0$
$(X+3)(X-2) = 0$

$X + 3 = 0$ $6(-3) - 12 = -2(-3)^2 + 4(-3)$
$X = -3$ $-18 - 12 = -2(9) - 12$
            $-30 = -18 - 12$
            $-30 = -30$ ok

$X - 2 = 0$ $6(2) - 12 = -2(2)^2 + 4(2)$
$X = 2$ $12 - 12 = -2(4) + 8$
             $0 = -8 + 8$
             $0 = 0$ ok

18. $\dfrac{X^2 - 3X - 4}{X^2 + X - 6} \div \dfrac{-6X - 6}{X^2 - 2X} =$
$\dfrac{X^2 - 3X - 4}{X^2 + X - 6} \times \dfrac{X^2 - 2X}{-6X - 6} =$
$\dfrac{(X-4)(X+1)}{(X+3)(X-2)} \times \dfrac{(X)(X-2)}{(-6)(X+1)} = \dfrac{X^2 - 4X}{-6X - 18}$

19. $5\sqrt{\dfrac{8}{3}} - 6\sqrt{\dfrac{18}{7}} = 5\left(\dfrac{\sqrt{8}}{\sqrt{3}}\right) - 6\left(\dfrac{\sqrt{18}}{\sqrt{7}}\right) =$
$\dfrac{5\sqrt{8}}{\sqrt{3}} - \dfrac{6\sqrt{18}}{\sqrt{7}} = \dfrac{5\sqrt{4}\sqrt{2}}{\sqrt{3}} - \dfrac{6\sqrt{9}\sqrt{2}}{\sqrt{7}} =$
$\dfrac{5(2)\sqrt{2}}{\sqrt{3}} - \dfrac{6(3)\sqrt{2}}{\sqrt{7}} = \dfrac{10\sqrt{2}}{\sqrt{3}} - \dfrac{18\sqrt{2}}{\sqrt{7}} =$
$\dfrac{10\sqrt{2}\sqrt{3}}{3} - \dfrac{18\sqrt{2}\sqrt{7}}{7} = \dfrac{10\sqrt{6}}{3} - \dfrac{18\sqrt{14}}{7} =$
$\dfrac{10\sqrt{6}(7)}{3(7)} - \dfrac{18\sqrt{14}(3)}{7(3)} = \dfrac{70\sqrt{6}}{21} - \dfrac{54\sqrt{14}}{21} =$
$\dfrac{70\sqrt{6} - 54\sqrt{14}}{21}$

20. $\dfrac{X^2 - \dfrac{X}{2}}{X + \dfrac{2X}{3}} = \dfrac{\dfrac{2X^2}{2} - \dfrac{X}{2}}{\dfrac{3X}{3} + \dfrac{2X}{3}} = \dfrac{\dfrac{2X^2 - X}{2}}{\dfrac{3X + 2X}{3}} =$
$\dfrac{\dfrac{2X^2 - X}{2} \times \dfrac{3}{3X + 2X}}{\dfrac{3X + 2X}{3} \times \dfrac{3}{3X + 2X}} = \dfrac{2X^2 - X}{2} \times \dfrac{3}{3X + 2X} =$
$\dfrac{(X)(2X-1)}{2} \times \dfrac{3}{(X)(3+2)} = \dfrac{6X - 3}{10}$

Systematic Review 9D

1. $(X+7)^2 = X^2 + 14X + 49$

2. $(3X-4)^2 = 9X^2 - 24X + 16$

3. $X^2 - 6X + 9 = (X-3)^2$

4. $4X^2 + 16X + 16 = (2X+4)^2$

5. $(X+2)^3 = (X)^3 + 3(X)^2(2) + 3(X)(2)^2 + (2)^3 =$
$X^3 + 6X^2 + 12X + 8$

6. $\left(X - \dfrac{1}{5}\right)^3 =$

$(X)^3 + 3(X)^2\left(-\dfrac{1}{5}\right) + 3(X)\left(-\dfrac{1}{5}\right)^2 + \left(-\dfrac{1}{5}\right)^3 =$

$X^3 - \dfrac{3}{5}X^2 + \dfrac{3}{25}X - \dfrac{1}{125}$

7. $(3X + 2)^3 =$

$(3X)^3 + 3(3X)^2(2) + 3(3X)(2)^2 + (2)^3 =$

$27X^3 + 54X^2 + 36X + 8$

8. $(2X - 3)^3 =$

$(2X)^3 + 3(2X)^2(-3) + 3(2X)(-3)^2 + (-3)^3 =$

$8X^3 - 36X^2 + 54X - 27$

9. $5 + \sqrt{-4}$ or $5 + 2i$

10. $2X^2 - 15 = \left(X\sqrt{2} - \sqrt{15}\right)\left(X\sqrt{2} + \sqrt{15}\right)$

11. $\dfrac{10\sqrt{15}}{3\sqrt{5} + 8} = \dfrac{10\sqrt{15}\left(3\sqrt{5} - 8\right)}{\left(3\sqrt{5} + 8\right)\left(3\sqrt{5} - 8\right)} =$

$\dfrac{30\sqrt{75} - 80\sqrt{15}}{9(5) - 64} = \dfrac{30\sqrt{25}\sqrt{3} - 80\sqrt{15}}{45 - 64} =$

$\dfrac{30(5)\sqrt{3} - 80\sqrt{15}}{-19} =$

$\dfrac{150\sqrt{3} - 80\sqrt{15}}{-19}$ or $-\dfrac{150\sqrt{3} - 80\sqrt{15}}{19}$

12. $\dfrac{5}{2 + 6i} = \dfrac{5(2 - 6i)}{(2 + 6i)(2 - 6i)} = \dfrac{10 - 30i}{4 - 36(-1)} =$

$\dfrac{10 - 30i}{4 + 36} = \dfrac{10 - 30i}{40} = \dfrac{1 - 3i}{4}$

13. $(7i)\left(3\sqrt{-8}\right) = (7i)3\sqrt{4}\sqrt{-1}\sqrt{2} =$

$(7)(3)(2)(i)(i)\sqrt{2} = 42i^2\sqrt{2} =$

$42(-1)\sqrt{2} = -42\sqrt{2}$

14. $i^3 = i^2 \cdot i = -1 \cdot i = -i$

15. $\left(16^{-\frac{1}{4}}\right)^{-3} = \left(2^{-1}\right)^{-3} = 2^{(-1)(-3)} = 2^3 = 8$

16. $\left(\sqrt{X}\right)^{-3} = \left(X^{\frac{1}{2}}\right)^{-3} = X^{\left(\frac{1}{2}\right)(-3)} = X^{-\frac{3}{2}}$

17. $2X^2 = \dfrac{5}{2}X + 3$

$4X^2 = 5X + 6$

$4X^2 - 5X - 6 = 0$

$(4X + 3)(X - 2) = 0$

$4X + 3 = 0 \qquad 2\left(-\dfrac{3}{4}\right)^2 = \dfrac{5}{2}\left(-\dfrac{3}{4}\right) + 3$

$4X = -3 \qquad 2\left(\dfrac{9}{16}\right) = -\dfrac{15}{8} + \dfrac{24}{8}$

$X = -\dfrac{3}{4} \qquad \dfrac{18}{16} = \dfrac{9}{8}$

$\dfrac{9}{8} = \dfrac{9}{8}$ ok

$X - 2 = 0 \qquad 2(2)^2 = \dfrac{5}{2}(2) + 3$

$X = 2 \qquad 2(4) = 5 + 3$

$8 = 8$ ok

18. $\dfrac{6X - 42}{X + 1} \times \dfrac{X^2 - 1}{-9X + 9} =$

$\dfrac{(6)(X - 7)}{(X + 1)} \times \dfrac{(X - 1)(X + 1)}{(-9)(X - 1)} = \dfrac{(6)(X - 7)}{(-9)} =$

$\dfrac{2(X - 7)}{-3} = \dfrac{2X - 14}{-3}$ or $-\dfrac{2X - 14}{3}$

19. $2\sqrt{\dfrac{3}{5}} - X\sqrt{20} = 2\dfrac{\sqrt{3}}{\sqrt{5}} - X\sqrt{4}\sqrt{5} =$

$\dfrac{2\sqrt{3}\sqrt{5}}{\sqrt{5}\sqrt{5}} - 2X\sqrt{5} = \dfrac{2\sqrt{15}}{5} - \dfrac{2X\sqrt{5}(5)}{(5)} =$

$\dfrac{2\sqrt{15} - 10X\sqrt{5}}{5}$

20. $\dfrac{2 - \dfrac{3X + 4}{5}}{3X + \dfrac{2 + X}{3}} = \dfrac{\dfrac{10}{5} - \dfrac{3X + 4}{5}}{\dfrac{9X}{3} + \dfrac{2 + X}{3}} =$

$\dfrac{\dfrac{10 - (3X + 4)}{5}}{\dfrac{9X + 2 + X}{3}} = \dfrac{\dfrac{6 - 3X}{5}}{\dfrac{10X + 2}{3}} =$

$\dfrac{\dfrac{6 - 3X}{5} \times \dfrac{3}{10X + 2}}{\dfrac{10X + 2}{3} \times \dfrac{3}{10X + 2}} =$

$\dfrac{6 - 3X}{5} \times \dfrac{3}{10X + 2} = \dfrac{18 - 9X}{50X + 10}$

Systematic Review 9E

1. $(X + 8)^2 = X^2 + 16X + 64$

2. $(4X + 1)^2 = 16X^2 + 8X + 1$

3. $X^2 + 8X + 16 = (X + 4)^2$

4. $9X^2 + 12X + 4 = (3X + 2)^2$

5. $(X + 6)^3 =$
$(X)^3 + 3(X)^2(6) + 3(X)(6)^2 + (6)^3 =$
$X^3 + 18X^2 + 108X + 216$

6. $(2X + 5)^3 =$
$(2X)^3 + 3(2X)^2(5) + 3(2X)(5)^2 + (5)^3 =$
$8X^3 + 60X^2 + 150X + 125$

7. $(X + 4)^3 =$
$(X)^3 + 3(X)^2(4) + 3(X)(4)^2 + (4)^3 =$
$X^3 + 12X^2 + 48X + 64$

8. $\left(X - \dfrac{2}{3}\right)^3 =$
$(X)^3 + 3(X)^2\left(-\dfrac{2}{3}\right) + 3(X)\left(-\dfrac{2}{3}\right)^2 + \left(-\dfrac{2}{3}\right)^3 =$
$X^3 - 2X^2 + \dfrac{4}{3}X - \dfrac{8}{27}$

9. $6 - 3\sqrt{-9}$ or $6 - 3\sqrt{9}\sqrt{-1} = 6 - 9i$

10. $100X^2 - 83 = \left(10X - \sqrt{83}\right)\left(10X + \sqrt{83}\right)$

11. $\dfrac{5\sqrt{7}}{2\sqrt{7} - 3} = \dfrac{5\sqrt{7}\left(2\sqrt{7} + 3\right)}{\left(2\sqrt{7} - 3\right)\left(2\sqrt{7} + 3\right)} =$
$\dfrac{10\sqrt{49} + 15\sqrt{7}}{4(7) - 9} =$
$\dfrac{10(7) + 15\sqrt{7}}{28 - 9} = \dfrac{70 + 15\sqrt{7}}{19}$

12. $\dfrac{4}{10 - 7i} = \dfrac{4(10 + 7i)}{(10 - 7i)(10 + 7i)} =$
$\dfrac{40 + 28i}{100 - 49i^2} = \dfrac{40 + 28i}{100 - (-49)} =$
$\dfrac{40 + 28i}{100 + 49} = \dfrac{40 + 28i}{149}$

13. $\left(10i^2\right)\left(\sqrt{-75}\right) = 10i^2\sqrt{25}\sqrt{-1}\sqrt{3} =$
$10(-1)(5)(i)\sqrt{3} = -50i\sqrt{3}$

14. $(2i)^3 = 8\left(i^3\right) = 8i^2i = 8(-1)i = -8i$

15. $\left(9^{\frac{1}{2}}\right)^{-5} = 3^{-5} = \dfrac{1}{3^5} = \dfrac{1}{243}$

16. $\left(\sqrt[3]{X^9}\right)^{-2} = \left[\left(X^9\right)^{\frac{1}{3}}\right]^{-2} =$
$X^{(9)\left(\frac{1}{3}\right)(-2)} = X^{-6}$ or $\dfrac{1}{X^6}$

17. $5X^2 - 3X = 0$
$(X)(5X - 3) = 0$

$X = 0 \qquad 5(0)^2 - 3(0) = 0$
$\qquad\qquad\qquad\qquad 0 = 0$ ok

$5X - 3 = 0 \quad 5\left(\dfrac{3}{5}\right)^2 - 3\left(\dfrac{3}{5}\right) = 0$
$5X = 3$
$X = \dfrac{3}{5} \qquad 5\left(\dfrac{9}{25}\right) - \dfrac{9}{5} = 0$
$\qquad\qquad\qquad \dfrac{45}{25} - \dfrac{9}{5} = 0$
$\qquad\qquad\qquad \dfrac{9}{5} - \dfrac{9}{5} = 0$
$\qquad\qquad\qquad\qquad 0 = 0$ ok

18. $\dfrac{X^2 + 5X}{25 - X^2} \div \dfrac{X + 5}{10X - 50} =$
$\dfrac{(X)(X + 5)}{(-1)\left(X^2 - 25\right)} \times \dfrac{(10)(X - 5)}{(X + 5)} =$
$\dfrac{(X)\cancel{(X + 5)}}{(-1)\cancel{(X - 5)}(X + 5)} \times \dfrac{(10)\cancel{(X - 5)}}{\cancel{(X + 5)}} = \dfrac{10X}{-X - 5}$

19. $3\sqrt{\dfrac{2}{7}} - 7\sqrt{\dfrac{3}{X}} = 3\left(\dfrac{\sqrt{2}}{\sqrt{7}}\right) - 7\left(\dfrac{\sqrt{3}}{\sqrt{X}}\right) =$
$\dfrac{3\sqrt{2}}{\sqrt{7}} - \dfrac{7\sqrt{3}}{\sqrt{X}} = \dfrac{3\sqrt{2}\sqrt{7}}{\sqrt{7}\sqrt{7}} - \dfrac{7\sqrt{3}\sqrt{X}}{\sqrt{X}\sqrt{X}} =$
$\dfrac{3\sqrt{14}}{7} - \dfrac{7\sqrt{3X}}{X} = \dfrac{3\sqrt{14}\,(X)}{7(X)} - \dfrac{7\sqrt{3X}\,(7)}{X(7)} =$
$\dfrac{3X\sqrt{14}}{7X} - \dfrac{49\sqrt{3X}}{7X} = \dfrac{3X\sqrt{14} - 49\sqrt{3X}}{7X}$

20. $\dfrac{1 + \dfrac{4 - 5X}{2}}{\dfrac{X + 3}{4}} = \dfrac{\dfrac{2}{2} + \dfrac{4 - 5X}{2}}{\dfrac{X + 3}{4}} =$

$\dfrac{\dfrac{6 - 5X}{2} \times \dfrac{4}{X + 3}}{\dfrac{X + 3}{4} \times \dfrac{4}{X + 3}} = \qquad \dfrac{6 - 5X}{2} \times \dfrac{4}{X + 3} =$

$\dfrac{6 - 5X}{1} \times \dfrac{2}{X + 3} = \dfrac{12 - 10X}{X + 3}$

Lesson Practice 10A

1. $6 + 1 = 7$ terms:

   $1A^6 7^0 + 6A^5 7^1 + 15A^4 7^2 + 20A^3 7^3 + 15A^2 7^4 + 6A^1 7^5 + 1A^0 7^6 = A^6 + 42A^5 + 735A^4 + 6,860A^3 + 36,015A^2 + 100,842A + 117,649$

2. $5 + 1 = 6$ terms: $1X^5(-2)^0 + 5X^4(-2)^1 + 10X^3(-2)^2 + 10X^2(-2)^3 + 5X^1(-2)^4 + 1X^0(-2)^5 = X^5 - 10X^4 + 40X^3 - 80X^2 + 80X - 32$

3. $4 + 1 = 5$ terms: $1(3X)^4 1^0 + 4(3X)^3 1^1 + 6(3X)^2 1^2 + 4(3X)^1 1^3 + 1(3X)^0 1^4 = 81X^4 + 108X^3 + 54X^2 + 12X + 1$

4. $6 + 1 = 7$ terms: $1R^6\left(-\frac{1}{2}\right)^0 + 6R^5\left(-\frac{1}{2}\right)^1 + 15R^4\left(-\frac{1}{2}\right)^2 + 20R^3\left(-\frac{1}{2}\right)^3 + 15R^2\left(-\frac{1}{2}\right)^4 +$

   $6R^1\left(-\frac{1}{2}\right)^5 + 1R^0\left(-\frac{1}{2}\right)^6 = R^6 - 3R^5 + \frac{15}{4}R^4 - \frac{5}{2}R^3 + \frac{15}{16}R^2 - \frac{3}{16}R + \frac{1}{64}$

5. fourth term would have $4 - 1$ or 3 factors

   exponent of 2B would be 3

   exponent of A would be $5 - 3 = 2$

   $\dfrac{5 \cdot \cancel{4}^2 \cdot \cancel{3}}{1 \cdot \cancel{2} \cdot \cancel{3}} = \dfrac{5 \cdot 2}{1} = 10 : \quad 10A^2(2B)^3 = 80A^2 B^3$

6. 3rd term $-$ 2 factors : $\dfrac{\cancel{6}^3 \cdot 5}{1 \cdot \cancel{2}} = 15 : 15X^4 2^2 = 60X^4$

7. 4 factors : $\dfrac{7 \cdot \cancel{6}^2 \cdot 5 \cdot \cancel{4}}{1 \cdot \cancel{2} \cdot \cancel{3} \cdot \cancel{4}} = 35 : 35(2X)^3(-2)^4 = 4,480X^3$

8. 1 factor : $\dfrac{4}{1} = 4 : 4X^3\left(-\frac{1}{3}\right)^1 = -\frac{4}{3}X^3$

9. 3 factors : $\dfrac{\cancel{6}^2 \cdot 5 \cdot \cancel{4}^2}{1 \cdot \cancel{2} \cdot \cancel{3}} = 20 : 20X^3 Y^3$

10. 5 factors : $\dfrac{\cancel{8}^4 \cdot 7 \cdot \cancel{6}^2 \cdot 5 \cdot \cancel{4}}{1 \cdot \cancel{2} \cdot \cancel{3} \cdot \cancel{4} \cdot \cancel{5}} = 56 : 56P^3(-Q)^5 = -56P^3 Q^5$

Lesson Practice 10B

1. 4 terms: $1B^3 4^0 + 3B^2 4^1 + 3B^1 4^2 + 1B^0 4^3 = B^3 + 12B^2 + 48B + 64$

2. 7 terms: $1(2X)^6 1^0 + 6(2X)^5 1^1 + 15(2X)^4 1^2 + 20(2X)^3 1^3 + 15(2X)^2 1^4 + 6(2X)^1 1^5 + 1(2X)^0 1^6 = 64X^6 + 192X^5 + 240X^4 + 160X^3 + 60X^2 + 12X + 1$

3. 6 terms: $1R^5(-T)^0 + 5R^4(-T)^1 + 10R^3(-T)^2 + 10R^2(-T)^3 + 5R^1(-T)^4 + 1R^0(-T)^5 = R^5 - 5R^4 T + 10R^3 T^2 - 10R^2 T^3 + 5RT^4 - T^5$

4. 5 terms: $1\left(\frac{1}{2}X\right)^4\left(\frac{1}{2}Y\right)^0 + 4\left(\frac{1}{2}X\right)^3\left(\frac{1}{2}Y\right)^1 + 6\left(\frac{1}{2}X\right)^2\left(\frac{1}{2}Y\right)^2 + 4\left(\frac{1}{2}X\right)^1\left(\frac{1}{2}Y\right)^3 + 1\left(\frac{1}{2}X\right)^0\left(\frac{1}{2}Y\right)^4 =$

   $\frac{1}{16}X^4 + \frac{1}{4}X^3 Y + \frac{3}{8}X^2 Y^2 + \frac{1}{4}XY^3 + \frac{1}{16}Y^4$

5. 0 factors: $1X^{10}(2Y)^0 = X^{10}$

6. 6 factors : $\dfrac{6 \cdot 5 \cdot 4 \cdot 3 \cdot 2 \cdot 1}{1 \cdot 2 \cdot 3 \cdot 4 \cdot 5 \cdot 6} = 1:\ 1A^0B^6 = B^6$

7. $\dfrac{\cancel{4}^2 \cdot 3}{1 \cdot \cancel{2}} = 6:\ 6(2X)^2 3^2 = 216X^2$

8. $\dfrac{\cancel{6}^3 \cdot 5 \cdot \cancel{4} \cdot \cancel{3}}{1 \cdot \cancel{2} \cdot \cancel{3} \cdot \cancel{4}} = 15:\ 15Q^2\left(\dfrac{2}{3}\right)^4 = 15Q^2\left(\dfrac{16}{81}\right) = \dfrac{80}{27}Q^2$

9. $\dfrac{5}{1} = 5:\ 5(3R)^4(-T)^1 = -405R^4T$

10. $\dfrac{4 \cdot 3 \cdot \cancel{2}}{1 \cdot \cancel{2} \cdot \cancel{3}} = 4:\ 4(2X)^1(-2Y)^3 = -64XY^3$

Systematic Review 10C

1. $5+1=6$

2. $1X^5 3^0 + 5X^4 3^1 + 10X^3 3^2 + 10X^2 3^3 + 5X^1 3^4 + 1X^0 3^5 = X^5 + 15X^4 + 90X^3 + 270X^2 + 405X + 243$

3. $6+1=7$

4. $1X^6(-4)^0 + 6X^5(-4)^1 + 15X^4(-4)^2 + 20X^3(-4)^3 + 15X^2(-4)^4 + 6X^1(-4)^5 + 1X^0(-4)^6 =$
   $X^6 - 24X^5 + 240X^4 - 1,280X^3 + 3,840X^2 - 6,144X + 4,096$

5. $\dfrac{4 \cdot \cancel{3} \cdot \cancel{2}}{1 \cdot \cancel{2} \cdot \cancel{3}} = 4:\ 4(2X)^1(Y)^3 = 8XY^3$

6. $\dfrac{4}{1} = 4:\ 4(2X)^3(Y)^1 = 32X^3Y$

7. $\dfrac{\cancel{6}^3 \cdot 5}{1 \cdot \cancel{2}} = 15:\ 15(X)^4(-2)^2 = 60X^4$

8. $\dfrac{\cancel{6} \cdot 5 \cdot \cancel{4} \cdot 3}{1 \cdot \cancel{2} \cdot \cancel{3} \cdot \cancel{4}} = 15:\ 15(X)^2(-2)^4 = 240X^2$

9. $(X+3A)^2 = X^2 + 6AX + 9A^2$

10. $9X^2 - 42X + 49 = (3X - 7)^2$

11. $(X+12)^3 = (X)^3 + 3(X)^2(12) + 3(X)(12)^2 + (12)^3 = X^3 + 36X^2 + 432X + 1,728$

12. $\left(2X - \dfrac{1}{3}\right)^3 = (2X)^3 + 3(2X)^2\left(-\dfrac{1}{3}\right)^1 + 3(2X)\left(-\dfrac{1}{3}\right)^2 + \left(-\dfrac{1}{3}\right)^3 =$
    $8X^3 - 4X^2 + \dfrac{2}{3}X - \dfrac{1}{27}$

13. $\dfrac{2\sqrt{2}+1}{3\sqrt{3}-1} = \dfrac{(2\sqrt{2}+1)(3\sqrt{3}+1)}{(3\sqrt{3}-1)(3\sqrt{3}+1)} = \dfrac{6\sqrt{6}+2\sqrt{2}+3\sqrt{3}+1}{9(3)-1} = \dfrac{6\sqrt{6}+2\sqrt{2}+3\sqrt{3}+1}{26}$

14. $\dfrac{-8i}{4+5i} = \dfrac{-8i(4-5i)}{(4+5i)(4-5i)} = \dfrac{-32i+40i^2}{16-25i^2} = \dfrac{-32i+40(-1)}{16-(-25)} = \dfrac{-32i-40}{41}$

15. $(-8i)(7i)(11i) = (-8)(7)(11)i^3 = -616i^3 = -616(-1)i = 616i$

16. $\left(-3\sqrt{-6}\right)\left(4\sqrt{6}\right) = (-3)(4)\sqrt{-1}\sqrt{6}\sqrt{6} = -12i\sqrt{36} = -12i(6) = -72i$

17. $6\sqrt{\dfrac{1}{2}} - \dfrac{9\sqrt{8}}{2} = 6\left(\dfrac{\sqrt{1}}{\sqrt{2}}\right) - \dfrac{9\sqrt{4}\sqrt{2}}{2} = \dfrac{6}{\sqrt{2}} - \dfrac{9(2)\sqrt{2}}{2} =$

$\dfrac{6\sqrt{2}}{\sqrt{2}\sqrt{2}} - \dfrac{18\sqrt{2}}{2} = \dfrac{6\sqrt{2}}{2} - \dfrac{18\sqrt{2}}{2} = -\dfrac{12\sqrt{2}}{2} = -6\sqrt{2}$

18. $\left(\sqrt[3]{64}\right)^{\frac{5}{2}} = 4^{\frac{5}{2}} = 2^5 = 32$

19. $\left[\left(2X^2 - 3X - 2\right) \div \left(X^3 - 2X^2\right)\right] \div \left[\left(4X^2 - 12X\right) \div \left(X^2 - 5X + 6\right)\right] =$

$\dfrac{2X^2 - 3X - 2}{X^3 - 2X^2} \div \dfrac{4X^2 - 12X}{X^2 - 5X + 6} = \dfrac{2X^2 - 3X - 2}{X^3 - 2X^2} \times \dfrac{X^2 - 5X + 6}{4X^2 - 12X} =$

$\dfrac{(2X + 1)\cancel{(X-2)}}{(X^2)\cancel{(X-2)}} \times \dfrac{\cancel{(X-3)}(X - 2)}{(4X)\cancel{(X-3)}} = \dfrac{2X + 1}{X^2} \times \dfrac{X - 2}{4X} = \dfrac{2X^2 - 3X - 2}{4X^3}$

20. $\dfrac{10X^1Y^0X^1}{X^1Y^1} - \dfrac{4^{-2}Y^2X^{-3}}{4^{-1}X^{-2}Y^1Y^0} - \dfrac{2Y^{-1}X^1X^{-2}}{Y^{-2}} =$

$10X^1Y^0X^1X^{-1}Y^{-1} - 4^{-2}Y^2X^{-3}4^1X^2Y^{-1}Y^0 - 2Y^{-1}X^1X^{-2}Y^2 =$

$10X^{1+1+(-1)}Y^{0+(-1)} - 4^{-2+1}X^{-3+2}Y^{2+(-1)+0} - 2X^{1+(-2)}Y^{-1+2} =$

$10XY^{-1} - 4^{-1}X^{-1}Y^1 - 2X^{-1}Y^1 =$

$\dfrac{10X}{Y} - \dfrac{Y}{4X} - \dfrac{2Y}{X} = \dfrac{10X(4X)}{Y(4X)} - \dfrac{Y(Y)}{4X(Y)} - \dfrac{2Y(4Y)}{X(4Y)} = \dfrac{40X^2 - 9Y^2}{4XY}$

Systematic Review 10D

1. $5 + 1 = 6$

2. $1(X)^5(A)^0 + 5(X)^4(A)^1 + 10(X)^3(A)^2 + 10(X)^2(A)^3 + 5(X)^1(A)^4 + 1(X)^0(A)^5 =$

$X^5 + 5X^4A + 10X^3A^2 + 10X^2A^3 + 5XA^4 + A^5$

3. $4 + 1 = 5$

4. $1\left(\dfrac{1}{3}X\right)^4(2)^0 + 4\left(\dfrac{1}{3}X\right)^3(2)^1 + 6\left(\dfrac{1}{3}X\right)^2(2)^2 + 4\left(\dfrac{1}{3}X\right)^1(2)^3 + 1\left(\dfrac{1}{3}X\right)^0(2)^4 =$

$\dfrac{1}{81}X^4 + \dfrac{8}{27}X^3 + \dfrac{8}{3}X^2 + \dfrac{32}{3}X + 16$

5. $\dfrac{\cancel{6}^2 \cdot 5 \cdot \cancel{4}^2}{1 \cdot \cancel{2} \cdot \cancel{3}}X^3\left(-\dfrac{2}{3}\right)^3 = -\dfrac{160}{27}X^3$

6. $\dfrac{6}{1}X^5\left(-\dfrac{2}{3}\right)^1 = -4X^5$

7. $\dfrac{5 \cdot 4}{1 \cdot 2}X^3(-3)^2 = 90X^3$

8. $\dfrac{5 \cdot 4 \cdot 3 \cdot 2 \cdot 1}{1 \cdot 2 \cdot 3 \cdot 4 \cdot 5}X^0(-3)^5 = -243$

9. $\left(X - 2A\right)^2 = X^2 - 4AX + 4A^2$

10. $16X^2 - 24X + 9 = \left(4X - 3\right)^2$

11. $(X-10)^3 = 1(X)^3(-10)^0 + 3(X)^2(-10)^1 + 3(X)^1(-10)^2 + 1(X)^0(-10)^3 = X^3 - 30X^2 + 300X - 1,000$

12. $\left(X+\dfrac{1}{2}\right)^3 = 1(X)^3\left(\dfrac{1}{2}\right)^0 + 3(X)^2\left(\dfrac{1}{2}\right)^1 + 3(X)^1\left(\dfrac{1}{2}\right)^2 + 1(X)^0\left(\dfrac{1}{2}\right)^3 = X^3 + \dfrac{3}{2}X^2 + \dfrac{3}{4}X + \dfrac{1}{8}$

13. $\dfrac{8\sqrt{11}-6}{4\sqrt{2}+3} = \dfrac{\left(8\sqrt{11}-6\right)\left(4\sqrt{2}-3\right)}{\left(4\sqrt{2}+3\right)\left(4\sqrt{2}-3\right)} = \dfrac{32\sqrt{22}-24\sqrt{11}-24\sqrt{2}+18}{16(2)-9} =$

$\dfrac{32\sqrt{22}-24\sqrt{11}-24\sqrt{2}+18}{32-9} = \dfrac{32\sqrt{22}-24\sqrt{11}-24\sqrt{2}+18}{23}$

14. $\dfrac{2i}{1+9i} = \dfrac{2i(1-9i)}{(1+9i)(1-9i)} = \dfrac{2i-18i^2}{1-81i^2} = \dfrac{2i-(18)(-1)}{1-81(-1)} = \dfrac{2i+18}{1+81} = \dfrac{2i+18}{82} = \dfrac{i+9}{41}$

15. $(5i)(9i)(12) = (5)(9)(12)(i^2) = 540(-1) = -540$

16. $\left(8\sqrt{-7}\right)\left(9\sqrt{-7}\right) = (8)(9)\sqrt{-7}\sqrt{-7} = 72(-7) = -504$

17. $\sqrt{\dfrac{5}{8}} - 2\sqrt{160} = \dfrac{\sqrt{5}}{\sqrt{8}} - 2\sqrt{16}\sqrt{10} = \dfrac{\sqrt{5}}{\sqrt{8}}\dfrac{\sqrt{2}}{\sqrt{2}} - \dfrac{2(4)\sqrt{10}}{1} =$

$\dfrac{\sqrt{10}}{\sqrt{16}} - \dfrac{8\sqrt{10}}{1} = \dfrac{\sqrt{10}}{4} - \dfrac{8\sqrt{10}(4)}{1(4)} = \dfrac{1\sqrt{10}-32\sqrt{10}}{4} = \dfrac{-31\sqrt{10}}{4}$

18. $\left(\sqrt[3]{125}\right)^3 = 5^3 = 125$

19. $\dfrac{4X^2-1}{2X+1} \div \dfrac{4X^2-4X+1}{8X} = \dfrac{4X^2-1}{2X+1} \times \dfrac{8X}{4X^2-4X+1} = \dfrac{\cancel{(2X-1)}(2X+1)}{\cancel{(2X+1)}} \times \dfrac{8X}{\cancel{(2X-1)}(2X-1)} = \dfrac{8X}{2X-1}$

20. $4X^1Y^0 + \dfrac{6X^1Y^{-2}}{X^1Y^{-3}} + \dfrac{9X^4Y^5}{X^3} = 4X + 6X^1Y^{-2}X^{-1}Y^3 + 9X^4Y^5X^{-3} = 4X + 6Y + 9XY^5$

Systematic Review 10E

1. $5+1=6$

2. $1(X)^5(-4)^0 + 5(X)^4(-4)^1 + 10(X)^3(-4)^2 + 10(X)^2(-4)^3 + 5(X)^1(-4)^4 + 1(X)^0(-4)^5 =$
   $X^5 - 20X^4 + 160X^3 - 640X^2 + 1,280X - 1,024$

3. $4+1=5$

4. $1(X)^4(2)^0 + 4(X)^3(2)^1 + 6(X)^2(2)^2 + 4(X)^1(2)^3 + 1(X)^0(2)^4 = X^4 + 8X^3 + 24X^2 + 32X + 16$

5. $\dfrac{5 \cdot 4 \cdot 3}{1 \cdot 2 \cdot 3}(2X)^2(3)^3 = (10)4X^2(27) = 1,080X^2$

6. $\dfrac{5}{1}(2X)^4(3)^1 = (5)16X^4(3) = 240X^4$

7. $1(2X)^4(1)^0 = 16X^4$

8. $\dfrac{4 \cdot 3}{1 \cdot 2}(2X)^2(1)^2 = (6)4X^2(1) = 24X^2$

9. $(X+A)^2 = X^2 + 2XA + A^2$

10. $36X^2 - 6X + \dfrac{1}{4} = \left(6X - \dfrac{1}{2}\right)^2$

11. $\left(X + \dfrac{4}{5}\right)^3 = 1(X)^3\left(\dfrac{4}{5}\right)^0 + 3(X)^2\left(\dfrac{4}{5}\right)^1 + 3(X)^1\left(\dfrac{4}{5}\right)^2 + \left(\dfrac{4}{5}\right)^3 = X^3 + \dfrac{12}{5}X^2 + \dfrac{48}{25}X + \dfrac{64}{125}$

12. $(3X+1)^3 = 1(3X)^3(1)^0 + 3(3X)^2(1)^1 + 3(3X)^1(1)^2 + 1(3X)^0(1)^3 = 27X^3 + 27X^2 + 9X + 1$

13. $\dfrac{\sqrt{8}}{5\sqrt{7}-4} = \dfrac{\sqrt{8}\left(5\sqrt{7}+4\right)}{\left(5\sqrt{7}-4\right)\left(5\sqrt{7}+4\right)} = \dfrac{5\sqrt{56}+4\sqrt{8}}{25(7)-16} =$

$\dfrac{5\sqrt{4}\sqrt{14}+4\sqrt{4}\sqrt{2}}{175-16} = \dfrac{5(2)\sqrt{14}+4(2)\sqrt{2}}{159} = \dfrac{10\sqrt{14}+8\sqrt{2}}{159}$

14. $\dfrac{-3i}{2-11i} = \dfrac{-3i(2+11i)}{(2-11i)(2+11i)} = \dfrac{-6i-33i^2}{4-121i^2} = \dfrac{-6i-33(-1)}{4-121(-1)} = \dfrac{-6i+33}{4+121} = \dfrac{-6i+33}{125}$

15. $(-5i)(6) = -30i$

16. $\left(5\sqrt{-8}\right)\left(-7\sqrt{-2}\right) = (5)(-7)\sqrt{-8}\sqrt{-2} = -35\sqrt{8}\sqrt{-1}\sqrt{2}\sqrt{-1} = -35\sqrt{16}i^2 = -35(4)(-1) = 140$

17. $\sqrt{\dfrac{1}{10}} + 3\sqrt{90} = \dfrac{\sqrt{1}}{\sqrt{10}} + 3\sqrt{9}\sqrt{10} = \dfrac{1}{\sqrt{10}} + \dfrac{3(3)\sqrt{10}}{1} =$

$\dfrac{\sqrt{10}}{\sqrt{10}\sqrt{10}} + \dfrac{9\sqrt{10}}{1} = \dfrac{\sqrt{10}}{10} + \dfrac{9\sqrt{10}\,(10)}{1(10)} = \dfrac{\sqrt{10}+90\sqrt{10}}{10} = \dfrac{91\sqrt{10}}{10}$

18. $\left(\sqrt{81}\right)^{\frac{3}{2}} = 9^{\frac{3}{2}} = 3^3 = 27$

19. $\dfrac{X^2+5X+6}{X^2-16} \div \dfrac{X^2+6X+9}{X^2+6X+8} = \dfrac{X^2+5X+6}{X^2-16} \times \dfrac{X^2+6X+8}{X^2+6X+9} =$

$\dfrac{(X+3)(X+2)}{(X-4)(X+4)} \times \dfrac{(X+2)(X+4)}{(X+3)(X+3)} = \dfrac{(X+2)(X+2)}{(X-4)(X+3)} = \dfrac{X^2+4X+4}{X^2-X-12}$

20. $\dfrac{12X^3X^2X^{-1}Y^{-2}}{Y^{-7}} + \dfrac{10X^2}{X^{-2}Y^5} + \dfrac{8XXYX^2}{Y^{-2}X^0Y^{-2}} = 12X^4Y^5 + \dfrac{10X^4}{Y^5} + 8X^4Y^5 = 20X^4Y^5 + \dfrac{10X^4}{Y^5}$

Lesson Practice 11A

1. $10 \div 2 = 5; \ 5^2 = 25$

2. $-8 \div 2 = -4; \ (-4)^2 = 16$

3. $\sqrt{4} = 2; \ 2(2) = 4; \ 4X$

4. $\sqrt{225} = 15; \ 2(15) = 30; \ 30A$

5. $X^2 + 2X + 3 = 0$
$X^2 + 2X + 1 = -2$
$(X + 1)^2 = -2$
$X + 1 = \pm\sqrt{-2}$
$X = -1 \pm i\sqrt{2}$

For most problems, the check for only one solution is shown. The check for the other solution should be similar.

$\left(-1 + i\sqrt{2}\right)^2 + 2\left(-1 + i\sqrt{2}\right) + 3 = 0$
$\left(-1 + i\sqrt{2}\right)\left(-1 + i\sqrt{2}\right) - 2 + 2i\sqrt{2} + 3 = 0$
$1 \cancel{-2i\sqrt{2}} + \left(i\sqrt{2}\right)^2 - 2 \cancel{+2i\sqrt{2}} + 3 = 0$
$1 + 2i^2 - 2 + 3 = 0$
$1 + 2(-1) - 2 + 3 = 0$
$1 - 2 - 2 + 3 = 0$

6. $X^2 - 5X + 4 = 0$
$(X - 4)(X - 1) = 0$

$\begin{array}{ll} X - 4 = 0 & (4)^2 - 5(4) + 4 = 0 \\ X = 4 & 16 - 20 + 4 = 0 \\ & 0 = 0 \end{array}$

$\begin{array}{ll} X - 1 = 0 & (1)^2 - 5(1) + 4 = 0 \\ X = 1 & 1 - 5 + 4 = 0 \\ & 0 = 0 \end{array}$

7. $2X^2 + 8X + 2 = 0$
$X^2 + 4X + 1 = 0$
$X^2 + 4X + 4 = 3$
$(X + 2)^2 = 3$
$X + 2 = \pm\sqrt{3}$
$X = -2 \pm \sqrt{3}$

$2\left(-2 + \sqrt{3}\right)^2 + 8\left(-2 + \sqrt{3}\right) + 2 = 0$
$2\left(-2 + \sqrt{3}\right)\left(-2 + \sqrt{3}\right) - 16 + 8\sqrt{3} + 2 = 0$
$2\left(4 - 4\sqrt{3} + 3\right) - 16 + 8\sqrt{3} + 2 = 0$
$8 \cancel{-8\sqrt{3}} + 6 - 16 \cancel{+8\sqrt{3}} + 2 = 0$
$8 + 6 - 16 + 2 = 0$
$0 = 0$

8. $X^2 + 4X - 7 = 0$
$X^2 + 4X + 4 = 7 + 4$
$(X + 2)^2 = 11$
$X + 2 = \pm\sqrt{11}$
$X = -2 \pm \sqrt{11}$

$\left(-2 + \sqrt{11}\right)^2 + 4\left(-2 + \sqrt{11}\right) - 7 = 0$
$\left(-2 + \sqrt{11}\right)\left(-2 + \sqrt{11}\right) - 8 + 4\sqrt{11} - 7 = 0$
$4 \cancel{-4\sqrt{11}} + 11 - 8 \cancel{+4\sqrt{11}} - 7 = 0$
$4 + 11 - 8 - 7 = 0$
$0 = 0$

9. $3X^2 - 9X + 3 = 0$
$X^2 - 3X + 1 = 0$
$X^2 - 3X + \dfrac{9}{4} = -1 + \dfrac{9}{4}$
$\left(X - \dfrac{3}{2}\right)^2 = \dfrac{5}{4}$
$X - \dfrac{3}{2} = \pm\sqrt{\dfrac{5}{4}}$
$X = \dfrac{3}{2} \pm \dfrac{\sqrt{5}}{2} = \dfrac{3 \pm \sqrt{5}}{2}$

$$3\left(\frac{3+\sqrt{5}}{2}\right)^2 - 9\left(\frac{3+\sqrt{5}}{2}\right) + 3 = 0$$

$$3\left(\frac{3+\sqrt{5}}{2}\right)\left(\frac{3+\sqrt{5}}{2}\right) - \frac{27+9\sqrt{5}}{2} + 3 = 0$$

$$3\left(\frac{9+6\sqrt{5}+5}{4}\right) - \frac{27+9\sqrt{5}}{2} + 3 = 0$$

$$\frac{27+18\sqrt{5}+15}{4} - \frac{27+9\sqrt{5}}{2} + 3 = 0$$

$$\frac{42+18\sqrt{5}}{4} - \frac{27+9\sqrt{5}}{2} + 3 = 0$$

$$\frac{\cancel{2}(21+9\sqrt{5})}{\cancel{4}_2} - \frac{27+9\sqrt{5}}{2} + 3 = 0$$

$$(2)\frac{21+9\sqrt{5}}{2} - (2)\frac{27+9\sqrt{5}}{2} + (2)3 = (2)0$$

$$21+9\sqrt{5} - (27+9\sqrt{5}) + 6 = 0$$

$$21\cancel{+9\sqrt{5}} - 27\cancel{-9\sqrt{5}} + 6 = 0$$

$$21-27+6 = 0$$

$$0 = 0$$

10. $X^2 - 2X - 11 = 0$

$$X^2 - 2X + 1 = 11 + 1$$

$$(X-1)^2 = 12$$

$$X - 1 = \pm\sqrt{12}$$

$$X = 1 \pm \sqrt{4}\sqrt{3}$$

$$X = 1 \pm 2\sqrt{3}$$

$$(1+2\sqrt{3})^2 - 2(1+2\sqrt{3}) - 11 = 0$$

$$(1+2\sqrt{3})(1+2\sqrt{3}) - 2 - 4\sqrt{3} - 11 = 0$$

$$1\cancel{+4\sqrt{3}} + 4(3) - 2\cancel{-4\sqrt{3}} - 11 = 0$$

$$1 + 12 - 2 - 11 = 0$$

$$0 = 0$$

5. $X^2 + 4X + 16 = 0$

$$X^2 + 4X + 4 = -16 + 4$$

$$(X+2)^2 = -12$$

$$X + 2 = \pm\sqrt{-12}$$

$$X = -2 \pm 2i\sqrt{3}$$

$$(-2+2i\sqrt{3})^2 + 4(-2+2i\sqrt{3}) + 16 = 0$$

$$(-2+2i\sqrt{3})(-2+2i\sqrt{3}) - 8 + 8i\sqrt{3} + 16 = 0$$

$$4\cancel{-8i\sqrt{3}} - 12 - 8\cancel{+8i\sqrt{3}} + 16 = 0$$

$$4 - 12 - 8 + 16 = 0$$

$$0 = 0$$

6. $2X^2 - 16X - 4 = 0$

$$2(X^2 - 8X - 2) = 2(0)$$

$$X^2 - 8X - 2 = 0$$

$$X^2 - 8X + 16 = 2 + 16$$

$$(X-4)^2 = 18$$

$$X - 4 = \pm\sqrt{18}$$

$$X = 4 \pm \sqrt{18} = 4 \pm 3\sqrt{2}$$

$$2(4+3\sqrt{2})^2 - 16(4+3\sqrt{2}) - 4 = 0$$

$$2(4+3\sqrt{2})(4+3\sqrt{2}) - 64 - 48\sqrt{2} - 4 = 0$$

$$2(16+24\sqrt{2}+9(2)) - 64 - 48\sqrt{2} - 4 = 0$$

$$32\cancel{+48\sqrt{2}} + 36 - 64\cancel{-48\sqrt{2}} - 4 = 0$$

$$32 + 36 - 64 - 4 = 0$$

$$0 = 0$$

Lesson Practice 11B

1. $-3 \div 2 = -\dfrac{3}{2}; \left(-\dfrac{3}{2}\right)^2 = \dfrac{9}{4}$

2. $\dfrac{1}{3} \div 2 = \dfrac{1}{6}; \left(\dfrac{1}{6}\right)^2 = \dfrac{1}{36}$

3. $\sqrt{1} = 1; 2(1) = 2; 2X$

4. $\sqrt{\dfrac{16}{100}} = \dfrac{4}{10}; 2\left(\dfrac{4}{10}\right) = \dfrac{8}{10} = \dfrac{4}{5}; \dfrac{4}{5}$ Y

7.

$$A^2 + 5A + \frac{1}{4} = 0$$

$$A^2 + 5A + \frac{25}{4} = \frac{-1}{4} + \frac{25}{4}$$

$$\left(A + \frac{5}{2}\right)^2 = \frac{24}{4}$$

$$\left(A + \frac{5}{2}\right)^2 = 6$$

$$A + \frac{5}{2} = \pm\sqrt{6}$$

$$A = -\frac{5}{2} \pm \sqrt{6}$$

$$\left(-\frac{5}{2} + \sqrt{6}\right)^2 + 5\left(-\frac{5}{2} + \sqrt{6}\right) + \frac{1}{4} = 0$$

$$\frac{25}{4} - 5\sqrt{6} + 6 - \frac{25}{2} + 5\sqrt{6} + \frac{1}{4} = 0$$

$$\frac{25}{4} + \frac{24}{4} - \frac{50}{4} + \frac{1}{4} = 0$$

$$\frac{0}{4} = 0$$

$$0 = 0$$

8.

$$X^2 + 8X - 10 = 0$$

$$X^2 + 8X + 16 = 10 + 16$$

$$(X + 4)^2 = 26$$

$$X + 4 = \pm\sqrt{26}$$

$$X = -4 \pm \sqrt{26}$$

$$\left(-4 + \sqrt{26}\right)^2 + 8\left(-4 + \sqrt{26}\right) - 10 = 0$$

$$16 - 8\sqrt{26} + 26 - 32 + 8\sqrt{26} - 10 = 0$$

$$16 + 26 - 32 - 10 = 0$$

$$0 = 0$$

9.

$$3X^2 + 18X + 3 = 0$$

$$3\left(X^2 + 6X + 1\right) = 3(0)$$

$$X^2 + 6X + 1 = 0$$

$$X^2 + 6X + 9 = -1 + 9$$

$$(X + 3)^2 = 8$$

$$X + 3 = \pm\sqrt{8}$$

$$X = -3 \pm 2\sqrt{2}$$

$$3\left(-3 + 2\sqrt{2}\right)^2 + 18\left(-3 + 2\sqrt{2}\right) + 3 = 0$$

$$3\left(9 - 12\sqrt{2} + 8\right) - 54 + 36\sqrt{2} + 3 = 0$$

$$27 - 36\sqrt{2} + 24 - 54 + 36\sqrt{2} + 3 = 0$$

$$27 + 24 - 54 + 3 = 0$$

$$0 = 0$$

10.

$$X^2 - 10X + 30 = 0$$

$$X^2 - 10X + 25 = -30 + 25$$

$$(X - 5)^2 = -5$$

$$X - 5 = \pm\sqrt{-5}$$

$$X = 5 \pm i\sqrt{5}$$

$$\left(5 + i\sqrt{5}\right)^2 - 10\left(5 + i\sqrt{5}\right) + 30 = 0$$

$$25 + 10i\sqrt{5} - 5 - 50 - 10i\sqrt{5} + 30 = 0$$

$$25 - 5 - 50 + 30 = 0$$

$$0 = 0$$

Systematic Review 11C

1. $\left(2X + \frac{1}{3}\right)^2 = 4X^2 + \frac{4}{3}X + \frac{1}{9}$

2. $(3X - 4)^2 = 9X^2 - 24X + 16$

3. 9

4. $2X^2 + 20X + \underline{}$

divide through by 2:

$X^2 + 10X + \underline{}$

complete the square:

$X^2 + 10X + 25$

multiply through by 2:

$2X^2 + 20X + \underline{\,50\,}$

5. $28X$

6. $\frac{3}{2}X$

7.

$$X^2 + 10X + 3 = 0$$

$$X^2 + 10X + 25 = -3 + 25$$

$$(X + 5)^2 = 22$$

$$X + 5 = \pm\sqrt{22}$$

$$X = -5 \pm \sqrt{22}$$

8.
$$\left(-5+\sqrt{22}\right)^2+10\left(-5+\sqrt{22}\right)+3=0$$
$$\left(-5+\sqrt{22}\right)\left(-5+\sqrt{22}\right)-50+10\sqrt{22}+3=0$$
$$25-10\sqrt{22}+22-50+10\sqrt{22}+3=0$$
$$25+22-50+3=0$$
$$0=0$$

$$\left(-5-\sqrt{22}\right)^2+10\left(-5-\sqrt{22}\right)+3=0$$
$$\left(-5-\sqrt{22}\right)\left(-5-\sqrt{22}\right)-50-10\sqrt{22}+3=0$$
$$25+10\sqrt{22}+22-50-10\sqrt{22}+3=0$$
$$25+22-50+3=0$$
$$0=0$$

9.
$$X^2-6X-6=0$$
$$X^2-6X+9=6+9$$
$$\left(X-3\right)^2=15$$
$$X-3=\pm\sqrt{15}$$
$$X=3\pm\sqrt{15}$$

10.
$$\left(3+\sqrt{15}\right)^2-6\left(3+\sqrt{15}\right)-6=0$$
$$9+6\sqrt{15}+15-18-6\sqrt{15}-6=0$$
$$9+15-18-6=0$$
$$0=0$$

$$\left(3-\sqrt{15}\right)^2-6\left(3-\sqrt{15}\right)-6=0$$
$$9-6\sqrt{15}+15-18+6\sqrt{15}-6=0$$
$$9+15-18-6=0$$
$$0=0$$

11.
$$\left(\frac{1}{2}X-3B\right)^4=1\left(\frac{1}{2}X\right)^4\left(-3B\right)^0+$$
$$4\left(\frac{1}{2}X\right)^3\left(-3B\right)^1+6\left(\frac{1}{2}X\right)^2\left(-3B\right)^2+$$
$$4\left(\frac{1}{2}X\right)^1\left(-3B\right)^3+1\left(\frac{1}{2}X\right)^0\left(-3B\right)^4=$$
$$\frac{1}{16}X^4-\frac{3}{2}X^3B+\frac{27}{2}X^2B^2-54XB^3+81B^4$$

12.
$$\left(X+1\right)^5=1X^5 1^0+5X^4 1^1+$$
$$10X^3 1^2+10X^2 1^3+5X^1 1^4+1X^0 1^5=$$
$$X^5+5X^4+10X^3+10X^2+5X+1$$

13.
$$\frac{5\cdot4}{1\cdot2}\left(2X\right)^3\left(-5\right)^2=10\cdot8X^3\cdot25=2{,}000X^3$$

14.
$$\frac{5\cdot4\cdot3\cdot2}{1\cdot2\cdot3\cdot4}\left(2X\right)^1\left(-5\right)^4=5\cdot2X\cdot625=6{,}250X$$

15.
$$\left(4X-6\right)^3=1\left(4X\right)^3\left(-6\right)^0+$$
$$3\left(4X\right)^2\left(-6\right)^1+3\left(4X\right)^1\left(-6\right)^2+1\left(4X\right)^0\left(-6\right)^3=$$
$$64X^3-288X^2+432X-216$$

16. The first term would need to be $\sqrt[3]{X^3}$, and the second term would need to be $\sqrt[3]{64}$. Expanding $\left(X+4\right)^3$ yields $X^3+12X^2+48X+64$, so the cube root is $\left(X+4\right)$.

17.
$$\frac{5-6\sqrt{-3}}{4i+7}=\frac{\left(5-6i\sqrt{3}\right)\left(4i-7\right)}{\left(4i+7\right)\left(4i-7\right)}=$$
$$\frac{20i-35-24i^2\sqrt{3}+42i\sqrt{3}}{16i^2-49}=$$
$$\frac{20i-35-24\left(-1\right)\sqrt{3}+42i\sqrt{3}}{16\left(-1\right)-49}=$$
$$\frac{20i-35+24\sqrt{3}+42i\sqrt{3}}{-16-49}=$$
$$\frac{20i-35+24\sqrt{3}+42i\sqrt{3}}{-65}$$

18.
$$\frac{3-\sqrt{2}}{3+\sqrt{2}}=\frac{\left(3-\sqrt{2}\right)\left(3-\sqrt{2}\right)}{\left(3+\sqrt{2}\right)\left(3-\sqrt{2}\right)}=$$
$$\frac{9-6\sqrt{2}+2}{9-2}=\frac{11-6\sqrt{2}}{7}$$

19.
$$\left(12i\right)\left(\sqrt{-5}-\sqrt{16}\right)=12i\sqrt{-5}-12i\sqrt{16}=$$
$$12i\sqrt{-1}\sqrt{5}-12i\left(4\right)=12i^2\sqrt{5}-48i=$$
$$12\left(-1\right)\sqrt{5}-48i=-12\sqrt{5}-48i$$

20.
$$\left(i^1\right)\left(i^3\right)\left(i^0\right)=i^{1+3+0}=i^4=\left(-1\right)^2=1$$

Systematic Review 11D

1. $\left(\frac{1}{2}X-5\right)^2=\frac{1}{4}X^2-5X+25$

2. $\left(2X-6\right)^2=4X^2-24X+36$

3. 49

4. 64

5. 6X

6. 16X

7.
$$X^2 - 4X + 5 = 0$$
$$X^2 - 4X + 4 = -5 + 4$$
$$(X - 2)^2 = -1$$
$$X - 2 = \pm\sqrt{-1}$$
$$X = 2 \pm i$$

8.
$$(2 + i)^2 - 4(2 + i) + 5 = 0$$
$$4 + 4i + i^2 - 8 - 4i + 5 = 0$$
$$4 + (-1) - 8 + 5 = 0$$
$$0 = 0$$

$$(2 - i)^2 - 4(2 - i) + 5 = 0$$
$$4 - 4i + i^2 - 8 + 4i + 5 = 0$$
$$4 + (-1) - 8 + 5 = 0$$
$$0 = 0$$

9.
$$X^2 + 12X + 11 = 0$$
$$(X + 11)(X + 1) = 0$$
$$X + 11 = 0 \qquad X + 1 = 0$$
$$X = -11 \qquad X = -1$$

10.
$$(-11)^2 + 12(-11) + 11 = 0$$
$$121 - 132 + 11 = 0$$
$$0 = 0$$

$$(-1)^2 + 12(-1) + 11 = 0$$
$$1 - 12 + 11 = 0$$
$$0 = 0$$

11.
$$\left(\frac{1}{3}X + 2\right)^4 = 1\left(\frac{1}{3}X\right)^4(2)^0 + 4\left(\frac{1}{3}X\right)^3(2)^1 +$$
$$6\left(\frac{1}{3}X\right)^2(2)^2 + 4\left(\frac{1}{3}X\right)^1(2)^3 + 1\left(\frac{1}{3}X\right)^0(2)^4 =$$
$$\frac{1}{81}X^4 + \frac{8}{27}X^3 + \frac{8}{3}X^2 + \frac{32}{3}X + 16$$

12.
$$(X - 2A)^5 = 1X^5(-2A)^0 + 5X^4(-2A)^1 +$$
$$10X^3(-2A)^2 + 10X^2(-2A)^3 + 5X^1(-2A)^4 +$$
$$1X^0(-2A)^5 =$$
$$X^5 - 10X^4A + 40X^3A^2 - 80X^2A^3 + 80XA^4 - 32A^5$$

13. $\frac{5}{1}(x)^4(2A)^1 = 10X^4A$

14. $\frac{5 \cdot 4 \cdot 3 \cdot 2 \cdot 1}{1 \cdot 2 \cdot 3 \cdot 4 \cdot 5}(x)^0(2A)^5 = 1 \cdot 1 \cdot (2A)^5 = 32A^5$

15.
$$(2X - 3)^3 = 1(2X)^3(-3)^0 + 3(2X)^2(-3)^1 +$$
$$3(2X)^1(-3)^2 + 1(2X)^0(-3)^3 =$$
$$8X^3 - 36X^2 + 54X - 27$$

16. $(X - 3)$

17.
$$\frac{4\sqrt{-6}}{8i - 9} = \frac{4\sqrt{-1}\sqrt{6}\,(8i + 9)}{(8i - 9)(8i + 9)} =$$
$$\frac{4i\sqrt{6}\,(8i + 9)}{64i^2 - 81} = \frac{32i^2\sqrt{6} + 36i\sqrt{6}}{64(-1) - 81} =$$
$$\frac{32(-1)\sqrt{6} + 36i\sqrt{6}}{-64 - 81} = \frac{-32\sqrt{6} + 36i\sqrt{6}}{-145}$$

18.
$$\frac{5 + \sqrt{-3}}{5 - \sqrt{-3}} = \frac{(5 + \sqrt{-3})(5 + \sqrt{-3})}{(5 - \sqrt{-3})(5 + \sqrt{-3})} =$$
$$\frac{25 + 10\sqrt{-3} + (-3)}{25 - (-3)} = \frac{22 + 10i\sqrt{3}}{28} =$$
$$\frac{11 + 5i\sqrt{3}}{14}$$

19.
$$(4i)(2i - \sqrt{-9}) =$$
$$(4i)(2i - 3i) = (4i)(-i) = -4i^2 = 4$$

20. $(i^4)(i^4) = i^8 = (-1)^4 = 1^2 = 1$

Systematic Review 11E

1. $\left(3X - \frac{1}{4}\right)^2 = 9X^2 - \frac{3}{2}X + \frac{1}{16}$

2. $(X + 11)^2 = X^2 + 22X + 121$

3. 16

4. 225

5. 12X

6. $4X^2 + \underline{\quad} + 9$

divide through by 4:

$X^2 + \underline{\quad} + \frac{9}{4}$

find middle term:

$X^2 + \frac{6}{2}X + \frac{9}{4}$

multiply through by 4:

$4X^2 + \underline{12X} + 9$

7. $X^2 - 3X - 9 = 0$

$X^2 - 3X + \dfrac{9}{4} = 9 + \dfrac{9}{4}$

$\left(X - \dfrac{3}{2}\right)^2 = \dfrac{45}{4}$

$X - \dfrac{3}{2} = \pm\sqrt{\dfrac{45}{4}}$

$X = \dfrac{3}{2} \pm \dfrac{\sqrt{45}}{\sqrt{4}}$

$X = \dfrac{3}{2} \pm \dfrac{\sqrt{9}\sqrt{5}}{2}$

$X + \dfrac{3 \pm 3\sqrt{5}}{2}$

8. $\left(\dfrac{3 + 3\sqrt{5}}{2}\right)^2 - 3\left(\dfrac{3 + 3\sqrt{5}}{2}\right) - 9 = 0$

$\dfrac{9 + 18\sqrt{5} + 9(5)}{4} - \dfrac{9 + 9\sqrt{5}}{2} - 9 = 0$

$\dfrac{9 + 18\sqrt{5} + 45}{4} - \dfrac{9 + 9\sqrt{5}}{2} - 9 = 0$

$\dfrac{54 + 18\sqrt{5}}{4} - \dfrac{9 + 9\sqrt{5}}{2} - 9 = 0$

$\dfrac{2(27 + 9\sqrt{5})}{2(2)} - \dfrac{9 + 9\sqrt{5}}{2} - 9 = 0$

$\dfrac{27 + 9\sqrt{5}}{2} - \dfrac{9 + 9\sqrt{5}}{2} - \dfrac{18}{2} = 0$

$\dfrac{27 + 9\sqrt{5} - (9 + 9\sqrt{5}) - 18}{2} = 0$

$\dfrac{27 + 9\sqrt{5} - 9 - 9\sqrt{5} - 18}{2} = 0$

$\dfrac{27 - 9 - 18}{2} = 0$

$\dfrac{0}{2} = 0$

$0 = 0$

The check for the other solution is not shown here, but it works out in a similar fashion.

9. $2X^2 + 3X - 2 = 0$

$(2X - 1)(X + 2) = 0$

$2X - 1 = 0 \qquad X + 2 = 0$

$2X = 1 \qquad\qquad X = -2$

$X = \dfrac{1}{2}$

10. $2\left(\dfrac{1}{2}\right)^2 + 3\left(\dfrac{1}{2}\right) - 2 = 0$

$2\left(\dfrac{1}{4}\right) + \dfrac{3}{2} - 2 = 0$

$\dfrac{2}{4} + \dfrac{3}{2} - 2 = 0$

$\dfrac{1}{2} + \dfrac{3}{2} - \dfrac{4}{2} = 0$

$0 = 0$

$2(-2)^2 + 3(-2) - 2 = 0$

$2(4) - 6 - 2 = 0$

$8 - 6 - 2 = 0$

$0 = 0$

11. $(X + 2)^5 = 1X^5 2^0 + 5X^4 2^1 + 10X^3 2^2 +$
$10X^2 2^3 + 5X^1 2^4 + 1X^0 2^5 =$
$X^5 + 10X^4 + 40X^3 + 80X^2 + 80X + 32$

12. $(2X - 1)^4 = 1(2X)^4(-1)^0 + 4(2X)^3(-1)^1 +$
$6(2X)^2(-1)^2 + 4(2X)^1(-1)^3 + 1(2X)^0(-1)^4 =$
$16X^4 - 32X^3 + 24X^2 - 8X + 1$

13. $\dfrac{6 \cdot 5}{1 \cdot 2}(X)^4(-1)^2 = 15X^4$

14. $\dfrac{6 \cdot 5 \cdot 4}{1 \cdot 2 \cdot 3}(X)^3(-1)^3 = -20X^3$

15. $(3X + 1)^3 =$
$1(3X)^3 1^0 + 3(3X)^2 1^1 + 3(3X)^1 1^2 + 1(3X)^0 1^3 =$
$27X^3 + 27X^2 + 9X + 1$

16. $(X + 5)$

17. $\dfrac{3 - 2\sqrt{-5}}{7i + 2} = \dfrac{(3 - 2\sqrt{-5})(7i - 2)}{(7i + 2)(7i - 2)} =$

$\dfrac{21i - 6 - 14i\sqrt{-5} + 4\sqrt{-5}}{49i^2 - 4} =$

$\dfrac{21i - 6 - 14i^2\sqrt{5} + 4i\sqrt{5}}{-49 - 4} =$

$\dfrac{21i - 6 - 14(-1)\sqrt{5} + 4i\sqrt{5}}{-53} =$

$\dfrac{21i - 6 + 14\sqrt{5} + 4i\sqrt{5}}{-53}$

18. $\dfrac{1 + \sqrt{X}}{2 - \sqrt{X}} = \dfrac{(1 + \sqrt{X})(2 + \sqrt{X})}{(2 - \sqrt{X})(2 + \sqrt{X})} =$

$\dfrac{2 + \sqrt{X} + 2\sqrt{X} + X}{4 - X} = \dfrac{2 + 3\sqrt{X} + X}{4 - X}$

19. $(18i)\left(\sqrt{-36}+7i\right)=(18i)(6i+7i)=$
$(18i)(13i)=234i^2=234(-1)=-234$

20. $(i^2)(i^1)(i^3)=i^{2+1+3}=i^6=(-1)^3=-1$

Lesson Practice 12A

1. $X=\dfrac{-(6)\pm\sqrt{(6)^2-4(1)(2)}}{2(1)}=\dfrac{-6\pm\sqrt{28}}{2}=$
$\dfrac{-6\pm2\sqrt{7}}{2}=-3\pm\sqrt{7}$

2. $X^2-5X+4=0$
$(X-4)(X-1)=0$

$X-4=0 \qquad X-1=0$
$\quad X=4 \qquad\quad X=1$

3. $X=\dfrac{-(7)\pm\sqrt{(7)^2-4(3)(-1)}}{2(3)}=\dfrac{-7\pm\sqrt{61}}{6}$

4. $A^2-10A-11=0$
$(A-11)(A+1)=0$

$A-11=0 \qquad A+1=0$
$\quad A=11 \qquad\quad A=-1$

5. $2Q^2+2=17Q$
$2Q^2-17Q+2=0$
$\dfrac{-(17)\pm\sqrt{(-17)^2-4(2)(2)}}{2(2)}=\dfrac{17\pm\sqrt{273}}{4}$

6. $5X^2+15X+10=0$
$(5)(X+1)(X+2)=0$

$X+1=0 \qquad X+2=0$
$\quad X=-1 \qquad\quad X=-2$

7. $\dfrac{1}{4}R^2-\dfrac{1}{2}R+\dfrac{3}{2}=0$
$(4)\dfrac{1}{4}R^2-(4)\dfrac{1}{2}R+(4)\dfrac{3}{2}=(4)0$
$R^2-2R+6=0$
$\dfrac{-(-2)\pm\sqrt{(-2)^2-4(1)(6)}}{2(1)}=$
$\dfrac{2\pm\sqrt{-20}}{2}=\dfrac{2\pm2i\sqrt{5}}{2}=1\pm i\sqrt{5}$

8. $16X^2=2X+4$
$8X^2=X+2$
$8X^2-X-2=0$
$X=\dfrac{-(-1)\pm\sqrt{(-1)^2-4(8)(-2)}}{2(8)}=\dfrac{1\pm\sqrt{65}}{16}$

9. $X=\dfrac{-(3)\pm\sqrt{(3)^2-4(2)(-8)}}{2(2)}=\dfrac{-3\pm\sqrt{73}}{4}$

10. $Y^2=\dfrac{3}{4}Y+2$
$(4)Y^2=(4)\dfrac{3}{4}Y+(4)2$
$4Y^2=3Y+8$
$4Y^2-3Y-8=0$
$X=\dfrac{-(-3)\pm\sqrt{(-3)^2-4(4)(-8)}}{2(4)}=\dfrac{3\pm\sqrt{137}}{8}$

Lesson Practice 12B

1. $X=\dfrac{-(-1)\pm\sqrt{(-1)^2-4(8)(-3)}}{2(8)}=\dfrac{1\pm\sqrt{97}}{16}$

2. $7=2X^2+X$
$0=2X^2+X-7$
$X=\dfrac{-(1)\pm\sqrt{(1)^2-4(2)(-7)}}{2(2)}=\dfrac{-1\pm\sqrt{57}}{4}$

3. $Q=\dfrac{-(-6)\pm\sqrt{(-6)^2-4(1)(3)}}{2(1)}=\dfrac{6\pm\sqrt{24}}{2}=$
$\dfrac{6\pm2\sqrt{6}}{2}=3\pm\sqrt{6}$

4. $2+3X+4X^2=0$
$4X^2+3X+2=0$
$X=\dfrac{-(3)\pm\sqrt{(3)^2-4(4)(2)}}{2(4)}=$
$\dfrac{-3\pm\sqrt{-23}}{8}=\dfrac{-3\pm i\sqrt{23}}{8}$

5. $P=P^2-2$
$0=P^2-P-2$
$0=(P-2)(P+1)$

$P-2=0 \qquad P+1=0$
$\quad P=2 \qquad\quad P=-1$

6.
$$X^2 + \frac{1}{5}X + 5 = 0$$
$$(5)X^2 + (5)\frac{1}{5}X + (5)5 = (5)0$$
$$5X^2 + X + 25 = 0$$
$$X = \frac{-(1) \pm \sqrt{(1)^2 - 4(5)(25)}}{2(5)} = \frac{-1 \pm i\sqrt{499}}{10}$$

7.
$$20X^2 + 40X = 30$$
$$2X^2 + 4X = 3$$
$$2X^2 + 4X - 3 = 0$$
$$X = \frac{-(4) \pm \sqrt{(4)^2 - 4(2)(-3)}}{2(2)} = \frac{-4 \pm \sqrt{40}}{4} =$$
$$\frac{-4 \pm 2\sqrt{10}}{4} = \frac{-2 \pm \sqrt{10}}{2}$$

8.
$$X = \frac{-(2) \pm \sqrt{(2)^2 - 4(5)(-1)}}{2(5)} = \frac{-2 \pm \sqrt{24}}{10} =$$
$$\frac{-2 \pm 2\sqrt{6}}{10} = \frac{-1 \pm \sqrt{6}}{5}$$

9.
$$3X^2 = -5X$$
$$3X^2 + 5X = 0$$
$$(X)(3X + 5) = 0$$
$$X = 0 \qquad 3X + 5 = 0$$
$$3X = -5$$
$$X = -\frac{5}{3}$$

10.
$$\frac{-(B) \pm \sqrt{(B)^2 - 4(A)(C)}}{2(A)} = \frac{-B \pm \sqrt{B^2 - 4AC}}{2A}$$

3.
$$X^2 - 3X + 1 = -6X$$
$$X^2 + 3X + 1 = 0$$
$$X = \frac{-(3) \pm \sqrt{(3)^2 - 4(1)(1)}}{2(1)} = \frac{-3 \pm \sqrt{5}}{2}$$

4.
$$X^2 + 4X - 12 = 0$$
$$(X + 6)(X - 2) = 0$$
$$X + 6 = 0 \qquad X - 2 = 0$$
$$X = -6 \qquad X = 2$$

5.
$$X = \frac{-(2) \pm \sqrt{(2)^2 - 4(2)(5)}}{2(2)} =$$
$$\frac{-2 \pm \sqrt{-36}}{4} = \frac{-2 \pm 6i}{4} = \frac{-1 \pm 3i}{2}$$

6.
$$X^2 + 8X = -16$$
$$X^2 + 8X + 16 = 0$$
$$(X + 4)(X + 4) = 0$$
$$X + 4 = 0 \qquad X = -4$$

7. 169

8. $2X^2 + 9X + \underline{}$
divide through by 2:
$$X^2 + \frac{9}{2}X + \underline{}$$
complete the square:
$$X^2 + \frac{9}{2}X + \frac{81}{16}$$
multiply through by 2:
$$2X^2 + 9X + \frac{81}{8}$$

9. $40X$

10. $2\sqrt{14}\,X$

Systematic Review 12C

1.
$$X^2 - 5X + 6 = 0$$
$$(X - 3)(X - 2) = 0$$
$$X - 3 = 0 \qquad X - 2 = 0$$
$$X = 3 \qquad X = 2$$

2.
$$X = \frac{-(4) \pm \sqrt{(4)^2 - 4(1)(2)}}{2(1)} =$$
$$\frac{-4 \pm 2\sqrt{2}}{2} = -2 \pm \sqrt{2}$$

11.

$$X^2 + \frac{1}{3}X - \frac{4}{3} = 0$$

$$X^2 + \frac{1}{3}X + \frac{1}{36} = \frac{4}{3} + \frac{1}{36}$$

$$\left(X + \frac{1}{6}\right)^2 = \frac{4(12)}{3(12)} + \frac{1}{36}$$

$$\left(X + \frac{1}{6}\right)^2 = \frac{49}{36}$$

$$X + \frac{1}{6} = \sqrt{\frac{49}{36}}$$

$$X = -\frac{1}{6} \pm \frac{7}{6}$$

$$X = -\frac{1}{6} + \frac{7}{6} \qquad X = -\frac{1}{6} - \frac{7}{6}$$

$$X = \frac{6}{6} = 1 \qquad X = -\frac{8}{6} = -\frac{4}{3}$$

12.

$$(1)^2 + \frac{1}{3}(1) - \frac{4}{3} = 0$$

$$1 + \frac{1}{3} - \frac{4}{3} = 0$$

$$\frac{3}{3} + \frac{1}{3} - \frac{4}{3} = 0$$

$$0 = 0$$

$$\left(-\frac{4}{3}\right)^2 + \frac{1}{3}\left(-\frac{4}{3}\right) - \frac{4}{3} = 0$$

$$\frac{16}{9} - \frac{4}{9} - \frac{12}{9} = 0$$

$$0 = 0$$

13.

$$(X - A)^6 = 1X^6(-A)^0 + 6X^5(-A)^1 +$$

$$15X^4(-A)^2 + 20X^3(-A)^3 + 15X^2(-A)^4 +$$

$$6X^1(-A)^5 + 1X^0(-A)^6 =$$

$$X^6 - 6X^5A + 15X^4A^2 - 20X^3A^3 +$$

$$15X^2A^4 - 6XA^5 + A^6$$

14. $\frac{4}{1}\left(\frac{1}{2}X\right)^3(-3A)^1 = 4\left(\frac{1}{8}\right)X^3(-3A) = -\frac{3}{2}X^3A$

15. $(5 - 2A)^3 =$

$$1(5)^3(-2A)^0 + 3(5)^2(-2A)^1 +$$

$$3(5)^1(-2A)^2 + 1(5)^0(-2A)^3 =$$

$$125 - 150A + 60A^2 - 8A^3$$

16. $(X - 2Y)$

17.

$$\frac{6 + 5i}{3i - 2} = \frac{(6 + 5i)(3i + 2)}{(3i - 2)(3i + 2)} =$$

$$\frac{18i + 12 + 15i^2 + 10i}{9i^2 - 4} = \frac{28i + 12 + 15(-1)}{9(-1) - 4} =$$

$$\frac{28i + 12 - 15}{-9 - 4} = \frac{28i - 3}{-13}$$

18.

$$\frac{2 + \sqrt{-49}}{2 - \sqrt{-49}} = \frac{(2 + \sqrt{-49})(2 + \sqrt{-49})}{(2 - \sqrt{-49})(2 + \sqrt{-49})} =$$

$$\frac{4 + 4\sqrt{-49} - 49}{4 - (-49)} = \frac{-45 + 4i(7)}{4 + 49} = \frac{-45 + 28i}{53}$$

19.

$$\frac{2}{3 - \sqrt{7}} = \frac{2(3 + \sqrt{7})}{(3 - \sqrt{7})(3 + \sqrt{7})} =$$

$$\frac{6 + 2\sqrt{7}}{9 - 7} = \frac{6 + 2\sqrt{7}}{2} = \frac{3 + \sqrt{7}}{1} = 3 + \sqrt{7}$$

20.

$$\frac{2 + \sqrt{5}}{2\sqrt{5} - 4} = \frac{(2 + \sqrt{5})(2\sqrt{5} + 4)}{(2\sqrt{5} - 4)(2\sqrt{5} + 4)} =$$

$$\frac{4\sqrt{5} + 8 + 4\sqrt{5} + 2\sqrt{25}}{4(5) - 16} = \frac{8\sqrt{5} + 8 + 2(5)}{20 - 16} =$$

$$\frac{8\sqrt{5} + 8 + 10}{4} = \frac{8\sqrt{5} + 18}{4} = \frac{4\sqrt{5} + 9}{2}$$

Systematic Review 12D

1. $X = \dfrac{-(-9) \pm \sqrt{(-9)^2 - 4(2)(-7)}}{2(2)} = \dfrac{9 \pm \sqrt{137}}{4}$

2. $X = \dfrac{-(5) \pm \sqrt{(5)^2 - 4(1)(-2)}}{2(1)} = \dfrac{-5 \pm \sqrt{33}}{2}$

3.

$$3X^2 + 7X + 4 = 0$$

$$(3X + 4)(X + 1) = 0$$

$$3X + 4 = 0 \qquad X + 1 = 0$$

$$3X = -4 \qquad X = -1$$

$$X = -\frac{4}{3}$$

4. $X = \dfrac{-(-6) \pm \sqrt{(-6)^2 - 4(1)(12)}}{2(1)} =$

$$\frac{6 \pm \sqrt{-12}}{2} = \frac{6 \pm 2i\sqrt{3}}{2} = 3 \pm i\sqrt{3}$$

5. $5X^2 - 3X - 2 = 0$

$(5X + 2)(X - 1) = 0$

$5X + 2 = 0 \qquad X - 1 = 0$

$5X = -2 \qquad\qquad X = 1$

$X = -\dfrac{2}{5}$

6. $4X^2 + 1 = 4X$

$4X^2 - 4X + 1 = 0$

$(2X - 1)(2X - 1) = 0$

$2X - 1 = 0$

$2X = 1$

$X = \dfrac{1}{2}$

7. $\dfrac{25}{4}$

8. $\dfrac{1}{16}$

9. $25X^2 + \underline{} + 1$

divide through by 25:

$X^2 + \underline{} + \dfrac{1}{25}$

complete the square:

$X^2 + \dfrac{2}{\underline{5}} X + \dfrac{1}{25}$

multiply through by 25:

$25X^2 + \underline{10X} + 1$

10. $49X^2 - \underline{} + 4$

divide through by 49:

$X^2 - \underline{} + \dfrac{4}{49}$

complete the square:

$X^2 - \dfrac{4}{\underline{7}} X + \dfrac{4}{49}$

multiply through by 49:

$49X^2 - \underline{28X} + 4$

11. $X^2 - 12X + 20 = 0$

$(X - 10)(X - 2) = 0$

$X - 10 = 0 \qquad X - 2 = 0$

$X = 10 \qquad\qquad X = 2$

12. $(10)^2 - 12(10) + 20 = 0$

$100 - 120 + 20 = 0$

$0 = 0$

$(2)^2 - 12(2) + 20 = 0$

$4 - 24 + 20 = 0$

$0 = 0$

13. $(X + 1)^4 =$

$1X^4 1^0 + 4X^3 1^1 + 6X^2 1^2 + 4X^1 1^3 + 1X^0 1^4 =$

$X^4 + 4X^3 + 6X^2 + 4X + 1$

14. $\dfrac{4 \cdot 3 \cdot 2 \cdot 1}{1 \cdot 2 \cdot 3 \cdot 4}\left(\dfrac{1}{2}X\right)^0 (-3A)^4 = 81A^4$

15. $\left(10 - \dfrac{1}{X}\right)^3 = 1(10)^3\left(-\dfrac{1}{X}\right)^0 +$

$3(10)^2\left(-\dfrac{1}{X}\right)^1 + 3(10)^1\left(-\dfrac{1}{X}\right)^2 + 1(10)^0\left(-\dfrac{1}{X}\right)^3 =$

$1{,}000 - \dfrac{300}{X} + \dfrac{30}{X^2} - \dfrac{1}{X^3}$

16. $(X + 2)$

17. $\dfrac{4 - 3i}{2i} = \dfrac{(4 - 3i)(i)}{2i(i)} = \dfrac{4i - 3i^2}{2i^2} =$

$\dfrac{4i - 3(-1)}{2(-1)} = \dfrac{4i + 3}{-2}$

18. $\dfrac{10 + \sqrt{-A}}{10 - \sqrt{-A}} = \dfrac{\left(10 + \sqrt{-A}\right)\left(10 + \sqrt{-A}\right)}{\left(10 - \sqrt{-A}\right)\left(10 + \sqrt{-A}\right)} =$

$\dfrac{100 + 20\sqrt{-A} - A}{100 - (-A)} = \dfrac{100 + 20i\sqrt{A} - A}{100 + A}$

19. $\dfrac{9}{7 + \sqrt{10}} = \dfrac{9\left(7 - \sqrt{10}\right)}{\left(7 + \sqrt{10}\right)\left(7 - \sqrt{10}\right)} =$

$\dfrac{63 - 9\sqrt{10}}{49 - 10} = \dfrac{63 - 9\sqrt{10}}{39} = \dfrac{21 - 3\sqrt{10}}{13}$

20. $\dfrac{4 - \sqrt{6}}{3\sqrt{7} + 5} = \dfrac{\left(4 - \sqrt{6}\right)\left(3\sqrt{7} - 5\right)}{\left(3\sqrt{7} + 5\right)\left(3\sqrt{7} - 5\right)} =$

$\dfrac{12\sqrt{7} - 20 - 3\sqrt{42} + 5\sqrt{6}}{9(7) - 25} =$

$\dfrac{12\sqrt{7} - 20 - 3\sqrt{42} + 5\sqrt{6}}{63 - 25} =$

$\dfrac{12\sqrt{7} - 20 - 3\sqrt{42} + 5\sqrt{6}}{38}$

Systematic Review 12E

1. $X^2 + 2X - 8 = 0$

$(X+4)(X-2) = 0$

$X + 4 = 0$ ⠀⠀⠀ $X - 2 = 0$
⠀⠀ $X = -4$ ⠀⠀⠀⠀ $X = 2$

2. $X^2 - 6X = -8$

$X^2 - 6X + 8 = 0$

$(X-4)(X-2) = 0$

$X - 4 = 0$ ⠀⠀⠀ $X - 2 = 0$
⠀⠀ $X = 4$ ⠀⠀⠀⠀ $X = 2$

3. $2X^2 - 15X + 7 = 0$

$(2X-1)(X-7) = 0$

$2X - 1 = 0$ ⠀⠀⠀ $X - 7 = 0$
⠀ $2X = 1$ ⠀⠀⠀⠀ $X = 7$
⠀⠀ $X = \dfrac{1}{2}$

4. $3X^2 + 4X = 7$

$3X^2 + 4X - 7 = 0$

$(3X+7)(X-1) = 0$

$3X + 7 = 0$ ⠀⠀⠀ $X - 1 = 0$
⠀ $3X = -7$ ⠀⠀⠀⠀ $X = 1$
⠀⠀ $X = -\dfrac{7}{3}$

5. $2 = 5X + X^2$

$0 = X^2 + 5X - 2$

$X = \dfrac{-(5) \pm \sqrt{(5)^2 - 4(1)(-2)}}{2(1)} = \dfrac{-5 \pm \sqrt{33}}{2}$

6. $X^2 + 2X - 15 = 0$

$(X+5)(X-3) = 0$

$X + 5 = 0$ ⠀⠀⠀ $X - 3 = 0$
⠀⠀ $X = -5$ ⠀⠀⠀⠀ $X = 3$

7. $4X^2 + 28X + \underline{}$

divide through by 4:

$X^2 + 7X + \underline{}$

complete the square:

$X^2 + 7X + \dfrac{49}{4}$

multiply through by 4:

$4X^2 + 28X + 49$

8. $9X^2 - 36X + \underline{}$

divide through by 9:

$X^2 - 4X + \underline{}$

complete the square:

$X^2 - 4X + \underline{4}$

multiply through by 9:

$9X^2 - 36X + 36$

9. $36X^2 + \underline{} + 25$

divide through by 36:

$X^2 + \underline{} + \dfrac{25}{36}$

complete the square:

$X^2 + \dfrac{10}{\underline{6}}X + \dfrac{25}{36}$

multiply through by 36:

$36X^2 + \underline{60X} + 25$

10. $81X^2 - \underline{} + 121$

divide through by 81:

$X^2 - \underline{} + \dfrac{121}{81}$

complete the square:

$X^2 - \dfrac{22}{\underline{9}}X + \dfrac{121}{81}$

multiply through by 81:

$81X^2 - \underline{198X} + 121$

11. $X^2 + 5X - 14 = 0$

$(X+7)(X-2) = 0$

$X + 7 = 0$ ⠀⠀⠀ $X - 2 = 0$
⠀⠀ $X = -7$ ⠀⠀⠀⠀ $X = 2$

12. $(-7)^2 + 5(-7) - 14 = 0$
⠀⠀ $49 - 35 - 14 = 0$
⠀⠀⠀⠀⠀⠀ $0 = 0$

$(2)^2 + 5(2) - 14 = 0$
⠀⠀ $4 + 10 - 14 = 0$
⠀⠀⠀⠀⠀ $0 = 0$

13. $(2X + 1)^5 =$

$1(2X)^5 1^0 + 5(2X)^4 1^1 + 10(2X)^3 1^2 +$

$10(2X)^2 1^3 + 5(2X)^1 1^4 + 1(2X)^0 1^5 =$

$32X^5 + 80X^4 + 80X^3 + 40X^2 + 10X + 1$

14. $\dfrac{5 \cdot 4}{1 \cdot 2} \left(\dfrac{1}{3}X\right)^3 (2)^2 = 10\left(\dfrac{1}{27}\right) X^3 (4) = \dfrac{40}{27} X^3$

15. $\left(X - \dfrac{3}{5}\right)^3 = 1X^3 \left(-\dfrac{3}{5}\right)^0 + 3X^2 \left(-\dfrac{3}{5}\right)^1 +$

$3X^1 \left(-\dfrac{3}{5}\right)^2 + 1X^0 \left(-\dfrac{3}{5}\right)^3 =$

$X^3 - \dfrac{9}{5}X^2 + \dfrac{27}{25}X - \dfrac{27}{125}$

16. $(2X + 1)$

17. $\dfrac{10 + i}{5i} = \dfrac{(10 + i)(i)}{5i(i)} = \dfrac{10i + i^2}{5i^2} = \dfrac{10i - 1}{-5}$

18. $\dfrac{10}{5 - \sqrt{8}} = \dfrac{10(5 + \sqrt{8})}{(5 - \sqrt{8})(5 + \sqrt{8})} = \dfrac{50 + 10\sqrt{8}}{25 - 8} =$

$\dfrac{50 + 10\sqrt{4}\sqrt{2}}{17} = \dfrac{50 + 10(2)\sqrt{2}}{17} = \dfrac{50 + 20\sqrt{2}}{17}$

19. $\dfrac{2 + 3\sqrt{6}}{1 - \sqrt{6}} = \dfrac{(2 + 3\sqrt{6})(1 + \sqrt{6})}{(1 - \sqrt{6})(1 + \sqrt{6})} =$

$\dfrac{2 + 2\sqrt{6} + 3\sqrt{6} + 3(6)}{1 - 6} = \dfrac{2 + 5\sqrt{6} + 18}{-5} =$

$\dfrac{20 + 5\sqrt{6}}{-5} = \dfrac{4 + \sqrt{6}}{-1} = -4 - \sqrt{6}$

20. $\dfrac{6 - \sqrt{2}}{10\sqrt{3} - 8} = \dfrac{(6 - \sqrt{2})(10\sqrt{3} + 8)}{(10\sqrt{3} - 8)(10\sqrt{3} + 8)} =$

$\dfrac{60\sqrt{3} + 48 - 10\sqrt{6} - 8\sqrt{2}}{100(3) - 64} =$

$\dfrac{2(30\sqrt{3} + 24 - 5\sqrt{6} - 4\sqrt{2})}{300 - 64} =$

$\dfrac{2(30\sqrt{3} + 24 - 5\sqrt{6} - 4\sqrt{2})}{236} =$

$\dfrac{30\sqrt{3} + 24 - 5\sqrt{6} - 4\sqrt{2}}{118}$

Lesson Practice 13A

1. $(6)^2 - 4(1)(9) = 0$

real, rational, equal (double root)

$X^2 + 6X + 9 = 0$
$(X + 3)(X + 3) = 0$
$X + 3 = 0$
$X = -3$

2. $(7)^2 - 4(2)(3) = 25$

real, rational, unequal

$2X^2 + 7X + 3 = 0$
$(2X + 1)(X + 3) = 0$

$2X + 1 = 0 \qquad\qquad X + 3 = 0$
$2X = -1 \qquad\qquad X = -3$
$X = -\dfrac{1}{2}$

3. $(3)^2 - 4(-2)(6) = 57$

real, irrational, unequal

$X = \dfrac{-(3) \pm \sqrt{(3)^2 - 4(-2)(6)}}{2(-2)} = \dfrac{-3 \pm \sqrt{57}}{-4}$

4. $(-2)^2 - 4(3)(5) = -56$

imaginary

$X = \dfrac{-(-2) \pm \sqrt{(-2)^2 - 4(3)(5)}}{2(3)} =$

$\dfrac{2 \pm 2i\sqrt{14}}{6} = \dfrac{1 \pm i\sqrt{14}}{3}$

5. $7X^2 - 3X = 20$

$7X^2 - 3X - 20 = 0$

$(-3)^2 - 4(7)(-20) = 569$

real, irrational, unequal

$X = \dfrac{-(-3) \pm \sqrt{(-3)^2 - 4(7)(-20)}}{2(7)} = \dfrac{3 \pm \sqrt{569}}{14}$

Lesson Practice 13B

1. $2R^2 = -5R + 3$

$2R^2 + 5R - 3 = 0$

$(5)^2 - 4(2)(-3) = 49$

real, rational, unequal

$2R^2 + 5R - 3 = 0$
$(2R - 1)(R + 3) = 0$

$2R - 1 = 0 \qquad\qquad R + 3 = 0$
$2R = 1 \qquad\qquad R = -3$
$R = \dfrac{1}{2}$

2. $\frac{1}{4}X^2 + 2X = -4$

$X^2 + 8X = -16$

$X^2 + 8X + 16 = 0$

$(8)^2 - 4(1)(16) = 0$

real, rational, equal (double roots)

$X^2 + 8X + 16 = 0$
$(X + 4)(X + 4) = 0$
$X + 4 = 0$
$X = -4$

3. $11 + 7Y = -6Y^2$

$6Y^2 + 7Y + 11 = 0$

$(7)^2 - 4(6)(11) = -215$

imaginary

$X = \frac{-(7) \pm \sqrt{(7)^2 - 4(6)(11)}}{2(6)} =$

$\frac{-7 \pm \sqrt{-215}}{12} = \frac{-7 \pm i\sqrt{215}}{12}$

4. $8X^2 + 10X + 2 = 0$

$4X^2 + 5X + 1 = 0$

$(5)^2 - 4(4)(1) = 9$

real, rational, unequal

$4X^2 + 5X + 1 = 0$
$(4X + 1)(X + 1) = 0$

$4X + 1 = 0 \qquad X + 1 = 0$
$4X = -1 \qquad X = -1$
$X = -\frac{1}{4}$

5. $(-5)^2 - 4(6)(-3) = 97$

real, irrational, unequal

$X = \frac{-(-5) \pm \sqrt{(-5)^2 - 4(6)(-3)}}{2(6)} = \frac{5 \pm \sqrt{97}}{12}$

2. $X = \frac{-(3) \pm \sqrt{(3)^2 - 4(1)(1)}}{2(1)} = \frac{-3 \pm \sqrt{5}}{2}$

3. $X^2 + 4X = -49$

$X^2 + 4X + 49 = 0$

$(4)^2 - 4(1)(49) = -180$

imaginary

4. $X = \frac{-(4) \pm \sqrt{(4)^2 - 4(1)(49)}}{2(1)} =$

$\frac{-4 \pm 6i\sqrt{5}}{2} = -2 \pm 3i\sqrt{5}$

5. $(-5)^2 - 4(1)(-9) = 61$

real, irrational, unequal

6. $X = \frac{-(-5) \pm \sqrt{(-5)^2 - 4(1)(-9)}}{2(1)} = \frac{5 \pm \sqrt{61}}{2}$

7. $(11)^2 - 4(2)(12) = 25$

real, rational, unequal

8. $2X^2 + 11X + 12 = 0$

$(2X + 3)(X + 4) = 0$

$2X + 3 = 0 \qquad X + 4 = 0$
$2X = -3 \qquad X = -4$
$X = -\frac{3}{2}$

9. $X = \frac{-(-8) \pm \sqrt{(-8)^2 - 4(1)(8)}}{2(1)} = \frac{8 \pm \sqrt{32}}{2} =$

$\frac{8 \pm 4\sqrt{2}}{2} = 4 \pm 2\sqrt{2}$

10. $X = \frac{-(7) \pm \sqrt{(7)^2 - 4(1)(12)}}{2(1)} = \frac{-7 \pm \sqrt{1}}{2} =$

$\frac{-7 + 1}{2} = \frac{-6}{2} = -3$

and $\frac{-7 - 1}{2} = \frac{-8}{2} = -4$

X = -3 and X = -4

Systematic Review 13C

1. $(3)^2 - 4(1)(1) = 5$

real, irrational, unequal

11. $X^2 - 7X + 1 = 0$

$X^2 - 7X + \dfrac{49}{4} = -1 + \dfrac{49}{4}$

$\left(X - \dfrac{7}{2}\right)^2 = \dfrac{-4}{4} + \dfrac{49}{4}$

$\left(X - \dfrac{7}{2}\right)^2 = \dfrac{45}{4}$

$X - \dfrac{7}{2} = \pm\sqrt{\dfrac{45}{4}}$

$X = \dfrac{7}{2} \pm \dfrac{\sqrt{45}}{\sqrt{4}}$

$X = \dfrac{7}{2} \pm \dfrac{3\sqrt{5}}{2} = \dfrac{7 \pm 3\sqrt{5}}{2}$

12. $\left(\dfrac{7 + 3\sqrt{5}}{2}\right)^2 - 7\left(\dfrac{7 + 3\sqrt{5}}{2}\right) + 1 = 0$

$\dfrac{(7 + 3\sqrt{5})(7 + 3\sqrt{5})}{4} - \dfrac{49 + 21\sqrt{5}}{2} + 1 = 0$

$\dfrac{49 + 42\sqrt{5} + 9(5)}{4} - \dfrac{49 + 21\sqrt{5}}{2} + 1 = 0$

$\dfrac{49 + 42\sqrt{5} + 45}{4} - \dfrac{49 + 21\sqrt{5}}{2} + 1 = 0$

$\dfrac{94 + 42\sqrt{5}}{4} - \dfrac{49 + 21\sqrt{5}}{2} + 1 = 0$

$\dfrac{47 + 21\sqrt{5}}{2} - \dfrac{49 + 21\sqrt{5}}{2} + \dfrac{2}{2} = 0$

$\dfrac{47 + 21\sqrt{5} - (49 + 21\sqrt{5}) + 2}{2} = 0$

$\dfrac{47 + 2\mathbf{1}\sqrt{5} - 49 - 2\mathbf{1}\sqrt{5} + 2}{2} = 0$

$\dfrac{47 - 49 + 2}{2} = 0$

$\dfrac{0}{2} = 0$

$0 = 0$

The check for the other solution is not shown here, but it works out in a similar fashion.

13. $\left(\dfrac{1}{2}X + 3\right)^6 = 1\left(\dfrac{1}{2}X\right)^6 (3)^0 +$

$6\left(\dfrac{1}{2}X\right)^5 (3)^1 + 15\left(\dfrac{1}{2}X\right)^4 (3)^2 +$

$20\left(\dfrac{1}{2}X\right)^3 (3)^3 + 15\left(\dfrac{1}{2}X\right)^2 (3)^4 +$

$6\left(\dfrac{1}{2}X\right)^1 (3)^5 + 1\left(\dfrac{1}{2}X\right)^0 (3)^6 =$

$\dfrac{1}{64}X^6 + \dfrac{9}{16}X^5 + \dfrac{135}{16}X^4 + \dfrac{135}{2}X^3 +$

$\dfrac{1{,}215}{4}X^2 + 729X + 729$

14. $\dfrac{4 \cdot 3 \cdot 2}{1 \cdot 2 \cdot 3}(4X)^1(-1)^3 = -16X$

15. $\left(\dfrac{1}{4}X + \dfrac{1}{5}\right)^3 =$

$1\left(\dfrac{1}{4}X\right)^3 \left(\dfrac{1}{5}\right)^0 + 3\left(\dfrac{1}{4}X\right)^2 \left(\dfrac{1}{5}\right)^1 +$

$3\left(\dfrac{1}{4}X\right)^1 \left(\dfrac{1}{5}\right)^2 + 1\left(\dfrac{1}{4}X\right)^0 \left(\dfrac{1}{5}\right)^3 =$

$\dfrac{1}{64}X^3 + \dfrac{3}{80}X^2 + \dfrac{3}{100}X + \dfrac{1}{125}$

16. $(X - 5)$

17. $\dfrac{4 - 3\sqrt{5}}{2} \cdot \dfrac{4 + 3\sqrt{5}}{2} =$

$\dfrac{(4 - 3\sqrt{5})(4 + 3\sqrt{5})}{(2)(2)} =$

$\dfrac{16 - 9(5)}{4} = \dfrac{16 - 45}{4} = \dfrac{-29}{4}$

18. $\dfrac{2 + \dfrac{1}{4}}{2 - \dfrac{1}{4}} = \dfrac{\dfrac{8}{4} + \dfrac{1}{4}}{\dfrac{8}{4} - \dfrac{1}{4}} = \dfrac{\dfrac{9}{4}}{\dfrac{7}{4}} = \dfrac{\dfrac{9}{4} \times \dfrac{4}{7}}{\dfrac{7}{4} \times \dfrac{4}{7}} =$

$\dfrac{9}{4} \times \dfrac{4}{7} = \dfrac{36}{28} = \dfrac{9}{7}$

19. $\dfrac{X}{5} - \dfrac{7}{2} - \dfrac{X}{6} = 0 \qquad \text{LCD} = 30$

$(30)\dfrac{X}{5} - (30)\dfrac{7}{2} - (30)\dfrac{X}{6} = 0(30)$

$6X - 105 - 5X = 0$

$X - 105 = 0$

$X = 105$

Check: $\dfrac{105}{5} - \dfrac{7}{2} - \dfrac{105}{6} = 0$

$6(105) - 105 - 5(105) = 0$

20. $(X - Ai)(X + Ai) = X^2 - A^2 i^2 =$
$X^2 - A^2(-1) = X^2 + A^2$

Systematic Review 13D

1. $(-2)^2 - 4(1)(-3) = 16$
real, rational, unequal

2. $X^2 - 2X - 3 = 0$
$(X - 3)(X + 1) = 0$

$X - 3 = 0$ $\qquad$ $X + 1 = 0$
$X = 3$ $\qquad\qquad$ $X = -1$

3. $(-2)^2 - 4(1)(5) = -16$
imaginary

4. $X = \dfrac{-(-2) \pm \sqrt{(-2)^2 - 4(1)(5)}}{2(1)} =$
$\dfrac{2 \pm \sqrt{-16}}{2} = \dfrac{2 \pm 4i}{2} = 1 \pm 2i$

5. $(-20)^2 - 4(4)(25) = 0$
real, rational, equal (double root)

6. $4X^2 - 20X + 25 = 0$
$(2X - 5)(2X - 5) = 0$
$2X - 5 = 0$
$2X = 5$
$X = \dfrac{5}{2}$

7. $(-2)^2 - 4(2)(5) = -36$
imaginary

8. $X = \dfrac{-(-2) \pm \sqrt{(-2)^2 - 4(2)(5)}}{2(2)} =$
$\dfrac{2 \pm 6i}{4} = \dfrac{1 \pm 3i}{2}$

9. $3X^2 + 6X = -2$
$3X^2 + 6X + 2 = 0$
$X = \dfrac{-(6) \pm \sqrt{(6)^2 - 4(3)(2)}}{2(3)} = \dfrac{-6 \pm \sqrt{12}}{6} =$
$\dfrac{-6 \pm 2\sqrt{3}}{6} = \dfrac{-3 \pm \sqrt{3}}{3}$

10. $X = \dfrac{-(2) \pm \sqrt{(2)^2 - 4(7)(1)}}{2(7)} = \dfrac{-2 \pm \sqrt{-24}}{14} =$
$\dfrac{-2 \pm 2i\sqrt{6}}{14} = \dfrac{-1 \pm i\sqrt{6}}{7}$

11. $X^2 - 6X = 2$
$X^2 - 6X + 9 = 2 + 9$
$(X - 3)^2 = 11$
$X - 3 = \pm\sqrt{11}$
$X = 3 \pm \sqrt{11}$

12. $(3 + \sqrt{11})^2 - 6(3 + \sqrt{11}) - 2 = 0$
$(3 + \sqrt{11})^2 - 6(3 + \sqrt{11}) = 2$
$9 \cancel{+ 6\sqrt{11}} + 11 - 18 \cancel{- 6\sqrt{11}} = 2$
$9 + 11 - 18 = 2$
$2 = 2$

$(3 - \sqrt{11})^2 - 6(3 - \sqrt{11}) - 2 = 0$
$9 \cancel{- 6\sqrt{11}} + 11 - 18 \cancel{+ 6\sqrt{11}} = 2$
$9 + 11 - 18 = 2$
$2 = 2$

13. $(X + A)^4 =$
$1X^4 A^0 + 4X^3 A^1 + 6X^2 A^2 + 4X^1 A^3 + 1X^0 A^4 =$
$X^4 + 4X^3 A + 6X^2 A^2 + 4XA^3 + A^4$

14. $\dfrac{6}{1}(4X)^5(-1)^1 = (-6)(1{,}024X^5) = -6{,}144X^5$

15. $\left(X - \dfrac{2}{9}\right)^3 = 1X^3\left(-\dfrac{2}{9}\right)^0 + 3X^2\left(-\dfrac{2}{9}\right)^1 +$
$3X^1\left(-\dfrac{2}{9}\right)^2 + 1X^0\left(-\dfrac{2}{9}\right)^3 =$
$X^3 - \dfrac{2}{3}X^2 + \dfrac{4}{27}X - \dfrac{8}{729}$

16. $(3X + 1)$

17. $\dfrac{7 + 2\sqrt{X}}{6} \cdot \dfrac{7 - 2\sqrt{X}}{6} =$
$\dfrac{(7 + 2\sqrt{X})(7 - 2\sqrt{X})}{(6)(6)} = \dfrac{49 - 4X}{36}$

18. $\dfrac{X - \dfrac{1}{X}}{3 + \dfrac{1}{3}} = \dfrac{\dfrac{X(X)}{1(X)} - \dfrac{1}{X}}{\dfrac{3(3)}{1(3)} + \dfrac{1}{3}} = \dfrac{\dfrac{X^2 - 1}{X}}{\dfrac{9 + 1}{3}} = \dfrac{\dfrac{X^2 - 1}{X}}{\dfrac{10}{3}} =$

$\dfrac{\dfrac{X^2 - 1}{X} \times \dfrac{3}{10}}{\dfrac{10}{3} \times \dfrac{3}{10}} = \dfrac{X^2 - 1}{X} \times \dfrac{3}{10} = \dfrac{3X^2 - 3}{10X}$

19. $\dfrac{4X + 1}{3} - 1 = X + \dfrac{3X - 8}{5} \qquad LCM = 15$

$\left(^5\cancel{15}\right)\dfrac{4X + 1}{3} - (15)1 = (15)X + \left(^3\cancel{15}\right)\dfrac{3X - 8}{5}$

$(5)(4X + 1) - 15 = 15X + (3)(3X - 8)$

$20X + 5 - 15 = 15X + 9X - 24$

$5 - 15 + 24 = 15X + 9X - 20X$

$14 = 4X$

$\dfrac{14}{4} = X = \dfrac{7}{2} \text{ or } 3\dfrac{1}{2}$

20. $(2X - 3i)(2X + 3i) = 4X^2 - 9i^2 =$

$4X^2 - 9(-1) = 4X^2 + 9$

Systematic Review 13E

1. $(7)^2 - 4(3)(2) = 25$

real, rational, unequal

2. $3X^2 + 7X + 2 = 0$

$(3X + 1)(X + 2) = 0$

$3X + 1 = 0 \qquad X + 2 = 0$

$\quad 3X = -1 \qquad \quad X = -2$

$\quad X = -\dfrac{1}{3}$

3. $(-5)^2 - 4(2)(4) = -7$

imaginary

4. $X = \dfrac{-(-5) \pm \sqrt{(-5)^2 - 4(2)(4)}}{2(2)} =$

$\dfrac{5 \pm \sqrt{-7}}{4} = \dfrac{5 \pm i\sqrt{7}}{4}$

5. $(-2)^2 - 4(4)(9) = -140$

imaginary

6. $X = \dfrac{-(-2) \pm \sqrt{(-2)^2 - 4(4)(9)}}{2(4)} =$

$\dfrac{2 \pm \sqrt{-140}}{8} = \dfrac{2 \pm 2i\sqrt{35}}{8} = \dfrac{1 \pm i\sqrt{35}}{4}$

7. $(-4)^2 - 4(2)(-7) = 72$

real, irrational, unequal

8. $X = \dfrac{-(-4) \pm \sqrt{(-4)^2 - 4(2)(-7)}}{2(2)} =$

$\dfrac{4 \pm \sqrt{72}}{4} = \dfrac{4 \pm 6\sqrt{2}}{4} = \dfrac{2 \pm 3\sqrt{2}}{2}$

9. $2X^2 + 6X = 3$

$2X^2 + 6X - 3 = 0$

$X = \dfrac{-(6) \pm \sqrt{(6)^2 - 4(2)(-3)}}{2(2)} = \dfrac{-6 \pm \sqrt{60}}{4} =$

$\dfrac{-6 \pm 2\sqrt{15}}{4} = \dfrac{-3 \pm \sqrt{15}}{2}$

10. $5X^2 + 4 = 8X$

$5X^2 - 8X + 4 = 0$

$X = \dfrac{-(-8) \pm \sqrt{(-8)^2 - 4(5)(4)}}{2(5)} =$

$\dfrac{8 \pm \sqrt{-16}}{10} = \dfrac{8 \pm 4i}{10} = \dfrac{4 \pm 2i}{5}$

11. $3X^2 + 8X - 3 = 0$

$\dfrac{3X^2 + 8X - 3}{3} = \dfrac{0}{3}$

$X^2 + \dfrac{8}{3}X - 1 = 0$

$X^2 + \dfrac{8}{3}X + \dfrac{16}{9} = 1 + \dfrac{16}{9}$

$\left(X + \dfrac{4}{3}\right)^2 = \dfrac{25}{9}$

$X + \dfrac{4}{3} = \pm\sqrt{\dfrac{25}{9}}$

$X = -\dfrac{4}{3} \pm \dfrac{5}{3}$

$X = -\dfrac{4}{3} + \dfrac{5}{3} \qquad X = -\dfrac{4}{3} - \dfrac{5}{3}$

$X = \dfrac{1}{3} \qquad\qquad X = -\dfrac{9}{3} = -3$

12. $3\left(\dfrac{1}{3}\right)^2 + 8\left(\dfrac{1}{3}\right) - 3 = 0$

$3\left(\dfrac{1}{9}\right) + \dfrac{8}{3} - \dfrac{9}{3} = 0$

$\dfrac{3}{9} + \dfrac{8}{3} - \dfrac{9}{3} = 0$

$\dfrac{1}{3} + \dfrac{8}{3} - \dfrac{9}{3} = 0$

$0 = 0$

$3(-3)^2 + 8(-3) - 3 = 0$

$3(9) - 24 - 3 = 0$

$27 - 24 - 3 = 0$

$0 = 0$

13. $(X + 2A)^5 =$

$1X^5(2A)^0 + 5X^4(2A)^1 + 10X^3(2A)^2 +$

$10X^2(2A)^3 + 5X^1(2A)^4 + 1X^0(2A)^5 =$

$X^5 + 10X^4A + 40X^3A^2 + 80X^2A^3 + 80XA^4 + 32A^5$

14. $\dfrac{6 \cdot 5 \cdot 4 \cdot 3 \cdot 2}{1 \cdot 2 \cdot 3 \cdot 4 \cdot 5}(X)^1(-4)^5 =$

$6X(-1,024) = -6,144X$

15. $(2X - A)^3 = 1(2X)^3(-A)^0 + 3(2X)^2(-A)^1 +$

$3(2X)^1(-A)^2 + 1(2X)^0(-A)^3 =$

$8X^3 - 12X^2A + 6XA^2 - A^3$

16. $(2X + 3Y)$

17. $\dfrac{10 - \sqrt{AX}}{4} \cdot \dfrac{10 + \sqrt{AX}}{4} =$

$\dfrac{\left(10 - \sqrt{AX}\right)\left(10 + \sqrt{AX}\right)}{(4)(4)} = \dfrac{100 - AX}{16}$

18. $\dfrac{2X^2 - \dfrac{1}{X^2}}{\dfrac{4}{X}} = \dfrac{\dfrac{2X^2(X^2)}{1(X^2)} - \dfrac{1}{X^2}}{\dfrac{4}{X}} = \dfrac{\dfrac{2X^4 - 1}{X^2}}{\dfrac{4}{X}} =$

$\dfrac{\dfrac{2X^4 - 1}{X^2} \times \dfrac{X}{4}}{\dfrac{4}{X} \times \dfrac{X}{4}} = \dfrac{2X^4 - 1}{X^2} \times \dfrac{X}{4} =$

$\dfrac{(2X^4 - 1)(X)}{(X)(X)(4)} = \dfrac{2X^4 - 1}{4X}$

19. $X + 3 - \dfrac{6X - 5}{2} = \dfrac{2X - 7}{6}$ $\text{LCM} = 6$

$(6)X + (6)3 - (6)\dfrac{6X - 5}{2} = (6)\dfrac{(2X - 7)}{6}$

$6X + 18 - \dfrac{6(6X - 5)}{2} = 2X - 7$

$6X + 18 - 3(6X - 5) = 2X - 7$

$6X + 18 - 18X + 15 = 2X - 7$

$6X - 18X - 2X = -7 - 18 - 15$

$-14X = -40$

$X = \dfrac{-40}{-14} = \dfrac{20}{7} \text{ or } 2\dfrac{6}{7}$

20. $(A + Bi)(A - Bi) = A^2 - B^2i^2 =$

$A^2 - B^2(-1) = A^2 + B^2$

Lesson Practice 14A

1. $45.00 - 33.75 = \$11.25$ saved

$WP \times 45 = 11.25$

$WP = \dfrac{11.25}{45} = .25 \text{ or } 25\%$

2. $25.00 - 15.00 = \$10.00$ profit

$WP \times 15 = 10$

$WP = \dfrac{10}{15} = \dfrac{2}{3} \approx 67\%$

3. $25.00 - 15.00 = \$10.00$ profit

$WP \times 25 = 10$

$WP = \dfrac{10}{25} = \dfrac{2}{5} = 40\%$

4. $.28 \times 32 = 8.96$

$32.00 + 8.96 = \$40.96$

5. $P = $ original cost

$2,500 = P + 15\%(P)$

$2,500 = P + .15P$

$2,500 = P(1 + .15)$

$2,500 = P(1.15)$

$\dfrac{2,500}{1.15} = P = \$2,174$

6. $.55 \times 195 = \$107.25$ discount

$195 - 107.25 = \$87.75$

7. P = original price

$32.45 = P + .06P$

$32.45 = P(1 + .06)$

$32.45 = P(1.06)$

$\dfrac{32.45}{1.06} = P = \$30.61$

8. $.054 \times 45.50 \approx \$2.46$ tax

$.15 \times 45.50 \approx \$6.83$ tip

$45.50 + 2.46 + 6.83 = \$54.79$

9. $Si = 28$; $O = 16$; $SiO_2 = 28 + 2(16) = 60$

$\dfrac{Si}{SiO_2} = \dfrac{28}{60} \approx 47\%$

10. $\dfrac{O_2}{SiO_2} = \dfrac{32}{60} \approx 53\%$

11. $Fe = 56$; $O = 16$;

$Fe_2O_3 = 2(56) + 3(16) = 160$

$\dfrac{Fe_2}{Fe_2O_3} = \dfrac{112}{160} = 70\%$

12. $\dfrac{O_3}{Fe_2O_3} = \dfrac{48}{160} = 30\%$

Lesson Practice 14B

1. $.13 \times .59 \approx \$.08$ markup

$.59 + .08 = \$.67$ new price

2. $.153 \times 8,500 = \$1,300.50$

3. $14.04 - 13.50 = \$.54$ in tax

$WP \times 13.50 = .54$

$WP = \dfrac{.54}{13.50} = .04 = 4\%$

4. $.20 \times 13.50 = \$2.70$

5. $15.00 - 9.50 = \$5.50$ markup

$WP \times 15.00 = 5.5$

$WP = \dfrac{5.5}{15} \approx .37 = 37\%$

6. $.12 \times 1.38 = \$.17$ increase

$1.38 + .17 = \$1.55$

7. $4.5 + 7 = 11.5\%$ taxes and fee

$955 = P + 11.5\%(P)$

$955 = P + .115P$

$955 = P(1 + .115)$

$955 = P(1.115)$

$\dfrac{955}{1.115} = P \approx \$856.50$ base price

$856.50 \times .045 = \$38.54$ sales tax

$856.50 \times .07 = \$59.96$ delivery fee

8. $.45 \times 75 = 33.75$ discount

$75 - 33.75 = \$41.25$ new price

or

$100\% - 45\% = 55\%$

$.55 \times 75 = \$41.25$

9. $K = 39$; $Cr = 52$; $O = 16$;

$K_2Cr_2O_7 = 2(39) + 2(52) + 7(16) = 294$

$\dfrac{K_2}{K_2Cr_2O_7} = \dfrac{78}{294} \approx 27\%$

10. $\dfrac{Cr_2}{K_2Cr_2O_7} = \dfrac{104}{294} \approx 35\%$

11. $Na = 23$; $O = 16$; $H = 1$;

$NaOH = 23 + 16 + 1 = 40$

$\dfrac{Na}{NaOH} = \dfrac{23}{40} \approx 58\%$

12. $\dfrac{H}{NaOH} = \dfrac{1}{40} \approx 3\%$

Systematic Review 14C

1. $Profit = 9.50 - 4.00 = \$5.50$

$WP \times wholesale = profit$

$WP \times 4 = \$5.50$

$WP = \dfrac{5.50}{4} \approx 1.38 = 138\%$

2. $WP \times 9.50 = \$5.50$

$WP = \dfrac{5.50}{9.50} \approx .58 = 58\%$

3. $.062 \times 23,600 = \$1,463.20$

4. $.0765 \times 23,600 = \$1,805.40$

5. $28.50 \times .15 \approx \$4.28$

6. $28.50 \times .0825 \approx \$2.35$

7. $C = 12; S = 32;$

$CS_2 = 12 + 2(32) = 76$

$\dfrac{C}{CS_2} = \dfrac{12}{76} \approx .16 = 16\%$

8. $\dfrac{S_2}{CS_2} = \dfrac{64}{76} \approx .84 = 84\%$

9. $(8)^2 - 4(2)(8) = 0$

real, rational, equal

10. $2X^2 + 8X + 8 = 0$

$2(X^2 + 4X + 4) = 0$

$2(X + 2)^2 = 0$

$X + 2 = 0$

$X = -2$

11. $(-5)^2 - 4(1)(-7) = 53$

real, irrational, unequal

12. $X = \dfrac{-(-5) \pm \sqrt{(-5)^2 - 4(1)(-7)}}{2(1)} = \dfrac{5 \pm \sqrt{53}}{2}$

13. $(4)^2 - 4(1)(6) = -8$

imaginary

14. $X = \dfrac{-(4) \pm \sqrt{(4)^2 - 4(1)(6)}}{2(1)} = \dfrac{-4 \pm \sqrt{-8}}{2} =$

$\dfrac{-4 \pm 2i\sqrt{2}}{2} = -2 \pm i\sqrt{2}$

15. $2X^2 - 1 = -3X$

$2X^2 + 3X - 1 = 0$

$= \dfrac{-(3) \pm \sqrt{(3)^2 - 4(2)(-1)}}{2(2)} = \dfrac{-3 \pm \sqrt{17}}{4}$

16. $X = \dfrac{-(-5) \pm \sqrt{(-5)^2 - 4(1)(5)}}{2(1)} = \dfrac{5 \pm \sqrt{5}}{2}$

17. $X^2 + 4X = 32$

$X^2 + 4X - 32 = 0$

$(X + 8)(X - 4) = 0$

$X + 8 = 0 \qquad X - 4 = 0$

$X = -8 \qquad X = 4$

18. $(-8)^2 + 4(-8) = 32$

$64 - 32 = 32$

$32 = 32$

$(4)^2 + 4(4) = 32$

$16 + 16 = 32$

$32 = 32$

19. $(X - 1)^5 =$

$1X^5(-1)^0 + 5X^4(-1)^1 + 10X^3(-1)^2 +$

$10X^2(-1)^3 + 5X^1(-1)^4 + 1X^0(-1)^5 =$

$X^5 - 5X^4 + 10X^3 - 10X^2 + 5X - 1$

20. $\dfrac{4 \cdot 3 \cdot 2}{1 \cdot 2 \cdot 3}(X)^1(3)^3 = 4X(27) = 108X$

Systematic Review 14D

1. $\text{Profit} = 19.95 - 12.50 = \$7.45$

$WP \times 12.50 = \$7.45$

$WP = \dfrac{7.45}{12.50} \approx .60 = 60\%$

2. $WP \times 19.95 = \$7.45$

$WP = \dfrac{7.45}{19.95} \approx .37 = 37\%$

3. $WP \times 19.95 = \$12.50$

$WP = \dfrac{12.50}{19.95} \approx .63 = 63\%$

4. $78.10 \times .055 \approx \$4.30$

5. $78.10 \times .15 \approx \$11.72$

6. $\text{Final cost} = 78.10 + 4.30 + 11.72 = \$94.12$

$\text{Tax and tip} = 4.30 + 11.72 = \$16.02$

$WP \times 94.12 = \$16.02$

$WP = \dfrac{16.02}{94.12} \approx 17\%$

7. $H = 1; S = 32; H_2S = 2(1) + 32 = 34$

$\dfrac{H_2}{H_2S} = \dfrac{2}{34} \approx .06 = 6\%$

8. $\dfrac{S}{H_2S} = \dfrac{32}{34} \approx .94 = 94\%$

9. $(8)^2 - 4(4)(20) = 64 - 320 = -256$

imaginary

10. $X = \dfrac{-(8) \pm \sqrt{(8)^2 - 4(4)(20)}}{2(4)} =$

$\dfrac{-8 \pm 16i}{8} = -1 \pm 2i$

11. $2X^2 - 7X = -4$

$2X^2 - 7X + 4 = 0$

$(-7)^2 - 4(2)(4) = 49 - 32 = 17$

real, irrational, unequal

12. $X = \dfrac{-(-7) \pm \sqrt{(-7)^2 - 4(2)(4)}}{2(2)} = \dfrac{7 \pm \sqrt{17}}{4}$

13. $3X^2 = 7X + 4$

$3X^2 - 7X - 4 = 0$

$(-7)^2 - 4(3)(-4) = 49 + 48 = 97$

real, irrational, unequal

14. $X = \dfrac{-(-7) \pm \sqrt{(-7)^2 - 4(3)(-4)}}{2(3)} = \dfrac{7 \pm \sqrt{97}}{6}$

15. $6X^2 - 5X = 0$

$6X^2 - 5X + 0 = 0$

$X = \dfrac{-(-5) \pm \sqrt{(-5)^2 - 4(6)(0)}}{2(6)} =$

$\dfrac{5 \pm \sqrt{25}}{12} = \dfrac{5 \pm 5}{12}$

$X = \dfrac{5+5}{12}$　　　　$X = \dfrac{5-5}{12}$

$X = \dfrac{10}{12} = \dfrac{5}{6}$　　　$X = \dfrac{0}{12} = 0$

16. $5X^2 - 4 = 0$

$5X^2 + 0X - 4 = 0$

$X = \dfrac{-(0) \pm \sqrt{(0)^2 - 4(5)(-4)}}{2(5)} =$

$\dfrac{\pm\sqrt{80}}{10} = \dfrac{\pm 4\sqrt{5}}{10} = \dfrac{\pm 2\sqrt{5}}{5}$

17. $5X^2 + 4X = 0$

$(X)(5X + 4) = 0$

$X = 0$　　　　　$5X + 4 = 0$

$5X = -4$

$X = -\dfrac{4}{5}$

18. $5(0)^2 + 4(0) = 0$

$0 + 0 = 0$

$0 = 0$

$5\left(-\dfrac{4}{5}\right)^2 + 4\left(-\dfrac{4}{5}\right) = 0$

$5\left(\dfrac{16}{25}\right) - \dfrac{16}{5} = 0$

$\dfrac{80}{25} - \dfrac{16}{5} = 0$

$\dfrac{16}{5} - \dfrac{16}{5} = 0$

$0 = 0$

19. $(X - 2A)^5 =$

$1X^5(-2A)^0 + 5X^4(-2A)^1 + 10X^3(-2A)^2 +$

$10X^2(-2A)^3 + 5X^1(-2A)^4 + 1X^0(-2A)^5 =$

$X^5 - 10X^4A + 40X^3A^2 -$

$80X^2A^3 + 80XA^4 - 32A^5$

20. $\dfrac{4 \cdot 3}{1 \cdot 2}(X)^2(3)^2 = 6X^2(9) = 54X^2$

Systematic Review 14E

1. discount $= 149.95 \times .60 = \$89.97$

new price $= 149.95 - 89.97 = \$59.98$

or

$100\% - 60\% = 40\%$

new price $= 149.95 \times .40 = \$59.98$

For this type of problem, the student may use either method.

2. discount $= 399 \times .60 = \$239.40$

new price $= 399 - 239.40 = \$159.60$

3. discount $= 21.90 \times .60 = \$13.14$

new price $= 21.90 - 13.14 = \$8.76$

4. $54.45 \times .0725 \approx \$3.95$

5. $54.45 \times .16 \approx \$8.71$

6. final cost $= 54.45 + 3.95 + 8.71 = \$67.11$

tax and tip $= 8.71 + 3.95 = \$12.66$

$WP \times 67.11 = \$12.66$

$WP = \dfrac{12.66}{67.11} \approx .19 = 19\%$

7. $H = 1; C = 12; CH_4 = 12 + 4(1) = 16$

$\dfrac{C}{CH_4} = \dfrac{12}{16} = .75 = 75\%$

8. $\dfrac{H_4}{CH_4} = \dfrac{4}{16} = .25 = 25\%$

9. $(3)^2 - 4(1)(-5) = 9 + 20 = 29$

real, irrational, unequal

10. $X = \dfrac{-(3) \pm \sqrt{(3)^2 - 4(1)(-5)}}{2(1)} = \dfrac{-3 \pm \sqrt{29}}{2}$

11. $3X^2 = X + 3$

$3X^2 - X - 3 = 0$

$(-1)^2 - 4(3)(-3) = 1 + 36 = 37$

real, irrational, unequal

12. $X = \dfrac{-(-1) \pm \sqrt{(-1)^2 - 4(3)(-3)}}{2(3)} = \dfrac{1 \pm \sqrt{37}}{6}$

13. $3X^2 - 5X = -2$

$3X^2 - 5X + 2 = 0$

$(-5)^2 - 4(3)(2) = 25 - 24 = 1$

real, rational, unequal

14. $3X^2 - 5X + 2 = 0$

$(3X - 2)(X - 1) = 0$

$3X - 2 = 0 \qquad X - 1 = 0$

$3X = 2 \qquad\qquad X = 1$

$X = \dfrac{2}{3}$

15. $4X^2 + 7X = 2$

$4X^2 + 7X - 2 = 0$

$X = \dfrac{-(7) \pm \sqrt{(7)^2 - 4(4)(-2)}}{2(4)} =$

$\dfrac{-7 \pm \sqrt{49 + 32}}{8} = \dfrac{-7 \pm \sqrt{81}}{8} = \dfrac{-7 \pm 9}{8}$

$X = \dfrac{-7 + 9}{8} \qquad X = \dfrac{-7 - 9}{8}$

$X = \dfrac{2}{8} = \dfrac{1}{4} \qquad X = \dfrac{-16}{8} = -2$

16. $3X^2 + 5 = 8X$

$3X^2 - 8X + 5 = 0$

$X = \dfrac{-(-8) \pm \sqrt{(-8)^2 - 4(3)(5)}}{2(3)} = \dfrac{8 \pm 2}{6} = \dfrac{4 \pm 1}{3}$

$X = \dfrac{4 + 1}{3} \qquad X = \dfrac{4 - 1}{3}$

$X = \dfrac{5}{3} \qquad\quad X = \dfrac{3}{3} = 1$

17. $X^2 - 8X + 9 = 0$

$X^2 - 8X + 16 = -9 + 16$

$(X - 4)^2 = 7$

$X - 4 = \pm\sqrt{7}$

$X = 4 \pm \sqrt{7}$

18. $(4 + \sqrt{7})^2 - 8(4 + \sqrt{7}) + 9 = 0$

$(4 + \sqrt{7})(4 + \sqrt{7}) - 32 - 8\sqrt{7} + 9 = 0$

$16 + 4\sqrt{7} + 4\sqrt{7} + 7 - 32 - 8\sqrt{7} + 9 = 0$

$16 + \cancel{8\sqrt{7}} + 7 - 32 - \cancel{8\sqrt{7}} + 9 = 0$

$16 + 7 - 32 + 9 = 0$

$0 = 0$

$(4 - \sqrt{7})^2 - 8(4 - \sqrt{7}) + 9 = 0$

$(4 - \sqrt{7})(4 - \sqrt{7}) - 32 + 8\sqrt{7} + 9 = 0$

$16 - 4\sqrt{7} - 4\sqrt{7} + 7 - 32 + 8\sqrt{7} + 9 = 0$

$16 - \cancel{8\sqrt{7}} + 7 - 32 + \cancel{8\sqrt{7}} + 9 = 0$

$16 + 7 - 32 + 9 = 0$

$0 = 0$

19. $(2X-1)^5 =$

$1(2X)^5(-1)^0 + 5(2X)^4(-1)^1 + 10(2X)^3(-1)^2 +$

$10(2X)^2(-1)^3 + 5(2X)^1(-1)^4 + 1(2X)^0(-1)^5 =$

$32X^5 - 80X^4 + 80X^3 - 40X^2 + 10X - 1$

20. $\dfrac{5\cdot 4\cdot 3\cdot 2\cdot 1}{1\cdot 2\cdot 3\cdot 4\cdot 5}(2X)^0(-3)^5 = \dfrac{1}{1}(1)(-243) = -243$

Lesson Practice 15A

The order of the steps may be changed in some of these solutions without changing the final results.

1. $AFG = H$

$\dfrac{A\cancel{FG}}{\cancel{FG}} = \dfrac{H}{FG}$

$A = \dfrac{H}{FG}$

2. $AB = GF$

$\dfrac{\cancel{A}B}{\cancel{A}} = \dfrac{GF}{A}$

$B = \dfrac{GF}{A}$

3. $\dfrac{X}{YZ} = \dfrac{P}{Q}$

$(\cancel{YZ})\dfrac{X}{\cancel{YZ}} = (YZ)\dfrac{P}{Q}$

$X = \dfrac{PYZ}{Q}$

4. $\dfrac{X}{YZ} = \dfrac{A}{B}$

$(B\cancel{YZ})\dfrac{X}{\cancel{YZ}} = (\cancel{B}YZ)\dfrac{A}{\cancel{B}}$

$BX = AYZ$

$\dfrac{BX}{AZ} = \dfrac{A\cancel{YZ}}{\cancel{AZ}}$

$\dfrac{BX}{AZ} = Y$

5. $C - A = D + B$

$-A = D + B - C$

$(-1)(-A) = (-1)(D + B - C)$

$A = -D - B + C$

6. $X + Y + Z = B + A$

$X = B + A - Y - Z$

7. $\dfrac{B}{C+D} = 0$

$(\cancel{C+D})\dfrac{B}{\cancel{C+D}} = (C+D)0$

$B = (C+D)0$

$B = 0$

8. $G(A+B) = D$

$\dfrac{G\cancel{(A+B)}}{\cancel{(A+B)}} = \dfrac{D}{A+B}$

$G = \dfrac{D}{A+B}$

9. $\dfrac{1}{Y} = \dfrac{X}{Z}$

$(\cancel{Y}Z)\dfrac{1}{\cancel{Y}} = (Y\cancel{Z})\dfrac{X}{\cancel{Z}}$

$Z = XY$

$\dfrac{Z}{X} = \dfrac{XY}{X}$

$\dfrac{Z}{X} = Y$

10. $Q = RS + RT$

$Q = R(S+T)$

$\dfrac{Q}{S+T} = \dfrac{R(S+T)}{S+T}$

$\dfrac{Q}{S+T} = R$

11. $R = \dfrac{2}{3}X + Y$

$(3)R = (3)\dfrac{2}{3}X + (3)Y$

$3R = 2X + 3Y$

$3R - 3Y = 2X$

$\dfrac{3R - 3Y}{2} = X$

12. $B = 2\pi rh$

$\dfrac{B}{2rh} = \dfrac{2\pi rh}{2rh}$

$\dfrac{B}{2rh} = \pi$

Lesson Practice 15B

1. $\dfrac{1}{Y} = \dfrac{1}{X} + \dfrac{1}{Z}$

$(XYZ)\dfrac{1}{Y} = (XYZ)\dfrac{1}{X} + (XYZ)\dfrac{1}{Z}$

$\dfrac{X\!Y\!Z}{Y} = \dfrac{XYZ}{X} + \dfrac{XY\!Z}{Z}$

$XZ = YZ + XY$

$XZ - XY = YZ$

$X(Z - Y) = YZ$

$X = \dfrac{YZ}{Z - Y}$

2. $\dfrac{B_1}{A_2} = \dfrac{A_1}{B_2}$

$\left(A_2 B_2\right)\dfrac{B_1}{A_2} = \left(A_2 B_2\right)\dfrac{A_1}{B_2}$

$B_2 B_1 = A_2 A_1$

$B_2 = \dfrac{A_2 A_1}{B_1}$

3. $R = BW(1 + X)$

$\dfrac{R}{B(1 + X)} = \dfrac{B\!W(1\!+\!X)}{B(1\!+\!X)}$

$\dfrac{R}{B(1 + X)} = W$ or $\dfrac{R}{B + BX} = W$

4. $2A = \dfrac{1}{2}A + B$

$(2)2A = (2)\dfrac{1}{2}A = (2)B$

$4A = A + 2B$

$3A = 2B$

$A = \dfrac{2B}{3}$

5. $XYZ = YZQ$

$\dfrac{X\!Y\!Z}{YZ} = \dfrac{Y\!Z\!Q}{YZ}$

$X = Q$

6. $X = \dfrac{XY}{4}$

$(4)X = (4)\dfrac{XY}{4}$

$4X = XY$

$\dfrac{4X}{X} = \dfrac{XY}{X}$

$4 = Y$

7. $D = RT$

$\dfrac{D}{R} = \dfrac{RT}{R}$

$\dfrac{D}{R} = T$

8. $TS = XT + XS$

$TS = X(T + S)$

$\dfrac{TS}{T + S} = \dfrac{X(T + S)}{T + S}$

$\dfrac{TS}{T + S} = X$

9. $3 - \dfrac{A}{B} = C$

$(B)3 - (B)\dfrac{A}{B} = (B)C$

$3B - A = BC$

$3B - BC = A$

$B(3 - C) = A$

$\dfrac{B(3 - C)}{3 - C} = \dfrac{A}{3 - C}$

$B = \dfrac{A}{3 - C}$

10. $Q(P + R) - S = 10$

$QP + QR - S = 10$

$QP = -QR + S + 10$

$P = \dfrac{-QR + S + 10}{Q}$

11. $AX + BX + CX = D$

$X(A + B + C) = D$

$X = \dfrac{D}{A + B + C}$

12. $X = \dfrac{Y}{W + i}$

$(W + i)X = (W + i)\dfrac{Y}{W + i}$

$WX + iX = Y$

$iX = Y - WX$

$i = \dfrac{Y - WX}{X}$ or $i = \dfrac{Y}{X} - \dfrac{WX}{X} = \dfrac{Y}{X} - W$

Systematic Review 15C

1. $\dfrac{B}{C} = \dfrac{A}{D}$

$(\cancel{C})\dfrac{B}{\cancel{C}} = (C)\dfrac{A}{D}$

$B = \dfrac{CA}{D}$

2. $D = RT$

$\dfrac{D}{T} = R$

3. $\dfrac{AB}{C} = \dfrac{Y}{X}$

$(XC)\dfrac{AB}{C} = (XC)\dfrac{Y}{X}$

$\dfrac{X\cancel{C}AB}{\cancel{C}} = \dfrac{\cancel{X}CY}{\cancel{X}}$

$XAB = CY$

$X = \dfrac{CY}{AB}$

4. $\dfrac{1}{A} = \dfrac{B}{C}$

$(\cancel{A}C)\dfrac{1}{\cancel{A}} = (A\cancel{C})\dfrac{B}{\cancel{C}}$

$C = AB$

$\dfrac{C}{B} = A$

5. $D = RT$

$\dfrac{D}{R} = T$

6. $\dfrac{A}{B} - \dfrac{D}{E} = 0$

$(B\cancel{E})\dfrac{A}{\cancel{B}} - (\cancel{B}E)\dfrac{D}{\cancel{E}} = (BE)0$

$EA - BD = 0$

$EA = BD$

$\dfrac{EA}{B} = D$

7. retail = wholesale + 40% of retail

$59 = W + (.40)(59)$

$59 = W + 23.60$

$59 - 23.60 = W$

$\$35.40 = W$

8. $WP \times 35.40 = 23.60$

$WP = \dfrac{23.60}{35.40} \approx .67 = 67\%$

9. $62.30 \times .15 \approx \$9.35$

10. $\text{tax} = 76.15 - (62.30 + 9.35)$

$\quad = 76.15 - (71.65)$

$\quad = \$4.50$

$\text{tax} = WP \times 62.30$

$4.50 = WP \times 62.30$

$\dfrac{4.50}{62.30} = WP \approx .07 = 7\%$

11. $H = 1; C = 12; 0 = 16$

$H_2C0 = (2)1 + 12 + 16 = 30$

$\dfrac{C}{H_2C0} = \dfrac{12}{30} = .4 = 40\%$

12. $\dfrac{0}{H_2CO} = \dfrac{16}{30} \approx .53 = 53\%$

13. $(9)^2 - 4(1)(20) = 81 - 80 = 1$

real, rational, unequal

14. $X^2 + 9X + 20 = 0$

$(X + 5)(X + 4) = 0$

$X + 5 = 0 \qquad X + 4 = 0$

$\quad X = -5 \qquad\quad X = -4$

15. $X^2 - 25 = 0$

$X^2 + 0X - 25 = 0$

$(0)^2 - 4(1)(-25) = 0 + 100 = 100$

real, rational, unequal

16. $X^2 - 25 = 0$

$(X - 5)(X + 5) = 0$

$X - 5 = 0 \qquad X + 5 = 0$

$\quad X = 5 \qquad\quad X = -5$

17. $9X^2 + 3X = 2$

$9X^2 + 3X - 2 = 0$

$X = \dfrac{-(3) \pm \sqrt{(3)^2 - 4(9)(-2)}}{2(9)} =$

$\dfrac{-3 \pm \sqrt{81}}{18} = \dfrac{-3 \pm 9}{18} = \dfrac{-1 \pm 3}{6}$

$X = \dfrac{-1 + 3}{6} \qquad X = \dfrac{-1 - 3}{6}$

$X = \dfrac{2}{6} = \dfrac{1}{3} \qquad X = \dfrac{-4}{6} = -\dfrac{2}{3}$

18.

$$3X^2 - 15X - 42 = 0$$

$$3(X^2 - 5X - 14) = 3(0)$$

$$X^2 - 5X - 14 = 0$$

$$X = \frac{-(-5) \pm \sqrt{(-5)^2 - 4(1)(-14)}}{2(1)} =$$

$$\frac{5 \pm \sqrt{81}}{2} = \frac{5 \pm 9}{2}$$

$$X = \frac{5+9}{2} \qquad X = \frac{5-9}{2}$$

$$X = \frac{14}{2} = 7 \qquad X = \frac{-4}{2} = -2$$

19.

$$4X^2 - X - 4 = 0$$

$$\frac{4X^2 - X - 4}{4} = \frac{0}{4}$$

$$X^2 - \frac{1}{4}X - 1 = 0$$

$$X^2 - \frac{1}{4}X + \frac{1}{64} = 1 + \frac{1}{64}$$

$$\left(X - \frac{1}{8}\right)^2 = \frac{65}{64}$$

$$X - \frac{1}{8} = \pm\sqrt{\frac{65}{64}}$$

$$X = \frac{1}{8} \pm \frac{\sqrt{65}}{\sqrt{64}}$$

$$X = \frac{1}{8} \pm \frac{\sqrt{65}}{8} = \frac{1 \pm \sqrt{65}}{8}$$

20.

$$4\left(\frac{1+\sqrt{65}}{8}\right)^2 - \left(\frac{1+\sqrt{65}}{8}\right) - 4 = 0$$

$$(4)\frac{(1+\sqrt{65})(1+\sqrt{65})}{(8)(8)} + \frac{-(1+\sqrt{65})}{8} - 4 = 0$$

$$(4)\frac{1 + 2\sqrt{65} + 65}{64} + \frac{-1 - \sqrt{65}}{8} - 4 = 0$$

$$\frac{1 + 2\sqrt{65} + 65}{16} + \frac{-2 - 2\sqrt{65}}{16} - \frac{64}{16} = 0$$

$$\frac{1 + 65}{16} + \frac{-2}{16} - \frac{64}{16} = 0$$

$$\frac{1 + 65 - 2 - 64}{16} = 0$$

$$\frac{0}{16} = 0$$

$$0 = 0$$

The other solution is not shown, but it works out in a similar fashion.

Systematic Review 15D

1. $V = LWH$

$$\frac{V}{LW} = H$$

2. $A = \frac{AB}{2}$

$$(2)A = (2)\frac{AB}{2}$$

$$2A = AB$$

$$\frac{2A}{A} = B$$

$$2 = B$$

3. $P = 2L + 2W$

$$P - 2W = 2L$$

$$\frac{P - 2W}{2} = L$$

4. $V = \pi R^2 H$

$$\frac{V}{\pi R^2} = H$$

5. $A = 2\pi RH$

$$\frac{A}{2\pi H} = R$$

6. $I = \frac{E}{R + r}$

$$(R + r)I = E$$

$$RI + rI = E$$

$$RI = E - rI$$

$$R = \frac{E - rI}{I} \text{ or } \frac{E}{I} - r$$

7. G = gross

$$G = 15.3\%G + 3\%G + \text{net}$$

$$G = .153G + .03G + 968.40$$

$$G = .183G + 968.40$$

$$G - .183G = 968.40$$

$$.817G = 968.40$$

$$G = \frac{968.40}{.817} \approx \$1,185.31$$

8. 90 hr per week for 2 weeks = 180 hours

$$\$1,185.31 \div 180 \approx \$6.59 \text{ per hour}$$

9. $12\% \times 6.59 = .12 \times 6.59 \approx \$.79$ raise

$$6.59 + .79 = \$7.38 \text{ new rate}$$

$$7.38 \times 180 = \$1,328.40 \text{ gross}$$

$$1,328.40 \times .153 \approx \$203.25 \text{ first tax}$$

$$1,328.40 \times .03 \approx \$39.85 \text{ second tax}$$

$$\$203.25 + 39.85 = \$243.10 \text{ total tax}$$

$$1,328.40 - 243.10 = \$1,085.30$$

10. $1,328.40

11. $Li = 7; S = 32; O = 16$

$Li_2SO_3 = (2)7 + 32 + (3)16 = 94$

$\dfrac{Li_2}{Li_2SO_3} = \dfrac{14}{94} \approx .15 = 15\%$

12. $\dfrac{S}{Li_2SO_3} = \dfrac{32}{94} \approx .34 = 34\%$

13. $(-4)^2 - 4(1)(13) = 16 - 52 = -36$

imaginary

14. $X = \dfrac{-(-4) \pm \sqrt{(-4)^2 - 4(1)(13)}}{2(1)} =$

$\dfrac{4 \pm \sqrt{-36}}{2} = \dfrac{4 \pm 6i}{2} = 2 \pm 3i$

15. $X^2 + 6X = 3$

$X^2 + 6X - 3 = 0$

$(6)^2 - 4(1)(-3) = 36 + 12 = 48$

real, irrational, unequal

16. $X \dfrac{-(6) \pm \sqrt{(6)^2 - 4(1)(-3)}}{2(1)} =$

$\dfrac{-6 \pm 4\sqrt{3}}{2} = -3 \pm 2\sqrt{3}$

17. $2X^2 + 6X = 3$

$2X^2 + 6X - 3 = 0$

$X = \dfrac{-(6) \pm \sqrt{(6)^2 - 4(2)(-3)}}{2(2)} =$

$\dfrac{-6 \pm \sqrt{60}}{4} = \dfrac{-6 \pm 2\sqrt{15}}{4} = \dfrac{-3 \pm \sqrt{15}}{2}$

18. $2X^2 + 13X = 2X$

$2X^2 + 13X - 2X = 0$

$2X^2 + 11X + 0 = 0$

$X = \dfrac{-(11) \pm \sqrt{(11)^2 - 4(2)(0)}}{2(2)} =$

$\dfrac{-11 \pm \sqrt{121}}{4} = \dfrac{-11 \pm 11}{4}$

$X = \dfrac{-11 + 11}{4} = \dfrac{0}{4} = 0$

$X = \dfrac{-11 - 11}{4} = \dfrac{-22}{4} = -\dfrac{11}{2}$

19. $\dfrac{4X^2Y^{-3}}{12A^3Y^{-1}} \cdot \dfrac{7AX^2}{A^{-2}} \cdot \dfrac{9AY^2}{14X^{-2}A} =$

$\dfrac{3X^2Y^{-3}AX^2Y^2}{A^3Y^{-1}A^{-2}2X^{-2}} = \dfrac{3AX^4Y^{-1}}{2A^1X^{-2}Y^{-1}} =$

$\dfrac{3X^4}{2X^{-2}} = \dfrac{3X^4X^2}{2} = \dfrac{3X^6}{2}$

20. $\dfrac{8A^2X}{132X^{-2}} \cdot \dfrac{12X^{-2}A}{11X^3A^{-2}} = \dfrac{8A^2X^1X^{-2}A^1}{11X^{-2}11X^3A^{-2}} =$

$\dfrac{8A^3X^{-1}}{121X^1A^{-2}} = \dfrac{8A^3X^{-1} \cdot X^{-1}A^2}{121} =$

$\dfrac{8A^5X^{-2}}{121} = \dfrac{8A^5}{121X^2}$

Systematic Review 15E

1. $F = \dfrac{9}{5}C + 32$

$F - 32 = \dfrac{9}{5}C$

$5(F - 32) = 9C$

$\dfrac{5}{9}(F - 32) = C$

2. $\dfrac{W_1}{W_2} = \dfrac{L_2}{L_1}$

$(W_2L_1)\dfrac{W_1}{W_2} = (W_2L_1)\dfrac{L_2}{L_1}$

$L_1W_1 = W_2L_2$

$\dfrac{L_1W_1}{L_2} = W_2$

3. $A = 2\pi r(H + r)$

$\dfrac{A}{2\pi r} = H + r$

$\dfrac{A}{2\pi r} - r = H$

4. $\dfrac{1}{F} = \dfrac{1}{A} - \dfrac{1}{B}$

$\dfrac{1}{F} + \dfrac{1}{B} = \dfrac{1}{A}$

$\dfrac{(B)}{F(B)} + \dfrac{(F)}{B(F)} = \dfrac{1}{A}$

$\dfrac{B+F}{BF} = \dfrac{1}{A}$

$(A)\left(\dfrac{B+F}{BF}\right) = (A)\dfrac{1}{A}$

$(A)\left(\dfrac{B+F}{BF}\right) = 1$

$A\left(\dfrac{\cancel{B+F}}{\cancel{BF}}\right)\left(\dfrac{\cancel{BF}}{\cancel{B+F}}\right) = 1\left(\dfrac{BF}{B+F}\right)$

$A = \dfrac{BF}{B+F}$

5. $F = K \cdot \dfrac{M_1 M_2}{D_2}$

$\dfrac{F}{K} = \dfrac{M_1 M_2}{D_2}$

$(D_2)\dfrac{F}{K} = (D_2)\dfrac{M_1 M_2}{D_2}$

$\dfrac{D_2 F}{K} = M_1 M_2$

$\dfrac{D_2 F}{K M_2} = M_1$

6. $A = 2\pi r h$

$\dfrac{A}{2rh} = \pi$

7. WP × games played = wins

$WP \times (56 + 25) = 56$

$WP \times 81 = 56$

$WP = \dfrac{56}{81} \approx .69 = 69\%$

8. $WP \times 81 = $ losses

$WP \times 81 = 25$

$WP = \dfrac{25}{81} \approx .31 = 31\%$

9. TW = total games won

FH = games won first half

SH = games won second half

TW − FH = SH

$105 - 56 = 49$ games

10. $\dfrac{49}{81} \approx .60 = 60\%$

11. $H = 1; N = 14$

$\dfrac{N}{NH_3} = \dfrac{14}{17} \approx .82 = 82\%$

12. $\dfrac{H_3}{NH_3} = \dfrac{3}{17} \approx .18 = 18\%$

13. $(-7)^2 - 4(3)(2) = 49 - 24 = 25$

real, rational, unequal

14. $3X^2 - 7X + 2 = 0$

$(3X - 1)(X - 2) = 0$

$\begin{array}{ll} 3X - 1 = 0 & X - 2 = 0 \\ 3X = 1 & X = 2 \\ X = \dfrac{1}{3} & \end{array}$

15. $5X^2 = 45$

$5X^2 + 0X - 45 = 0$

$(0)^2 - 4(5)(-45) = 900$

real, rational, unequal

16. $5X^2 = 45$

$5X^2 - 45 = 0$

$(5)(X^2 - 9) = 0$

$(5)(X - 3)(X + 3) = 0$

$\begin{array}{ll} X - 3 = 0 & X + 3 = 0 \\ X = 3 & X = -3 \end{array}$

17. $3X^2 + 2X = 0$

$3X^2 + 2X + 0 = 0$

$X = \dfrac{-(2) \pm \sqrt{(2)^2 - 4(3)(0)}}{2(3)} =$

$\dfrac{-2 \pm \sqrt{4}}{6} = \dfrac{-2 \pm 2}{6} = \dfrac{-1 \pm 1}{3} = 0, -\dfrac{2}{3}$

18.
$$4X^2 + 3 = 12X$$
$$4X^2 - 12X + 3 = 0$$
$$X = \frac{-(-12) \pm \sqrt{(-12)^2 - 4(4)(3)}}{2(4)} =$$
$$\frac{12 \pm 4\sqrt{6}}{8} = \frac{3 \pm \sqrt{6}}{2}$$

19.
$$\frac{2X+1}{5} - X = \frac{4-3X}{4} - 2$$
$$(20)\frac{2X+1}{5} - (20)X = (20)\frac{4-3X}{4} - (20)2$$
$$(4)(2X+1) - 20X = (5)(4-3X) - 40$$
$$8X + 4 - 20X = 20 - 15X - 40$$
$$8X - 20X + 15X = 20 - 40 - 4$$
$$3X = -24$$
$$X = \frac{-24}{3} = -8$$

20.
$$\frac{4X}{9} - 1 = \frac{-5X}{12} + X$$
$$(36)\frac{4X}{9} - (36)1 = (36)\frac{-5X}{12} + (36)X$$
$$16X - 36 = -15X + 36X$$
$$16X + 15X - 36X = 36$$
$$-5X = 36$$
$$X = \frac{36}{-5} \text{ or } -\frac{36}{5}$$

Lesson Practice 16A

1. A = apples; R = oranges; T = total
$$\frac{A}{R} = \frac{6}{5}; \frac{A}{T} = \frac{6}{11}; \frac{R}{T} = \frac{5}{11}$$
$$\frac{6}{5} = \frac{12}{R}$$
$$R = \frac{60}{6} = 10 \text{ oranges}$$

2. C = cloudy; S = sunny; T = total
$$\frac{C}{S} = \frac{1}{2}; \frac{C}{T} = \frac{1}{3}; \frac{S}{T} = \frac{2}{3}$$
$$\frac{1}{3} = \frac{C}{30}$$
$$\frac{30}{3} = C = 10 \text{ cloudy}$$
$$30 - 10 = 20 \text{ sunny}$$

3.
$$\frac{A}{B} = \frac{2}{5}; \frac{A}{T} = \frac{2}{7}; \frac{B}{T} = \frac{5}{7}$$
$$\frac{B}{490,000} = \frac{5}{7}$$
$$B = \frac{5 \cdot 490,000}{7}$$
$$B = 5 \cdot 70,000 = 350,000 \text{ votes}$$

4.
$$\frac{S}{R} = \frac{8}{7}; \frac{S}{T} = \frac{8}{15}; \frac{R}{T} = \frac{7}{15}$$
$$\frac{S}{56} = \frac{8}{7}$$
$$S = \frac{56 \cdot 8}{7} = 8 \cdot 8 = 64 \text{ squirrels}$$

5.
$$\frac{R}{M} = \frac{3}{5}; \frac{R}{T} = \frac{3}{8}; \frac{M}{T} = \frac{5}{8}$$
$$\frac{M}{24} = \frac{5}{8}$$
$$M = \frac{5 \cdot 24}{8} = 5 \cdot 3 = 15 \text{ students like math}$$

6.
$$\frac{Na}{NaCl} = \frac{23}{58}$$
$$\frac{Na}{406} = \frac{23}{58}$$
$$Na = \frac{406 \cdot 23}{58} = 161 \text{ grams}$$

7.
$$\frac{Cl}{NaCl} = \frac{35}{58}$$
$$\frac{Cl}{406} = \frac{35}{58}$$
$$Cl = \frac{35 \cdot 406}{58} = 245 \text{ grams}$$

8.
$$\frac{H_2}{H_2CO_2} = \frac{2}{46}$$
$$\frac{H_2}{352} = \frac{2}{46}$$
$$H_2 = \frac{2 \cdot 352}{46} \approx 15.30 \text{ grams}$$

9.
$$\frac{C}{352} = \frac{12}{46}$$
$$C = \frac{12 \cdot 352}{46} \approx 91.83 \text{ grams}$$

10.
$$\frac{O_2}{352} = \frac{32}{46}$$
$$O_2 = \frac{32 \cdot 352}{46} \approx 244.87 \text{ grams}$$

Lesson Practice 16B

1. N = amount of nitrogen

    G = amount of other ingredients

    T = total amount of fertilizer

    $\dfrac{N}{G} = \dfrac{5}{10}$; $\dfrac{N}{T} = \dfrac{5}{15}$; $\dfrac{G}{T} = \dfrac{10}{15}$

    $\dfrac{N}{135} = \dfrac{5}{15}$

    $N = \dfrac{5 \times 135}{15} = 45$ lb

2. S_1 = homeschoolers

    S_2 = all others

    $\dfrac{S_1}{S_2} = \dfrac{3}{7}$; $\dfrac{S_1}{T} = \dfrac{3}{10}$; $\dfrac{S_2}{T} = \dfrac{7}{10}$

    $\dfrac{90}{T} = \dfrac{3}{10}$

    $T = \dfrac{90 \times 10}{3} = 300$ students

3. S = sports

    W = school work

    $\dfrac{S}{W} = \dfrac{2}{3}$; $\dfrac{S}{T} = \dfrac{2}{5}$; $\dfrac{W}{T} = \dfrac{3}{5}$

    $\dfrac{4}{W} = \dfrac{2}{3}$

    $W = \dfrac{4 \times 3}{2} = 6$ hours

4. 8 gal = 32 qt

    $\dfrac{F}{A} = \dfrac{3}{1}$; $\dfrac{F}{T} = \dfrac{3}{4}$; $\dfrac{A}{T} = \dfrac{1}{4}$

    $\dfrac{A}{32} = \dfrac{1}{4}$

    $A = \dfrac{32 \times 1}{4} = 8$ qt antique ivory

    $32 - 8 = 24$ qt forest green

5. $\dfrac{T_1}{T_2} = \dfrac{4}{5}$; $\dfrac{T_1}{T} = \dfrac{4}{9}$; $\dfrac{T_2}{T} = \dfrac{5}{9}$

    $\dfrac{22}{T_2} = \dfrac{4}{5}$

    $T_2 = \dfrac{5 \times 22}{4} = 27.5$ in

6. $\dfrac{C}{CF_2Cl_2} = \dfrac{12}{120}$

    $\dfrac{C}{480} = \dfrac{12}{120}$

    $C = \dfrac{12(480)}{120} = 48$ g

7. $\dfrac{F_2}{480} = \dfrac{38}{120}$

    $F_2 = \dfrac{38(480)}{120} = 152$ g

8. $\dfrac{Cl_2}{480} = \dfrac{70}{120}$

    $\dfrac{70(480)}{120} = 280$ g

9. $\dfrac{K_2}{K_2S} = \dfrac{78}{110}$

    $\dfrac{K_2}{550} = \dfrac{78}{110}$

    $K_2 = \dfrac{78(550)}{110} = 390$ g

10. $\dfrac{S}{550} = \dfrac{32}{110}$

    $S = \dfrac{32(550)}{110} = 160$ g

Systematic Review 16C

1. $\dfrac{b}{g} = \dfrac{4}{3}$; $\dfrac{b}{t} = \dfrac{4}{7}$; $\dfrac{g}{t} = \dfrac{3}{7}$

2. $\dfrac{b}{t} = \dfrac{4}{7}$; We need to know the number of

    boys, and we are given the total

    number of students

3. $\dfrac{b}{t} = \dfrac{4}{7}$

    $\dfrac{b}{21} = \dfrac{4}{7}$

    $b = \dfrac{21(4)}{7} = 12$

4. $\dfrac{C_2}{C_2H_2} = \dfrac{24}{26}$; $\dfrac{H_2}{C_2H_2} = \dfrac{2}{26}$; $\dfrac{H_2}{C_2} = \dfrac{2}{24}$

5. $\dfrac{C_2}{C_2H_2} = \dfrac{24}{26}$

    $\dfrac{C_2}{234} = \dfrac{24}{26}$

    $C_2 = \dfrac{24(234)}{26} = 216$ g

6. $\dfrac{H_2}{C_2H_2} = \dfrac{2}{26}$

$\dfrac{H_2}{234} = \dfrac{2}{26}$

$H_2 = \dfrac{2(234)}{26} = 18 \text{ g}$

check : $216 + 18 = 234 \text{ g}$

7. $\dfrac{Fe}{FeCl_3} = \dfrac{56}{161}$; $\dfrac{Cl_3}{FeCl_3} = \dfrac{105}{161}$; $\dfrac{Fe}{Cl_3} = \dfrac{56}{105}$

8. $\dfrac{Fe}{FeCl_3} = \dfrac{56}{161}$

$\dfrac{Fe}{805} = \dfrac{56}{161}$

$Fe = \dfrac{56(805)}{161} = 280 \text{ g}$

9. $\dfrac{Cl_3}{FeCl_3} = \dfrac{105}{161}$

$\dfrac{Cl_3}{805} = \dfrac{105}{161}$

$Cl_3 = \dfrac{105(805)}{161} = 525 \text{ g}$

10. $\dfrac{Y}{Z} = X$

$\dfrac{Y}{X} = Z$

11. $\dfrac{R}{S} = \dfrac{T}{QW}$

$\dfrac{RQW}{S} = T$

12. $\dfrac{13}{18} \approx .72 = 72\%$

13. $\dfrac{5}{18} \approx .28 = 28\%$

14. $\dfrac{C}{CF_2Cl_2} = \dfrac{12}{12 + 38 + 70} = \dfrac{12}{120} = .1 = 10\%$

15. $\dfrac{F_2}{CF_2Cl_2} = \dfrac{38}{120} \approx .3167 \approx 32\%$

16. $\dfrac{Cl_2}{CF_2Cl_2} = \dfrac{70}{120} \approx .5833 \approx 58\%$

17. $2X^2 + X = -\dfrac{1}{2}$

$2X^2 + X + \dfrac{1}{2} = 0$

$b^2 - 4ac \Rightarrow (1)^2 - 4(2)\left(\dfrac{1}{2}\right) = 1 - 4 = -3$

imaginary

18. $\dfrac{-(1) \pm \sqrt{(1)^2 - 4(2)\left(\dfrac{1}{2}\right)}}{2(2)} = \dfrac{-1 \pm i\sqrt{3}}{4}$

19. $X^2 + \dfrac{7}{4}X = \dfrac{1}{2}$

$X^2 + \dfrac{7}{4}X + \dfrac{49}{64} = \dfrac{1}{2} + \dfrac{49}{64}$

$\left(X + \dfrac{7}{8}\right)^2 = \dfrac{32}{64} + \dfrac{49}{64}$

$\left(X + \dfrac{7}{8}\right)^2 = \dfrac{81}{64}$

$X + \dfrac{7}{8} = \pm\sqrt{\dfrac{81}{64}}$

$X + \dfrac{7}{8} = \pm\dfrac{9}{8}$

$X = -\dfrac{7}{8} \pm \dfrac{9}{8}$

$X = -\dfrac{7}{8} + \dfrac{9}{8} = \dfrac{2}{8} = \dfrac{1}{4}$

$X = -\dfrac{7}{8} - \dfrac{9}{8} = -\dfrac{16}{8} = -2$

20. $\left(\dfrac{1}{4}\right)^2 + \dfrac{7}{4}\left(\dfrac{1}{4}\right) = \dfrac{1}{2}$ $(-2)^2 + \dfrac{7}{4}(-2) = \dfrac{1}{2}$

$\dfrac{1}{16} + \dfrac{7}{16} = \dfrac{1}{2}$ $4 - \dfrac{14}{4} = \dfrac{1}{2}$

$\dfrac{8}{16} = \dfrac{1}{2}$ $\dfrac{16}{4} - \dfrac{14}{4} = \dfrac{1}{2}$

$\dfrac{1}{2} = \dfrac{1}{2}$ $\dfrac{2}{4} = \dfrac{1}{2}$

$\dfrac{1}{2} = \dfrac{1}{2}$

Systematic Review 16D

1. R = orange juice

   C = cranberry juice

   $\dfrac{R}{C} = \dfrac{4}{3}$; $\dfrac{R}{T} = \dfrac{4}{7}$; $\dfrac{C}{T} = \dfrac{3}{7}$

2. $\dfrac{C}{T} = \dfrac{3}{7}$; We need to know total, and we are given cranberry.

3. $\dfrac{C}{T} = \dfrac{3}{7}$

   $\dfrac{165}{T} = \dfrac{3}{7}$

   $T = \dfrac{165(7)}{3} = 385$

4. $\dfrac{K_2}{K_2O} = \dfrac{78}{94}$; $\dfrac{O}{K_2O} = \dfrac{16}{94}$; $\dfrac{K_2}{O} = \dfrac{78}{16}$

5. $\dfrac{K_2}{K_2O} = \dfrac{78}{94}$

   $\dfrac{K_2}{752} = \dfrac{78}{94}$

   $K_2 = \dfrac{78(752)}{94} = 624$ g

6. $\dfrac{O}{K_2O} = \dfrac{16}{94}$

   $\dfrac{O}{752} = \dfrac{16}{94}$

   $O = \dfrac{752(16)}{94} = 128$ g

7. $\dfrac{C}{CHF_3} = \dfrac{12}{70}$; $\dfrac{H}{CHF_3} = \dfrac{1}{70}$; $\dfrac{F_3}{CHF_3} = \dfrac{57}{70}$

8. $\dfrac{C}{CHF_3} = \dfrac{12}{70}$

   $\dfrac{C}{840} = \dfrac{12}{70}$

   $C = \dfrac{840(12)}{70} = 144$ g

9. $\dfrac{F_3}{CHF_3} = \dfrac{57}{70}$

   $\dfrac{F_3}{840} = \dfrac{57}{70}$

   $F_3 = \dfrac{840(57)}{70} = 684$ g

10. $r = \dfrac{1}{3}\pi r^2 H$

    $\dfrac{3r}{\pi r^2} = H$

    $\dfrac{3}{\pi r} = H$

11. $S = N\left(\dfrac{A+L}{T}\right)$

    $\dfrac{ST}{A+L} = N$

12. $\dfrac{932}{1,650} \approx .565 = 56.5\%$

13. $\dfrac{718}{1,650} \approx .435 = 43.5\%$

14. $\dfrac{C}{H_2CO} = \dfrac{12}{30} = .4 = 40\%$

15. $\dfrac{H_2}{H_2CO} = \dfrac{2}{30} \approx .066 \approx 7\%$

16. $\dfrac{O}{H_2CO} = \dfrac{16}{30} \approx .533 \approx 53\%$

17. $X^2 + 16 = -8X$

    $X^2 + 8X + 16 = 0$

    $(8)^2 - 4(1)(16) = 64 - 64 = 0$

    real, rational, equal (double root)

18. $X^2 + 8X + 16 = 0$

    $(X+4)^2 = 0$

    $X + 4 = 0$

    $X = -4$

19. $\dfrac{8X-3}{6} + 1 = \dfrac{X-5}{3} - \dfrac{2-3X}{8}$

    $(24)\dfrac{8X-3}{6} + (24)1 = (24)\dfrac{X-5}{3} - (24)\dfrac{2-3X}{8}$

    $(4)\dfrac{8X-3}{1} + 24 = (8)\dfrac{X-5}{1} - (3)\dfrac{2-3X}{1}$

    $4(8X-3) + 24 = 8(X-5) - 3(2-3X)$

    $32X - 12 + 24 = 8X - 40 - 6 + 9X$

    $32X + 12 = 17X - 46$

    $15X = -58$

    $X = -\dfrac{58}{15} = -3\dfrac{13}{15}$

20. $\dfrac{3X}{7} - X = \dfrac{5X}{3} - 2$

$(21)\dfrac{3X}{7} - (21)X = (21)\dfrac{5X}{3} - (21)2$

$3(3X) - 21X = 7(5X) - 42$

$9X - 21X = 35X - 42$

$-12X = 35X - 42$

$42 = 35X + 12X$

$42 = 47X$

$\dfrac{42}{47} = X$

Systematic Review 16E

1. $\dfrac{F}{S} = \dfrac{3}{1}; \dfrac{F}{T} = \dfrac{3}{4}; \dfrac{S}{T} = \dfrac{1}{4}$

2. $\dfrac{S}{T} = \dfrac{1}{4}$: We need to know total fans, and we are given soccer fans.

3. $\dfrac{S}{T} = \dfrac{1}{4}$

$\dfrac{11,300}{T} = \dfrac{1}{4}$

$T = 4(11,300) = 45,200$ fans

4. $\dfrac{H_2}{H_2S} = \dfrac{2}{34}; \dfrac{S}{H_2S} = \dfrac{32}{34}; \dfrac{H_2}{S} = \dfrac{2}{32}$

5. $\dfrac{S}{H_2S} = \dfrac{32}{34}$

$\dfrac{S}{442} = \dfrac{32}{34}$

$S = \dfrac{32(442)}{34} = 416$ g

6. $\dfrac{H_2}{H_2S} = \dfrac{2}{34}$

$\dfrac{H_2}{442} = \dfrac{2}{34}$

$H_2 = \dfrac{2(442)}{34} = 26$ g

7. $\dfrac{Fe_2}{NFe_2} = \dfrac{112}{126}; \dfrac{N}{NFe_2} = \dfrac{14}{126}; \dfrac{N}{Fe_2} = \dfrac{14}{112}$

8. $\dfrac{Fe_2}{NFe_2} = \dfrac{112}{126}$

$\dfrac{Fe_2}{882} = \dfrac{112}{126}$

$Fe_2 = \dfrac{882(112)}{126} = 784$ g

9. $\dfrac{N}{NFe_2} = \dfrac{14}{126}$

$\dfrac{N}{882} = \dfrac{14}{126}$

$N = \dfrac{882(14)}{126} = 98$ g

10. $\dfrac{1}{F} = \dfrac{1}{A} + \dfrac{1}{B}$

$\dfrac{1}{F} = \dfrac{1(B)}{A(B)} + \dfrac{1(A)}{B(A)}$

$\dfrac{1}{F} = \dfrac{B}{AB} + \dfrac{A}{AB}$

$\dfrac{1}{F} = \dfrac{A+B}{AB}$

$F = \dfrac{AB}{A+B}$

11. $\dfrac{AB}{XY} + \dfrac{CD}{E} = 0$

$\dfrac{CD}{E} = -\dfrac{AB}{XY}$

$CD = E\left(-\dfrac{AB}{XY}\right)$

$-\dfrac{CDXY}{AB} = E$

12. $(.45)(5,435,960) = 2,446,182$

13. $\dfrac{.55}{.45} = \dfrac{11}{9}$

14. $\dfrac{Na}{NaOH} = \dfrac{23}{23+16+1} = \dfrac{23}{40} = .575 \approx 58\%$

15. $\dfrac{O}{NaOH} = \dfrac{16}{40} = .40 = 40\%$

16. $\dfrac{H}{NaOH} = \dfrac{1}{40} = .025 \approx 3\%$

17.
$$X^2 - \frac{2}{3}X = \frac{4}{3}$$
$$3(X^2) - 3\left(\frac{2}{3}X\right) = 3\left(\frac{4}{3}\right)$$
$$3X^2 - 2X = 4$$
$$3X^2 - 2X - 4 = 0$$
$$(-2)^2 - 4(3)(-4) = 4 + 48 = 52$$
real, irrational, unequal

18.
$$\frac{-(-2) \pm \sqrt{(-2)^2 - 4(3)(-4)}}{2(3)} =$$
$$\frac{-(-2) \pm 2\sqrt{13}}{6} = \frac{1 \pm \sqrt{13}}{3}$$

19.
$$3^2 - X = 1.25X - 8.4$$
$$9 - X = 1.25X - 8.4$$
$$(100)9 - (100)X = (100)1.25X - (100)8.4$$
$$900 - 100X = 125X - 840$$
$$900 + 840 = 125X + 100X$$
$$1{,}740 = 225X$$
$$\frac{1{,}740}{225} = X = \frac{116}{15} \text{ or } 7\frac{11}{15}$$

20.
$$\frac{X}{2} + 15 = \frac{X}{3} + X$$
$$(6)\frac{X}{2} + (6)15 = (6)\frac{X}{3} + (6)X$$
$$3X + 90 = 2X + 6X$$
$$3X + 90 = 8X$$
$$90 = 5X$$
$$18 = X$$

Lesson Practice 17A

1. $\frac{156 \text{ in}}{1} \times \frac{1 \text{ ft}}{12 \text{ in}} = 13 \text{ ft}$

2. $\frac{8 \text{ lb}}{1} \times \frac{16 \text{ oz}}{1 \text{ lb}} = 128 \text{ oz}$

3. $\frac{7 \text{ cm}}{1} \times \frac{1 \text{ m}}{100 \text{ cm}} = .07 \text{ m}$

4. $\frac{15 \text{ in}^2}{1} \times \frac{1 \text{ ft}}{12 \text{ in}} \times \frac{1 \text{ ft}}{12 \text{ in}} = \frac{15}{144} = .104 \text{ ft}^2$

5. $\frac{25 \text{ gal}}{1} \times \frac{4 \text{ qt}}{1 \text{ gal}} \times \frac{2 \text{ pt}}{1 \text{ qt}} = 200 \text{ pt}$

6. $\frac{10 \text{ mi}^2}{1} \times \frac{5{,}280 \text{ ft}}{1 \text{ mi}} \times \frac{5{,}280 \text{ ft}}{1 \text{ mi}} =$
278,784,000 ft^2

7. $\frac{13 \text{ oz}}{1} \times \frac{28 \text{ g}}{1 \text{ oz}} = 364 \text{ g}$

8. $\frac{9 \text{ liters}}{1} \times \frac{1.06 \text{ qt}}{1 \text{ liter}} = 9.54 \text{ qt}$

9. $\frac{350 \text{ cm}}{1} \times \frac{.4 \text{ in}}{1 \text{ cm}} = 140 \text{ in}$

10. $\frac{17 \text{ yd}}{1} \times \frac{.9 \text{ m}}{1 \text{ yd}} = 15.3 \text{ m}$

11. $\frac{4 \text{ km}}{1} \times \frac{.62 \text{ mi}}{1 \text{ km}} \times \frac{5{,}280 \text{ ft}}{1 \text{ mi}} = 13{,}094.4 \text{ ft}$

12. $\frac{50 \text{ gal}}{1} \times \frac{4 \text{ qt}}{1 \text{ gal}} \times \frac{.95 \text{ liters}}{1 \text{ qt}} = 190 \text{ liters}$

Lesson Practice 17B

1. $\frac{7 \text{ mi}}{1} \times \frac{5{,}280 \text{ ft}}{1 \text{ mi}} = 36{,}960 \text{ ft}$

2. $\frac{6{,}342 \text{ lb}}{1} \times \frac{1 \text{ ton}}{2{,}000 \text{ lb}} = 3.171 \text{ tons}$

3. $\frac{7{,}040 \text{ yd}}{1} \times \frac{3 \text{ ft}}{1 \text{ yd}} \times \frac{1 \text{ mi}}{5{,}280 \text{ ft}} = 4 \text{ mi}$

4. $\frac{852 \text{ ft}^2}{1} \times \frac{12 \text{ in}}{1 \text{ ft}} \times \frac{12 \text{ in}}{1 \text{ ft}} = 122{,}688 \text{ in}^2$

5. $\frac{95 \text{ km}^2}{1} \times \frac{1{,}000 \text{ m}}{1 \text{ km}} \times \frac{1{,}000 \text{ m}}{1 \text{ km}} =$
95,000,000 m^2

6. $\frac{580 \text{ g}}{1} \times \frac{1 \text{ kg}}{1{,}000 \text{ g}} = .58 \text{ kg}$

7. $\frac{87 \text{ in}}{1} \times \frac{1 \text{ yd}}{36 \text{ in}} \times \frac{.9 \text{ m}}{1 \text{ yd}} = 2.175 \text{ m}$

8. $\frac{106 \text{ mi}}{1} \times \frac{1.6 \text{ km}}{1 \text{ mi}} = 169.6 \text{ km}$

9. $\frac{45 \text{ kg}}{1} \times \frac{2.2 \text{ lb}}{1 \text{ kg}} \times \frac{16 \text{ oz}}{1 \text{ lb}} = 1{,}584 \text{ oz}$

10. $\dfrac{9 \text{ lb}}{1} \times \dfrac{.45 \text{ kg}}{1 \text{ lb}} \times \dfrac{1{,}000 \text{ g}}{1 \text{ kg}} = 4{,}050 \text{ g}$

OR :

$\dfrac{9 \text{ lb}}{1} \times \dfrac{16 \text{ oz}}{1 \text{ lb}} \times \dfrac{28 \text{ g}}{1 \text{ oz}} = 4{,}032 \text{ g}$

Differences due to rounding of metric/imperial conversion factors.

11. $\dfrac{3 \text{ liters}}{1} \times \dfrac{1.06 \text{ qt}}{1 \text{ liter}} \times \dfrac{2 \text{ pt}}{1 \text{ qt}} = 6.36 \text{ pt}$

12. $\dfrac{14 \text{ kg}}{1} \times \dfrac{2.2 \text{ lb}}{1 \text{ kg}} = 30.8 \text{ lb}$

Systematic Review 17C

1. $\dfrac{3 \text{ tons}}{1} \times \dfrac{2{,}000 \text{ lb}}{1 \text{ ton}} \times \dfrac{16 \text{ oz}}{1 \text{ lb}} = 96{,}000 \text{ oz}$

2. $\dfrac{24 \text{ pt}}{1} \times \dfrac{1 \text{ qt}}{2 \text{ pt}} \times \dfrac{1 \text{ gal}}{4 \text{ qt}} = 3 \text{ gal}$

3. $\dfrac{5 \text{ yd}^2}{1} \times \dfrac{3 \text{ ft}}{1 \text{ yd}} \times \dfrac{3 \text{ ft}}{1 \text{ yd}} = 45 \text{ ft}^2$

4. $\dfrac{2 \text{ ft}^3}{1} \times \dfrac{1 \text{ yd}}{3 \text{ ft}} \times \dfrac{1 \text{ yd}}{3 \text{ ft}} \times \dfrac{1 \text{ yd}}{3 \text{ ft}} =$

$\dfrac{2}{27} \text{ yd}^3 \text{ or} \approx .074 \text{ yd}^3$

5. $\dfrac{4 \text{ mi}^2}{1} \times \dfrac{1{,}760 \text{ yd}}{1 \text{ mi}} \times \dfrac{1{,}760 \text{ yd}}{1 \text{ mi}} =$

$12{,}390{,}400 \text{ yd}^2$

6. $\dfrac{10 \text{ mi}}{1} \times \dfrac{1.6 \text{ km}}{1 \text{ mi}} = 16 \text{ km}$

7. $\dfrac{25 \text{ oz}}{1} \times \dfrac{28 \text{ g}}{1 \text{ oz}} = 700 \text{ g}$

8. $\dfrac{5 \text{ lb}}{1} \times \dfrac{.45 \text{ kg}}{1 \text{ lb}} = 2.25 \text{ kg}$

9. $\dfrac{20 \text{ m}}{1} \times \dfrac{1.1 \text{ yd}}{1 \text{ m}} = 22 \text{ yd}$

10. $\dfrac{12 \text{ liters}}{1} \times \dfrac{1.06 \text{ qt}}{1 \text{ liter}} = 12.72 \text{ qt}$

11. $\dfrac{H_2}{H_2O} = \dfrac{2}{18} ; \dfrac{O}{H_2O} = \dfrac{16}{18} ; \dfrac{H_2}{O} = \dfrac{2}{16}$

12. $\dfrac{O}{H_2O} = \dfrac{16}{18}$

$\dfrac{O}{234} = \dfrac{16}{18}$

$O = \dfrac{234(16)}{18} = 208 \text{ g}$

13. $\dfrac{H_2}{H_2O} = \dfrac{2}{18}$

$\dfrac{H_2}{234} = \dfrac{2}{18}$

$H_2 = \dfrac{234(2)}{18} = 26 \text{ g}$

14. $\dfrac{\text{pecan}}{\text{oak}} = \dfrac{7}{2} ; \dfrac{\text{oak}}{\text{total}} = \dfrac{2}{9} ; \dfrac{\text{pecan}}{\text{total}} = \dfrac{7}{9}$

15. given total, looking for pecan:

$\dfrac{P}{T} = \dfrac{7}{9}$

16. $\dfrac{P}{T} = \dfrac{7}{9}$

$\dfrac{P}{36} = \dfrac{7}{9}$

$P = \dfrac{36(7)}{9} = 28 \text{ pecan trees}$

17. $\dfrac{1}{X} = \dfrac{2Y}{3} + A$

$\dfrac{1}{X} = \dfrac{2Y}{3} + \dfrac{A(3)}{1(3)}$

$\dfrac{1}{X} = \dfrac{2Y + 3A}{3}$

$1 = (X)\left(\dfrac{2Y + 3A}{3} \right)$

$\dfrac{3}{2Y + 3A} = X$

18. $\dfrac{A}{C} = 5 - \dfrac{1}{B}$

$\dfrac{A}{C} = \dfrac{5(B)}{1(B)} - \dfrac{1}{B}$

$\dfrac{A}{C} = \dfrac{5B - 1}{B}$

$A = C\left(\dfrac{5B - 1}{B}\right)$

$\dfrac{A}{1}\left(\dfrac{B}{5B - 1}\right) = C$

$\dfrac{AB}{5B - 1} = C$

19. $\dfrac{O}{MgO} = \dfrac{16}{40} = .4 = 40\%$

20. $\dfrac{Mg}{MgO} = \dfrac{24}{40} = .6 = 60\%$

Systematic Review 17D

1. $\dfrac{65 \text{ yd}}{1} \times \dfrac{3 \text{ ft}}{1 \text{ yd}} \times \dfrac{12 \text{ in}}{1 \text{ ft}} = 2{,}340 \text{ in}$

2. $\dfrac{10.6 \text{ m}}{1} \times \dfrac{1 \text{ km}}{1{,}000 \text{ m}} = .0106 \text{ km}$

3. $\dfrac{50 \text{ ft}^2}{1} \times \dfrac{1 \text{ yd}}{3 \text{ ft}} \times \dfrac{1 \text{ yd}}{3 \text{ ft}} =$

$\dfrac{50}{9} \text{ yd}^2 \text{ or } 5\dfrac{5}{9} \text{ yd}^2 \text{ or } \approx 5.56 \text{ yd}^2$

4. $\dfrac{1{,}860 \text{ in}^3}{1} \times \dfrac{1 \text{ yd}}{36 \text{ in}} \times \dfrac{1 \text{ yd}}{36 \text{ in}} \times \dfrac{1 \text{ yd}}{36 \text{ in}} \approx$

$.04 \text{ yd}^3$

5. $\dfrac{4 \text{ cm}^2}{1} \times \dfrac{1 \text{ m}}{100 \text{ cm}} \times \dfrac{1 \text{ m}}{100 \text{ cm}} = .0004 \text{ m}^2$

6. $\dfrac{45 \text{ in}}{1} \times \dfrac{2.5 \text{ cm}}{1 \text{ in}} = 112.5 \text{ cm}$

7. $\dfrac{7 \text{ qt}}{1} \times \dfrac{.95 \text{ liters}}{1 \text{ qt}} = 6.65 \text{ liters}$

8. $\dfrac{200 \text{ cm}}{1} \times \dfrac{.4 \text{ in}}{1 \text{ cm}} = 80 \text{ in}$

9. $\dfrac{3 \text{ km}}{1} \times \dfrac{.62 \text{ mi}}{1 \text{ km}} = 1.86 \text{ mi}$

$\dfrac{1.86 \text{ mi}}{1} \times \dfrac{5{,}280 \text{ ft}}{1 \text{ mi}} = 9{,}820.8 \text{ ft}$

10. $\dfrac{9 \text{ kg}}{1} \times \dfrac{2.2 \text{ lb}}{1 \text{ kg}} = 19.8 \text{ lb}$

11. $\dfrac{H}{HCl} = \dfrac{1}{36}; \dfrac{Cl}{HCl} = \dfrac{35}{36}; \dfrac{H}{Cl} = \dfrac{1}{35}$

12. $\dfrac{Cl}{HCl} = \dfrac{35}{36}$

$\dfrac{Cl}{612} = \dfrac{35}{36}$

$Cl = \dfrac{612(35)}{36} = 595 \text{ g}$

13. $\dfrac{H}{HCl} = \dfrac{1}{36}$

$\dfrac{H}{612} = \dfrac{1}{36}$

$H = \dfrac{612(1)}{36} = 17 \text{ g}$

14. $\dfrac{S}{T} = \dfrac{2}{7}; \dfrac{D}{T} = \dfrac{5}{7}; \dfrac{S}{D} = \dfrac{2}{5}$

15. given sunny, looking for dreary:

$\dfrac{S}{D} = \dfrac{2}{5}$

16. $\dfrac{S}{D} = \dfrac{2}{5}$

$\dfrac{32}{D} = \dfrac{2}{5}$

$2D = 160$

$D = 80 \text{ dreary days}$

17. $\dfrac{A}{a} = \dfrac{B}{b}$

$A = \dfrac{Ba}{b}$

18. $\dfrac{A}{a} = \dfrac{B}{b}$

$A = a\left(\dfrac{B}{b}\right)$

$A\left(\dfrac{b}{B}\right) = a$

$\dfrac{Ab}{B} = a$

19. $\dfrac{O_2}{CO_2} = \dfrac{16 + 16}{12 + 16 + 16} = \dfrac{32}{44} \approx .727 = 72.7\%$

20. $\dfrac{C}{CO_2} = \dfrac{12}{44} \approx .273 = 27.3\%$

Systematic Review 17E

1. $\dfrac{28\ ft}{1} \times \dfrac{12\ in}{1\ ft} = 336\ in$

2. $\dfrac{.28\ km}{1} \times \dfrac{1{,}000\ m}{1\ km} = 280\ m$

3. $\dfrac{8{,}000\ yd^2}{1} \times \dfrac{1\ mi}{1{,}760\ yd} \times \dfrac{1\ mi}{1{,}760\ yd} \approx$ $.0026\ mi^2$

4. $\dfrac{39\ km^3}{1} \times \dfrac{1{,}000\ m}{1\ km} \times \dfrac{1{,}000\ m}{1\ km} \times \dfrac{1{,}000\ m}{1\ km} =$ $3.9 \times 10^{10}\ m^3$

5. $\dfrac{4.8\ m^2}{1} \times \dfrac{1{,}000\ mm}{1\ m} \times \dfrac{1{,}000\ mm}{1\ m} =$ $4.8 \times 10^6\ mm^2$

6. $\dfrac{12\ yd}{1} \times \dfrac{.9\ m}{1\ yd} = 10.8\ m$

7. $\dfrac{10{,}000\ ft}{1} \times \dfrac{1\ mi}{5{,}280\ ft} \times \dfrac{1.6\ km}{1\ mi} \approx 3.03\ km$

8. $\dfrac{3\ lb}{1} \times \dfrac{.45\ kg}{1\ lb} \times \dfrac{1{,}000\ g}{1\ kg} = 1{,}350\ g$

9. $\dfrac{36\ m}{1} \times \dfrac{1.1\ yd}{1\ m} \times \dfrac{3\ ft}{1\ yd} = 118.8\ ft$

10. $\dfrac{340\ g}{1} \times \dfrac{.035\ oz}{1\ g} = 11.9\ oz$

11. $\dfrac{C}{CH_4} = \dfrac{12}{16}$; $\dfrac{H_4}{CH_4} = \dfrac{4}{16}$; $\dfrac{C}{H_4} = \dfrac{12}{4}$

12. $\dfrac{C}{CH_4} = \dfrac{12}{16}$

    $\dfrac{C}{208} = \dfrac{12}{16}$

    $C = \dfrac{12(208)}{16} = 156\ g$

13. $\dfrac{H_4}{CH_4} = \dfrac{4}{16}$

    $\dfrac{H_4}{208} = \dfrac{4}{16}$

    $H_4 = \dfrac{4(208)}{16} = 52\ g$

14. D = decorated
    N = not decorated
    T = total
    $\dfrac{D}{T} = \dfrac{7}{12}$; $\dfrac{N}{T} = \dfrac{5}{12}$; $\dfrac{D}{N} = \dfrac{7}{5}$

15. given not decorated, looking for decorated:
    $\dfrac{D}{N} = \dfrac{7}{5}$

16. $\dfrac{D}{N} = \dfrac{7}{5}$

    $\dfrac{D}{35} = \dfrac{7}{5}$

    $D = \dfrac{7(35)}{5} = 49$ decorated

17. $Y = X(A - B)$

    $\dfrac{Y}{A - B} = X$

18. $Y = X(A - B)$
    $Y = AX - BX$
    $BX = AX - Y$
    $B = \dfrac{AX - Y}{X}$ or $\dfrac{AX}{X} - \dfrac{Y}{X} = A - \dfrac{Y}{X} = B$

19. $\dfrac{O}{H_2O} = \dfrac{16}{2 + 16} = \dfrac{16}{18} \approx .89 = 89\%$

20. $\dfrac{H_2}{H_2O} = \dfrac{2}{18} \approx .11 = 11\%$

Lesson Practice 18A

1. $D = RT \Rightarrow D = 45\ mph \times 4\ hr = 180\ mi$

2. $D = RT \Rightarrow R = \dfrac{D}{T} = \dfrac{36\ yd}{72\ min} = \dfrac{1}{2}\ ypm$

3. $D = RT \Rightarrow T = \dfrac{D}{R} = \dfrac{36\ yd}{12\ ypm} = 3\ min$

4.

$D_B = D_S$

$R_B T_B = R_S T_S$

$(60)(9) = (50)(T_S) \begin{cases} R_B = 60 \\ T_B = 9 \\ R_S = 60 - 10 = 50 \end{cases}$

$540 = 50 T_S$

$10.8 = T_S = 10\frac{8}{10} = 10\frac{48}{60} = 10$ hr 48 min

6:00 AM + 10:48 = 4:48 PM

5.

$D_1 = D_2$

$R_1 T_1 = R_2 T_2$

$(45)(T_1) = (35)(T_1 + .4) \begin{cases} R_1 = 45 \\ R_2 = 35 \\ T_2 = T_1 + .4 \end{cases}$

$45 T_1 = 35 T_1 + 14$

$10 T_1 = 14$

$T_1 = 1.4$ hr

$D = RT \Rightarrow 45(1.4) = 63$ mi

6.

$D_G = D_J$

$R_G T_G = R_J T_J$

$(55)(5) = (R_J)(4) \begin{cases} R_G = 55 \\ T_G = 5 \\ T_J = 4 \end{cases}$

$275 = 4 R_J$

$R_J = 68.75$ mph

4.

$D_C = D_W$

$R_C T_C = R_W T_W$

$(50)(.3) = (R_W)(3.75) \begin{cases} T_C = .3 \\ R_C = 50 \\ T_W = 3.75 \end{cases}$

$15 = 3.75 R_W$

$R_W = 4$ mph

$D = R_W T_W \Rightarrow (4)(3.75) = 15$ mi

5.

$D_{Jo} = D_{Je}$

$R_{Jo} T_{Jo} = R_{Je} T_{Je}$

$(4)(2) = (3)(T_{Je}) \begin{cases} R_{Jo} = 4 \\ R_{Je} = 3 \\ T_{Jo} = 2 \end{cases}$

$8 = 3 T_{Je}$

$T_{Je} = \frac{8}{3}$ hr or $2\frac{2}{3}$ hr or 2 hr 40 min

$D = RT \Rightarrow (4)(2) = 8$ mi

6.

$D_1 = D_2$

$R_1 T_1 = R_2 T_2$

$9 = (12)(T_2) \begin{cases} D_1 = R_1 T_1 = 9 \\ R_2 = 15 - 3 = 12 \end{cases}$

$T_2 = .75$ hr or 45 min

Lesson Practice 18B

1. $D = RT \Rightarrow R = \frac{D}{T} = \frac{336 \text{ mi}}{6 \text{ hr}} = 56$ mph

2. $D = RT \Rightarrow T = \frac{D}{R} = \frac{21 \text{ mi}}{3.5 \text{ mph}} = 6$ hr

3. $D = RT \Rightarrow D = 4.5 \text{ fpm} \times 120 \text{ min} = 540$ ft

Systematic Review 18C

1. $D = RT \Rightarrow T = \frac{D}{R} = \frac{10.5 \text{ km}}{3 \text{ km/hr}} = 3.5$ hr

2. $D = RT \Rightarrow D = 7 \text{ km/hr} \times 4\frac{1}{4} \text{ hr} = 29\frac{3}{4}$ km

3. $D = RT \Rightarrow R = \frac{D}{T} = \frac{10 \text{ km}}{1.5 \text{ hr}} = 6\frac{2}{3}$ km/hr

4.
$$\frac{420 \text{ mi}}{D_L = D_V} \longrightarrow$$

$$R_L T_L = R_V T_V$$

$$(60)(7) = (50)(T_V) \begin{cases} R_V = R_L - 10 \\ R_V = (60) - 10 = 50 \\ T_L = 3:30 - 8:30 = 7 \text{ hr} \end{cases}$$

$$420 = 50T_V$$

$$8.4 \text{ hr} = T_V$$

$$T_V = 8 \text{ hr} + (.4 \times 60) \text{ min}$$

$$T_V = 8 \text{ hr} + 24 \text{ min}$$

5. $8:00 \text{ AM} + 8:24 = 4:24 \text{ PM arrival}$

6.
$$\frac{?}{D_L = D_V} \longrightarrow$$

$$D_L = D_V$$

$$R_L T_L = R_V T_V$$

$$T_V = 5:00 \text{ PM} - 8:40 \text{ AM} = 8\frac{1}{3} \text{ hr}$$

$$T_L = 5:00 \text{ PM} - 9:30 \text{ AM} = 7\frac{1}{2} \text{ hr}$$

$$R_V = R_L - 6$$

$$R_L\left(7\frac{1}{2}\right) = \left(R_L - 6\right)\left(8\frac{1}{3}\right)$$

$$6\left(\frac{15}{2}R_L = \frac{25}{3}R_L - 50\right)$$

$$45R_L = 50R_L - 300$$

$$-5R_L = -300 \Rightarrow R_L = 60$$

$$R_V = 60 - 6 = 54$$

7. $D = (60)(7.5) = 450 \text{ mi}$
or $(54)(8.33) = 450 \text{ mi}$

8. $\dfrac{12,500 \text{ lb}}{1} \times \dfrac{1 \text{ ton}}{2,000 \text{ lb}} = 6.25 \text{ tons}$

9. $\dfrac{3.4 \text{ m}}{1} \times \dfrac{100 \text{ cm}}{1 \text{ m}} = 340 \text{ cm}$

10. $\dfrac{500 \text{ in}^2}{1} \times \dfrac{1 \text{ ft}}{12 \text{ in}} \times \dfrac{1 \text{ ft}}{12 \text{ in}} \approx 3.47 \text{ ft}^2$

11. $\dfrac{14,000 \text{ mm}^2}{1} \times \dfrac{1 \text{ m}}{1,000 \text{ mm}} \times \dfrac{1 \text{ m}}{1,000 \text{ mm}} =$
.014 m^2

12. $\dfrac{4 \text{ gal}}{1} \times \dfrac{4 \text{ qt}}{1 \text{ gal}} \times \dfrac{.95 \text{ liters}}{1 \text{ qt}} = 15.2 \text{ liters}$

13. $\dfrac{75 \text{ cm}}{1} \times \dfrac{.4 \text{ in}}{1 \text{ cm}} = 30 \text{ in}$

14. $\dfrac{C}{CO} = \dfrac{12}{28}; \dfrac{O}{CO} = \dfrac{16}{28}; \dfrac{C}{O} = \dfrac{12}{16}$

15. $\dfrac{O}{1,204} = \dfrac{16}{28}$

$$O = \frac{16(1,204)}{28} = 688 \text{ g}$$

16. $\dfrac{C}{1,204} = \dfrac{12}{28}$

$$C = \frac{12(1,204)}{28} = 516 \text{ g}$$

17. $C = \dfrac{5}{9}(F - 32) \Rightarrow \dfrac{9}{5}C = F - 32$

$$\frac{9}{5}C + 32 = F$$

18. $\dfrac{Na}{NaCl} = \dfrac{23}{23 + 35} = \dfrac{23}{58} \approx .397 = 39.7\%$

19. $\dfrac{Cl}{NaCl} = \dfrac{35}{58} \approx .603 = 60.3\%$

20. $\left[(64)^{\frac{1}{2}}\right]^{\frac{2}{3}} = 8^{\frac{2}{3}} = 2^2 = 4$

Systematic Review 18D

1. $D = RT \Rightarrow T = \dfrac{D}{R} = \dfrac{1,750 \text{ yd}}{1,000 \text{ yd/h}} = 1\dfrac{3}{4} \text{ hr}$
   If $R = 500 \text{ yd}/30 \text{ min}$, then $R = 1,000 \text{ yd/ hr}$

2. $D = RT \Rightarrow D = 12 \text{ mph} \times \dfrac{9}{4} \text{ h} = 27 \text{ mi}$

3. $D = RT \Rightarrow T = \dfrac{D}{R} = \dfrac{16}{2.5 \text{ mph}} = 6.4 \text{ hr}$

4. jet ski

$D_A = D_M$

$R_A T_A = R_M T_M$

$$(24)(T_A) = (20)\left(T_A + \frac{1}{4}\right) \begin{cases} R_A = 24 \text{ kph} \\ R_M = 20 \text{ kph} \\ T_M = T_A + \frac{1}{4} \text{ hr} \\ \left(15 \text{ min} = \frac{1}{4} \text{ hr}\right) \end{cases}$$

$24T_A = 20T_A + 5$

$4T_A = 5$

$T_A = \frac{5}{4} = 1\frac{1}{4}$

$T_M = T_A + \frac{1}{4} \Rightarrow T_M = 1\frac{1}{4} + \frac{1}{4} = 1\frac{1}{2} \text{ hr}$

5. $D = (24)\left(\frac{5}{4}\right) = 30 \text{ km}$

or $D = (20)\left(\frac{3}{2}\right) = 30 \text{ km}$

6. canoe

$D_A = D_M$

$R_A(T_M + 1) = R_M T_M$

$$6(T_M + 1) = 7.5T_M \begin{cases} R_A = 6 \text{ kph} \\ R_M = 7.5 \text{ kph} \\ T_A = T_M + 1 \end{cases}$$

$6T_M + 6 = 7.5T_M$

$6 = 1.5T_M$

$\frac{6}{1.5} = T_M = 4$

$T_A = T_M + 1 \Rightarrow T_A = 4 + 1 = 5 \text{ hr}$

7. $D = (6)(5) = 30 \text{ km}$

or $(7.5)(4) = 30 \text{ km}$

8. $\frac{78 \text{ qt}}{1} \times \frac{1 \text{ gal}}{4 \text{ qt}} = 19.5 \text{ gal}$

9. $\frac{105,600 \text{ oz}}{1} \times \frac{1 \text{ lb}}{16 \text{ oz}} \times \frac{1 \text{ ton}}{2,000 \text{ lb}} =$

3.3 tons

10. $\frac{7 \text{ yd}^2}{1} \times \frac{3 \text{ ft}}{1 \text{ yd}} \times \frac{3 \text{ ft}}{1 \text{ yd}} = 63 \text{ ft}^2$

11. $\frac{100 \text{ km}^2}{1} \times \frac{1,000 \text{ m}}{1 \text{ km}} \times \frac{1,000 \text{ m}}{1 \text{ km}} =$

$100,000,000 \text{ m}^2$ or $1 \times 10^8 \text{ m}^2$

12. $\frac{23 \text{ m}}{1} \times \frac{100 \text{ cm}}{1 \text{ m}} \times \frac{.4 \text{ in}}{1 \text{ cm}} = 920 \text{ in}$

13. $\frac{15 \text{ m}}{1} \times \frac{1.1 \text{ yd}}{1 \text{ m}} = 16.5 \text{ yd}$

14. $\frac{H}{NH} = \frac{2}{1}; \frac{H}{T} = \frac{2}{3}; \frac{NH}{T} = \frac{1}{3}$

15. $\frac{H}{NH} = \frac{2}{1}$

We need H, and are given NH.

16. $\frac{H}{NH} = \frac{2}{1} \Rightarrow \frac{H}{32} = \frac{2}{1}$

$H = 2(32) = 64$ days of homework

17. $\frac{1}{X} = \frac{2}{Y} + \frac{3}{Z}$

$\frac{1}{X} = \frac{2(Z)}{Y(Z)} + \frac{3(Y)}{Z(Y)}$

$\frac{1}{X} = \frac{2Z + 3Y}{YZ}$

$(X)\frac{1}{X} = (X)\frac{2Z + 3Y}{YZ}$

$1 = (X)\frac{2Z + 3Y}{YZ}$

$\frac{YZ}{2Z + 3Y} = X$

18. $\frac{K}{KCN} = \frac{39}{39 + 12 + 14} = \frac{39}{65} = .6 = 60\%$

19. $\frac{N}{KCN} = \frac{14}{65} \approx .215 = 21.5\%$

20. $(i^2)(i^3) = i^{2+3} = i^5 = i^4 i^1 = (-1)^2 i = 1i = i$

Systematic Review 18E

1. $D = RT \Rightarrow T = \frac{D}{R} = \frac{180 \text{ yd}}{15 \text{ ypm}} = 12 \text{ min}$

2. $D = RT \Rightarrow D = 18 \text{ ypm} \times 15 \text{ min} = 270 \text{ yd}$

3. $D = RT \Rightarrow R = \frac{D}{T} = \frac{525 \text{ yd}}{21 \text{ min}} = 25 \text{ ypm}$

4. $\longrightarrow$

$D_F = D_G$

$R_F T_F = R_G T_G$

$(500)(3.5) = (R_G)(5) \begin{cases} R_F = 500 \\ T_F = 3.5 \text{ hr} \\ T_G = 5 \text{ hr} \end{cases}$

$1{,}750 = 5R_G$

$\dfrac{1{,}750}{5} = R_G = 350 \text{ mph}$

5. $D = R_F T_F = (500)(3.5) = 1{,}750 \text{ mi}$

or $D = R_G T_G = (350)(5) = 1{,}750 \text{ mi}$

6. $R_T = R_B + 20 = 70 \text{ mph}$

7. $\longrightarrow$

$D_B = D_T$

$R_B T_B = R_T T_T$

$(50)(T_T + 10) = (70)(T_T) \begin{cases} R_B = 50 \\ T_B = T_T + 10 \\ R_T = R_B + 20 \\ R_T = 70 \end{cases}$

$50T_T + 500 = 70T_T$

$500 = 20T_T$

$\dfrac{500}{20} = T_T = 25 \text{ hr}$

$T_B = T_T + 10 = 25 + 10 = 35 \text{ hr}$

8. $\dfrac{17 \text{ gal}}{1} \times \dfrac{4 \text{ qt}}{1 \text{ gal}} \times \dfrac{2 \text{ pt}}{1 \text{ qt}} = 136 \text{ pt}$

9. $\dfrac{1 \text{ qt}}{1} \times \dfrac{2 \text{ pt}}{1 \text{ qt}} = 2 \text{ pt}$

10. $\dfrac{2 \text{ yd}^3}{1} \times \dfrac{36 \text{ in}}{1 \text{ yd}} \times \dfrac{36 \text{ in}}{1 \text{ yd}} \times \dfrac{36 \text{ in}}{1 \text{ yd}} = 93{,}312 \text{ i}$

11. $\dfrac{5.6 \text{ cm}^3}{1} \times \dfrac{10 \text{ mm}}{1 \text{ cm}} \times \dfrac{10 \text{ mm}}{1 \text{ cm}} \times \dfrac{10 \text{ mm}}{1 \text{ cm}}$

$5{,}600 \text{ mm}^3$

12. $\dfrac{18 \text{ oz}}{1} \times \dfrac{28 \text{ g}}{1 \text{ oz}} = 504$

13. $\dfrac{17 \text{ liters}}{1} \times \dfrac{1.06 \text{ qt}}{1 \text{ liter}} = 18.02 \text{ qt}$

14. $\dfrac{N}{NF_3} = \dfrac{14}{71}; \dfrac{F_3}{NF_3} = \dfrac{57}{71}; \dfrac{N}{F_3} = \dfrac{14}{57}$

15. $\dfrac{N}{2{,}059} = \dfrac{14}{71}$

$N = \dfrac{14(2{,}059)}{71} = 406 \text{ g}$

16. $\dfrac{F}{2{,}059} = \dfrac{57}{71}$

$F = \dfrac{57(2{,}059)}{71} = 1{,}653 \text{ g}$

17. $(F + 32) = \dfrac{9}{5}C \Rightarrow \dfrac{5}{9}(F + 32) = C$

18. $\dfrac{C}{CF_2Cl_2} = \dfrac{12}{12 + 38 + 70} = \dfrac{12}{120} = .1 = 10\%$

19. $\dfrac{Cl_2}{CF_2Cl_2} = \dfrac{70}{120} \approx .583 = 58.3\%$

20. $7 - 5i$

Lesson Practice 19A

1. $\overset{\displaystyle 350 \text{ mi}}{\underset{D_1 \qquad D_2}{\vdash\!\!-\!\!-\!\!\cdot\!\!-\!\!-\!\!\dashv}}$

$D_1 + D_2 = 350$

$R_1 T_1 + R_2 T_2 = 350$

$(50)(T) + (90)(T) = 350 \begin{cases} R_1 = 50 \\ R_2 = 90 \\ T_1 = T_2 \end{cases}$

$140T = 350$

$T = \dfrac{350}{140} = 2.5 \text{ hr}$

$8{:}30 \text{ AM} + 2{:}30 = 11{:}00 \text{ AM}$

2.

$$\overset{\text{39 mi}}{\underset{D_T \qquad D_W}{\longmapsto\!\!\!\!\longrightarrow\!\!\!\!\dashv}}$$

$D_T + D_W = 39$

$R_T T_T + R_W T_W = 39$

$$(12)(5 - T_W) + (5)(T_W) = 39 \begin{cases} R_T = 12 \\ R_W = 5 \\ T_T + T_W = 5 \\ T_T = 5 - T_W \end{cases}$$

$60 - 12T_W + 5T_W = 39$

$-7T_W = -21$

$T_W = \dfrac{-21}{-7} = 3$ hr walking

$T_T = 5 - T_W \Rightarrow T_T = 5 - 3 = 2$ hr trotting

3.

$$\overset{\text{1 mi}}{\underset{D_F \qquad D_S}{\longmapsto\!\!\!\!\longleftarrow\!\!\!\!\dashv}}$$

$D_F + D_S = 1$

$R_F T_F + R_S T_S = 1$

$$(10)(T) + (5)(T) = 1 \begin{cases} R_F = 10 \\ R_S = 5 \\ T_F = T_S \end{cases}$$

$15T = 1$

$T = \dfrac{1}{15}$ hr $= \dfrac{1}{15}(60)$ min $= 4$ min

4.

$$\overset{\text{19 mi}}{\underset{D_M \qquad D_L}{\longmapsto\!\!\!\!\longleftarrow\!\!\!\!\dashv}}$$

$D_M + D_L = 19$

$R_M T_M + R_L T_L = 19$

$$(3)(T_M) + (5)(T_M - 1) = 19 \begin{cases} R_M = 3 \\ R_L = 5 \\ T_L = T_M - 1 \end{cases}$$

$3T_M + 5T_M - 5 = 19$

$8T_M = 24$

$T_M = \dfrac{24}{8} = 3$ hr

2:30 + 3:00 = 5:30 PM

5.

$$\overset{\text{360 mi}}{\underset{D_1 \qquad D_2}{\longmapsto\!\!\!\!\longrightarrow\!\!\!\!\dashv}}$$

$D_1 + D_2 = 360$

$R_1 T_1 + R_2 T_2 = 360$

$$(40)(7 - T_2) + (65)(T_2) = 360 \begin{cases} R_1 = 40 \\ R_2 = 65 \\ T_1 + T_2 = 7 \\ T_1 = 7 - T_2 \end{cases}$$

$280 - 40T_2 + 65T_2 = 360$

$25T_2 = 80$

$T_2 = \dfrac{80}{25} = 3.2$ hr at 65 mph

$T_1 = 7 - T_2 \Rightarrow$

$T_1 = 7 - 3.2 = 3.8$ hr at 40 mph

Lesson Practice 19B

1.

$$\overset{\text{20 mi}}{\underset{D_S \qquad D_E}{\longmapsto\!\!\!\!\longrightarrow\!\!\!\!\dashv}}$$

$D_S + D_E = 20$

$R_S T_S + R_E T_E = 20$

$$(2)(T_S) + (13)(T_S + 1) = 20 \begin{cases} R_S = 2 \\ R_E = 13 \\ T_E = T_S + 1 \end{cases}$$

$2T_S + 13T_S + 13 = 20$

$15T_S = 7$

$T_S = \dfrac{7}{15}$ hr $= \dfrac{7}{15}(60)$ min $= 28$ min

2.

$$1,320 \text{ steps}$$
$$\overleftrightarrow{\quad D_R \quad D_S \quad}$$

$$D_R + D_S = 1,320$$
$$R_R T_R + R_S T_S = 1,320$$

$$(30)(T) + (25)(T) = 1,320 \begin{cases} R_R = 30 \\ R_S = 25 \\ T_R = T_S \end{cases}$$

$$55T = 1,320$$
$$T = \frac{1,320}{55} = 24 \text{ min}$$

3.

$$500 \text{ mi}$$
$$\overrightarrow{\quad D_1 \quad D_2 \quad}$$

$$D_1 + D_2 = 500$$
$$R_1 T_1 + R_2 T_2 = 500$$

$$(130)(3T_2) + (110)(T_2) = 500 \begin{cases} R_1 = 130 \\ R_2 = 110 \\ T_1 = 3T_2 \end{cases}$$

$$390T_2 + 110T_2 = 500$$
$$500T_2 = 500$$
$$T_2 = 1 \text{ hr}$$
$$T_1 = 3T_2 \Rightarrow T_1 = 3(1) = 3 \text{ hr}$$

4.

$$3050 \text{ mi}$$
$$\overleftrightarrow{\quad D_M \quad D_L \quad}$$

$$D_M + D_L = 3,050$$
$$R_M T_M + R_L T_L = 3,050$$

$$(400)(T_L + 2) + (500)(T_L) = 3,050 \begin{cases} R_M = 400 \\ T_M = T_L + 2 \\ R_L = 500 \end{cases}$$

$$400T_L + 800 + 500T_L = 3,050$$
$$900T_L = 2,250$$
$$T_L = \frac{2,250}{900} = 2.5 \text{ days}$$

5.

$$600 \text{ mi}$$
$$\overleftrightarrow{\quad D_L \quad D_M \quad}$$

$$D_M + D_L = 600$$
$$R_M T_M + R_L T_L = 600$$

$$(65)(T_M) + (70)(T_M - 3) = 600 \begin{cases} R_M = 65 \\ R_L = 70 \\ T_L = T_M - 3 \end{cases}$$

$$65T_M + 70T_M - 210 = 600$$
$$135T_M = 810$$
$$T_M = \frac{810}{135} = 6 \text{ hr}$$
$$9:00 \text{ AM} + 6:00 = 3:00 \text{ PM}$$

Systematic Review 19C

1. Jumps $= RT \Rightarrow T = \frac{J}{R} = \frac{100}{40} = 2\frac{1}{2} \text{ min}$

2. $J = RT \Rightarrow R = \frac{J}{T} = \frac{129}{3} = 43 \text{ jumps per min}$

3. $J = RT \Rightarrow J = \frac{60 \text{ jumps}}{\text{min}} \times 2\frac{1}{4} \text{ min}$

$J = \frac{60 \text{ jumps}}{\text{min}} \times \frac{9}{4} \text{ min} = 135 \text{ jumps}$

4.

$$\overrightarrow{\qquad\qquad}$$

$$D_H = D_M$$
$$R_H T_H = R_M T_M$$

$$(20)\left(\frac{1}{10}\right) = R_M\left(\frac{1}{6}\right) \begin{cases} R_H = 20 \\ T_H = \frac{6}{60} = \frac{1}{10} \text{ hr} \\ T_M = \frac{10}{60} = \frac{1}{6} \text{ hr} \end{cases}$$

$$\frac{20}{10} = \frac{1}{6} R_M$$

$$2 = \frac{1}{6} R_M$$

$$R_M = 2\left(\frac{6}{1}\right) = 12 \text{ mph}$$

5. $D = RT = (12)\left(\frac{1}{6}\right) \text{ or } (20)\left(\frac{1}{10}\right)$

$D = 2 \text{ mi}$

6.

$$\overset{\text{4,000 yds}}{\overset{\longleftrightarrow}{\underset{D_A \qquad D_E}{}}}$$

$D_A + D_E = 4{,}000$

$R_A T_A + R_E T_E = 4{,}000$

$(240)(T_E + 2) + (200)(T_E) = 4{,}000$ $\begin{cases} R_A = 240 \\ R_E = 200 \\ T_A = T_E + 2 \end{cases}$

$240 T_E + 480 + 200 T_E = 4{,}000$

$440 T_E = 3{,}520$

$T_E = \dfrac{3{,}520}{440} = 8 \text{ min}$

7. $T_A = T_E + 2 \Rightarrow T_A = 8 + 2 = 10 \text{ min}$

8. $D_A = (240)(10) = 2{,}400 \text{ yd}$

$D_E = (200)(8) = 1{,}600 \text{ yd}$

9.

$$\overset{\text{600'}}{\overset{\longleftrightarrow}{\underset{D_P \qquad D_E}{}}}$$

$D_P + D_E = 600$

$R_P T_P + R_E T_E = 600$

$R_P(11) + (27)(10) = 600$ $\begin{cases} R_E = 27 \text{ fps} \\ T_E = 10 \text{ sec} \\ T_P = 10 + 1 = 11 \text{ sec} \end{cases}$

$11 R_P + 270 = 600$

$11 R_P = 330$

$R_P = 30 \text{ fps}$

10. $D_E = (27)(10) = 270 \text{ ft}$

11. $D_P = (30)(11) = 330 \text{ ft}$

12. $\dfrac{100 \text{ in}}{1} \times \dfrac{1 \text{ yd}}{36 \text{ in}} \approx 2.78 \text{ yd or } 2\dfrac{7}{9} \text{ yd}$

13. $\dfrac{4{,}320 \text{ in}^3}{1} \times \dfrac{1 \text{ ft}}{12 \text{ in}} \times \dfrac{1 \text{ ft}}{12 \text{ in}} \times \dfrac{1 \text{ ft}}{12 \text{ in}} =$

2.5 ft^3

14. $\dfrac{5 \text{ m}}{1} \times \dfrac{100 \text{ cm}}{1 \text{ m}} = 500 \text{ cm}$

$\dfrac{500 \text{ cm}}{1} \times \dfrac{.4 \text{ in}}{1 \text{ cm}} = 200 \text{ in}$

15. $\dfrac{80 \text{ qt}}{1} \times \dfrac{.95 \text{ liters}}{1 \text{ qt}} = 76 \text{ liters}$

16. $\dfrac{Mg}{MgO} = \dfrac{24}{24 + 16} = \dfrac{24}{40}; \dfrac{O}{MgO} = \dfrac{16}{40};$

$\dfrac{O}{Mg} = \dfrac{16}{24}$

17. $\dfrac{O}{720} = \dfrac{16}{40}$

$O = \dfrac{720(16)}{40} = 288 \text{ g}$

18. $\dfrac{Mg}{720} = \dfrac{24}{40}$

$Mg = \dfrac{720(24)}{40} = 432 \text{ g}$

19. $\sqrt{\sqrt[3]{X}} = \left(X^{\frac{1}{3}}\right)^{\frac{1}{2}} = X^{\left(\frac{1}{3}\right)\left(\frac{1}{2}\right)} = X^{\frac{1}{6}}$

20. $3\sqrt{-18} - 5\sqrt{-8} =$

$3\sqrt{-1}\sqrt{9}\sqrt{2} - 5\sqrt{-1}\sqrt{4}\sqrt{2} =$

$3i(3)\sqrt{2} - 5i(2)\sqrt{2} =$

$9i\sqrt{2} - 10i\sqrt{2} = -i\sqrt{2}$

Systematic Review 19D

1. Words = RT

$W = 30 \text{ wpm} \times 7 \text{ min} = 210 \text{ words}$

2. $W = RT \Rightarrow R = \dfrac{W}{T} = \dfrac{385}{11} = 35 \text{ wpm}$

3. $W = RT \Rightarrow R = \dfrac{W}{T}$

$R = \dfrac{820 \text{ w}}{40 \text{ wpm}} = 20.5 \text{ min or } 20 \text{ min } 30 \text{ sec}$

4.

$$\longrightarrow$$

$D_B = D_K$

$R_B T_B = R_K T_K$

$(6)\left(4\dfrac{1}{2}\right) = R_K(4)$ $\begin{cases} R_B = 6 \text{ mph} \\ T_B = 4\dfrac{1}{2} \text{ hr} \\ T_K = 4 \text{ hr} \end{cases}$

$$(6)\left(\frac{9}{2}\right) = 4R_K$$

$$27 = 4R_K$$

$$R_K = \frac{27}{4} = 6\frac{3}{4} \text{ mph}$$

5. $D = RT \Rightarrow (6)(4.5) \text{ or } (6.75)(4) = 27 \text{ mi}$

6.
```
|      400 mi      |
|------►----◄------|
    D_C      D_D
```

$$D_C + D_D = 400$$

$$R_C T_C + R_D T_D = 400$$

$$(60)\left(T_C\right) + (55)\left(T_C + 1\right) = \begin{cases} R_C = 60 \\ R_D = 60 - 5 = 55 \\ T_D = T_C + 1 \end{cases}$$

$$60T_C + 55T_C + 55 = 400$$

$$115T_C = 345$$

$$T_C = \frac{345}{115} = 3 \text{ hr}$$

7. $T_D = T_C + 1 \Rightarrow T_D = 3 + 1 = 4 \text{ hr}$

8. 7:00 AM + 4 = 11:00 AM or
   8:00 AM + 3 = 11:00 AM

9.
```
|     1,031 mi     |
|------►----◄------|
    D_S      D_C
```

$$D_S + D_C = 1,031$$

$$R_S T_S + R_C T_C = 1,031$$

$$52\left(T_C + 2\right) + 51\left(T_C\right) = 1,031 \begin{cases} R_S = 52 \\ R_C = 51 \\ T_S = T_C + 2 \end{cases}$$

$$52T_C + 104 + 51T_C = 1,031$$

$$103T_C = 927$$

$$T_C = \frac{927}{103} = 9 \text{ hr}$$

10. $T_S = T_C + 2 \Rightarrow T_S = 9 + 2 = 11 \text{ hr}$

11. 5:30 AM + 11 = 4:30 PM or
    7:30 AM + 9 = 4:30 PM

12. $\dfrac{8,200 \text{ mm}}{1} \times \dfrac{1 \text{ m}}{1,000 \text{ mm}} = 8.2 \text{ m}$

13. $\dfrac{790 \text{ m}^2}{1} \times \dfrac{1 \text{ km}}{1,000 \text{ m}} \times \dfrac{1 \text{ km}}{1,000 \text{ m}} =$

    $.00079 \text{ km}^2 = 7.9 \times 10^{-4} \text{ km}^2$

14. $\dfrac{4 \text{ km}}{1} \times \dfrac{1,000 \text{ m}}{1 \text{ km}} \times \dfrac{1.1 \text{ yd}}{1 \text{ m}} \times \dfrac{3 \text{ ft}}{1 \text{ yd}} =$

    13,200 ft or

    $\dfrac{4 \text{ km}}{1} \times \dfrac{.62 \text{ mi}}{1 \text{ km}} \times \dfrac{5,280 \text{ ft}}{1 \text{ mi}} = 13,094.4 \text{ ft}$

    The difference is due to rounding.

15. $\dfrac{20 \text{ in}}{1} \times \dfrac{2.5 \text{ cm}}{1 \text{ in}} = 50 \text{ cm}$

16. $\dfrac{W}{B} = \dfrac{6}{5} ; \dfrac{W}{T} = \dfrac{6}{11} ; \dfrac{B}{T} = \dfrac{5}{11}$

17. $\dfrac{B}{T} = \dfrac{5}{11}$

    Total given, looking for black.

18. $\dfrac{B}{T} = \dfrac{5}{11} \Rightarrow \dfrac{B}{253} = \dfrac{5}{11}$

    $B = \dfrac{253(5)}{11} = 115 \text{ black walls}$

19. $\dfrac{3i}{8+5i} = \dfrac{3i(8-5i)}{(8+5i)(8-5i)} = \dfrac{24i - 15i^2}{64 - 25i^2} =$

    $\dfrac{24i - (15)(-1)}{64 - (25)(-1)} = \dfrac{24i + 15}{64 + 25} = \dfrac{24i + 15}{89}$

20. $(X+10)^3 = X^3 + 3X^2 10^1 + 3X^1 10^2 + 10^3 =$
    $X^3 + 30X^2 + 300X + 1,000$

Systematic Review 19E

1. Pushups $= RT \Rightarrow R = \dfrac{P}{T} = \dfrac{50}{.5} = 100 \text{ per min}$

2. $P = RT \Rightarrow R = \dfrac{P}{T} = \dfrac{20}{1/3} = 60 \text{ per min}$

3. $P = RT \Rightarrow T = \dfrac{P}{R} = \dfrac{35P}{70 \text{ ppm}} = \dfrac{1}{2} \text{ min or } 30 \text{ sec}$

4.

$D_W = D_H$

$R_W T_W = R_H T_H$

$(R_H + 15)\left(\dfrac{11}{4}\right) = R_H(4)$ $\begin{cases} T_W = 2\dfrac{3}{4} \text{ hr} \\ T_H = 4 \text{ hr} \\ T_W = R_H + 15 \end{cases}$

$(R_H + 15)\left(\dfrac{11}{4}\right)(4) = R_H(4)(4)$

$(R_H + 15)(11) = 16 R_H$

$11 R_H + 165 = 16 R_H$

$165 = 5 R_H$

$R_H = \dfrac{165}{5} = 33 \text{ mph}$

5. $D = RT \Rightarrow D = (33)(4) = 132 \text{ mi}$

6.

1600 mi

$D_1 \qquad D_2$

$D_1 + D_2 = 1,600$

$R_1 T_1 + R_2 D_2 = 1,600$

$40 T_1 + 60(T_1 + 3) = 1,600$ $\begin{cases} R_1 = 40 \\ R_2 = 40 + 20 = 60 \\ T_2 = T_1 + 3 \end{cases}$

$40 T_1 + 60 T_1 + 180 = 1,600$

$100 T_1 = 1,420$

$T_1 = \dfrac{1,420}{100} = 14.2 \text{ hr Gerry drove}$

7. $T_2 = 14.2 + 3 = 17.2 \text{ hr for the team}$

8. $40 + 20 = 60 \text{ mph is rate on second day}$

9-11.

20 mi

$D_W \qquad D_R$

$D_W + D_R = 20$

$R_W T_W + R_R T_R = 20$

$(4)(3) + (6)(T_R) = 20$ $\begin{cases} R_W = 4 \\ T_W = 3 \\ R_R = 4 + 2 = 6 \end{cases}$

$12 + 6 T_R = 20$

$6 T_R = 8$

$T_R = \dfrac{8}{6} = 1\dfrac{1}{3} \text{ hr or 1 hr and 20 min}$

$R_R = 4 + 2 = 6 \text{ mph}$

12. $\dfrac{9 \text{ yd}}{1} \times \dfrac{3 \text{ ft}}{1 \text{ yd}} = 27 \text{ ft}$

13. $\dfrac{35 \text{ m}^2}{1} \times \dfrac{100 \text{ cm}}{1 \text{ m}} \times \dfrac{100 \text{ cm}}{1 \text{ m}} =$

$350,000 \text{ cm}^2 = 3.5 \times 10^5 \text{ cm}^2$

14. $\dfrac{300 \text{ g}}{1} \times \dfrac{.035 \text{ oz}}{1 \text{ g}} = 10.5 \text{ oz}$

15. $\dfrac{16 \text{ mi}}{1} \times \dfrac{1.6 \text{ km}}{1 \text{ mi}} = 25.6 \text{ km}$

16. $\dfrac{N}{NH_3} = \dfrac{14}{17}$; $\dfrac{H_3}{NH_3} = \dfrac{3}{17}$; $\dfrac{H_3}{N} = \dfrac{3}{14}$

17. $\dfrac{N}{646} = \dfrac{14}{17}$

$N = \dfrac{646(14)}{17} = 532 \text{ g}$

18. $\dfrac{H}{646} = \dfrac{3}{17}$

$H = \dfrac{646(3)}{17} = 114 \text{ g}$

19. 5 terms

$1X^4(-3)^0 + 4X^3(-3)^1 + 6X^2(-3)^2 +$

$4X^1(-3)^3 + 1X^0(-3)^4 =$

$X^4 - 12X^3 + 54X^2 - 108X + 81$

20. $\dfrac{5 \times 4}{1 \times 2}(2X)^3(5)^2 =$

$\dfrac{20}{2}(8X^3)(25) = 2,000X^3$

Lesson Practice 20A

1. positive; $\dfrac{\text{up } 3}{\text{over } 1} = 3$

2. $b = -2$

3. negative; $\dfrac{\text{down } 2}{\text{over } 1}$ or $\dfrac{-2}{1} = -2$

4. $b = 1$

5. $3X + 2Y = 9$

  $2Y = -3X + 9$

  $Y = -\dfrac{3}{2}X + \dfrac{9}{2}$

6. $Y = 5X + 1$

 $-5X + Y = 1$

  or

 $5X - Y = -1$

7. $2X + \dfrac{1}{2}Y = 3$

  $\dfrac{1}{2}Y = -2X + 3$

  $Y = -4X + 6$

8. $Y = X + 8$

 $-X + Y = 8$

  or

 $X - Y = -8$

9. $m = \dfrac{Y_2 - Y_1}{X_2 - X_1} = \dfrac{(-4) - (1)}{(-3) - (2)} = \dfrac{-5}{-5} = 1$

10. $m = \dfrac{Y_2 - Y_1}{X_2 - X_1} = \dfrac{(1) - (2)}{(5) - (-3)} = \dfrac{-1}{8} = -\dfrac{1}{8}$

11. $m = \dfrac{Y_2 - Y_1}{X_2 - X_1} = \dfrac{(2) - (-6)}{(5) - (1)} = \dfrac{8}{4} = 2$

12. $m = \dfrac{Y_2 - Y_1}{X_2 - X_1} = \dfrac{(-2) - (4)}{(1) - (-1)} = \dfrac{-6}{2} = -3$

13. $y = mx + b \Rightarrow (-5) = (2)(-2) + b$

      $-5 = -4 + b$

      $-1 = b$

  $Y = 2X - 1$: slope-intercept

 $-2X + Y = -1$ or

  $2X - Y = 1$: standard form

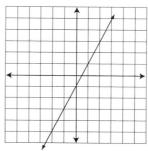

14. $m = \dfrac{Y_2 - Y_1}{X_2 - X_1} = \dfrac{(-4) - (2)}{(3) - (-2)} = \dfrac{-6}{5} = -\dfrac{6}{5}$

 $y = mx + b \Rightarrow (2) = \left(-\dfrac{6}{5}\right)(-2) + b$

      $2 = \dfrac{12}{5} + b$

    $\dfrac{10}{5} - \dfrac{12}{5} = b$

     $-\dfrac{2}{5} = b$

 $Y = -\dfrac{6}{5}X - \dfrac{2}{5}$: slope-intercept

   $\dfrac{6}{5}X + Y = -\dfrac{2}{5}$

 $(5)\left(\dfrac{6}{5}\right)(X) + (5)(Y) = (5)\left(-\dfrac{2}{5}\right)$

   $6X + 5Y = -2$: standard form

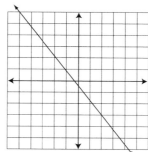

15. $y = mx + b \Rightarrow (5) = (-1)(5) + b$

$\qquad 5 = -5 + b$

$\qquad 10 = b$

$Y = -X + 10$: slope-intercept

$X + Y = 10$: standard form

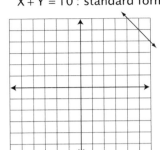

Lesson Practice 20B

1. positive; $\dfrac{\text{up } 1}{\text{over } 3} = \dfrac{1}{3}$

2. $b \approx 1$

3. positive; $\dfrac{\text{up } 1}{\text{over } 1} = \dfrac{1}{1} = 1$

4. $b \approx -4$

5. $7Y = -X + 14$

$\qquad Y = -\dfrac{1}{7}X + 2$

6. $Y = \dfrac{2}{3}X + 6$

$\qquad 3Y = 2X + 18$

$\qquad -2X + 3Y = 18 \text{ or } 2X - 3Y = -18$

7. $2Y = -\dfrac{1}{3}X + 2$

$\qquad Y = -\dfrac{1}{6}X + 1$

8. $Y = 5X - 4$

$\qquad -5X + Y = -4 \text{ or } 5X - Y = 4$

9. $m = \dfrac{Y_2 - Y_1}{X_2 - X_1} = \dfrac{(-3) - (4)}{(-2) - (6)} = \dfrac{-7}{-8} = \dfrac{7}{8}$

10. $m = \dfrac{Y_2 - Y_1}{X_2 - X_1} = \dfrac{(5) - (1)}{(6) - (-2)} = \dfrac{4}{8} = \dfrac{1}{2}$

11. $m = \dfrac{Y_2 - Y_1}{X_2 - X_1} = \dfrac{(8) - (-3)}{(1) - (2)} = \dfrac{11}{-1} = -11$

12. $m = \dfrac{Y_2 - Y_1}{X_2 - X_1} = \dfrac{(-4) - (3)}{(1) - (-5)} = \dfrac{-7}{6} = -\dfrac{7}{6}$

13. $m = \dfrac{Y_2 - Y_1}{X_2 - X_1} = \dfrac{(0) - (-6)}{(0) - (3)} = \dfrac{6}{-3} = -2$

$(0) = (-2)(0) + b$

$0 = 0 + b$

$0 = b$

$Y = -2X$: y-intercept form

$2X + Y = 0$: standard form

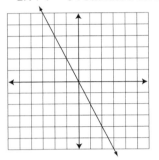

14. $(6) = (-5)(2) + b$

$\qquad 6 = -10 + b$

$\qquad 16 = b$

$Y = -5X + 16$: y-intercept form

$5X + Y = 16$: standard form

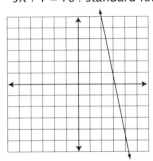

15. $m = \dfrac{Y_2 - Y_1}{X_2 - X_1} = \dfrac{(5) - (0)}{(1) - (-4)} = \dfrac{5}{5} = 1$

$(5) = (1)(1) + b$

$5 = 1 + b$

$4 = b$

$Y = X + 4$: slope-intercept form

$-X + Y = 4 \text{ or } X - Y = -4$: standard form

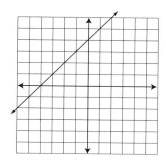

Systematic Review 20C

1. positive; $\dfrac{\text{up } 12}{\text{over } 6} = 2$

2. $b \approx 1$

3. $Y = mX + b$

   $(2) = \left(\dfrac{1}{2}\right)(6) + b$

   $2 = 3 + b$

   $-1 = b$

4. $Y = \dfrac{1}{2}X - 1$

5. $(2)Y = (2)\left(\dfrac{1}{2}\right)X - (2)1$

   $2Y = X - 2$

   $-X + 2Y = -2$ or $X - 2Y = 2$

6. on the graph

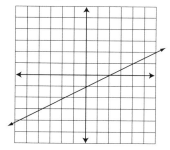

7. $m = \dfrac{Y_2 - Y_1}{X_2 - X_1} = \dfrac{(5) - (-1)}{(1) - (5)} = \dfrac{6}{-4} = -\dfrac{3}{2}$

   $Y = mX + b$

   $(-1) = \left(-\dfrac{3}{2}\right)(5) + b$

   $-1 = -\dfrac{15}{2} + b$

   $\dfrac{-2}{2} + \dfrac{15}{2} = b$

   $\dfrac{13}{2} = b$

8. $Y = -\dfrac{3}{2}X + \dfrac{13}{2}$

9. $(2)Y = (2)\left(-\dfrac{3}{2}\right)X + (2)\left(\dfrac{13}{2}\right)$

   $2Y = -3X + 13$

   $3X + 2Y = 13$

10. on the graph

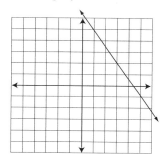

11.

    $\overset{\displaystyle 360}{\underset{D_C \qquad D_M}{\rule{0pt}{0pt}}}$

    $D_C + D_M = 360$

    $R_C T_C + R_M T_M = 360$

    $(R_M + 2)(15) + R_M(15) = 360 \quad \begin{cases} R_M = R_C - 2 \text{ or} \\ R_C = R_M + 2 \\ T_M = 15 \\ T_C = 15 \end{cases}$

    $15R_M + 30 + 15R_M = 360$

    $30R_M = 330$

    $R_M = \dfrac{330}{30} = 11$ ft per sec

    $R_C = R_M + 2 \Rightarrow R_C = 11 + 2 = 13$ ft per sec

12. $D_M = R_M T_M \Rightarrow D_M = (11)(15) = 165$ ft

13. $D_C = R_C T_C \Rightarrow D_C = (13)(15) = 195$ ft

14.

$$\vdash \underset{D_W \quad D_R}{\overset{14}{\longrightarrow}} \dashv$$

$$D_W + D_R = 14$$

$$R_W T_W + R_R T_R = 14$$

$$(5)(4T_R) + (8)(T_R) = 14 \begin{cases} T_W = 4T_R \\ R_W = 5 \\ R_R = R_W + 3 = 8 \end{cases}$$

$$20T_R + 8T_R = 14$$

$$28T_R = 14$$

$$T_R = \frac{14}{28} = \frac{1}{2} \text{ hr}$$

$$T_W = 4T_R \Rightarrow T_W = 4\left(\frac{1}{2}\right) = 2 \text{ hr}$$

15. $D_W = R_W T_W \Rightarrow D_W = (5)(2) = 10 \text{ mi}$

$D_R = R_R T_R \Rightarrow D_R = (8)\left(\frac{1}{2}\right) = 4 \text{ mi}$

16. $\dfrac{5,000 \text{ ft}^3}{1} \times \dfrac{12 \text{ in}}{1 \text{ ft}} \times \dfrac{12 \text{ in}}{1 \text{ ft}} \times \dfrac{12 \text{ in}}{1 \text{ ft}} =$

$8,640,000 \text{ or } 8.64 \times 10^6 \text{ in}^3$

17. $\dfrac{100 \text{ oz}}{1} \times \dfrac{28 \text{ g}}{1 \text{ oz}} = 2,800 \text{ or } 2.8 \times 10^3 \text{ g}$

18. $\dfrac{C}{CHF_3} = \dfrac{12}{12+1+57} = \dfrac{12}{70} \approx .171 = 17.1\%$

19. $\dfrac{H}{CHF_3} = \dfrac{1}{70} \approx .014 = 1.4\%$

20. $\dfrac{F_3}{CHF_3} = \dfrac{3(19)}{70} \approx .814 = 81.4\%$

Systematic Review 20D

1. negative; $\dfrac{\text{up } 10}{\text{over } -10} = \dfrac{10}{-10} = -1$

2. $b = 2$

3. $Y = mX + b \Rightarrow (2) = \left(-\dfrac{4}{3}\right)(3) + b$

$$2 = -4 + b$$

$$6 = b$$

4. $Y = -\dfrac{4}{3}X + 6$

5. $(3)Y = (3)\left(-\dfrac{4}{3}\right)X + (3)6$

$$3Y = -4X + 18$$

$$4X + 3Y = 18$$

6. on the graph

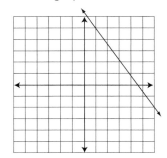

7. $m = \dfrac{Y_2 - Y_1}{X_2 - X_1} = \dfrac{(4)-(2)}{(-3)-(1)} = \dfrac{2}{-4} = -\dfrac{1}{2}$

$Y = mX + b \Rightarrow \quad (2) = \left(-\dfrac{1}{2}\right)(1) + b$

$$2 = -\dfrac{1}{2} + b$$

$$2 + \dfrac{1}{2} = b$$

$$b = 2\dfrac{1}{2} \text{ or } \dfrac{5}{2}$$

8. $Y = -\dfrac{1}{2}X + \dfrac{5}{2}$

9. $(2)Y = (2)\left(-\dfrac{1}{2}\right)X + (2)\left(\dfrac{5}{2}\right)$

$$2Y = -1X + 5$$

$$X + 2Y = 5$$

10. on the graph

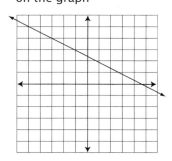

11.

$$\overset{12}{\underset{D_C \quad D_M}{\vdash\!\!\!\dashrightarrow\!\!\!\dashv}}$$

$D_C + D_M = 12$

$$(9)\left(T_M + \frac{1}{3}\right) + (45)(T_M) = 12 \begin{cases} R_C = 9 \\ R_M = 45 \\ T_C = T_M + \frac{1}{3} \text{ hr} \end{cases}$$

$9T_M + 3 + 45T_M = 12$

$54T_M = 9$

$T_M = \dfrac{9}{54} = \dfrac{1}{6} \text{ hr} = \dfrac{1}{6}(60) \text{ min} = 10 \text{ min}$

$6:20 + :10 = 6:30 \text{ PM}$

$D_C = R_C T_C \Rightarrow D_C = (9)\left(\dfrac{1}{2}\right) = \dfrac{9}{2} = 4\dfrac{1}{2} \text{ mi}$

$D_M = R_M T_M \Rightarrow$

$D_M (45)\left(\dfrac{1}{6}\right) = \dfrac{45}{6} = \dfrac{15}{2} = 7\dfrac{1}{2} \text{ mi}$

14.

$$\text{L.A.} \xrightarrow{\qquad D_1 \qquad\qquad D_2 \qquad} \text{Balt.}$$
$$2{,}880 \text{ mi}$$

$D_1 + D_2 = 2{,}880$

$R_1 T_1 + R_2 T_2 = 2{,}880$

$$(R_2 + 12)(32) + (R_2)(16) = 2{,}880 \begin{cases} R_1 = R_2 + 12 \\ T_1 = 32 \\ T_2 = 16 \end{cases}$$

$32R_2 + 384 + 16R_2 = 2{,}880$

$48R_2 = 2{,}496$

$R_2 = \dfrac{2{,}496}{48} = 52 \text{ mph}$

$R_1 = R_2 + 12 \Rightarrow R_1 = 52 + 12 = 64 \text{ mph}$

15. $D_1 = R_1 T_1 \Rightarrow D_1 = (64)(32) = 2{,}048 \text{ mi}$

$D_2 = R_2 T_2 \Rightarrow D_2 = (52)(16) = 832 \text{ mi}$

16. $\dfrac{1{,}300 \text{ ft}^2}{1} \times \dfrac{12 \text{ in}}{1 \text{ ft}} \times \dfrac{12 \text{ in}}{1 \text{ ft}} = 1{,}300 \times 144 =$

$187{,}200 \text{ in}^2 \text{ or } 1.872 \times 10^5 \text{ in}^2$

17. $\dfrac{20 \text{ lb}}{1} \times \dfrac{.45 \text{ kg}}{1 \text{ lb}} = 9 \text{ kg}$

18. $\dfrac{C}{350} = \dfrac{12}{70} \Rightarrow C = \dfrac{12(350)}{70} = 60 \text{ g}$

19. $\dfrac{H}{350} = \dfrac{1}{70} \Rightarrow H = \dfrac{1(350)}{70} = 5 \text{ g}$

20. $\dfrac{F}{350} = \dfrac{57}{70} \Rightarrow F = \dfrac{57(350)}{70} = 285 \text{ g}$

Systematic Review 20E

1. positive; $\dfrac{\text{up } 6}{\text{over } 12} = \dfrac{6}{12} = \dfrac{1}{2}$

2. $b \approx -3$

3. $Y = mX + b \Rightarrow (-1) = \left(\dfrac{5}{2}\right)(-2) + b$

$\qquad\qquad -1 = -5 + b$

$\qquad\qquad 4 = b$

4. $Y = mX + b \Rightarrow Y = \dfrac{5}{2}X + 4$

5. $\qquad Y = \dfrac{5}{2}X + 4$

$\qquad (2)Y = (2)\left(\dfrac{5}{2}\right)X + (2)(4)$

$\qquad 2Y = 5X + 8$

$-5X + 2Y = 8 \text{ or } 5X - 2Y = -8$

6. on the graph

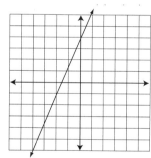

7. $m = \dfrac{Y_2 - Y_1}{X_2 - X_1} = \dfrac{(5) - (-3)}{(1) - (-1)} = \dfrac{8}{2} = 4$

$Y = mX + b \Rightarrow (5) = (4)(1) + b$

$\qquad\qquad 5 = 4 + b$

$\qquad\qquad 1 = b$

8. $Y = 4X + 1$

9. $Y = 4X + 1$
$-4X + Y = 1$ or $4X - Y = -1$

10. on the graph

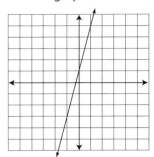

11.
$$\xrightarrow{\quad D_R \quad} \text{post office}$$
$$\text{home} \xleftarrow{\quad D_W \quad}$$

$D_R = D_W$
$R_R T_R = R_W T_W$

$$(R_W + 6)\left(\frac{1}{2}\right) = R_W(2) \begin{cases} T_R = \dfrac{1}{2} \\ T_W = 2 \\ R_R = R_W + 6 \end{cases}$$

$\dfrac{1}{2}R_W + 3 = 2R_W$

$3 = \dfrac{4}{2}R_W - \dfrac{1}{2}R_W$

$3 = \dfrac{3}{2}R_W$

$\left(\dfrac{2}{3}\right)(3) = \left(\dfrac{2}{3}\right)\left(\dfrac{3}{2}\right)R_W$

$2 \text{ mph} = R_W$

12. $R_R = R_W + 6 \Rightarrow R_R = 2 + 6 = 8 \text{ mph}$

13. $D = R_R T_R \Rightarrow D = (8)\left(\dfrac{1}{2}\right) = \dfrac{8}{2} = 4 \text{ mi}$
or $D = R_W T_W \Rightarrow D = (2)(2) = 4 \text{ mi}$

14.
$$\xrightarrow{\quad D_W \quad} \text{mail box}$$
$$\text{home} \xleftarrow{\quad D_R \quad}$$

$D_W = D_R$
$R_W T_W = R_R T_R$

$$(225)(3.6 + T_R) = 900T_R \begin{cases} R_W = 225 \text{ fpm} \\ R_R = 4R_W = \\ (4)(225) = 900 \text{ fpm} \\ T_W = 3.6 + T_R \end{cases}$$

$810 + 225T_R = 900T_R$

$810 = 675T_R$

$\dfrac{810}{675} = T_R = 1.2 \text{ min to run back}$

15. $T_W = 3.6 + 1.2 = 4.8 \text{ min}$
$D_W = (4.8)(225) = 1,080 \text{ ft to the mailbox}$

16. $\dfrac{400 \text{ ft}^2}{1} \times \dfrac{1 \text{ yd}}{3 \text{ ft}} \times \dfrac{1 \text{ yd}}{3 \text{ ft}} \approx 44.4 \text{ yd}^2$

17. $\dfrac{.75 \text{ kg}}{1} \times \dfrac{2.2 \text{ lb}}{1 \text{ kg}} = 1.65 \text{ lb}$

18. $\dfrac{Na}{Na_3PO_4} = \dfrac{(3)(23)}{69 + 31 + 64} = \dfrac{69}{164} \approx .42 = 42\%$

19. $\dfrac{P}{Na_3PO_4} = \dfrac{31}{164} \approx .19 = 19\%$

20. $\dfrac{O}{Na_3PO_4} = \dfrac{(4)(16)}{164} = \dfrac{64}{164} \approx .39 = 39\%$

Lesson Practice 21A

1. same; different
2. negative reciprocal
3. $Y = 3X + 2; m = 3$
$Y = mX + b \Rightarrow (0) = (3)(0) + b$
$0 = 0 + b$
$0 = b$
4. $Y = 3X + 0$ or $Y = 3X$

5. on the graph

given line

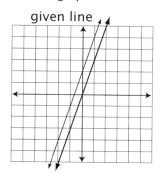

6. $Y = 2X - 1; \ m = 2$

$Y = mX + b \Rightarrow (1) = (2)(3) + b$
$$1 = 6 + b$$
$$-5 = b$$

7. $Y = 2X - 5$

8. on the graph

given line

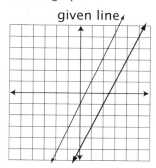

9. $Y = -X + 4; \ m = -1;$ slope of

perpendicular line is $-\left(\dfrac{1}{-1}\right) = 1$

$Y = mX + b \Rightarrow (5) = (1)(-1) + b$
$$5 = -1 + b$$
$$6 = b$$

10. $Y = X + 6$

11. on the graph

given line

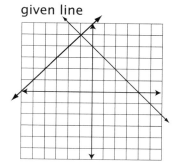

12. graph $Y = X + 3$

Choose one test point from each side of the line:

$(0, 0), (-1, 3)$

Other test points may be used.

$Y \le X + 3 \Rightarrow (0) \le (0) + 3$
$$0 \le 3 \text{ true}$$

$Y \le X + 3 \Rightarrow (3) \le (-1) + 3$
$$3 \le 2 \text{ false}$$

line is solid

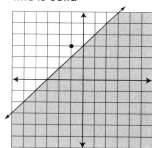

13. graph $Y = -2X - 1$

test points: $(0, 0), (-1, -2)$

$Y < -2X - 1 \Rightarrow \quad (0) < (-2)(0) - 1$

$\qquad\qquad\qquad\qquad 0 < -1$ false

$Y < -2X - 1 \Rightarrow \quad (-2) < (-2)(-1) - 1$

$\qquad\qquad\qquad\qquad -2 < 2 - 1$

$\qquad\qquad\qquad\qquad -2 < 1$ true

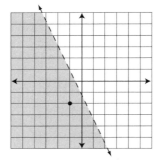

Lesson Practice 21B

1. parallel

2. perpendicular

3. $Y = \dfrac{1}{2}X - 3$; $m = -2$ (negative reciprocal)

$Y = mX + b \Rightarrow \quad (-1) = (-2)(4) + b$

$\qquad\qquad\qquad\qquad -1 = -8 + b$

$\qquad\qquad\qquad\qquad 7 = b$

4. $Y = -2X + 7$

5. on the graph

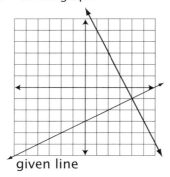

given line

6. $Y = -3X$; $m = \dfrac{1}{3}$ (negative reciprocal)

$Y = mX + b \Rightarrow \quad (3) = \left(\dfrac{1}{3}\right)(-1) + b$

$\qquad\qquad\qquad\qquad 3 = -\dfrac{1}{3} + b$

$\qquad\qquad\qquad 3 + \dfrac{1}{3} = b = \dfrac{10}{3}$

7. $Y = \dfrac{1}{3}X + \dfrac{10}{3}$

8. on the graph

given line

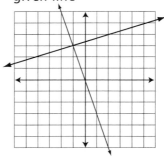

9. $Y = \dfrac{1}{3}X + 4$; $m = \dfrac{1}{3}$

$Y = mX + b \Rightarrow \quad (-2) = \left(\dfrac{1}{3}\right)(-2) + b$

$\qquad\qquad\qquad\qquad -2 = -\dfrac{2}{3} + b$

$\qquad\qquad\qquad -2 + \dfrac{2}{3} = b$

$\qquad\qquad\qquad -\dfrac{6}{3} + \dfrac{2}{3} = b$

$\qquad\qquad\qquad\qquad -\dfrac{4}{3} = b$

10. $Y = \dfrac{1}{3}X - \dfrac{4}{3}$

11. on the graph

given line

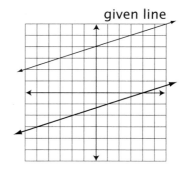

12. graph $Y = 2X + 4$

   test points: $(0, 0), (-2, 2)$

   $Y > 2X + 4 \Rightarrow (0) > (2)(0) + 4$
   $\qquad\qquad 0 > 0 + 4$
   $\qquad\qquad 0 > +4:$ false

   $Y > 2X + 4 \Rightarrow (2) > (2)(-2) + 4$
   $\qquad\qquad 2 > -4 + 4$
   $\qquad\qquad 2 > 0:$ true

   dotted line

   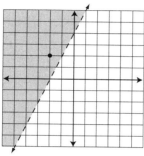

13. graph $Y = -4X - 2$

   test points: $(0, 0), (-3, 0)$

   $Y \leq -4X - 2 \Rightarrow (0) \leq (-4)(0) - 2$
   $\qquad\qquad 0 \leq 0 - 2$
   $\qquad\qquad 0 \leq -2$ false

   $Y \leq -4X - 2 \Rightarrow (0) \leq (-4)(-3) - 2$
   $\qquad\qquad 0 \leq 12 - 2$
   $\qquad\qquad 0 \leq 10$ true

   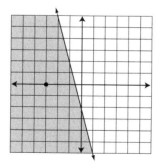

Systematic Review 21C

1. $Y = X - 3; m = 1$

   $Y = mX + b \Rightarrow (0) = (1)(1) + b$
   $\qquad\qquad\qquad 0 = 1 + b$
   $\qquad\qquad\qquad -1 = b$

2. $Y = X - 1$

3. $-X + Y = -1$ or $X - Y = 1$

4. on the graph

   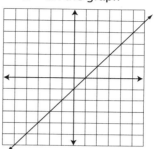

5. $2Y = -3X + 4$

   $Y = -\dfrac{3}{2}X + \dfrac{4}{2}$

   $Y = -\dfrac{3}{2}X + 2$

6. $(0, 0), (4, 0)$ (Other points may be used.)

7. $Y > -\dfrac{3}{2}X + 2 \Rightarrow (0) > \left(-\dfrac{3}{2}\right)(0) + 2$
   $\qquad\qquad\qquad 0 > 0 + 2$
   $\qquad\qquad\qquad 0 > 2$ false

   $Y > -\dfrac{3}{2}X + 2 \Rightarrow (0) > \left(-\dfrac{3}{2}\right)(4) + 2$

   $\qquad\qquad\qquad 0 > -\dfrac{12}{2} + 2$
   $\qquad\qquad\qquad 0 > -6 + 2$
   $\qquad\qquad\qquad 0 > -4$ true

8. dotted line: see graph

   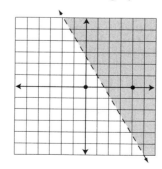

9. $m = \dfrac{Y_2 - Y_1}{X_2 - X_1} = \dfrac{(0) - (3)}{(0) - (2)} = \dfrac{-3}{-2} = \dfrac{3}{2}$

$Y = mX + b \Rightarrow (0) = \left(\dfrac{3}{2}\right)(0) + b$

$0 = 0 + b$

$0 = b$

10. $Y = \dfrac{3}{2}X$

11. $Y = \dfrac{3}{2}X$

$2Y = (2)\left(\dfrac{3}{2}\right)X$

$2Y = 3X$

$-3X + 2Y = 0$ or $3X - 2Y = 0$

12. on the graph

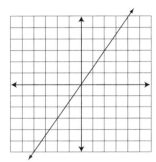

13.

$T_E + T_G = 5 \Rightarrow (T_G + 1) + T_G = 5$

$2T_G + 1 = 5$

$2T_G = 4$

$T_G = 2$

$D_G + D_E = 260$

$R_G T_G + R_E T_E = 260$

$R(2) + R(3) = 260 \begin{cases} R_G = R_E \\ T_G = 2 \\ T_E = T_G + 1 = 2 + 1 = 3 \end{cases}$

$5R = 260$

$R = \dfrac{260}{5} = 52$ mph

14. $D_E = R_E T_E \Rightarrow D_E = (52)(3) = 156$ mi

15. $D_G = R_G T_G \Rightarrow D_G = (52)(2) = 104$ mi

16.

$\dfrac{3 \text{ mi}^2}{1} \times \dfrac{5{,}280 \text{ ft}}{1 \text{ mi}} \times \dfrac{5{,}280 \text{ ft}}{1 \text{ mi}} \times \dfrac{1 \text{ yd}}{3 \text{ ft}} \times \dfrac{1 \text{ yd}}{3 \text{ ft}} =$

$9{,}292{,}800 \text{ yd}^2$

17. $\dfrac{88 \text{ gal}}{1} \times \dfrac{4 \text{ qt}}{1 \text{ gal}} \times \dfrac{.95 \text{ liters}}{1 \text{ qt}} =$

334.4 liters

18. $\dfrac{Mg}{MgCrO_4} = \dfrac{24}{24 + 52 + 64} =$

$\dfrac{24}{140} \approx .17 = 17\%$

19. $\dfrac{Cr}{MgCrO_4} = \dfrac{52}{140} \approx .37 = 37\%$

20. $\dfrac{O}{MgCrO_4} = \dfrac{(4)(16)}{140} = \dfrac{64}{140} \approx .46 = 46\%$

Systematic Review 21D

1. $2Y = -6X + 10$

$Y = -\dfrac{6}{2}X + \dfrac{10}{2}$

$Y = -3X + 5;\ m = \dfrac{1}{3}$

$Y = mX + b \Rightarrow (0) = \left(\dfrac{1}{3}\right)(-1) + b$

$0 = -\dfrac{1}{3} + b$

$\dfrac{1}{3} = b$

2. $Y = \dfrac{1}{3}X + \dfrac{1}{3}$

3. $(3)Y = (3)\left(\dfrac{1}{3}\right)X + (3)\left(\dfrac{1}{3}\right)$

$3Y = X + 1$

$-X + 3Y = 1$ or $X - 3Y = -1$

4. see graph

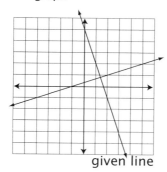

given line

5. $Y = 3X + 1$

6. $(0, 0), (-2, 0)$

7. $Y < 3X + 1 \Rightarrow (0) < (3)(0) + 1$
$$0 < 0 + 1$$
$$0 < 1 \text{ true}$$

$Y < 3X + 1 \Rightarrow (0) < (3)(-2) + 1$
$$0 < -6 + 1$$
$$0 < -5 \text{ false}$$

8. dotted line: see graph

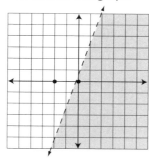

9. $m = \dfrac{Y_2 - Y_1}{X_2 - X_1} = \dfrac{(2) - (-2)}{(5) - (-3)} = \dfrac{4}{8} = \dfrac{1}{2}$

$Y = mX + b \Rightarrow (2) = \left(\dfrac{1}{2}\right)(5) + b$

$$2 = \dfrac{5}{2} + b$$

$$2 - \dfrac{5}{2} = b = -\dfrac{1}{2}$$

10. $Y = \dfrac{1}{2}X - \dfrac{1}{2}$

11. $(2)Y = (2)\left(\dfrac{1}{2}\right)X - (2)\left(\dfrac{1}{2}\right)$

$$2Y = X - 1$$

$$-X + 2Y = -1 \text{ or } X - 2Y = 1$$

12. see graph

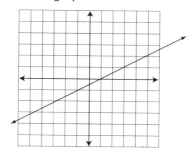

13.

$$\overset{94}{\underset{D_C \quad D_J}{\rule{3cm}{0.4pt}}}$$

$$D_C + D_J = 94$$

$$R_C T_C + R_J T_J = 94$$

$$(3R_J)(5) + (R_J)\left(\dfrac{2}{3}\right) = 94 \quad \begin{cases} T_C = 5 \\ T_J = \dfrac{2}{3} \\ R_C = 3R_J \end{cases}$$

$$15R_J + \dfrac{2}{3}R_J = 94$$

$$\dfrac{45}{3}R_J + \dfrac{2}{3}R_J = 94$$

$$\dfrac{47}{3}R_J = 94$$

$$R_J = (94)\left(\dfrac{3}{47}\right) = \dfrac{282}{47} = 6 \text{ mph}$$

$$R_C = 3R_J \Rightarrow R_C = (3)(6) = 18 \text{ mph}$$

16.

$$\dfrac{14{,}500 \text{ in}^3}{1} \times \dfrac{1 \text{ ft}}{12 \text{ in}} \times \dfrac{1 \text{ ft}}{12 \text{ in}} \times \dfrac{1 \text{ ft}}{12 \text{ in}} \approx 8.39 \text{ ft}^2$$

17. $\dfrac{50 \text{ oz}}{1} \times \dfrac{1 \text{ lb}}{16 \text{ oz}} \times \dfrac{.45 \text{ kg}}{1 \text{ lb}} \approx 1.4 \text{ kg}$

18. $\dfrac{Mg}{MgCrO_4} = \dfrac{24}{24 + 52 + 64} = \dfrac{24}{140}$

$$\dfrac{Mg}{1{,}260} = \dfrac{24}{140}$$

$$Mg = \dfrac{1{,}260(24)}{140} = 216 \text{ g}$$

19. $\dfrac{Cr}{MgCrO_4} = \dfrac{52}{140}$

$$\dfrac{Cr}{1{,}260} = \dfrac{52}{140}$$

$$Cr = \dfrac{1{,}260(52)}{140} = 468 \text{ g}$$

20. $\dfrac{O}{MgCrO_4} = \dfrac{64}{140}$

$\dfrac{O}{1,260} = \dfrac{64}{140}$

$O = \dfrac{1,260(64)}{140} = 576\ g$

Systematic Review 21E

1. $2Y = X$

$Y = \dfrac{1}{2}X;\ m = \dfrac{1}{2}$

$Y = mX + b \Rightarrow (-3) = \left(\dfrac{1}{2}\right)(-2) + b$

$-3 = -1 + b$

$-2 = b$

2. $Y = \dfrac{1}{2}X - 2$

3. $Y = \dfrac{1}{2}X - 2$

$2Y = X - 4$

$-X + 2Y = -4$ or $X - 2Y = 4$

4. on the graph

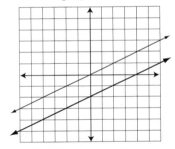

5. $-Y = 2X \Rightarrow Y = -2X$

6. $(1, 0), (-1, 0)$

7. $-Y \geq 2X \Rightarrow -(1) \geq 2(0)$

$-1 \geq 0$ false

$-Y \geq 2X \Rightarrow -(0) \geq 2(-1)$

$0 \geq -2$ true

8. solid line: see graph

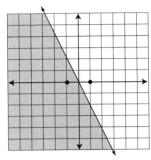

9. $m = \dfrac{Y_2 - Y_1}{X_2 - X_1} = \dfrac{(4) - (-3)}{(4) - (-1)} = \dfrac{7}{5}$

$Y = mX + b \Rightarrow (4) = \left(\dfrac{7}{5}\right)(4) + b$

$4 = \dfrac{28}{5} + b$

$\dfrac{20}{5} - \dfrac{28}{5} = b$

$-\dfrac{8}{5} = b$

10. $Y = \dfrac{7}{5}X - \dfrac{8}{5}$

11. $Y = \dfrac{7}{5}X - \dfrac{8}{5}$

$5Y = 7X - 8$

$-7X + 5Y = -8$ or $7X - 5Y = 8$

12. see graph

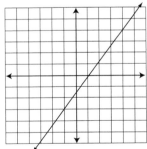

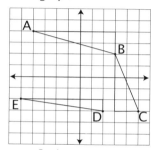

$D_P + D_F = 60$

13-15.

$D_P + D_F = 60$

$R_P T_P + R_F T_F = 60$

$(4)(12) + (8)(T_F) = 60 \begin{cases} R_P = 4 \\ T_P = 12 \\ R_F = 8 \end{cases}$

$48 + 8T_F = 60$

$8T_F = 12$

$T_F = \dfrac{12}{8} = \dfrac{3}{2}$ or $1\dfrac{1}{2}$ hours

16. $\dfrac{7.6 \ m^3}{1} \times \dfrac{100 \ cm}{1 \ m} \times \dfrac{100 \ cm}{1 \ m} \times \dfrac{100 \ cm}{1 \ m} =$

$7.6 \times 10^6 \ cm^3$

17. $\dfrac{620 \ km}{1} \times \dfrac{.62 \ mi}{1 \ km} = 384.4 \ mi$

18. $\dfrac{C}{C_2H_5Cl} = \dfrac{2(12)}{24 + 5 + 35} = \dfrac{24}{64} = .375 = 37.5\%$

19. $\dfrac{H}{C_2H_5Cl} = \dfrac{1(5)}{64} \approx .078 = 7.8\%$

20. $\dfrac{Cl}{C_2H_5Cl} = \dfrac{35}{64} \approx .547 = 54.7\%$

Lesson Practice 22A

1. $d = \sqrt{\Delta X^2 + \Delta Y^2}$ or

$d = \sqrt{(X_2 - X_1)^2 + (Y_2 - Y_1)^2}$

2. $\left(\dfrac{X_1 + X_2}{2}, \dfrac{Y_1 + Y_2}{2} \right)$

3. see graph

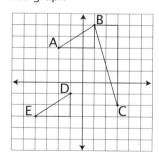

4. $(AB)^2 = \left[(1) - (-2) \right]^2 + \left[(5) - (3) \right]^2$

$(AB)^2 = (3)^2 + (2)^2 = 9 + 4 = 13$

$AB = \sqrt{13}$

5. $(BC)^2 = \left[(3) - (1) \right]^2 + \left[(-2) - (5) \right]^2$

$(BC)^2 = (2)^2 + (-7)^2 = 4 + 49 = 53$

$BC = \sqrt{53}$

6. $(DE)^2 = \left[(-4) - (-1) \right]^2 + \left[(-3) - (-1) \right]^2$

$(DE)^2 = (-3)^2 + (-2)^2 = 9 + 4 = 13$

$DE = \sqrt{13}$

7. see graph

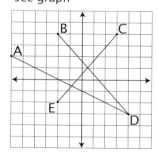

8. $(AB)^2 = \left[(3) - (-4) \right]^2 + \left[(2) - (4) \right]^2$

$(AB)^2 = (7)^2 + (-2)^2 = 49 + 4 = 53$

$AB = \sqrt{53}$

9. $(BC)^2 = \left[(5) - (3) \right]^2 + \left[(-3) - (2) \right]^2$

$(BC)^2 = (2)^2 + (-5)^2 = 4 + 25 = 29$

$BC = \sqrt{29}$

10. $(DE)^2 = \left[(-5) - (2) \right]^2 + \left[(-2) - (-3) \right]^2$

$(DE)^2 = (-7)^2 + (1)^2 = 49 + 1 = 50$

$DE = \sqrt{50} = 5\sqrt{2}$

11. see graph

12. $\text{midpoint} = \left(\dfrac{(-6)+(4)}{2}, \dfrac{(2)+(-3)}{2} \right) =$
$\left(\dfrac{-2}{2}, \dfrac{-1}{2} \right) = \left(-1, -\dfrac{1}{2} \right)$

13. $\text{midpoint} = \left(\dfrac{(-2)+(4)}{2}, \dfrac{(4)+(-3)}{2} \right) =$
$\left(\dfrac{2}{2}, \dfrac{1}{2} \right) = \left(1, \dfrac{1}{2} \right)$

14. $\text{midpoint} = \left(\dfrac{(-2)+(3)}{2}, \dfrac{(-2)+(4)}{2} \right) =$
$\left(\dfrac{1}{2}, \dfrac{2}{2} \right) = \left(\dfrac{1}{2}, 1 \right)$

15. see graph

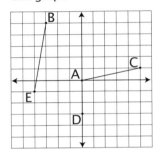

16. $\text{midpoint} = \left(\dfrac{(0)+(5)}{2}, \dfrac{(0)+(1)}{2} \right) = \left(\dfrac{5}{2}, \dfrac{1}{2} \right)$

17. $\text{midpoint} = \left(\dfrac{(-3)+(-4)}{2}, \dfrac{(5)+(-1)}{2} \right) =$
$\left(\dfrac{-7}{2}, \dfrac{4}{2} \right) = \left(-\dfrac{7}{2}, 2 \right)$

18. $\text{midpoint} = \left(\dfrac{(0)+(0)}{2}, \dfrac{(0)+(-3)}{2} \right) =$
$\left(\dfrac{0}{2}, \dfrac{-3}{2} \right) = \left(0, -\dfrac{3}{2} \right)$

Lesson Practice 22B

1. $d = \sqrt{\Delta X^2 + \Delta Y^2}$ or
$d = \sqrt{(X_2 - X_1)^2 + (Y_2 - Y_1)^2}$

2. $\left(\dfrac{X_1 + X_2}{2}, \dfrac{Y_1 + Y_2}{2} \right)$

3. see graph

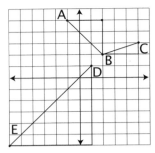

4. $(AB)^2 = \left[(2) - (-1) \right]^2 + \left[(2) - (5) \right]^2$
$(AB)^2 = (3)^2 + (-3)^2 = 9 + 9 = 18$
$AB = \sqrt{18} = \sqrt{9}\sqrt{2} = 3\sqrt{2}$

5. $(BC)^2 = \left[(5) - (2) \right]^2 + \left[(3) - (2) \right]^2$
$(BC)^2 = (3)^2 + (1)^2 = 9 + 1 = 10$
$BC = \sqrt{10}$

6. $(DE)^2 = \left[(-6) - (1) \right]^2 + \left[(-6) - (1) \right]^2$
$(DE)^2 = (-7)^2 + (-7)^2 = 49 + 49 = 98$
$DE = \sqrt{98} = \sqrt{49}\sqrt{2} = 7\sqrt{2}$

7. see graph

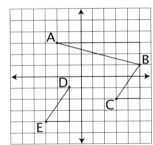

8. $(AB)^2 = \left[(5) - (-2) \right]^2 + \left[(1) - (3) \right]^2$
$(AB)^2 = (7)^2 + (-2)^2 = 49 + 4 = 53$
$AB = \sqrt{53}$

9. $(BC)^2 = \left[(3) - (5) \right]^2 + \left[(-2) - (1) \right]^2$
$(BC)^2 = (-2)^2 + (-3)^2 = 4 + 9 = 13$
$BC = \sqrt{13}$

10. $(DE)^2 = \left[(-3) - (-1) \right]^2 + \left[(-4) - (-1) \right]^2$
$(DE)^2 = (-2)^2 + (-3)^2 = 4 + 9 = 13$
$DE = \sqrt{13}$

11. see graph

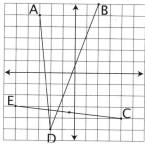

12. midpoint $= \left(\dfrac{(-3)+(-2)}{2}, \dfrac{(5)+(-5)}{2} \right) =$

$\left(\dfrac{-5}{2}, \dfrac{0}{2} \right) = \left(-\dfrac{5}{2}, 0 \right)$

13. midpoint $= \left(\dfrac{(2)+(-2)}{2}, \dfrac{(6)+(-5)}{2} \right) =$

$\left(\dfrac{0}{2}, \dfrac{1}{2} \right) = \left(0, \dfrac{1}{2} \right)$

14. midpoint $= \left(\dfrac{(-5)+(4)}{2}, \dfrac{(-3)+(-4)}{2} \right) =$

$\left(\dfrac{-1}{2}, \dfrac{-7}{2} \right) = \left(-\dfrac{1}{2}, -\dfrac{7}{2} \right)$

15. see graph

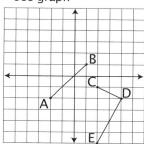

16. midpoint $= \left(\dfrac{(-2)+(1)}{2}, \dfrac{(-2)+(1)}{2} \right) =$

$\left(\dfrac{-1}{2}, \dfrac{-1}{2} \right) = \left(-\dfrac{1}{2}, -\dfrac{1}{2} \right)$

17. midpoint $= \left(\dfrac{(2)+(4)}{2}, \dfrac{(-1)+(-2)}{2} \right) =$

$\left(\dfrac{6}{2}, \dfrac{-3}{2} \right) = \left(3, -\dfrac{3}{2} \right)$

18. midpoint $= \left(\dfrac{(4)+(2)}{2}, \dfrac{(-2)+(-6)}{2} \right) =$

$\left(\dfrac{6}{2}, \dfrac{-8}{2} \right) = (3, -4)$

Systematic Review 22C

1. see graph

2. see graph

3. see graph

4. 3; 5; see graph

5. $3^2 + 5^2 = (AE)^2$

$9 + 25 = (AE)^2$

$34 = (AE)^2; \sqrt{34} = AE$

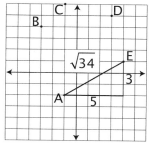

6. $(AB)^2 = \left[(-1) - (-3) \right]^2 + \left[(-2) - (4) \right]^2$

$(AB)^2 = (2)^2 + (-6)^2 = 4 + 36 = 40$

$AB = \sqrt{40} = \sqrt{4}\sqrt{10} = 2\sqrt{10}$

7. $(BC)^2 = \left[(-1) - (-3) \right]^2 + \left[(6) - (4) \right]^2$

$(BC)^2 = (2)^2 + (2)^2 = 4 + 4 = 8$

$BC = \sqrt{8} = \sqrt{4}\sqrt{2} = 2\sqrt{2}$

8. $(CE)^2 = \left[(-1) - (4) \right]^2 + \left[(6) - (1) \right]^2$

$(CE)^2 = (-5)^2 + (5)^2 = 25 + 25 = 50$

$CE = \sqrt{50} = \sqrt{25}\sqrt{2} = 5\sqrt{2}$

9. midpoint $= \left(\dfrac{(-3)+(4)}{2}, \dfrac{(4)+(1)}{2} \right) =$

$\left(\dfrac{1}{2}, \dfrac{5}{2} \right)$ or $\left(\dfrac{1}{2}, 2\dfrac{1}{2} \right)$

10. midpoint $= \left(\dfrac{(-3)+(3)}{2}, \dfrac{(4)+(5)}{2} \right) =$

$\left(\dfrac{0}{2}, \dfrac{9}{2} \right)$ or $\left(0, 4\dfrac{1}{2} \right)$

11. $\text{midpoint} = \left(\dfrac{(-1)+(-1)}{2}, \dfrac{(-2)+(6)}{2} \right) =$

$\left(\dfrac{-2}{2}, \dfrac{4}{2} \right) = (-1, 2)$

12. $m = \dfrac{Y_2 - Y_1}{X_2 - X_1} = \dfrac{(-1)-(3)}{(-5)-(4)} = \dfrac{-4}{-9} = \dfrac{4}{9}$

$Y = mX + b \Rightarrow \qquad (3) = \left(\dfrac{4}{9} \right)(4) + b$

$3 = \dfrac{16}{9} + b$

$\dfrac{27}{9} - \dfrac{16}{9} = b$

$\dfrac{11}{9} = b$

$Y = \dfrac{4}{9}X + \dfrac{11}{9}$

13. see graph

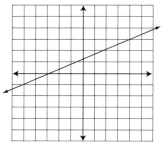

14. $Y = 3X - \dfrac{1}{2}; \; m = 3$

$Y = mX + b \Rightarrow \; (-2) = (3)(2) + b$

$-2 = 6 + b$

$-8 = b$

$Y = 3X - 8$

15. see graph

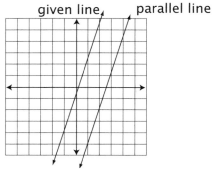

given line   parallel line

16. $Y = -\dfrac{2}{5}X + \dfrac{1}{3}; \; m = \dfrac{5}{2}$

$Y = mX + b \Rightarrow \qquad (-3) = \left(\dfrac{5}{2} \right)(-3) + b$

$-3 = -\dfrac{15}{2} + b$

$-\dfrac{6}{2} + \dfrac{15}{2} = b$

$\dfrac{9}{2} = b$

$Y = \dfrac{5}{2}X + \dfrac{9}{2}$

17. see graph

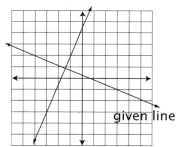

perpendicular line

given line

18. see graph

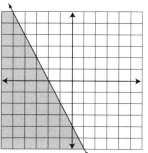

19. $(0, 0), (-4, 0)$

$$Y \leq -2X - 4 \Rightarrow \quad (0) \leq -2(0) - 4$$
$$0 \leq 0 - 4$$
$$0 \leq -4 \text{ false}$$

$$Y \leq -2X - 4 \Rightarrow \quad (0) \leq -2(-4) - 4$$
$$0 \leq 8 - 4$$
$$0 \leq 4 \text{ true}$$

20. solid; see graph

Systematic Review 22D

1. see graph
2. see graph
3. see graph
4. 4; 3; see graph
5. $3^2 + 4^2 = (CD)^2$

$$9 + 16 = (CD)^2$$
$$25 = (CD)^2$$
$$\sqrt{25} = CD = 5$$

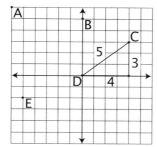

6. $(AD)^2 = \left[(6) - (0)\right]^2 + \left[(-6) - (0)\right]^2$

$$(AD)^2 = 36 + 36 = 72$$
$$AD = \sqrt{72} = \sqrt{36}\sqrt{2} = 6\sqrt{2}$$

7. $(CE)^2 = \left[(4) - (-5)\right]^2 + \left[(3) - (-2)\right]^2$

$$(CE)^2 = (9)^2 + (5)^2 = 81 + 25 = 106$$
$$CE = \sqrt{106}$$

8. $(BD)^2 = \left[(0) - (0)\right]^2 + \left[(0) - (5)\right]^2$

$$(BD)^2 = (0)^2 + (-5)^2 = 0 + 25 = 25$$
$$BD = \sqrt{25} = 5$$

9. $\text{midpoint} = \left(\dfrac{(-6) + (0)}{2}, \dfrac{(6) + (5)}{2} \right) =$

$$\left(-\frac{6}{2}, \frac{11}{2} \right) \text{ or} \left(-3, 5\frac{1}{2} \right)$$

10. $\text{midpoint} = \left(\dfrac{(0) + (-5)}{2}, \dfrac{(5) + (-2)}{2} \right) =$

$$\left(-\frac{5}{2}, \frac{3}{2} \right) \text{ or } \left(-2\frac{1}{2}, 1\frac{1}{2} \right)$$

11. $\text{midpoint} = \left(\dfrac{(4) + (0)}{2}, \dfrac{(3) + (0)}{2} \right) =$

$$\left(\frac{4}{2}, \frac{3}{2} \right) = \left(2, 1\frac{1}{2} \right)$$

12. $m = \dfrac{Y_2 - Y_1}{X_2 - X_1} = \dfrac{(-3) - (3)}{(1) - (-4)} = \dfrac{-6}{5} = -\dfrac{6}{5}$

$$Y = mX + b \Rightarrow \quad (3) = \left(-\frac{6}{5} \right)(-4) + b$$
$$3 = \frac{24}{5} + b$$
$$\frac{15}{5} - \frac{24}{5} = b$$
$$-\frac{9}{5} = b$$

$$Y = -\frac{6}{5}X - \frac{9}{5}$$

13. see graph

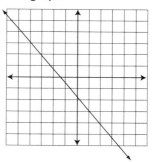

14. $Y = \frac{2}{3}X - \frac{1}{3}$; $m = \frac{2}{3}$

$Y = mX + b \Rightarrow (4) = \left(\frac{2}{3}\right)(3) + b$

$4 = 2 + b$

$2 = b$

$Y = \frac{2}{3}X + 2$

15. see graph

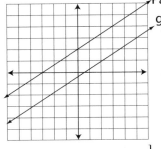

Parallel line
given line

16. $Y = -4X - 1$; $m = \frac{1}{4}$

$Y = mX + b \Rightarrow (3) = \left(\frac{1}{4}\right)(-2) + b$

$3 = -\frac{2}{4} + b$

$3 + \frac{1}{2} = b = \frac{7}{2}$

$Y = \frac{1}{4}X + \frac{7}{2}$

17. see graph

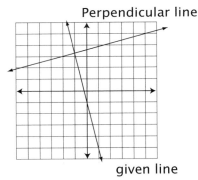

Perpendicular line

given line

18. $5Y \geq 3X - 15$; $Y \geq \frac{3}{5}X - 3$; see graph

19. $(0, 0), (0, -5)$

$5Y \geq 3X - 15 \Rightarrow 5(0) \geq 3(0) - 15$

$0 \geq 0 - 15$

$0 \geq -15$ true

$5Y \geq 3X - 15 \Rightarrow 5(-5) \geq 3(0) - 15$

$-25 \geq 0 - 15$

$-25 \geq -15$ false

20. solid; see graph

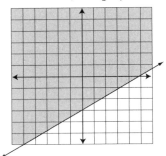

Systematic Review 22E

1. see graph

2. see graph

3. see graph

4. 4; 5; see graph

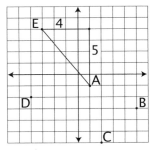

5. $4^2 + 5^2 = (AE)^2$

$16 + 25 = (AE)^2$

$41 = (AE)^2$

$\sqrt{41} = AE$

6. $(BD)^2 = \left[(5) - (-4)\right]^2 + \left[(-3) - (-2)\right]^2$

$(BD)^2 = (9)^2 + (-1)^2 = 81 + 1 = 82$

$BD = \sqrt{82}$

7. $(BC)^2 = \left[(5) - (2)\right]^2 + \left[(-3) - (-6)\right]^2$

$(BC)^2 = (3)^2 + (3)^2 = 9 + 9 = 18$

$BC = \sqrt{18} = \sqrt{9}\sqrt{2} = 3\sqrt{2}$

8. $(CE)^2 = \left[(2) - (-3)\right]^2 + \left[(-6) - (4)\right]^2$

$(CE)^2 = (5)^2 + (-10)^2 = 25 + 100 = 125$

$CE = \sqrt{125} = \sqrt{25}\sqrt{5} = 5\sqrt{5}$

9. $\text{midpoint} = \left(\dfrac{(-4) + (2)}{2}, \dfrac{(-6) + (-2)}{2}\right) =$

$\left(\dfrac{-2}{2}, \dfrac{-8}{2}\right) = (-1, -4)$

10. $\text{midpoint} = \left(\dfrac{(5) + (-3)}{2}, \dfrac{(-3) + (4)}{2}\right) =$

$\left(\dfrac{2}{2}, \dfrac{1}{2}\right) = \left(1, \dfrac{1}{2}\right)$

11. $\text{midpoint} = \left(\dfrac{(2) + (-3)}{2}, \dfrac{(-6) + (4)}{2}\right) =$

$\left(\dfrac{-1}{2}, \dfrac{-2}{2}\right) = \left(-\dfrac{1}{2}, -1\right)$

12. $m = \dfrac{Y_2 - Y_1}{X_2 - X_1} = \dfrac{(-2) - (1)}{(-1) - (4)} = \dfrac{-3}{-5} = \dfrac{3}{5}$

$Y = mX + b; \quad (1) = \left(\dfrac{3}{5}\right)(4) + b$

$1 = \dfrac{12}{5} + b; \ b = -\dfrac{7}{5}; \ Y = \dfrac{3}{5}X - \dfrac{7}{5}$

13. see graph

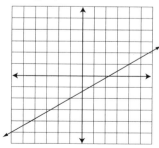

14. $Y = \dfrac{3}{4}X + 2; \ m = \dfrac{3}{4}$

$Y = mX + b; \quad (-3) = \left(\dfrac{3}{4}\right)(1) + b$

$-3 = \dfrac{3}{4} + b; \ b = -\dfrac{15}{4}; \ Y = \dfrac{3}{4}X - \dfrac{15}{4}$

15. see graph

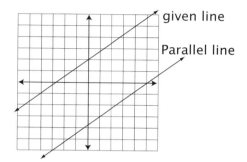

given line

Parallel line

16. $Y = -\dfrac{1}{2}X + 1; \ m = 2$

$Y = mX + b \Rightarrow (3) = (2)(0) + b$

$3 = 0 + b$

$3 = b$

$Y = 2X + 3$

17. see graph

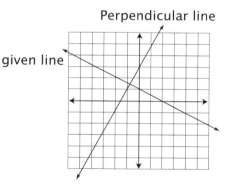

Perpendicular line

given line

18. $-4Y < 3X + 2$

Dividing by -4 reverses the inequality.

$Y > \dfrac{3}{-4}X + \dfrac{2}{-4}$

$Y = -\dfrac{3}{4}X - \dfrac{1}{2}$ use for graph

see graph

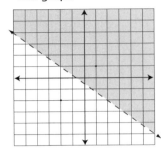

19. $(1, 1), (-2, -2)$

$-4Y < 3X + 2 \Rightarrow -4(1) < 3(1) + 2$

$-4 < 3 + 2$

$-4 < 5$ true

$-4Y < 3X + 2 \Rightarrow -4(-2) < 3(-2) + 2$

$8 < -6 + 2$

$8 < -4$ false

20. dotted; see graph

Lesson Practice 23A

1. $X^2 + Y^2 = 36$

$(X - 0)^2 + (Y - 0)^2 = 6^2$

center: $(0, 0)$

radius: 6

2. see graph

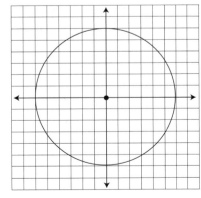

3. center: $(3, -3)$

radius: 4

$(X - (3))^2 + (Y - (-3))^2 = 4^2$

$(X - 3)^2 + (Y + 3)^2 = 16$

4. see graph

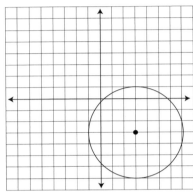

5. $X^2 + 2X + Y^2 + 4Y = 4$

$X^2 + 2X + 1 + Y^2 + 4Y + 4 = 4 + 1 + 4$

$(X + 1)^2 + (Y + 2)^2 = 9$

center: $(-1, -2)$

radius: 3

6. see graph

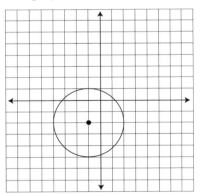

7.
$$4X^2 + Y^2 = 36$$
$$4(X+0)^2 + (Y+0)^2 = 6^2$$
center: $(0, 0)$

$4X^2 + Y^2 = 36$	$4X^2 + Y^2 = 36$
$4X^2 + (0)^2 = 36$	$4(0)^2 + Y^2 = 36$
$4X^2 = 36$	$Y^2 = 36$
$X^2 = 9$	$Y = \pm 6$
$X = \pm 3$	

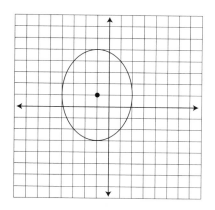

8. see graph

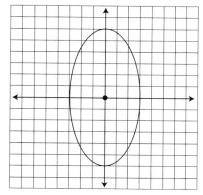

9. $16(X+1)^2 + 9(Y-1)^2 = 144$
center: $(-1, 1)$

$$16(X+1)^2 + 9(Y-1)^2 = 144$$
$$16(X+1)^2 + 9((1)-1)^2 = 144$$
$$16(X+1)^2 = 144$$
$$(X+1)^2 = 9$$
$$X+1 = \pm 3$$

$X+1 = 3$	$X+1 = -3$
$X = 2$	$X = -4$

$$16(X+1)^2 + 9(Y-1)^2 = 144$$
$$16((-1)+1)^2 + 9(Y-1)^2 = 144$$
$$9(Y-1)^2 = 144$$
$$(Y-1)^2 = 16$$
$$Y-1 = \pm 4$$

X	Y
-1	5
-1	-3
2	1
-4	1

$Y-1 = 4$	$Y-1 = -4$
$Y = 5$	$Y = -3$

10. see graph

Lesson Practice 23B

1. $(X+4)^2 + (Y+4)^2 = 5$
center: $(-4, -4)$
radius: $\sqrt{5} \approx 2.24$

2. see graph

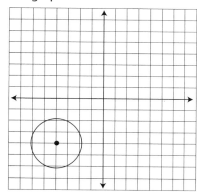

3. $(X-2)^2 + (Y-1)^2 = (4.5)^2 = 20.25$

4. see graph

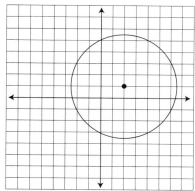

5. $X^2 - 8x + 16 + Y^2 + 12Y + 36 = -48 + 52$

$(X - 4)^2 + (Y + 6)^2 = 4$

center: $(4, -6)$

radius: 2

6. see graph

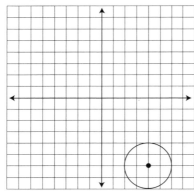

7. $4(X - 2)^2 + 16(Y + 1)^2 = 64$

center: $(2, -1)$

$4(0)^2 + 16(Y + 1)^2 = 64$ $4(X - 2)^2 + 16(0)^2 = 64$

$16(Y + 1)^2 = 64$ $4(X - 2)^2 = 64$

$(Y + 1)^2 = 4$ $(X - 2)^2 = 16$

$Y + 1 = \pm 2$ $X - 2 = \pm 4$

$Y = 1, -3$ $X = 6, -2$

X	Y
2	1
2	-3
6	-1
-2	-1

8. see graph

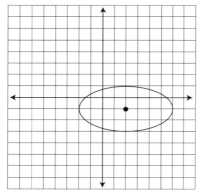

9. $(X - 1)^2 + 9(Y + 1)^2 = 9$

center: $(1, -1)$

$(0)^2 + 9(Y + 1)^2 = 9$ $(X - 1)^2 + 9(0)^2 = 9$

$9(Y + 1)^2 = 9$ $(X - 1)^2 = 9$

$(Y + 1)^2 = 1$ $X - 1^2 = \pm 3$

$Y + 1 = \pm 1$ $X = 4, -2$

$Y = 0, -2$

X	Y
1	0
1	-2
4	-1
-2	-1

10. see graph

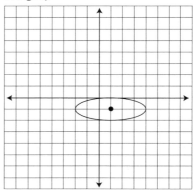

Systematic Review 23C

1. $X^2 + Y^2 = 9$

$(X - 0)^2 + (Y - 0)^2 = 9$

center: $(0, 0)$

radius: $\sqrt{9} = 3$

2. see graph

3. center: $(1, 1)$

radius: 3

$(X - 1)^2 + (Y - 1)^2 = 9$

4. see graph

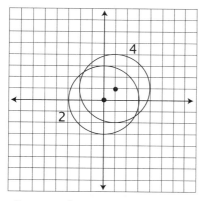

5. $X^2 + 6X + Y^2 + 6Y = -2$

$X^2 + 6X + 9 + Y^2 + 6Y + 9 = -2 + 9 + 9$

$(X + 3)^2 + (Y + 3)^2 = 16$

center: $(-3, -3)$

radius: $\sqrt{16} = 4$

6. see graph

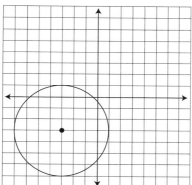

7. $6X^2 + 4Y^2 = 24$

$6(X - 0)^2 + 4(Y - 0)^2 = 24$

center: $(0, 0)$

8. $6X^2 + 4Y^2 = 24$

$6X^2 + 4(0)^2 = 24$

$6X^2 = 24$

$X^2 = 4$

$X = \pm 2$

9. $6X^2 + 4Y^2 = 24$

$6(0)^2 + 4Y^2 = 24$

$4Y^2 = 24$

$Y^2 = 6$

$Y = \pm\sqrt{6} \approx \pm 2.5$

X	Y
0	2.5
0	−2.5
2	0
−2	0

$\sqrt{6}$ is given in rounded decimal form to make it easier to graph the ellipse.

10. see graph

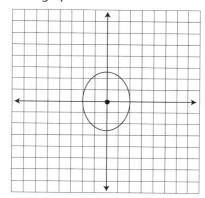

11. $(BC)^2 = \left[(2) - (-2)\right]^2 + \left[(3) - (-4)\right]^2$

$(BC)^2 = (4)^2 + (7)^2 = 16 + 49 = 65$

$BC = \sqrt{65}$

12. $(AB)^2 = \left[(5) - (2)\right]^2 + \left[(-6) - (3)\right]^2$

$(AB)^2 = (3)^2 + (-9)^2 = 9 + 81 = 90$

$AB = \sqrt{90} = \sqrt{9}\sqrt{10} = 3\sqrt{10}$

13. midpoint $= \left(\dfrac{(2) + (-2)}{2}, \dfrac{(3) + (-4)}{2}\right) =$

$\left(\dfrac{0}{2}, \dfrac{-1}{2}\right) = \left(0, -\dfrac{1}{2}\right)$

14. midpoint $= \left(\dfrac{(5) + (-2)}{2}, \dfrac{(-6) + (-4)}{2}\right) =$

$\left(\dfrac{3}{2}, \dfrac{-10}{2}\right) = \left(1\dfrac{1}{2}, -5\right)$

15. $3Y = X - 6$

$Y = \dfrac{1}{3}X - 2; \ m = \dfrac{1}{3}$

$Y = mX + b \Rightarrow (4) = \dfrac{1}{3}(-3) + b$

$\qquad\qquad\qquad 4 = -1 + b$

$\qquad\qquad\qquad 5 = b$

$Y = \dfrac{1}{3}X + 5$

16. see graph

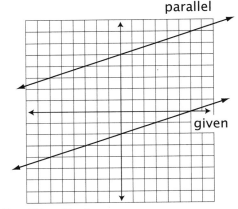

17. $5Y = -X - 5$

$Y = -\dfrac{1}{5}X - 1; \ m = 5$

$Y = mX + b \Rightarrow (-3) = (5)(-1) + b$

$\qquad\qquad\qquad -3 = -5 + b$

$\qquad\qquad\qquad 2 = b$

$Y = 5X + 2$

18. see graph

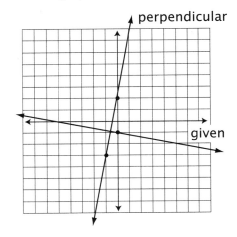

19. $2Y - 2X = 3$

$2Y = 2X + 3$

$Y = X + \dfrac{3}{2} :$ see graph

$(0, 0), (-2, 2)$

$2Y - 2X \geq 3 \Rightarrow 2(0) - 2(0) \geq 3$

$\qquad\qquad\qquad\qquad 0 - 0 \geq 3$

$\qquad\qquad\qquad\qquad 0 \geq 3 \text{ false}$

$2Y - 2X \geq 3 \Rightarrow 2(2) - 2(-2) \geq 3$

$\qquad\qquad\qquad\qquad 4 - (-4) \geq 3$

$\qquad\qquad\qquad\qquad 8 \geq 3 \text{ true}$

see graph

20. solid: see graph

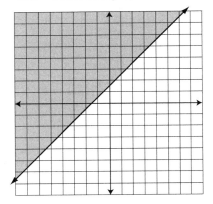

Systematic Review 23D

1. $(X - 2)^2 + (Y + 3)^2 = 36$

center: $(2, -3)$

radius: $\sqrt{36} = 6$

2. see graph

3. center: $(-2, 0)$

radius: 5

$[X - (-2)]^2 + [Y - (0)]^2 = 5^2$

$(X + 2)^2 + Y^2 = 25$

4. see graph

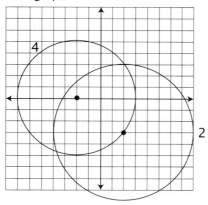

5. $X^2 - 6X + Y^2 = 16$

$X^2 - 6X + 9 + Y^2 = 16 + 9$

$(X - 3)^2 + (Y^2 - 0)^2 = 25$

center: $(3, 0)$

radius: $\sqrt{25} = 5$

6. see graph

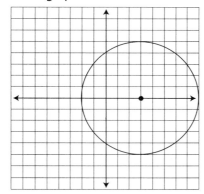

7. $\dfrac{(X + 3)^2}{4} + \dfrac{(Y - 1)^2}{16} = 1$

$(16)\left(\dfrac{(X + 3)^2}{4}\right) + (16)\left(\dfrac{(Y - 1)^2}{16}\right) = (16)(1)$

$4(X + 3)^2 + (Y - 1)^2 = 16$

center: $(-3, 1)$

8. $4(X + 3)^2 + (Y - 1)^2 = 16$

$4(X + 3)^2 + ((1) - 1)^2 = 16$

$4(X + 3)^2 = 16$

$(X + 3)^2 = 4$

$X + 3 = \pm 2$

$X + 3 = 2 \qquad X + 3 = -2$

$X = -1 \qquad X = -5$

9. $4(X + 3)^2 + (Y - 1)^2 = 16$

$4((-3) + 3)^2 + (Y - 1)^2 = 16$

$(Y - 1)^2 = 16$

$Y - 1 = \pm 4$

$Y - 1 = 4 \qquad Y - 1 = -4$

$Y = 5 \qquad Y = -3$

10. see graph

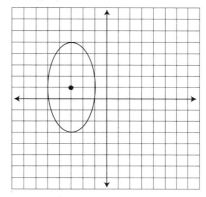

11. $(AB)^2 = \left[(6) - (-2)\right]^2 + \left[(1) - (-1)\right]^2$

$(AB)^2 = (8)^2 + (2)^2 = 64 + 4 = 68$

$AB = \sqrt{68} = \sqrt{4}\sqrt{17} = 2\sqrt{17}$

12. $(AC)^2 = \left[(6)-(-4)\right]^2 + \left[(1)-(4)\right]^2$

$(AC)^2 = (10)^2 + (-3)^2 = 100 + 9 = 109$

$AC = \sqrt{109}$

13. $\text{midpoint} = \left(\dfrac{(-4)+(-2)}{2}, \dfrac{(4)+(-1)}{2}\right) =$

$\left(\dfrac{-6}{2}, \dfrac{3}{2}\right) = \left(-3, 1\dfrac{1}{2}\right)$

14. $\text{midpoint} = \left(\dfrac{(6)+(-4)}{2}, \dfrac{(1)+(4)}{2}\right) =$

$\left(\dfrac{2}{2}, \dfrac{5}{2}\right) = \left(1, 2\dfrac{1}{2}\right)$

15. $4Y + X = -2$

$4Y = -X - 2$

$Y = -\dfrac{1}{4}X - \dfrac{2}{4}$

$Y = -\dfrac{1}{4}X - \dfrac{1}{2}; \ m = -\dfrac{1}{4}$

$Y = mX + b \Rightarrow (-4) = \left(-\dfrac{1}{4}\right)(0) + b$

$\qquad\qquad\qquad -4 = b$

$Y = -\dfrac{1}{4}X - 4$

16. see graph

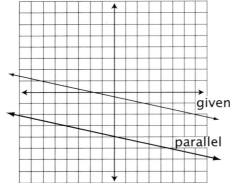

given

parallel

17. $2Y = X + 5$

$Y = \dfrac{1}{2}X + \dfrac{5}{2}; \ m = -2$

$Y = mX + b \Rightarrow (3) = (-2)(2) + b$

$\qquad\qquad\qquad 3 = -4 + b$

$\qquad\qquad\qquad 7 = b$

$Y = -2X + 7$

18. see graph

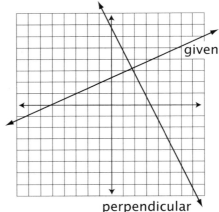

given

perpendicular

19. $5Y < 2X + \dfrac{5}{2}$

$Y < \left(\dfrac{1}{5}\right)(2X) + \left(\dfrac{1}{5}\right)\left(\dfrac{5}{2}\right)$

$Y < \dfrac{2}{5}X + \dfrac{1}{2}$

$(0, 0), (0, 2)$ on the graph

$5Y < 2X + \dfrac{5}{2} \Rightarrow 5(0) < 2(0) + \dfrac{5}{2}$

$\qquad\qquad\qquad 0 < 0 + \dfrac{5}{2}$

$\qquad\qquad\qquad 0 < \dfrac{5}{2} \ \text{true}$

$5Y < 2X + \dfrac{5}{2} \Rightarrow 5(2) < 2(0) + \dfrac{5}{2}$

$\qquad\qquad\qquad 10 < 0 + \dfrac{5}{2}$

$\qquad\qquad\qquad 10 < \dfrac{5}{2} \ \text{false}$

20. dotted line: see graph

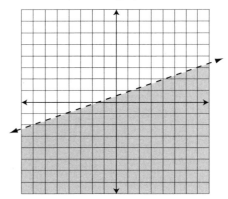

Systematic Review 23E

1. $3X^2 + 3Y^2 = 75$

   $X^2 + Y^2 = 25$

   $(X - 0)^2 + (Y - 0)^2 = 25$

   center: $(0, 0)$

   radius: $\sqrt{25} = 5$

2. see graph

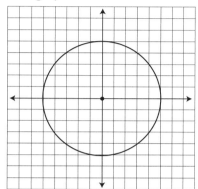

3. center: $(2, -2)$

   radius: 2

   $(X - 2)^2 + (Y + 2)^2 = 2^2$

   $(X - 2)^2 + (Y + 2)^2 = 4$

4. see graph

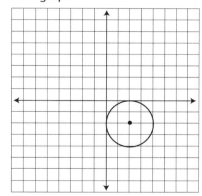

5. $X^2 + 2X + Y^2 + 2Y = 34$

   $X^2 + 2X + 1 + Y^2 + 2Y + 1 = 34 + 1 + 1$

   $(X + 1)^2 + (Y + 1)^2 = 36$

   center: $(-1, -1)$

   radius: $\sqrt{36} = 6$

6. see graph

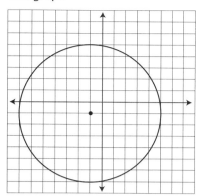

7. $\dfrac{X^2}{25} + \dfrac{Y^2}{9} = 1$

   $(225)\left(\dfrac{X^2}{25}\right) + (225)\left(\dfrac{Y^2}{9}\right) = (225)(1)$

   $9X^2 + 25Y^2 = 225$

   center: $(0, 0)$

8. $9X^2 + 25Y^2 = 225$

   $9X^2 + 25(0)^2 = 225$

   $9X^2 = 225$

   $X^2 = 25$

   $X = \pm 5$

9. $9X^2 + 25Y^2 = 225$

   $9(0)^2 + 25Y^2 = 225$

   $25Y^2 = 225$

   $Y^2 = 9$

   $Y = \pm 3$

10. see graph

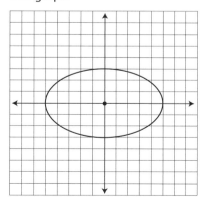

11. $(AC)^2 = \left[(3)-(6)\right]^2 + \left[(6)-(3)\right]^2$

$(AC)^2 = (-3)^2 + (3)^2 = 9+9 = 18$

$AC = \sqrt{18} = \sqrt{9}\sqrt{2} = 3\sqrt{2}$

12. $(BC)^2 = \left[(0)-(6)\right]^2 + \left[(0)-(3)\right]^2$

$(BC)^2 = (-6)^2 + (-3)^2 = 36+9 = 45$

$BC = \sqrt{45} = \sqrt{9}\sqrt{5} = 3\sqrt{5}$

13. $\text{midpoint} = \left(\dfrac{(3)+(0)}{2}, \dfrac{(6)+(0)}{2}\right) =$

$\left(\dfrac{3}{2}, \dfrac{6}{2}\right) = \left(1\dfrac{1}{2}, 3\right)$

14. $\text{midpoint} = \left(\dfrac{(3)+(6)}{2}, \dfrac{(6)+(3)}{2}\right) =$

$\left(\dfrac{9}{2}, \dfrac{9}{2}\right) = \left(4\dfrac{1}{2}, 4\dfrac{1}{2}\right)$

15. $2Y + 2X = -3$

$2Y = -2X - 3$

$Y = -X - \dfrac{3}{2} : m = -1$

$Y = mX + b \Rightarrow \ (4) = (-1)(-2) + b$

$\qquad\qquad\qquad 4 = 2 + b$

$\qquad\qquad\qquad 2 = b$

$\qquad\qquad\qquad Y = -X + 2$

16. see graph

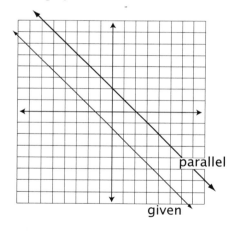

parallel

given

17. $Y = -3X - 2 : m = \dfrac{1}{3}$

$Y = mX + b \Rightarrow \qquad (3) = \left(\dfrac{1}{3}\right)(1) + b$

$\qquad\qquad\qquad\qquad 3 = \dfrac{1}{3} + b$

$\qquad\qquad\qquad\qquad 3 - \dfrac{1}{3} = b$

$\qquad\qquad\qquad\qquad \dfrac{9}{3} - \dfrac{1}{3} = b = \dfrac{8}{3}$

$\qquad Y = \dfrac{1}{3}X + \dfrac{8}{3}$

18. see graph

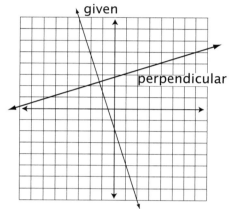

given

perpendicular

19. $5Y + 4X \leq 0$

$5Y \leq -4X$

$Y \leq -\dfrac{4}{5}X$

see graph

test points: $(-2, -2), (2, 2)$

$5Y + 4X \leq 0 \Rightarrow \ 5(-2) + 4(-2) \leq 0$

$\qquad\qquad\qquad\qquad -10 + -8 \leq 0$

$\qquad\qquad\qquad\qquad\qquad -18 \leq 0 \text{ true}$

$5Y + 4X \leq 0 \Rightarrow \ 5(2) + 4(2) \leq 0$

$\qquad\qquad\qquad\qquad 10 + 8 \leq 0$

$\qquad\qquad\qquad\qquad\qquad 18 \leq 0 \text{ false}$

20. solid: see graph

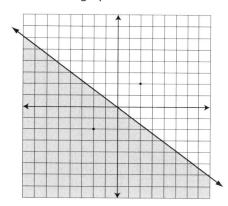

3. $Y = \frac{1}{3}X^2$

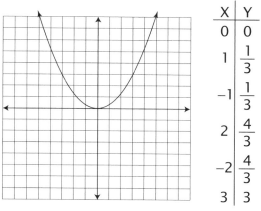

X	Y
0	0
1	$\frac{1}{3}$
−1	$\frac{1}{3}$
2	$\frac{4}{3}$
−2	$\frac{4}{3}$
3	3

Lesson Practice 24A

1. $Y = 3X^2$

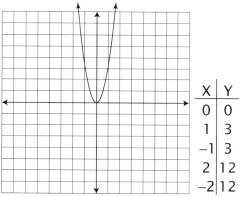

X	Y
0	0
1	3
−1	3
2	12
−2	12

4. $X = 4Y^2$

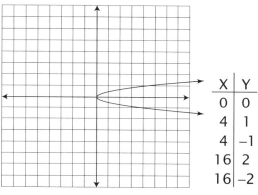

X	Y
0	0
4	1
4	−1
16	2
16	−2

5. $X = -3Y^2 + 1$

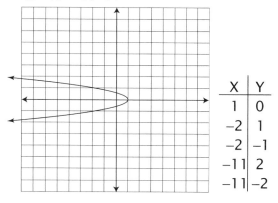

X	Y
1	0
−2	1
−2	−1
−11	2
−11	−2

2. $Y = -X^2$

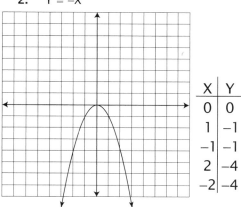

X	Y
0	0
1	−1
−1	−1
2	−4
−2	−4

ALGEBRA 2

6. $Y = X^2 - 4$

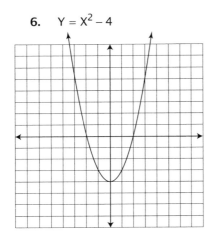

X	Y
0	−4
1	−3
−1	−3
2	0
−2	0

3. $X = \frac{1}{2}Y^2 + 3$

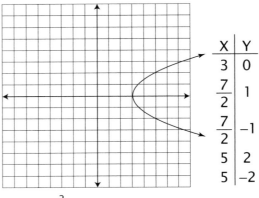

X	Y
3	0
$\frac{7}{2}$	1
$\frac{7}{2}$	−1
5	2
5	−2

4. $3X^2 = Y + 1$

$-Y = -3X^2 + 1$

$Y = 3X^2 - 1$

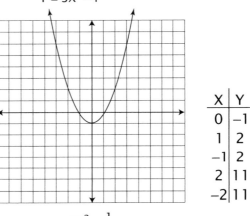

X	Y
0	−1
1	2
−1	2
2	11
−2	11

Lesson Practice 24B

1. $Y = 2X^2 + 5$

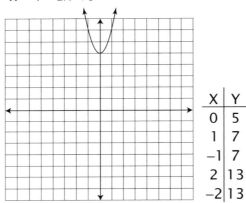

X	Y
0	5
1	7
−1	7
2	13
−2	13

2. $2X = -4Y^2 + 8$

$X = -2Y^2 + 4$

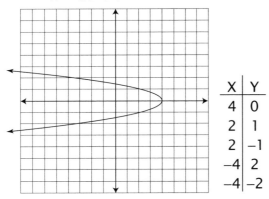

X	Y
4	0
2	1
2	−1
−4	2
−4	−2

5. $Y = -6X^2 + \frac{1}{2}$

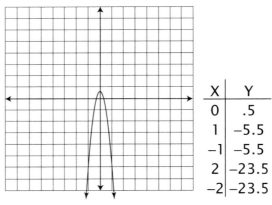

X	Y
0	.5
1	−5.5
−1	−5.5
2	−23.5
−2	−23.5

6. $X = Y^2 + 5$

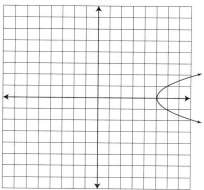

3. see graph

4.

X	Y
0	4
1	1
−1	1
2	−8
−2	−8

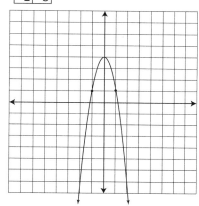

Systematic Review 24C

1. see graph

2.

X	Y
0	−3
1	−2
−1	−2
2	1
−2	1

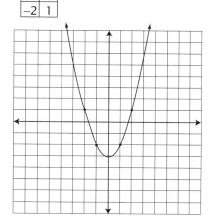

5. 1

6. −2

7. $\dfrac{1}{2}$

8. 1

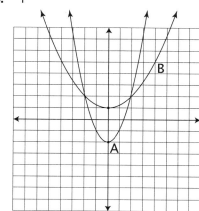

9. $2X^2 + 2Y^2 = 8$

$X^2 + Y^2 = 4$

$(X - 0)^2 + (Y - 0)^2 = 2^2$

center: $(0, 0)$

radius: 2

10. see graph

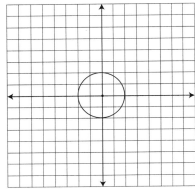

11. $(X - (1))^2 + (Y - (-1))^2 = 3^2$

$(X - 1)^2 + (Y + 1)^2 = 9$

12. see graph

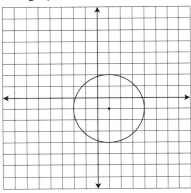

13. $\dfrac{X^2}{4} + \dfrac{Y^2}{9} = 1$

$(36)\dfrac{X^2}{4} + (36)\dfrac{Y^2}{9} = (36)1$

$9X^2 + 4Y^2 = 36$

$9(X - 0)^2 + 4(Y - 0)^2 = 36$

center: $(0, 0)$

14. $9X^2 + 4Y^2 = 36$

$9(0)^2 + 4Y^2 = 36$

$4Y^2 = 36$

$Y^2 = 9$

$Y = \pm\sqrt{9} = \pm3$

$9X^2 + 4Y^2 = 36$

$9X^2 + 4(0)^2 = 36$

$9X^2 = 36$

$X^2 = 4$

$X = \pm\sqrt{4} = \pm2$

see graph

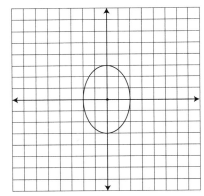

15. $(AB)^2 = [(4) - (2)]^2 + [(4) - (-1)]^2$

$(AB)^2 = (2)^2 + (5)^2 = 4 + 25 = 29$

$AB = \sqrt{29}$

16. $(BC)^2 = [(-1) - (4)]^2 + [(2) - (4)]^2$

$(BC)^2 = (-5)^2 + (-2)^2 = 25 + 4 = 29$

$BC = \sqrt{29}$

17. midpoint $= \left(\dfrac{(2) + (-1)}{2}, \dfrac{(-1) + (2)}{2}\right) = \left(\dfrac{1}{2}, \dfrac{1}{2}\right)$

18. midpoint $= \left(\dfrac{(2) + (4)}{2}, \dfrac{(-1) + (4)}{2}\right) =$

$\left(\dfrac{6}{2}, \dfrac{3}{2}\right) = \left(3, 1\dfrac{1}{2}\right)$

19. $2Y + \frac{2}{3}X = 1$

$2Y = -\frac{2}{3}X + 1$

$Y = -\frac{1}{3}X + \frac{1}{2} : m = 3$

$Y = mX + b \Rightarrow (3) = (3)(2) + b$

$\qquad\qquad\qquad 3 = 6 + b$

$\qquad\qquad\qquad -3 = b$

$Y = 3X - 3$

20. $5Y > 3X + 5$

$Y > \frac{3}{5}X + 1$

$(0, 0):$

$5Y > 3X + 5 \Rightarrow 5(0) > 3(0) + 5$

$\qquad\qquad\qquad 0 > 0 + 5$

$\qquad\qquad\qquad 0 > 5 \text{ false}$

$(0, 2):$

$5Y > 3X + 5 \Rightarrow 5(2) > 3(0) + 5$

$\qquad\qquad\qquad 10 > 0 + 5$

$\qquad\qquad\qquad 10 > 5 \text{ true}$

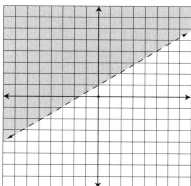

Systematic Review 24D

1. $2Y = X^2 + 4$

$Y = \frac{1}{2}X^2 + 2$

see graph

2.

X	Y
0	2
1	$2\frac{1}{2}$
-1	$2\frac{1}{2}$
2	4

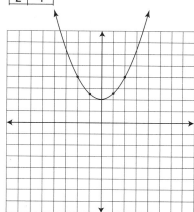

3. see graph

4.

X	Y
0	-1
1	$-1\frac{1}{2}$
-1	$-1\frac{1}{2}$
2	-3
-2	-3

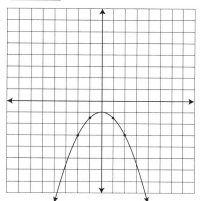

5. -2

6. 3

7. 3

8. 0

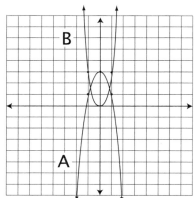

9.
$$X^2 - 4X + Y^2 - 2Y = 4$$
$$X^2 - 4X + 4 + Y^2 - 2Y + 1 = 4 + 4 + 1$$
$$(X - 2)^2 + (Y - 1)^2 = 9$$

center: $(2, 1)$

radius: $\sqrt{9} = 3$

10. see graph

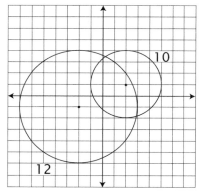

11.
$$[X - (-2)]^2 + [Y - (-1)]^2 = 5^2$$
$$(X + 2)^2 + (Y + 1)^2 = 25$$

12. see graph

13.
$$(X + 2)^2 + 4(Y - 1)^2 = 16$$
$$(X - (-2))^2 + 4(Y - (1))^2 = 4^2$$

center: $(-2, 1)$

14.
$$(X + 2)^2 + 4(Y - 1)^2 = 16$$
$$((-2) + 2)^2 + 4(Y - 1)^2 = 16$$
$$4(Y - 1)^2 = 16$$
$$(Y - 1)^2 = 4$$
$$Y - 1 = \pm 2$$
$$Y - 1 = 2 \text{ and } Y - 1 = -2$$

$Y = 3$ and $Y = (-1)$ when $X = (-2)$

$$(X + 2)^2 + (Y - 1)^2 = 16$$
$$(X + 2)^2 + ((1) - 1)^2 = 16$$
$$(X + 2)^2 = 16$$
$$X + 2 = \pm 4$$
$$X + 2 = 4 \text{ and } X + 2 = -4$$

$X = 2$ and $X = -6$ when $Y = 1$

see graph

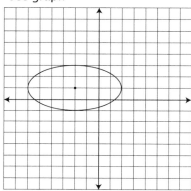

15.
$$(AB)^2 = [(6) - (0)]^2 + [(3) - (0)]^2$$
$$(AB)^2 = (6)^2 + (3)^2 = 36 + 9 = 45$$
$$AB = \sqrt{45} = \sqrt{9}\sqrt{5} = 3\sqrt{5}$$

16.
$$(AC)^2 = [(-1) - (0)]^2 + [(-3) - (0)]^2$$
$$(AC)^2 = (-1)^2 + (-3)^2 = 1 + 9 = 10$$
$$AC = \sqrt{10}$$

17. center $= \left(\dfrac{(6) + (-1)}{2}, \dfrac{(3) + (-3)}{2} \right) =$
$$\left(\dfrac{5}{2}, \dfrac{0}{2} \right) = \left(2\tfrac{1}{2}, 0 \right)$$

18. center $= \left(\dfrac{(6)+(0)}{2}, \dfrac{(3)+(0)}{2} \right) =$

$\left(\dfrac{6}{2}, \dfrac{3}{2} \right) = \left(3, 1\dfrac{1}{2} \right)$

19. $5Y = X + 10$

$Y = \dfrac{1}{5}X + 2 \quad m = -5$

$Y = mX + b \Rightarrow (-2) = (-5)(1) + b$

$\qquad -2 = -5 + b$

$\qquad 3 = b$

$Y = -5X + 3$

20. $X - Y > \dfrac{1}{2}$

graph $X - Y = \dfrac{1}{2}$:

$-Y = -X + \dfrac{1}{2}$

$Y = X - \dfrac{1}{2}$

$(0, 0)$:

$X - Y > \dfrac{1}{2} \Rightarrow (0) - (0) > \dfrac{1}{2}$

$\qquad\qquad 0 > \dfrac{1}{2}$ false

$(3, 0)$:

$X - Y > \dfrac{1}{2} \Rightarrow (3) - (0) > \dfrac{1}{2}$

$\qquad\qquad 3 > \dfrac{1}{2}$ true

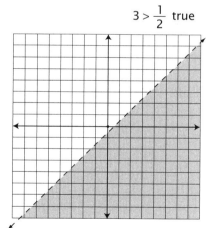

Systematic Review 24E

1. $2X^2 = -Y$

$Y = -2X^2$

see graph

2.

X	Y
0	0
1	-2
-1	-2
2	-8
-2	-8

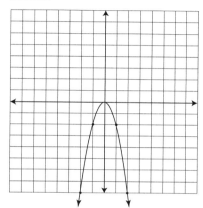

3. $Y + 1 = 2X^2$

$Y = 2X^2 - 1$

see graph

4.

X	Y
0	-1
1	1
-1	1
2	7
-2	7

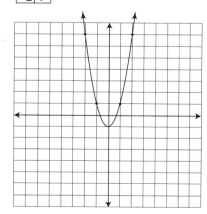

5. −2

6. 1

7. $-\dfrac{1}{2}$

8. 2

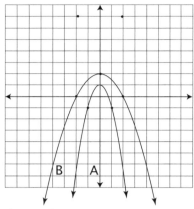

9. $\dfrac{1}{2}X^2 + \dfrac{1}{2}Y^2 = 8$

$X^2 + Y^2 = 16$

$(X - 0)^2 + (Y - 0)^2 = 16$

center: $(0, 0)$ radius: 4

10. see graph

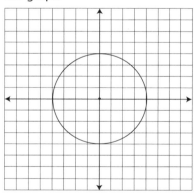

11. $\left(X - (0)\right)^2 + \left(Y - (2)\right)^2 = 3^2$

$(X)^2 + (Y - 2)^2 = 9$

12. see graph

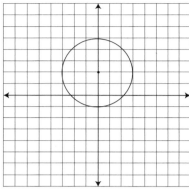

13. $\dfrac{(X-1)^2}{16} + \dfrac{(Y+1)^2}{4} = 1$

$(64)\dfrac{(X-1)^2}{16} + (64)\dfrac{(Y+1)^2}{4} = (64)1$

$4(X-1)^2 + 16(Y+1)^2 = 64$

center: $(1, -1)$

14. $4(X-1)^2 + 16(Y+1)^2 = 64$

$4\left((1) - 1\right)^2 + 16(Y+1)^2 = 64$

$16(Y+1)^2 = 64$

$(Y+1)^2 = 4$

$Y + 1 = \pm\sqrt{4} = \pm 2$

$Y + 1 = 2 \qquad Y + 1 = -2$

$Y = 1$ and $\qquad Y = -3$ when $X = 1$

$4(X-1)^2 + 16(Y+1)^2 = 64$

$4(X-1)^2 + 16\left((-1) + 1\right) = 64$

$4(X-1)^2 = 64$

$(X-1)^2 = 16$

$X - 1 = \pm\sqrt{16} = \pm 4$

$X - 1 = 4 \qquad X - 1 = -4$

$X = 5$ and $\qquad X = -3$ when $Y = (-1)$

see graph

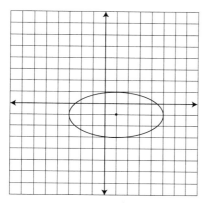

15. $(AB)^2 = \left[(0) - (-5)\right]^2 + \left[(4) - (5)\right]^2$

$(AB)^2 = [5]^2 + [-1]^2 = 25 + 1 = 26$

$AB = \sqrt{26}$

16. $(AC)^2 = \left[(4) - (-5)\right]^2 + \left[(-3) - (5)\right]^2$

$(AC)^2 = [9]^2 + [-8]^2 = 81 + 64 = 145$

$AC = \sqrt{145}$

17. midpoint $= \left(\dfrac{(4) + (0)}{2}, \dfrac{(-3) + (4)}{2}\right) =$

$\left(\dfrac{4}{2}, \dfrac{1}{2}\right) = \left(2, \dfrac{1}{2}\right)$

18. midpoint $= \left(\dfrac{(-5) + (0)}{2}, \dfrac{(5) + (4)}{2}\right) =$

$\left(\dfrac{-5}{2}, \dfrac{9}{2}\right) = \left(-2\dfrac{1}{2}, 4\dfrac{1}{2}\right)$

19. $4Y - X - 6 = 0$

$4Y = X + 6$

$Y = \dfrac{1}{4}X + \dfrac{6}{4} : m = -4$

$Y = mX + b \Rightarrow (3) = (-4)(0) + b$

$\qquad\qquad\qquad 3 = 0 + b$

$\qquad\qquad\qquad 3 = b$

$Y = -4X + 3$

20. graph $Y = 4X - \dfrac{3}{2}$

$(0, 0)$:

$Y > 4X - \dfrac{3}{2} \Rightarrow (0) > 4(0) - \dfrac{3}{2}$

$\qquad\qquad\qquad 0 > -\dfrac{3}{2}$ true

$(2, 0)$:

$Y > 4X - \dfrac{3}{2} \Rightarrow (0) > 4(2) - \dfrac{3}{2}$

$\qquad\qquad\qquad 0 > 8 - \dfrac{3}{2}$

$\qquad\qquad\qquad 0 > \dfrac{16}{2} - \dfrac{3}{2}$

$\qquad\qquad\qquad 0 > \dfrac{13}{2}$ false

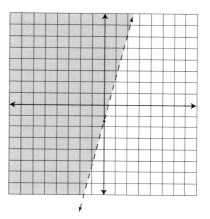

Lesson Practice 25A

You should be able to sketch the graph using the vertex and the original equation. If you wish, you may chart a few points for a more accurate graph. Estimate the graph first, so that you can choose the most useful values of X to try.

1. $Y = 3X^2 - 6X + 2$

$X = \dfrac{-B}{2A} = \dfrac{-(-6)}{2(3)} = \dfrac{6}{6} = 1$

$Y = 3(1)^2 - 6(1) + 2$

$Y = 3 - 6 + 2 = -1$

vertex $= (1, -1)$

X	Y
2	2
3	11

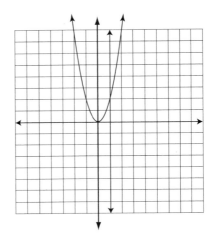

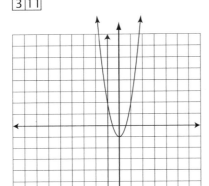

2. $4Y = 4X^2 + 8X + 4$

$Y = X^2 + 2X + 1$

$X = \dfrac{-B}{2A} = \dfrac{-(2)}{2(1)} = \dfrac{-2}{2} = -1$

$Y = (-1)^2 + 2(-1) + 1 = 1 - 2 + 1 = 0$

vertex $= (-1, 0)$

X	Y
-2	1
-4	9

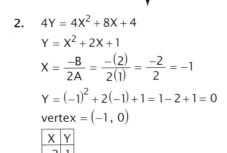

3. $Y = -X^2 + 6X - 4$

$X = \dfrac{-B}{2A} = \dfrac{-(6)}{2(-1)} = \dfrac{-6}{-2} = 3$

$Y = -(3)^2 + 6(3) - 4 = -9 + 18 - 4 = 5$

vertex $= (3, 5)$

X	Y
0	-4
2	4
6	-4

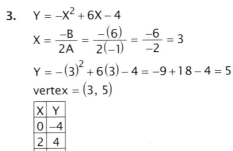

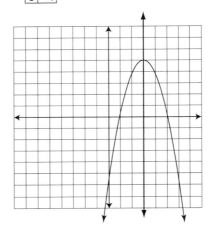

4. $Y = -4X^2 + 4X$

$X = \dfrac{-B}{2A} = \dfrac{-(4)}{2(-4)} = \dfrac{-4}{-8} = \dfrac{1}{2}$

$Y = -4\left(\dfrac{1}{2}\right)^2 + 4\left(\dfrac{1}{2}\right)$

$Y = -4\left(\dfrac{1}{4}\right) + \dfrac{4}{2}$

$Y = -1 + 2 = 1$

vertex $= \left(\dfrac{1}{2},\ 1\right)$

X	Y
1	0
2	-8

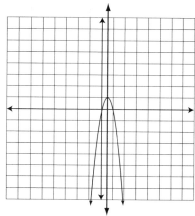

5. $Y = \dfrac{1}{4}X^2 - 3$

$Y = \dfrac{1}{4}X^2 + 0X - 3$

$X = \dfrac{-B}{2A} = \dfrac{-0}{2\left(\dfrac{1}{4}\right)} = 0$

$Y = \dfrac{1}{4}(0)^2 - 3 = 0 - 3 = -3$

vertex $= (0,\ -3)$

X	Y
2	-2
4	1
6	6

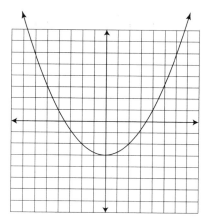

6. $Y = 3X^2 + 30X + 78$

$X = \dfrac{-B}{2A} = \dfrac{-(30)}{2(3)} = \dfrac{-30}{6} = -5$

$Y = 3(-5)^2 + 30(-5) + 78$

$Y = 3(25) - 150 + 78 =$

$\qquad 75 - 150 + 78 = 3$

vertex $= (-5,\ 3)$

X	Y
-4	6
-6	6

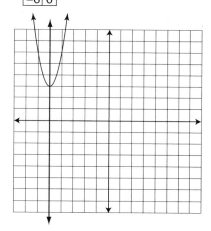

7. $\text{area} = X(18 - X)$

$\qquad = 18X - X^2$

$\qquad = -X^2 + 18X$

axis of symmetry =

$\dfrac{-B}{2A} = \dfrac{-(18)}{2(-1)} = \dfrac{-18}{-2} = 9$

$\text{area} = -(9)^2 + 18(9) = -81 + 162 = 81$

$\text{vertex (maxima)} = (9,\ 81)$

$\text{maximum area} = 9 \times 9 = 81 \text{ ft}^2$

$(36 - 2X) \div 2$

X ⎾   ⏋ X

2. $Y = X^2 + X - 6$

$X = \dfrac{-B}{2A} = \dfrac{-(1)}{2(1)} = \dfrac{-1}{2} = -\dfrac{1}{2}$

$Y = \left(-\dfrac{1}{2}\right)^2 + \left(-\dfrac{1}{2}\right) - 6$

$Y = \dfrac{1}{4} - \dfrac{1}{2} - 6 = \dfrac{1}{4} - \dfrac{2}{4} - \dfrac{24}{4} =$

$\qquad -\dfrac{25}{4} = -6\dfrac{1}{4}$

$\text{vertex} = \left(-\dfrac{1}{2},\ -6\dfrac{1}{4}\right)$

X	Y
2	0
3	6

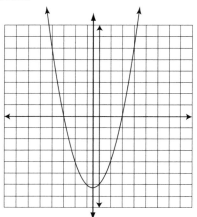

Lesson Practice 25B

1. $Y = 2X^2 - 4X + 1$

$X = \dfrac{-B}{2A} = \dfrac{-(-4)}{2(2)} = \dfrac{4}{4} = 1$

$Y = 2(1)^2 - 4(1) + 1 = 2 - 4 + 1 = -1$

$\text{vertex} = (1,\ -1)$

X	Y
2	1
3	7

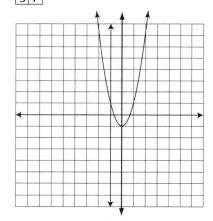

3. $5Y = -15X^2 + 30X$

$Y = -3X^2 + 6X$

$X = \dfrac{-B}{2A} = \dfrac{-(6)}{2(-3)} = \dfrac{-6}{-6} = 1$

$Y = -3(1)^2 + 6(1) = -3 + 6 = 3$

vertex $= (1, 3)$

X	Y
2	0
3	-9

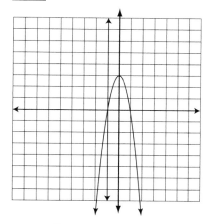

4. $Y = -2X^2 - 12X - 20$

$X = \dfrac{-B}{2A} = \dfrac{-(-12)}{2(-2)} = \dfrac{12}{-4} = -3$

$Y = -2(-3)^2 - 12(-3) - 20$

$Y = -2(9) + 36 - 20 = -18 + 36 - 20 = -2$

vertex $= (-3, -2)$

X	Y
-2	-4
-1	-10

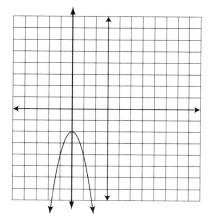

5. $Y = \dfrac{1}{4}X^2 + 3X + 3$

$X = \dfrac{-B}{2A} = \dfrac{-(3)}{2\left(\dfrac{1}{4}\right)} = \dfrac{-3}{\dfrac{1}{2}} = -3(2) = -6$

$Y = \dfrac{1}{4}(-6)^2 + 3(-6) + 3$

$Y = \dfrac{1}{4}(36) - 18 + 3 = 9 - 18 + 3 = -6$

vertex $= (-6, -6)$

X	Y
-2	-2
0	3
1	$6\dfrac{1}{4}$

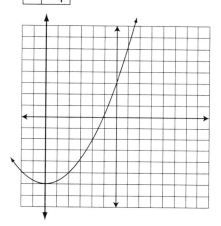

6. $X = 2Y^2 + 8Y$

    $Y = \dfrac{-B}{2A} = \dfrac{-(8)}{2(2)} = \dfrac{-8}{4} = -2$

    $X = 2(-2)^2 + 8(-2)$

    $X = 2(4) - 16 = 8 - 16 = -8$

    vertex $= (-8, -2)$

X	Y
-6	-3
-6	-1
0	0

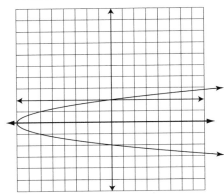

7. area $= X(260 - 2X)$

    $= 260X - 2X^2$

    $= -2X^2 + 260X$

    axis of symmetry $=$

    $\dfrac{-B}{2A} = \dfrac{-(260)}{2(-2)} = \dfrac{-260}{-4} = 65$

    area $= -2(65)^2 + 260(65) = 8,450$

    vertex (maxima) $= (65, 8450)$

    maximum area $= 65 \times 130 = 8,450 \text{ ft}^2$

    $$260 - 2X$$

    X [   ] X

Systematic Review 25C

1. $6Y = 3X^2 + 24X$

    $Y = \dfrac{3}{6}X^2 + \dfrac{24}{6}X$

    $Y = \dfrac{1}{2}X^2 + 4X$

    $X = \dfrac{-B}{2A} = \dfrac{-(4)}{2\left(\frac{1}{2}\right)} = \dfrac{-4}{1} = -4$

2. $Y = \dfrac{1}{2}(-4)^2 + 4(-4)$

    $Y = \dfrac{1}{2}(16) - 16 = 8 - 16 = -8$

    vertex is $(-4, -8)$

3. see graph

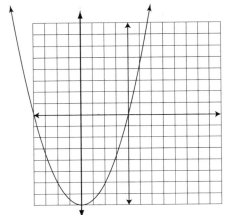

4. $Y = -2X^2 - 4X - 3$

    $X = \dfrac{-B}{2A} = \dfrac{-(-4)}{2(-2)} = \dfrac{4}{-4} = -1$

5. $Y = -2(-1)^2 - 4(-1) - 3$

    $Y = -2(1) + 4 - 3 = -2 + 4 - 3 = -1$

    vertex is $(-1, -1)$

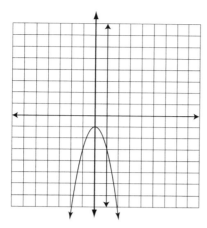

6. see graph

7. points up

8. Y-axis

9-10. area $= X(200 - 2X)$

$\qquad = 200X - 2X^2$

$\qquad = -2X^2 + 200X$

axis of symmetry =

$X = \dfrac{-B}{2A} = \dfrac{-(200)}{2(-2)} = \dfrac{-200}{-4} = 50$

area $= -2(50)^2 + 200(50)$

$\qquad = -2(2,500) + 10,000$

$\qquad = -5,000 + 10,000 = 5,000 \text{ ft}^2$

vertex $= (50, 5000)$

$X = 50$

$200 - 2X \Rightarrow 200 - 2(50) = 200 - 100 = 100$

dimensions $= 50 \text{ ft} \times 100 \text{ ft}$

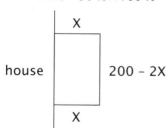

11. $Y = 2X^2 + 2$

$Y = 2X^2 + 0X + 2$

$X = \dfrac{-B}{2A} = \dfrac{-(0)}{2(2)} = 0$

$Y = 2(0)^2 + 2$

$Y = 0 + 2 = 2$

vertex $= (0, 2)$

see graph

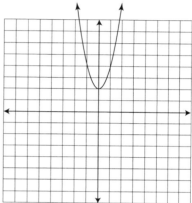

12. $2Y - X^2 = -2$

$2Y = X^2 - 2$

$Y = \dfrac{1}{2}X^2 - 1$

$Y = \dfrac{1}{2}X^2 + 0X - 1$

$X = \dfrac{-B}{2A} = \dfrac{-(0)}{2A} = 0$

$Y = \dfrac{1}{2}(0)^2 - 1 = 0 - 1 = -1$

vertex $= (0, -1)$

see graph

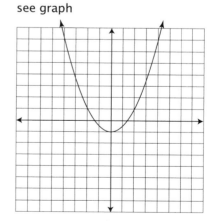

13. $2(X-3)^2 + 2(Y+2)^2 = 72$

$(X-3)^2 + (Y+2)^2 = 36$

center $= (3, -2)$

radius $= \sqrt{36} = 6$

14. center $= (1, 4)$; radius $= 5$

$(X-(1))^2 + (Y-(4))^2 = 5^2$

$(X-1)^2 + (Y-4)^2 = 25$

15-16. $X^2 + Y^2 = 4X + 5$

$X^2 - 4X + Y^2 = 5$

$X^2 - 4X + 4 + Y^2 = 5 + 4$

$(X-2)^2 + Y^2 = 9$

center $= (2, 0)$

radius $= \sqrt{9} = 3$

17. $(AB)^2 = [(-4)-(3)]^2 + [(-1)-(5)]^2$

$(AB)^2 = [-7]^2 + [-6]^2 = 49 + 36 = 85$

$AB = \sqrt{85}$

18. midpoint $= \left(\dfrac{(3)+(3)}{2}, \dfrac{(5)+(-2)}{2} \right) =$

$\left(\dfrac{6}{2}, \dfrac{3}{2} \right) = \left(3, 1\dfrac{1}{2} \right)$

19. $\dfrac{(X+2)^2}{16} + \dfrac{(Y+3)^2}{9} = 1$

$(144)\dfrac{(X+2)^2}{16} + (144)\dfrac{(Y+3)^2}{9} = (144)1$

$9(X+2)^2 + 16(Y+3)^2 = 144$

center $= (-2, -3)$

$9(X+2)^2 + 16(Y+3)^2 = 144$

$9((-2)+2)^2 + 16(Y+3)^2 = 144$

$16(Y+3)^2 = 144$

$(Y+3)^2 = 9$

$Y+3 = \pm 3$

$Y+3 = 3 \qquad Y+3 = -3$

$Y = 0$ and $\qquad Y = -6$ when $X = (-2)$

$9(X+2)^2 + 16(Y+3)^2 = 144$

$9(X+2)^2 + 16((-3)+3)^2 = 144$

$9(X+2)^2 = 144$

$(X+2)^2 = 16$

$X+2 = \pm 4$

$X+2 = 4 \qquad X+2 = -4$

$X = 2$ and $\qquad X = -6$ when $Y = (-3)$

20. see graph

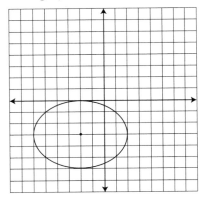

Systematic Review 25D

1. $Y = X^2 - 4X + 3$

$X = \dfrac{-B}{2A} = \dfrac{-(-4)}{2(1)} = \dfrac{4}{2} = 2$

2. $Y = (2)^2 - 4(2) + 3 = 4 - 8 + 3 = -1$

vertex $= (2, -1)$

3. see graph

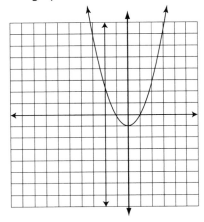

4. $2Y + X^2 = 8X - 4$

$2Y = -X^2 + 8X - 4$

$Y = -\frac{1}{2}X^2 + 4X - 2$

$X = \frac{-B}{2A} = \frac{-(4)}{2\left(-\frac{1}{2}\right)} = \frac{-4}{-1} = 4$

5. $Y = -\frac{1}{2}(4)^2 + 4(4) - 2$

$Y = -\frac{1}{2}(16) + 16 - 2 = -8 + 16 - 2 = 6$

vertex $= (4, 6)$

6. see graph

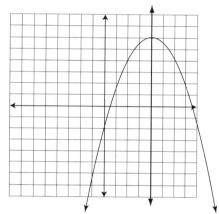

7. steeper

8. left and right

9. area $= X(6 - X)$

$= 6X - X^2$

$= -X^2 + 6X$

axis of symmetry $= \frac{-B}{2A} = \frac{-(6)}{2(-1)} = \frac{-6}{-2} = 3$

area $= -X^2 + 6X$

$= -(3)^2 + 6(3)$

$= -9 + 18 = 9 \text{ ft}^2$

vertex $= (3, 9)$

$X = 3$

dimensions $= 3 \text{ ft} \times 3 \text{ ft}$

10.

6 - X

X ☐ X

$1/2\ (12 - 2X) =$

6 - X

11. $3Y + 6X^2 = 6$

$3Y = -6X^2 + 6$

$Y = -2X^2 + 2$

$Y = -2X^2 + 0X + 2$

$X = \frac{-B}{2A} = \frac{-(0)}{2(-2)} = 0$

$Y = -2(0)^2 + 2 = 0 + 2 = 2$

vertex $= (0, 2)$

see graph

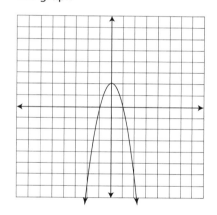

12. $Y + X^2 = 0$

$Y = -X^2$

$Y = -X^2 + 0X$

$X = \dfrac{-B}{2A} = \dfrac{-(0)}{2(-1)} = 0$

$Y = -X^2$

$Y = -(0)^2 = 0$

vertex $= (0, 0)$

see graph

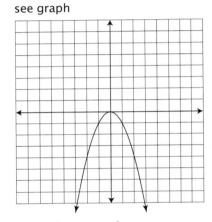

13. $3(X+1)^2 + 3(Y+4)^2 = 147$

$(X+1)^2 + (Y+4)^2 = 49$

center $= (-1, -4)$; radius $= \sqrt{49} = 7$

14. center $= (0, -3)$; radius $= 6$

$(X - (0))^2 + (Y - (-3))^2 = 6^2$

$X^2 + (Y+3)^2 = 36$

15-16. $X^2 + 4X - 3 = -Y^2 - 6Y$

$X^2 + 4X + Y^2 + 6Y = 3$

$X^2 + 4X + 4 + Y^2 + 6Y + 9 = 3 + 4 + 9$

$(X+2)^2 + (Y+3)^2 = 16$

center $= (-2, -3)$; radius $= \sqrt{16} = 4$

17. $(AB)^2 = \left[(3) - (-2)\right]^2 + \left[(-2) - (-2)\right]^2$

$(AB)^2 = (5)^2 + (0)^2 = 25 + 0 = 25$

$AB = \sqrt{25} = 5$

18. midpoint $= \left(\dfrac{(-2) + (3)}{2}, \dfrac{(-2) + (2)}{2} \right) =$

$\left(\dfrac{1}{2}, \dfrac{0}{2} \right) = \left(\dfrac{1}{2}, 0 \right)$

19. $-3Y + 2X \leq 3$

$-3Y \leq -2X + 3$

$Y \geq \dfrac{-2}{-3}X + \dfrac{3}{-3}$

$\left(\begin{array}{l} \text{multiplying by a negative number} \\ \text{reverses the sign of the inequality} \end{array} \right)$

$Y \geq \dfrac{2}{3}X - 1$

20. $(0, 0)$:

$Y \geq \dfrac{2}{3}X - 1 \Rightarrow (0) \geq \dfrac{2}{3}(0) - 1$

$0 \geq 0 - 1$

$0 \geq -1$ true

$(3, 0)$:

$Y \geq \dfrac{2}{3}X - 1 \Rightarrow (0) \geq \dfrac{2}{3}(3) - 1$

$0 \geq 2 - 1$

$0 \geq 1$ false

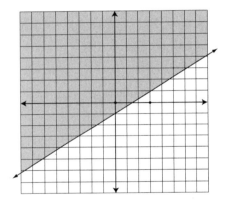

Systematic Review 25E

1. $Y + 2X^2 + 5 = 3X$

$Y = -2X^2 + 3X - 5$

$X = \dfrac{-B}{2A} = \dfrac{-(3)}{2(-2)} = \dfrac{-3}{-4} = \dfrac{3}{4}$

2. $Y = -2X^2 + 3X - 5$

$$Y = -2\left(\frac{3}{4}\right)^2 + 3\left(\frac{3}{4}\right) - 5$$

$$Y = -2\left(\frac{9}{16}\right) + \frac{9}{4} - 5$$

$$Y = -\frac{9}{8} + \frac{18}{8} - \frac{40}{8} = -\frac{31}{8} = -3\frac{7}{8}$$

$$\text{vertex} = \left(\frac{3}{4}, -3\frac{7}{8}\right)$$

3. see graph

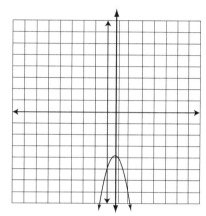

4. $Y + 2X = 3X^2 - 1$

$$Y = 3X^2 - 2X - 1$$

$$X = \frac{-B}{2A} = \frac{-(-2)}{2(3)} = \frac{2}{6} = \frac{1}{3}$$

5. $Y = 3\left(\frac{1}{3}\right)^2 - 2\left(\frac{1}{3}\right) - 1$

$$Y = 3\left(\frac{1}{9}\right) - \frac{2}{3} - 1$$

$$Y = \frac{1}{3} - \frac{2}{3} - \frac{3}{3} = \frac{-4}{3} = -1\frac{1}{3}$$

$$\text{vertex} = \left(\frac{1}{3}, -1\frac{1}{3}\right)$$

6. see graph

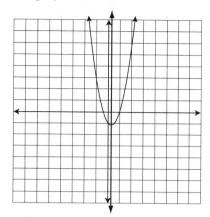

7. points down

8. $\dfrac{-B}{2A}$

9-10. $\text{area} = X(120 - 2X)$

$$= 120X - 2X^2$$

$$= -2X^2 + 120X$$

$\text{axis of symmetry} =$

$$\frac{-B}{2A} = \frac{-120}{2(-2)} = \frac{-120}{-4} = 30$$

$$-2X^2 + 120X \Rightarrow -2(30)^2 + 120(30)$$

$$= -2(900) + 3,600$$

$$= -1,800 + 3,600$$

$$= 1,800 \text{ ft}^2$$

$\text{dimensions} = 30 \text{ ft} \times 60 \text{ ft}$

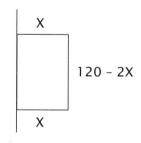

11. $4Y = -2X^2 - 8$

$Y = -\dfrac{2}{4}X^2 - \dfrac{8}{4}$

$Y = -\dfrac{1}{2}X^2 - 2$

$Y = -\dfrac{1}{2}X^2 + 0X - 2$

$X = \dfrac{-B}{2A} = \dfrac{-(0)}{2\left(-\dfrac{1}{2}\right)} = 0$

$Y = -\dfrac{1}{2}(0)^2 - 2 = 0 - 2 = -2$

vertex $= (0, -2)$; see graph

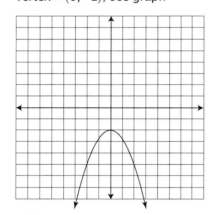

12. $3X^2 - Y = -1$

$-Y = -3X^2 - 1$

$Y = 3X^2 + 1$

$Y = 3X^2 + 0X + 1$

$X = \dfrac{-B}{2A} = \dfrac{-(0)}{2(3)} = 0$

$Y = 3(0)^2 + 1 = 0 + 1 = 1$

vertex $= (0, 1)$; see graph

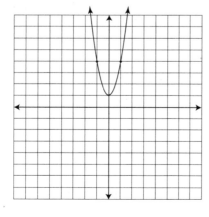

13. $(X - 2)^2 + (Y - 3)^2 = 6^2$

center $= (2, 3)$; radius $= 6$

14. center $= (-1, -1)$; radius $= 7$

$(X - (-1))^2 + (Y - (-1))^2 = 7^2$

$(X + 1)^2 + (Y + 1)^2 = 49$

15-16. $X^2 + Y^2 = 6Y - 5$

$X^2 + Y^2 - 6Y = -5$

$X^2 + (Y^2 - 6Y + 9) = -5 + 9$

$(X - 0)^2 + (Y - 3)^2 = 2^2$

center $= (0, 3)$ radius $= 2$

17. $(AC)^2 = \left[(-6) - (2)\right]^2 + \left[(-4) - (-6)\right]^2$

$(AC)^2 = [-8]^2 + [2]^2 = 64 + 4 = 68$

$AC = \sqrt{68} = \sqrt{4}\sqrt{17} = 2\sqrt{17}$

18. midpoint $= \left(\dfrac{(-2) + (2)}{2}, \dfrac{(-4) + (-6)}{2}\right) =$

$\left(\dfrac{0}{2}, \dfrac{-10}{2}\right) = (0, -5)$

19. $\dfrac{(X - 1)^2}{20} + \dfrac{(Y + 2)^2}{25} = 1$

$(100)\dfrac{(X - 1)^2}{20} + (100)\dfrac{(Y + 2)^2}{25} = (100)1$

$5(X - 1)^2 + 4(Y + 2)^2 = 100$

center $= (1, -2)$

$5((1) - 1)^2 + 4(Y + 2)^2 = 100$

$4(Y + 2)^2 = 100$

$(Y + 2)^2 = 25$

$Y + 2 = \pm 5$

$Y + 2 = 5 \qquad Y + 2 = -5$

$Y = 3$ and $\qquad Y = -7$ when $X = 1$

$5(X - 1)^2 + 4((-2) + 2)^2 = 100$

$5(X - 1)^2 = 100$

$(X - 1)^2 = 20$

$X - 1 = \pm\sqrt{20}$

$X - 1 = \pm\sqrt{4}\sqrt{5} = \pm 2\sqrt{5}$

continued on the next page

$$X - 1 = 2\sqrt{5}$$
$$X = 1 + 2\sqrt{5} \approx 5.5 \text{ when } Y = (-2)$$
$$X - 1 = -2\sqrt{5}$$
$$X = -1 - 2\sqrt{5} \approx -3.5 \text{ when } Y = (-2)$$

20. see graph

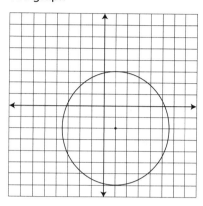

Lesson Practice 26A
Most decimal amounts are approximate.
In some cases, it seems better to choose
a value for Y first and then solve for X.

1. $XY = 8$

X	Y
1	8
2	4
8	1
−1	−8
−2	−4
−4	−2
−8	−1

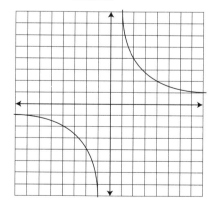

2. $XY - 12 = 0$

$XY = 12$

X	Y
2	6
3	4
6	2
−2	−6
−3	−4
−6	−2

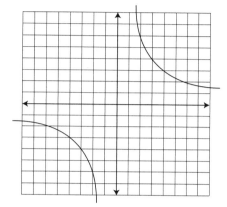

3. $-XY = -5$

$XY = 5$

X	Y
1	5
2	$\frac{5}{2}$
4	$\frac{5}{4}$
5	1
−1	−5
−2	$-\frac{5}{2}$
−4	$-\frac{5}{4}$
−5	−1

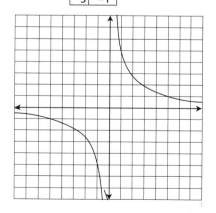

4. $X^2 - 5Y^2 = 25$

X	Y
±5	0
±5.5	±1
±6.7	±2

(We chose Y values and solved for X.)

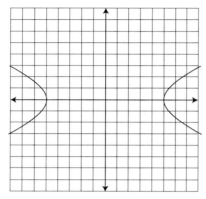

5. $3X^2 - Y^2 = 6$

X	Y
±1.4	0
±1.8	±2
±2.7	±4
±3.7	±6

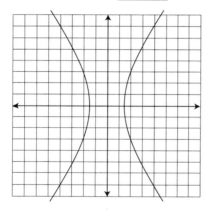

6. $5X^2 - 25 = Y^2$

X	Y
±2.2	0
±2.6	±3
±3.2	±5
±4.2	±8

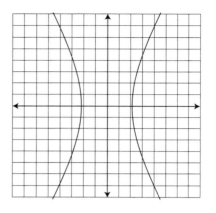

Lesson Practice 26B

1. $Y = \dfrac{-3}{X}$

$XY = -3$

X	Y
−3	1
−1	3
1	−3
3	−1

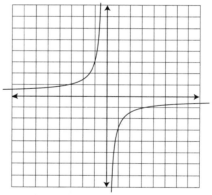

2. $0 = 4 - XY$

$-4 = -XY$

$(-1)(-4) = (-1)(-XY)$

$4 = XY$

X	Y
1	4
2	2
4	1
−1	−4
−2	−2
−4	−1

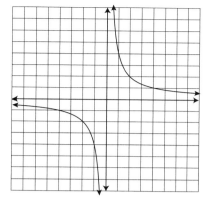

3. $-3XY = 18$

$XY = -6$

X	Y
1	−6
2	−3
6	−1
−1	6
−2	3
−6	1

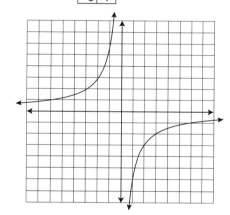

4. $\dfrac{1}{10}X^2 - \dfrac{1}{5}Y^2 = 1$

$(10)\dfrac{1}{10}X^2 - (10)\dfrac{1}{5}Y^2 = (10)(1)$

$X^2 - 2Y^2 = 10$

X	Y
±3.2	0
±4.2	±2
±6.5	±4

5. $2Y^2 - 3X^2 = 8$

X	Y
0	±2
±2	±3.2
±4	±5.3

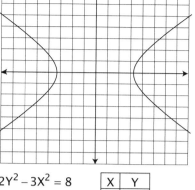

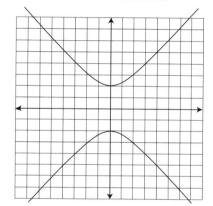

6. $Y^2 - 2X^2 = 16$

X	Y
0	±4
±2	±4.9
±4	±6.9

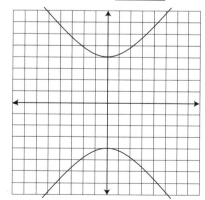

Systematic Review 26C

1. $XY - 4 = 0$

$XY = 4$

X	Y
1	4
2	2
4	1
−1	−4
−2	−2
−4	−1

2. see graph

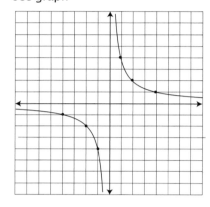

3. $Y = \dfrac{-6}{X}$

$XY = -6$

X	Y
1	−6
2	−3
3	−2
−1	6
−2	3
−3	2

4. see graph

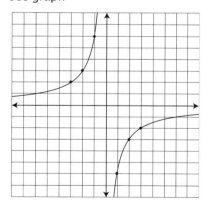

5. $2X^2 = 2Y^2 + 6$

$2X^2 - 2Y^2 = 6$

$X^2 - Y^2 = 3$

X	Y
±1.7	0
±2	1
±2.6	±2
±4.4	±4

6. see graph

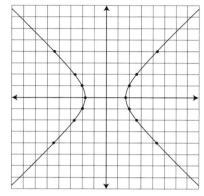

7. $Y = 2X^2 - 3X + 2$

$X = \dfrac{-B}{2A} = \dfrac{-(-3)}{2(2)} = \dfrac{3}{4}$

8. $Y = 2X^2 - 3X + 2$

$Y = 2\left(\dfrac{3}{4}\right)^2 - 3\left(\dfrac{3}{4}\right) + 2$

$Y = 2\left(\dfrac{9}{16}\right) - \dfrac{9}{4} + 2$

$Y = \dfrac{9}{8} - \dfrac{18}{8} + \dfrac{16}{8} = \dfrac{7}{8}$

$vertex = \left(\dfrac{3}{4}, \dfrac{7}{8}\right)$

9. see graph

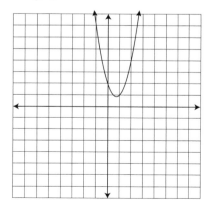

10. $\dfrac{1}{3}Y = \dfrac{2}{3}X^2 + \dfrac{1}{3}$

$Y = 2X^2 + 1$

see graph

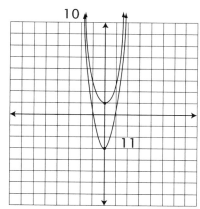

11. $2Y - 6X^2 = -6$

$2Y = 6X^2 - 6$

$Y = 3X^2 - 3$

see graph

12. $5(X-1)^2 + 5(Y+2)^2 = 500$

$(X-1)^2 + (Y+2)^2 = 100$

center $= (1, -2)$; radius $= \sqrt{100} = 10$

13. center $= (5, -2)$; radius $= 3$

$(X-(5))^2 + (Y-(-2))^2 = 3^2$

$(X-5)^2 + (Y+2)^2 = 9$

14-15. $9X^2 + 9Y^2 - 36X - 36Y = 252$

$X^2 + Y^2 - 4X - 4Y = 28$

$X^2 - 4X + Y^2 - 4Y = 28$

$X^2 - 4X + 4 + Y^2 - 4Y + 4 = 28 + 4 + 4$

$(X-2)^2 + (Y-2)^2 = 36$

center $= (2, 2)$; radius $= \sqrt{36} = 6$

16. $(AC)^2 = \left[(-2)-(-2)\right]^2 + \left[(-2)-(5)\right]^2$

$(AC)^2 = [0]^2 + [-7]^2 = 0 + 49 = 49$

$AC = \sqrt{49} = 7$

17. midpoint $= \left(\dfrac{(-2)+(-2)}{2}, \dfrac{(5)+(-2)}{2}\right) =$

$\left(\dfrac{-4}{2}, \dfrac{3}{2}\right) = \left(-2, 1\dfrac{1}{2}\right)$

18. $Y = X + 3$; $m = 1$

$Y = mX + b \Rightarrow \quad (3) = (1)(-2) + b$

$3 = -2 + b$

$5 = b$

$Y = X + 5$

19. $\dfrac{(X-1)^2}{9} + \dfrac{(Y+1)^2}{1} = 1$

$(9)\dfrac{(X-1)^2}{9} + (9)\dfrac{(Y+1)^2}{1} = (9)(1)$

$(X-1)^2 + 9(Y+1)^2 = 9$

center $= (1, -1)$

$((1)-1)^2 + 9(Y+1)^2 = 9$

$9(Y+1)^2 = 9$

$(Y+1)^2 = 1$

$Y + 1 = \pm\sqrt{1} = \pm 1$

$Y + 1 = 1 \qquad Y + 1 = -1$

$Y = 0$ and $\qquad Y = -2$ when $X = 1$

$(X-1)^2 + 9((-1)+1)^2 = 9$

$(X-1)^2 = 9$

$X - 1 = \pm 3$

$X - 1 = 3 \qquad X - 1 = -3$

$X = 4$ and $\qquad X = -2$ when $Y = (-1)$

20. see graph

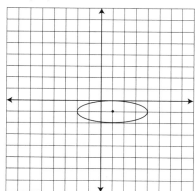

Systematic Review 26D

1. $X = \dfrac{-10}{Y}$

$XY = -10$

X	Y
1	−10
2	−5
5	−2
−1	10
−2	5
−5	2

2. see graph

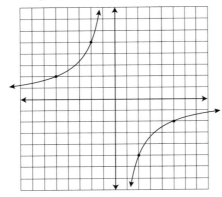

3. $Y = \dfrac{8}{X}$

$XY = 8$

X	Y
1	8
2	4
4	2
8	1
−2	−4
−4	−2

4. see graph

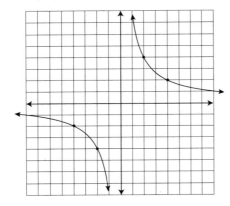

5. $-Y^2 = -X^2 + 2$

$X^2 - Y^2 = 2$

X	Y
±1.4	0
±1.7	±1
±3.3	±3
±5.2	±5

6. see graph

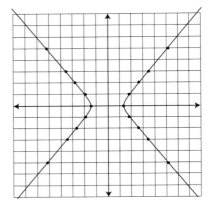

7. $Y = -\dfrac{3}{4}X^2 + 2X$

$X = \dfrac{-B}{2A} = \dfrac{-(2)}{2\left(-\dfrac{3}{4}\right)} = \dfrac{-2}{-\dfrac{3}{2}} =$

$(-2)\left(-\dfrac{2}{3}\right) = \dfrac{4}{3}$

8. $Y = -\dfrac{3}{4}X^2 + 2X \Rightarrow Y = -\dfrac{3}{4}\left(\dfrac{4}{3}\right)^2 + 2\left(\dfrac{4}{3}\right)$

$Y = -\dfrac{3}{4}\left(\dfrac{16}{9}\right) + \dfrac{8}{3}$

$Y = -\dfrac{48}{36} + \dfrac{8}{3}$

$Y = -\dfrac{4}{3} + \dfrac{8}{3} = \dfrac{4}{3}$

vertex $= \left(\dfrac{4}{3}, \dfrac{4}{3}\right)$

9. see graph

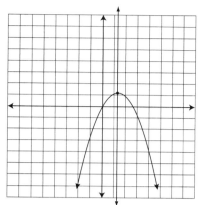

10. $8X - 4Y = 4X^2 + 1$

$-4Y = 4X^2 - 8X + 1$

$Y = -X^2 + 2X - \dfrac{1}{4}$

$X = \dfrac{-B}{2A} = \dfrac{-(2)}{2(-1)} = \dfrac{-2}{-2} = 1$

$Y = -X^2 + 2X - \dfrac{1}{4} \Rightarrow Y = -(1)^2 + 2(1) - \dfrac{1}{4}$

$Y = -1 + 2 - \dfrac{1}{4}$

$Y = 1 - \dfrac{1}{4} = \dfrac{4}{4} - \dfrac{1}{4} = \dfrac{3}{4}$

vertex $= \left(1, \dfrac{3}{4}\right)$

see graph

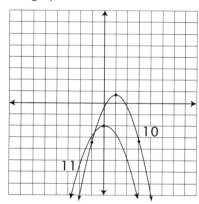

11. $X^2 = -Y - 2$

$Y = -X^2 - 2$; vertex $= (0, -2)$

see graph

12. $\dfrac{2}{3}(X+2)^2 + \dfrac{2}{3}(Y+2)^2 = 54$

$(X+2)^2 + (Y+2)^2 = \left(\dfrac{3}{2}\right)54$

$(X+2)^2 + (Y+2)^2 = 81$

center $= (-2, -2)$; radius $= \sqrt{81} = 9$

13. center $= (4, 3)$; radius $= 8$

$(X - (4))^2 + (Y - (3))^2 = 8^2$

$(X - 4)^2 + (Y - 3)^2 = 64$

14-15. $X^2 + Y^2 - 2Y - 3 = 0$

$X^2 + Y^2 - 2Y = 3$

$X^2 + Y^2 - 2Y + 1 = 3 + 1$

$X^2 + (Y - 1)^2 = 4$

center $= (0, 1)$; radius $= \sqrt{4} = 2$

16. $(BC)^2 = \left[(-2) - (3)\right]^2 + \left[(-2) - (1)\right]^2$

$(BC)^2 = [-5]^2 + [-3]^2 = 25 + 9 = 34$

$BC = \sqrt{34}$

17. midpoint $= \left(\dfrac{(-2) + (3)}{2}, \dfrac{(-2) + (1)}{2}\right) =$

$\left(\dfrac{1}{2}, \dfrac{-1}{2}\right) = \left(\dfrac{1}{2}, -\dfrac{1}{2}\right)$

18. $2Y + 5X = 3$

$2Y = -5X + 3$

$Y = -\dfrac{5}{2}X + \dfrac{3}{2}$; m of perpendicular $= \dfrac{2}{5}$

$Y = mX + b \Rightarrow (2) = \left(\dfrac{2}{5}\right)(1) + b$

$2 = \dfrac{2}{5} + b$

$2 - \dfrac{2}{5} = b$

$\dfrac{10}{5} - \dfrac{2}{5} = b$

$\dfrac{8}{5} = b = 1\dfrac{3}{5}$

$Y = \dfrac{2}{5}X + 1\dfrac{3}{5}$

19. $4(X-2)^2 + 16(Y+1)^2 = 64$

$(X-2)^2 + 4(Y+1)^2 = 16$

center = $(2, -1)$

$((2)-2)^2 + 4(Y+1)^2 = 16$

$4(Y+1)^2 = 16$

$(Y+1)^2 = 4$

$Y+1 = \pm\sqrt{4} = \pm 2$

$Y+1 = 2 \qquad Y+1 = -2$
$Y = 1 \qquad\quad Y = -3$

$(X-2)^2 + 4((-1)+1)^2 = 16$

$(X-2)^2 = 16$

$X-2 = \pm\sqrt{16} = \pm 4$

$X-2 = 4 \qquad X-2 = -4$
$X = 6 \qquad\quad X = -2$

extremities:
$(2, 1), (2, -3), (6, -1), (-2, -1)$

20. see graph

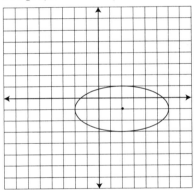

Systematic Review 26E

1. $XY = -12$

X	Y
2	-6
3	-4
6	-2
-2	6
-3	4
-6	2

2. see graph

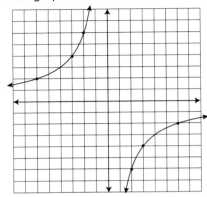

3. $-XY = -3$

$XY = 3$

X	Y
$\frac{1}{2}$	6
1	3
3	1
$-\frac{1}{2}$	-6
-1	-3
-3	-1

4. see graph

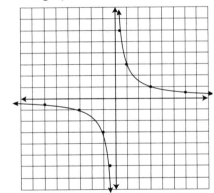

5. $9X^2 = 6Y^2 + 18$

$9X^2 - 6Y^2 = 18$

$3X^2 - 2Y^2 = 6$

X	Y
±1.4	0
±2.2	±2
±3.6	±4
±6.7	±8

6. see graph

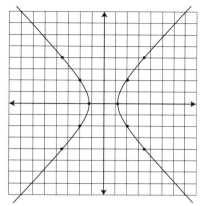

7. $Y = -X^2 + 4X - 4$

$X = \dfrac{-B}{2A} = \dfrac{-(4)}{2(-1)} = \dfrac{-4}{-2} = 2$

8. $Y = -X^2 + 4X - 4$

$Y = -(2)^2 + 4(2) - 4$

$Y = -4 + 8 - 4$

$Y = 0$; vertex is $(2, 0)$

9. see graph

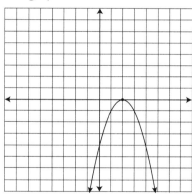

10. $3Y = -\dfrac{3}{2}X^2 + 3$

$Y = -\dfrac{1}{2}X^2 + 1$

see graph

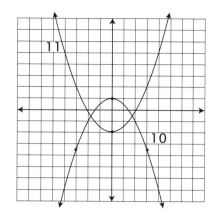

11. $\dfrac{1}{2} + \dfrac{1}{4}Y = \dfrac{1}{8}X^2$

$(4)\left(\dfrac{1}{2}\right) + (4)\left(\dfrac{1}{4}\right)Y = (4)\left(\dfrac{1}{8}\right)X^2$

$2 + Y = \dfrac{1}{2}X^2$

$Y = \dfrac{1}{2}X^2 - 2$

see graph

12. $\dfrac{1}{2}(X + 2)^2 + \dfrac{1}{2}(Y - 3)^2 = 32$

$(X + 2)^2 + (Y - 3)^2 = 64$

center $= (-2, 3)$; radius $= \sqrt{64} = 8$

13. center $= (-2, -2)$; radius $= 5$

$\left(X - (-2)\right)^2 + \left(Y - (-2)\right)^2 = 5^2$

$(X + 2)^2 + (Y + 2)^2 = 25$

14-15. $4X^2 - 32X + 64 + 4Y^2 - 24Y = 0$

$X^2 - 8X + 16 + Y^2 - 6Y = 0$

$X^2 - 8X + 16 + Y^2 - 6Y + 9 = 0 + 9$

$(X - 4)^2 + (Y - 3)^2 = 9$

center $= (4, 3)$; radius $= \sqrt{9} = 3$

16. $(CD)^2 = \left[(-2) - (4)\right]^2 + \left[(-2) - (-4)\right]^2$

$(CD)^2 = [-6]^2 + [2]^2 = 36 + 4 = 40$

$CD = \sqrt{40} = \sqrt{4}\sqrt{10} = 2\sqrt{10}$

17. midpoint $= \left(\dfrac{(-2) + (4)}{2}, \dfrac{(-2) + (-4)}{2}\right) =$

$\left(\dfrac{2}{2}, \dfrac{-6}{2}\right) = (1, -3)$

18. $3Y = X - 9$

$Y = \frac{1}{3}X - 3; \; m = \frac{1}{3}$

$Y = mX + b \Rightarrow \qquad (1) = \left(\frac{1}{3}\right)(4) + b$

$\qquad\qquad\qquad\qquad 1 = \frac{4}{3} + b$

$\qquad\qquad\qquad 1 - \frac{4}{3} = b$

$\qquad\qquad\qquad \frac{3}{3} - \frac{4}{3} = b$

$\qquad\qquad\qquad -\frac{1}{3} = b$

$Y = \frac{1}{3}X - \frac{1}{3}$

19. $\dfrac{(X+1)^2}{4} + \dfrac{(Y-2)^2}{9} = 1$

$(36)\dfrac{(X+1)^2}{4} + (36)\dfrac{(Y-2)^2}{9} = (36)(1)$

$9(X+1)^2 + 4(Y-2)^2 = 36$

center $= (-1, 2)$

$9\big((-1)+1\big)^2 + 4(Y-2)^2 = 36$

$\qquad\qquad 4(Y-2)^2 = 36$

$\qquad\qquad (Y-2)^2 = 9$

$\qquad\qquad Y - 2 = \pm\sqrt{9} = \pm 3$

$\qquad Y - 2 = 3 \qquad Y - 2 = -3$

$\qquad\quad Y = 5 \qquad\quad Y = -1$

$9(X+1)^2 + 4\big((2)-2\big)^2 = 36$

$\qquad\qquad 9(X+1)^2 = 36$

$\qquad\qquad (X+1)^2 = 4$

$\qquad\qquad X + 1 = \pm\sqrt{4} = \pm 2$

$\qquad X + 1 = 2 \qquad X + 1 = -2$

$\qquad\quad X = 1 \qquad\quad X = -3$

extremities:

$(-1, 5), (-1, -1), (1, 2), (-3, 2)$

20. see graph

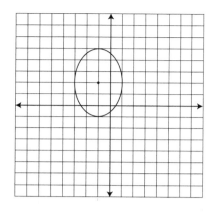

Lesson Practice 27A

1. $\begin{cases} XY = 12 \\ Y = 2X - 5 \end{cases}$

$XY = 12 \Rightarrow \qquad X(2X - 5) = 12$

$\qquad\qquad\qquad\qquad 2X^2 - 5X = 12$

$\qquad\qquad\qquad 2X^2 - 5X - 12 = 0$

$\qquad\qquad\qquad (2X + 3)(X - 4) = 0$

$\qquad\quad 2X + 3 = 0 \qquad X - 4 = 0$

$\qquad\qquad 2X = -3 \qquad\quad X = 4$

$\qquad\qquad\quad X = -\frac{3}{2}$

$XY = 12 \Rightarrow \left(-\frac{3}{2}\right)Y = 12$

$\qquad\qquad\qquad Y = 12\left(-\frac{2}{3}\right) = -8$

$XY = 12 \Rightarrow (4)Y = 12$

$\qquad\qquad\qquad Y = \frac{12}{4} = 3$

solutions: $\left(-\frac{3}{2}, -8\right), (4, 3)$

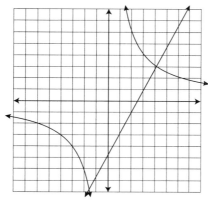

2. $\begin{cases} (X-4)^2 + (Y-4)^2 = 4 \\ 5X = 20 \Rightarrow X = 4 \end{cases}$

$(X-4)^2 + (Y-4)^2 = 4 \Rightarrow$

$((4)-4)^2 + (Y-4)^2 = 4$

$(Y-4)^2 = 4$

$Y - 4 = \pm\sqrt{4} = \pm 2$

$\begin{array}{ll} Y - 4 = 2 & Y - 4 = -2 \\ \quad Y = 6 & \quad Y = 2 \end{array}$

solutions:

$(4, 6), (4, 2)$

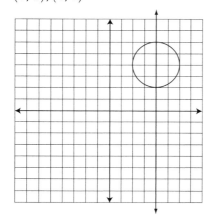

3. $\begin{cases} 4X^2 + 9Y^2 = 36 \\ Y = -X^2 - 2 \end{cases}$

$Y = -X^2 - 2$

$X^2 + Y + 2 = 0$

$-4(X^2 + Y + 2 = 0) \Rightarrow -4X^2 - 4Y - 8 = 0$

$\begin{array}{l} \quad 4X^2 + 9Y^2 \quad\quad - 36 = 0 \\ + \ \underline{\ -4X^2 \quad\quad - 4Y - \ \ 8 = 0} \\ \quad\quad\quad 9Y^2 - 4Y - 44 = 0 \end{array}$

find Y using quadratic formula:

$\dfrac{-(-4) \pm \sqrt{(-4)^2 - 4(9)(-44)}}{2(9)} =$

$\dfrac{4 \pm \sqrt{16 - (-1{,}584)}}{18} = \dfrac{4 \pm \sqrt{1{,}600}}{18} =$

$\dfrac{4 \pm 40}{18} = \dfrac{2(2 \pm 20)}{2(9)} = \dfrac{2 \pm 20}{9}$

$Y = \dfrac{2 + 20}{9} = \dfrac{22}{9} \approx 2.44$

$Y = \dfrac{2 - 20}{9} = \dfrac{-18}{9} = -2$

$Y = -X^2 - 2 \Rightarrow \ (2.44) = -X^2 - 2$

$\quad\quad\quad\quad\quad\quad 4.44 = -X^2$

$\quad\quad\quad\quad\quad -4.44 = X^2 \text{ no real solution}$

$Y = -X^2 - 2 \Rightarrow \ (-2) = -X^2 - 2$

$\quad\quad\quad\quad\quad\quad X^2 = -2 + 2$

$\quad\quad\quad\quad\quad\quad X^2 = 0$

$\quad\quad\quad\quad\quad\quad \ X = 0$

solution: $(0, -2)$

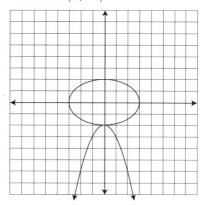

4. $\begin{cases} X^2 - 4Y^2 = 16 \Rightarrow -X^2 + 4Y^2 = -16 \\ X^2 + Y^2 = 49 \end{cases}$

$$X^2 + Y^2 = 49$$
$$+ \ \underline{-X^2 + 4Y^2 = -16}$$
$$5Y^2 = 33$$
$$Y^2 = \pm\sqrt{6.6}$$
$$Y \approx \pm 2.57$$

$$X^2 + (\pm 2.57)^2 = 49$$
$$X^2 + 6.6 = 49$$
$$X^2 = 42.4$$
$$X = \pm 6.51$$

solution:

$(6.51, 2.57), (6.51, -2.57),$
$(-6.51, -2.57), (-6.51, 2.57)$

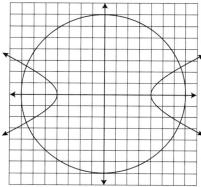

5. $\begin{cases} X = 2Y^2 - 5 \\ X = -2Y^2 - 1 \end{cases}$

$$X = 2Y^2 - 5$$
$$+ \ \underline{X = -2Y^2 - 1}$$
$$2X = 0 - 6$$

$$2X = -6$$
$$X = -3$$

$$X = -2Y^2 - 1 \Rightarrow (-3) = -2Y^2 - 1$$
$$-2 = -2Y^2$$
$$1 = Y^2$$
$$Y = \pm 1$$

solution: $(-3, 1), (-3, -1)$

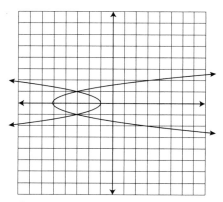

6. $\begin{cases} Y = \dfrac{1}{2}X^2 + 3 \\ Y = X + 4 \end{cases}$

$$Y = X + 4 \Rightarrow \left(\frac{1}{2}X^2 + 3\right) = X + 4$$

$$\frac{1}{2}X^2 - X - 1 = 0$$

$$X = \frac{-(-1) \pm \sqrt{(-1)^2 - 4\left(\frac{1}{2}\right)(-1)}}{2\left(\frac{1}{2}\right)} = \frac{1 \pm \sqrt{1+2}}{1} =$$

$$1 \pm \sqrt{3} \approx 1 \pm 1.73$$
$$X = 2.73, \ -.73$$

$$Y = X + 4 \Rightarrow \ Y = 2.73 + 4$$
$$Y = 6.73$$

$$Y = X + 4 \Rightarrow \ Y = -.73 + 4$$
$$Y = 3.27$$

solutions: $(2.73, 6.73), (-.73, 3.27)$

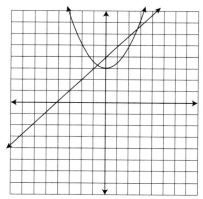

Lesson Practice 27B

1. $\begin{cases} Y = X + 4 \\ X^2 + Y^2 = 25 \end{cases}$

$X^2 + Y^2 = 25 \Rightarrow \quad X^2 + (X+4)^2 = 25$

$\qquad\qquad X^2 + X^2 + 8X + 16 = 25$

$\qquad\qquad\quad 2X^2 + 8X + 16 = 25$

$\qquad\qquad\quad 2X^2 + 8X - 9 = 0$

$X = \dfrac{-(8) \pm \sqrt{(8)^2 - 4(2)(-9)}}{2(2)} =$

$\dfrac{-8 \pm \sqrt{64 + 72}}{4} = \dfrac{-8 \pm \sqrt{136}}{4} =$

$\dfrac{-8 \pm \sqrt{4}\sqrt{34}}{4} = \dfrac{-8 \pm 2\sqrt{34}}{4} = \dfrac{-4 \pm \sqrt{34}}{2} \approx$

$\dfrac{-4 \pm 5.83}{2} \approx .92 \text{ or } -4.92$

$Y = X + 4 \Rightarrow \quad Y = (.92) + 4$

$\qquad\qquad\qquad\quad Y = 4.92$

$Y = X + 4 \Rightarrow \quad Y = (-4.92) + 4$

$\qquad\qquad\qquad\quad Y = -.92$

solutions: $(.92, 4.92), (-4.92, -.92)$

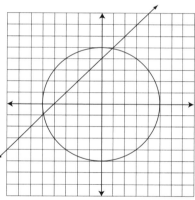

2. $\begin{cases} Y = -3 \\ Y = -2X^2 \end{cases}$

$Y = -2X^2 \Rightarrow \quad (-3) = -2X^2$

$\qquad\qquad\qquad \dfrac{-3}{-2} = X^2$

$\qquad\qquad\quad \pm\sqrt{1.5} = X^2 \approx \pm 1.22$

solutions: $(1.22, -3), (-1.22, -3)$

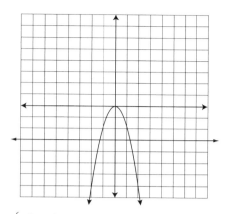

3. $\begin{cases} X^2 + Y^2 = 16 \\ X^2 + (Y-6)^2 = 4 \end{cases}$

$\qquad\qquad X^2 + (Y-6)^2 = 4$

$\qquad\quad X^2 + Y^2 - 12Y + 36 = 4$

$\qquad\quad X^2 + Y^2 - 12Y + 32 = 0$

$(-1)(X^2 + Y^2 = 16) \Rightarrow -X^2 - Y^2 = -16$

$\qquad\quad -X^2 - Y^2 \qquad\qquad = -16$

$+ \quad \underline{X^2 + Y^2 - 12Y + 32 = \quad 0}$

$\qquad\qquad\qquad -12Y + 32 = -16$

$\qquad\qquad\qquad\qquad -12Y = -48$

$\qquad\qquad\qquad\qquad\quad Y = 4$

$X^2 + Y^2 = 16 \Rightarrow \quad X^2 + (4)^2 = 16$

$\qquad\qquad\qquad\qquad X^2 + 16 = 16$

$\qquad\qquad\qquad\qquad\quad X^2 = 0$

$\qquad\qquad\qquad\qquad\quad X = 0$

solution: $(0, 4)$

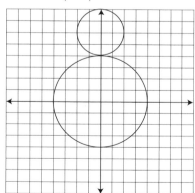

4. $\begin{cases} 4X^2 + Y^2 = 4 \\ X = Y^2 \end{cases}$

$4X^2 + Y^2 = 4 \Rightarrow \quad 4X^2 + (X) = 4$

$\qquad\qquad\qquad\qquad 4X^2 + X - 4 = 0$

$X = \dfrac{-(1) \pm \sqrt{(1)^2 - 4(4)(-4)}}{2(4)} = \dfrac{-1 \pm \sqrt{1 + 64}}{8} =$

$\dfrac{-1 \pm \sqrt{65}}{8} \approx \dfrac{-1 \pm 8.06}{8} \approx .88 \text{ or } -1.13$

$X = Y^2 \Rightarrow \quad (.88) = Y^2$

$\qquad\qquad\quad \pm\sqrt{.88} = Y \approx \pm.94$

$X = Y^2 \Rightarrow \quad (-1.13) = Y^2$

$\qquad\qquad\quad \pm\sqrt{-1.13} = Y \text{ (no real solution)}$

solution: $(.88, .94), (.88, -.94)$

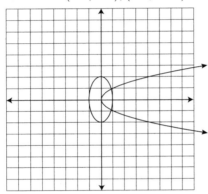

5. $\begin{cases} 4X^2 - 4Y^2 = 36 \\ Y = X - 1 \end{cases}$

$4X^2 - 4Y^2 = 36 \Rightarrow 4X^2 - 4(X-1)^2 = 36$

$\qquad\qquad 4X^2 - 4(X-1)^2 = 36$

$4X^2 - 4(X^2 - 2X + 1) = 36$

$4X^2 - 4X^2 + 8X - 4 = 36$

$\qquad\qquad\qquad 8X - 4 = 36$

$\qquad\qquad\qquad\qquad 8X = 40$

$\qquad\qquad\qquad\qquad\quad X = 5$

$Y = X - 1 \Rightarrow \quad Y = (5) - 1$

$\qquad\qquad\qquad\quad Y = 4$

solution: $(5, 4)$

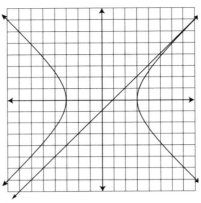

6. $\begin{cases} Y = 3X^2 \\ X^2 + Y^2 = 16 \end{cases}$

$(-3)(X^2 + Y^2 = 16) \Rightarrow -3X^2 - 3Y^2 = -48$

$Y = 3X^2 \Rightarrow 0 = 3X^2 - Y$

$\qquad -3X^2 - 3Y^2 \qquad\quad = -48$

$+ \quad \underline{3X^2 \qquad\quad - Y = \quad 0}$

$\qquad\qquad -3Y^2 - Y = -48$

$\qquad\quad -3Y^2 - Y + 48 = 0$

$Y = \dfrac{-(-1) \pm \sqrt{(-1)^2 - 4(-3)(48)}}{2(-3)} =$

$\dfrac{1 \pm \sqrt{1 - (-576)}}{-6} = \dfrac{1 \pm \sqrt{577}}{-6} \approx \dfrac{1 \pm 24.02}{-6} =$

$-4.17 \text{ or } \approx 3.84$

$Y = 3X^2 \Rightarrow \quad (-4.17) = 3X^2$

$\qquad\qquad\qquad\quad -1.39 = X^2$

$\qquad\qquad\quad \pm\sqrt{-1.39} = X \text{ (no real solution)}$

$Y = 3X^2 \Rightarrow \quad (3.84) = 3X^2$

$\qquad\qquad\qquad\quad 1.28 = X^2$

$\qquad\qquad\quad \pm\sqrt{1.28} = X \approx \pm1.13$

solutions: $(1.13, 3.84), (-1.13, 3.84)$

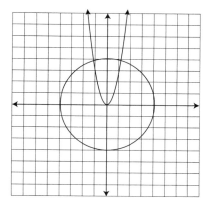

Systematic Review 27C

1. hyperbola and line
2. see graph

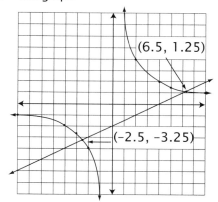

(6.5, 1.25)

(−2.5, −3.25)

3. $XY = 8 \Rightarrow \quad X\left(\frac{1}{2}X - 2\right) = 8$

$$\frac{1}{2}X^2 - 2X = 8$$

$$\frac{1}{2}X^2 - 2X - 8 = 0$$

4. $X = \dfrac{-(-2) \pm \sqrt{(-2)^2 - 4\left(\frac{1}{2}\right)(-8)}}{2\left(\frac{1}{2}\right)} =$

$\dfrac{2 \pm \sqrt{4 - (-16)}}{1} = 2 \pm \sqrt{20} = 2 \pm \sqrt{4}\sqrt{5} =$

$2 \pm 2\sqrt{5} \approx 6.48 \text{ or } -2.48$

5. $Y = \frac{1}{2}X - 2 \Rightarrow \quad Y = \frac{1}{2}(6.48) - 2$

$$Y = 3.24 - 2$$
$$Y = 1.24$$

$Y = \frac{1}{2}X - 2 \Rightarrow \quad Y = \frac{1}{2}(-2.48) - 2$

$$Y = -1.24 - 2$$
$$Y = -3.24$$

6. $(6.48, 1.24), (-2.48, -3.24)$
7. parabola and line
8. see graph

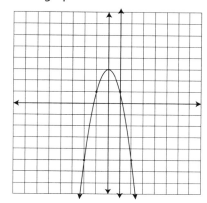

9. $X = 1$
10. $Y = -2X^2 + 3 \Rightarrow \quad Y = -2(1)^2 + 3$

$$Y = -2 + 3$$
$$Y = 1$$

11. $X = 1$
12. $(1, 1)$
13. circle and line
14. see graph

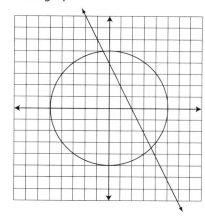

15. $75 = 3X^2 + 3Y^2$

$25 = X^2 + Y^2$

$X^2 + Y^2 = 25$

$X^2 + (-2X + 4)^2 = 25$

16.

$X^2 + 4X^2 - 16X + 16 = 25$

$5X^2 - 16X - 9 = 0$

$X = \dfrac{-(-16) \pm \sqrt{(-16)^2 - 4(5)(-9)}}{2(5)} =$

$\dfrac{16 \pm \sqrt{256 - (-180)}}{10} = \dfrac{16 \pm \sqrt{436}}{10} \approx$

$\dfrac{16 \pm 20.88}{10} \approx 3.69 \text{ or } -.49$

17. $Y = -2X + 4 \Rightarrow \quad Y = -2(3.69) + 4$

$Y = -7.38 + 4$

$Y = -3.38$

$Y = -2X + 4 \Rightarrow \quad Y = -2(-.49) + 4$

$Y = .98 + 4$

$Y = 4.98$

18. $(3.69, -3.38), (-.49, 4.98)$

19. $(AB)^2 = \left[(-5) - (0)\right]^2 + \left[(0) - (3)\right]^2$

$(AB)^2 = [-5]^2 + [-3]^2 = 25 + 9 = 34$

$AB = \sqrt{34}$

20. $\text{midpoint} = \left(\dfrac{(-5) + (3)}{2}, \dfrac{(0) + (-2)}{2} \right) =$

$\left(\dfrac{-2}{2}, \dfrac{-2}{2} \right) = (-1, -1)$

Systematic Review 27D
Don't forget that decimal amounts may be approximate.
In some cases, we chose a value for Y first and then solved for X.

1. hyperbola and circle

2. hyperbola:

X	Y
±4.2	0
≈ ±6	≈ ±4
≈ ±9	≈ ±8

circle: $X^2 + Y^2 = 32$

center $= (0, 0)$, radius $= \sqrt{32} \approx 5.7$

see graph

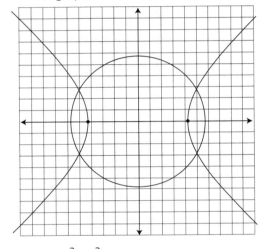

$X^2 - Y^2 = 18$

3. $+ \quad \dfrac{X^2 + Y^2 = 32}{2X^2 \qquad = 50}$

4. $X^2 = 25$

$X = \pm 5$

5. $X^2 + Y^2 = 32 \Rightarrow (\pm 5)^2 + Y^2 = 32$

$25 + Y^2 = 32$

$Y^2 = 7$

$Y = \pm\sqrt{7} \approx \pm 2.65$

6. $(5, 2.65), (5, -2.65), (-5, 2.65), (-5, -2.65)$

7. parabola and line

8. parabola:

axis of symmetry $= \dfrac{-B}{2A} = \dfrac{-(2)}{2(-1)} = 1$

$Y = -X^2 + 2X \Rightarrow Y = -(1)^2 + 2(1)$
$\qquad\qquad\qquad\quad Y = -1 + 2$
$\qquad\qquad\qquad\quad Y = 1$

vertex: $(1, 1)$
see graph

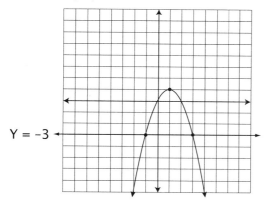

$Y = -3$

9. $Y = -3$

10. $Y = -X^2 + 2X \Rightarrow \qquad (-3) = -X^2 + 2X$
$\qquad\qquad\qquad\qquad\qquad\quad X^2 - 2X - 3 = 0$

$(X + 1)(X - 3) = 0$

$X + 1 = 0 \qquad\qquad\qquad X - 3 = 0$
$\qquad X = -1 \qquad\qquad\qquad\quad X = 3$

11. $Y = -3$

12. $(3, -3), (-1, -3)$

13. hyperbola and line

14. $3Y = -X + 3 \Rightarrow Y = -\dfrac{1}{3}X + 1$

$X^2 - Y^2 = 4 \Rightarrow X^2 - (0)^2 = 4$
$\qquad\qquad\qquad\qquad\quad X^2 = 4$
$\qquad\qquad\qquad\qquad\quad X = \pm 2$

X	Y
±2	0
±3	±2.24
±4	±3.46
±5	±4.58

14. graph

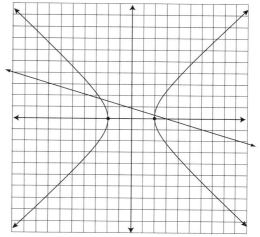

15. $3Y = -X + 3$

$X = -3Y + 3$

$X^2 - Y^2 = 4 \Rightarrow \qquad (-3Y + 3)^2 - Y^2 = 4$
$\qquad\qquad\qquad\qquad\qquad 9Y^2 - 18Y + 9 - Y^2 = 4$
$\qquad\qquad\qquad\qquad\qquad\quad 8Y^2 - 18Y + 5 = 0$

16. $Y = \dfrac{-(-18) \pm \sqrt{(-18)^2 - 4(8)(5)}}{2(8)} =$

$\dfrac{18 \pm \sqrt{324 - 160}}{16} = \dfrac{18 \pm \sqrt{164}}{16} \approx$

$\dfrac{18 \pm 12.81}{16} \approx 1.93 \text{ or } .32$

17. $X = -3Y + 3 \Rightarrow X = -3(1.93) + 3$
$\qquad\qquad\qquad\qquad\quad X = -5.79 + 3$
$\qquad\qquad\qquad\qquad\quad X = -2.79$

$X = -3Y + 3 \Rightarrow X = -3(.32) + 3$
$\qquad\qquad\qquad\qquad\quad X = -.96 + 3$
$\qquad\qquad\qquad\qquad\quad X = 2.04$

18. $(-2.79, 1.93), (2.04, .32)$

19. $(BD)^2 = \left[(-2) - (0)\right]^2 + \left[(-4) - (3)\right]^2$

$(BD)^2 = [-2]^2 + [-7]^2 = 4 + 49 = 53$

$BD = \sqrt{53}$

20. midpoint $= \left(\dfrac{(-2) + (0)}{2}, \dfrac{(-4) + (3)}{2}\right) =$

$\left(\dfrac{-2}{2}, \dfrac{-1}{2}\right) = \left(-1, -\dfrac{1}{2}\right)$

Systematic Review 27E

1. hyperbola and line

X	Y
$\frac{1}{2}$	6
1	3
3	1
6	$\frac{1}{2}$

2. see graph

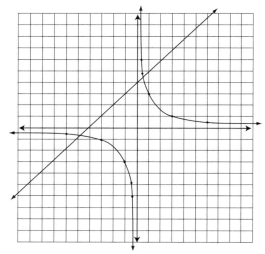

3. $XY = 3 \Rightarrow \quad X(X+4) = 3$

$\qquad\qquad\qquad X^2 + 4X = 3$

$\qquad\qquad\qquad X^2 + 4X - 3 = 0$

4. $X = \dfrac{-(4) \pm \sqrt{(4)^2 - 4(1)(-3)}}{2(1)} =$

$\dfrac{-4 \pm \sqrt{16+12}}{2} = \dfrac{-4 \pm \sqrt{28}}{2} =$

$\dfrac{-4 \pm \sqrt{4}\sqrt{7}}{2} = \dfrac{-4 \pm 2\sqrt{7}}{2} =$

$-2 \pm \sqrt{7} \approx -2 \pm 2.65$

$X = .65 \text{ or } -4.65$

5. $Y = X + 4 \Rightarrow \quad Y = (.65) + 4$

$\qquad\qquad\qquad Y = 4.65$

$Y = X + 4 \Rightarrow \quad Y = (-4.65) + 4$

$\qquad\qquad\qquad Y = -.65$

6. $(.65, 4.65), (-4.65, -.65)$

7. two parabolas

8. see graph

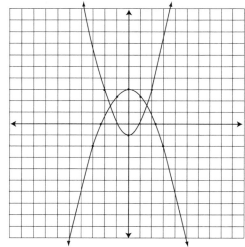

9. $Y = X^2 - 1$

$Y - X^2 = -1 \Rightarrow \qquad -Y \quad + X^2 = 1$

$\qquad\qquad\qquad + \quad Y + \dfrac{1}{2}X^2 = 3$

$\qquad\qquad\qquad\qquad\overline{\qquad \dfrac{3}{2}X^2 = 4}$

10. $X^2 = 4\left(\dfrac{2}{3}\right)$

$X^2 = \dfrac{8}{3}$

$X = \pm\sqrt{\dfrac{8}{3}} = \pm\dfrac{\sqrt{8}}{\sqrt{3}} = \pm\dfrac{2\sqrt{2}}{\sqrt{3}}\dfrac{\sqrt{3}}{\sqrt{3}} = \pm\dfrac{2\sqrt{6}}{3} \approx \pm1.63$

11. $Y = X^2 - 1 \Rightarrow \quad Y = (\pm1.63)^2 - 1$

$\qquad\qquad\qquad Y = 2.6569 - 1 \approx 1.66$

12. $(1.63, 1.66), (-1.63, 1.66)$

13. two circles

14. $\frac{1}{4}X^2 + \frac{1}{4}Y^2 = 4 \Rightarrow X^2 + Y^2 = 16$

center: $(0, 0)$; radius: $\sqrt{16} = 4$

$X^2 + (Y+6)^2 = 9$

center: $(0, -6)$; radius: $\sqrt{9} = 3$

see graph

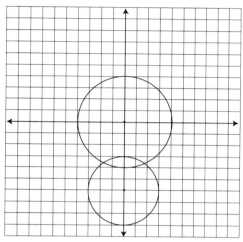

15. $X^2 + Y^2 = 16 \Rightarrow \quad -X^2 \qquad -Y^2 = -16$

$\underline{+ \quad X^2 + (Y+6)^2 \qquad = \quad 9}$

$\qquad\qquad (Y+6)^2 - Y^2 = -7$

16. $\qquad (Y+6)^2 - Y^2 = -7$

$Y^2 + 12Y + 36 - Y^2 = -7$

$12Y + 36 = -7$

$12Y = -43$

$Y = -\frac{43}{12} \approx -3.58$

17. $X^2 + (Y+6)^2 = 9$

$X^2 + ((-3.58) + 6)^2 = 9$

$X^2 + (2.42)^2 = 9$

$X^2 + 5.8564 = 9$

$X^2 = 9 - 5.8564$

$X^2 = 3.1436$

$X = \pm\sqrt{3.1436} \approx \pm 1.77$

18. $(1.77, -3.58), (-1.77, -3.58)$

19. $(AC)^2 = \left[(-5) - (3)\right]^2 + \left[(0) - (-2)\right]^2$

$(AC)^2 = [-8]^2 + [2]^2 = 64 + 4 = 68$

$AC = \sqrt{68} = \sqrt{4}\sqrt{17} = 2\sqrt{17}$

20. $\text{midpoint} = \left(\frac{(-5)+(3)}{2}, \frac{(0)+(-2)}{2}\right) =$

$\left(\frac{-2}{2}, \frac{-2}{2}\right) = (-1, -1)$

Lesson Practice 28A

1. $\qquad -5(N+D=7) \Rightarrow \quad -5N \quad -5D = -35$

$100(.05N + .10D = .45) \Rightarrow \quad \underline{5N + 10D = \quad 45}$

$\qquad\qquad\qquad\qquad\qquad 5D = \quad 10$

$\qquad\qquad\qquad\qquad\qquad D = \quad 2$

$N + D = 7 \qquad N + (2) = 7$

$\qquad\qquad\qquad N = 5$

2. $\qquad -1(P+N=24) \Rightarrow \quad -P \quad -N = -24$

$100(.01P + .05N = .56) \Rightarrow \quad \underline{P + 5N = \quad 56}$

$\qquad\qquad\qquad\qquad\qquad 4N = \quad 32$

$\qquad\qquad\qquad\qquad\qquad N = \quad 8$

$P + N = 24 \Rightarrow P + (8) = 24$

$\qquad\qquad\qquad P = 16$

3.

$\qquad -10(Q+D=11) \Rightarrow \quad -10Q - 10D = -110$

$100(.25Q + .10D = 2.15) \Rightarrow \quad \underline{25Q + 10D = \quad 215}$

$\qquad\qquad\qquad\qquad\qquad\qquad 15Q = \quad 105$

$\qquad\qquad\qquad\qquad\qquad\qquad Q = \quad 7$

$Q + D = 11 \Rightarrow (7) + D = 11$

$\qquad\qquad\qquad\qquad D = 4$

4. $2N + 3(N+1) - (N+2) = 21$

$2N + 3N + 3 - N - 2 = 21$

$4N + 1 = 21$

$4N = 20$

$N = 5$

integers = 5, 6, 7

5. $8N - 2(N+4) = 10 + (N+2)$

$8N - 2N - 8 = 10 + N + 2$

$8N - 2N - N = 10 + 2 + 8$

$5N = 20$

$N = 4$

integers = 4, 6, 8

6. $3N + 4(N + 2) = -13(N + 4)$

$3N + 4N + 8 = -13N - 52$

$3N + 4N + 13N = -52 - 8$

$20N = -60$

$N = -3$

integers $= -3, -1, 1$

7. $-80(M_1 + M_2 = 10) \Rightarrow \quad -80M_1 - 80M_2 = -800$

$100(.80M_1 + .90M_2 = .85(10)) \Rightarrow \quad \underline{80M_1 + 90M_2 = 850}$

$10M_2 = 50$

$M_2 = 5 \text{ lb}$

$M_1 + M_2 = 10 \Rightarrow \quad M_1 + (5) = 10$

$M_1 = 5 \text{ lb}$

8. $-1(S_1 + S_2 = 65) \Rightarrow \quad -S_1 - S_2 = -65$

$100(.10S_1 + .01S_2 = .09(65)) \Rightarrow \quad \underline{10S_1 + S_2 = 585}$

$9S_1 = 520$

$S_1 \approx 57.78 \text{ ml}$

$S_1 + S_2 = 65 \Rightarrow \quad 57.78 + S_2 = 65$

$S_2 = 7.22 \text{ ml}$

Lesson Practice 28B

1. $-5(N + D = 15) \Rightarrow \quad -5N - 5D = -75$

$100(.05N + .10D = 1.15) \Rightarrow \quad \underline{5N + 10D = 115}$

$5D = 40$

$D = 8$

$N + D = 15 \Rightarrow \quad N + (8) = 15$

$N = 7$

2. $-5(Q + N = 30) \Rightarrow \quad -5Q - 5N = -150$

$100(.25Q + .05N = 4.30) \Rightarrow \quad \underline{25Q + 5N = 430}$

$20Q = 280$

$Q = 14$

$Q + N = 30 \Rightarrow \quad (14) + N = 30$

$N = 16$

3. $-1(D + P = 14) \Rightarrow \quad -D - P = -14$

$100(.10D + .01P = .68) \Rightarrow \quad \underline{10D + P = 68}$

$9D = 54$

$D = 6$

$D + P = 14 \Rightarrow \quad (6) + P = 14$

$P = 8$

4. $4N + 8(N + 1) + 64 = 4(N + 2)$

$4N + 8N + 8 + 64 = 4N + 8$

$4N + 8N - 4N = 8 - 8 - 64$

$8N = -64$

$N = -8$

integers $= -8, -7, -6$

5. $2(N + 2) - N + 26 = 3(N + 4)$

$2N + 4 - N + 26 = 3N + 12$

$2N - N - 3N = 12 - 4 - 26$

$-2N = -18$

$N = 9$

integers $= 9, 11, 13$

6. $3N + 6(N + 2) = 8(N + 4) - 14$

$3N + 6N + 12 = 8N + 32 - 14$

$3N + 6N - 8N = 32 - 14 - 12$

$N = 6$

integers $= 6, 8, 10$

7.

$$-45\left(M_1 + M_2 = 60\right) \Rightarrow \quad -45M_1 - 45M_2 = -2{,}700$$

$$100\left(.65M_1 + .45M_2 = .50(60)\right) \Rightarrow \quad \underline{\quad 65M_1 + 45M_2 = 3{,}000\quad}$$

$$20M_1 = 300$$

$$M_1 = 15 \text{ lb}$$

$$M_1 + M_2 = 60 \Rightarrow (15) + M_2 = 60$$

$$M_2 = 45 \text{ lbs}$$

8.

$$-5\left(S_1 + S_2 = 80\right) \Rightarrow \quad -5S_1 - 5S_2 = -400$$

$$100\left(.15S_1 + .05S_2 = .07(80)\right) \Rightarrow \quad \underline{\quad 15S_1 + 5S_2 = 560\quad}$$

$$10S_1 = 160$$

$$S_1 = 16 \text{ liters}$$

$$S_1 + S_2 = 80 \Rightarrow (16) + S_2 = 80$$

$$S_2 = 64 \text{ liters}$$

Systematic Review 28C

1.

$$-1\left(N + P = 27\right) \Rightarrow \quad -N - P = -27$$

$$100\left(.05N + .01P = .75\right) \Rightarrow \quad \underline{\quad 5N + P = 75\quad}$$

$$4N = 48$$

$$N = 12$$

2. $N + P = 27 \Rightarrow (12) + P = 27$

$$P = 15$$

3. $4(N + 2) - 5(N) - (N + 1) = 5$

$$4N + 8 - 5N - N - 1 = 5$$

$$4N - 5N - N = 5 - 8 + 1$$

$$-2N = -2$$

$$N = 1$$

4. 1, 2, 3

5. $5(N + 4) - 6(N) = (N + 2) - 14$

$$5N + 20 - 6N = N + 2 - 14$$

$$5N - 6N - N = 2 - 14 - 20$$

$$-2N = -32$$

$$N = 16$$

6. 16, 18, 20

7. $9(N + 4) - 3(N + 2) = 8(N) + 4$

$$9N + 36 - 3N - 6 = 8N + 4$$

$$9N - 3N - 8N = 4 - 36 + 6$$

$$-2N = -26$$

$$N = 13$$

8. 13, 15, 17

9. $B_T = 2\%$; $B_F = 5\%$; $B_H = 3\%$

$$-2\left(B_T + B_F = 50\right) \Rightarrow \quad -2B_T - 2B_F = -100$$
$$100\left(.02B_T + .05B_F = .03(50)\right) \Rightarrow \quad \underline{2B_T + 5B_F = \quad 150}$$
$$3B_F = \quad 50$$
$$B_F = \quad 16\tfrac{2}{3} \text{ gal}$$

10. $B_T + B_F = 50 \Rightarrow B_T + \left(16\tfrac{2}{3}\right) = 50$

$$B_T = 33\tfrac{1}{3} \text{ gal}$$

11. $W_F = 95\%$; $W_E = 98\%$; $W_S = 97\%$

$$-95\left(W_E + W_F = 50\right) \Rightarrow \quad -95W_E - 95W_F = -4{,}750$$
$$100\left(.98W_E + .95W_F = .97(50)\right) \Rightarrow \quad \underline{98W_E + 95W_F = \quad 4{,}850}$$
$$3W_E = \quad 100$$
$$W_E = \quad 33\tfrac{1}{3} \text{ gal}$$

12. $W_E + W_F = 50 \Rightarrow \left(33\tfrac{1}{3}\right) + W_F = 50$

$$W_F = 16\tfrac{2}{3}$$

13. $X^2 + Y^2 = 8$ circle

$2Y - X^2 = 0$

$2Y = X^2$ parabola

14. see graph

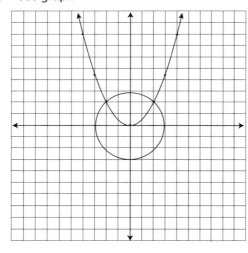

15.
$$X^2 + Y^2 \qquad = 8$$
$$2Y - X^2 = 0 \Rightarrow \underline{-X^2 + \qquad 2Y = 0}$$
$$Y^2 + 2Y = 8$$

16.
$$Y^2 + 2Y = 8$$
$$Y^2 + 2Y - 8 = 0$$
$$(Y + 4)(Y - 2) = 0$$
$$Y + 4 = 0 \qquad Y - 2 = 0$$
$$Y = -4 \qquad Y = 2$$

17. $2Y - X^2 = 0 \Rightarrow 2(-4) - X^2 = 0$
$$-8 = X^2$$
$$\sqrt{-8} = X: \text{ no real solution}$$

$2Y - X^2 = 0 \Rightarrow 2(2) - X^2 = 0$
$$4 = X^2$$
$$\pm\sqrt{4} = X = \pm 2$$

18. $(2, 2); (-2, 2)$

Systematic Review 28D

1.
$$-5(N+D=11) \Rightarrow -5N- 5D = -55$$
$$100(.05N+.10D=.7) \Rightarrow \underline{5N+10D = 70}$$
$$5D = 15$$
$$D = 3$$

2. $N+D=11 \Rightarrow N+(3)=11$
$$N=8$$

3. $5(N+1)-7(N)-3(N+2)=19$
$$5N+5-7N-3N-6=19$$
$$5N-7N-3N=19-5+6$$
$$-5N=20$$
$$N=-4$$

4. $-4, -3, -2$

5. $6(N)-8(N+2)=12$
$$6N-8N-16=12$$
$$6N-8N=12+16$$
$$-2N=28$$
$$N=-14$$

6. $-14, -12, -10$

7. $(N)+(N+2)+(N+4)=-15$
$$3N+6=-15$$
$$3N=-21$$
$$N=-7$$

8. $-7, -5, -3$

9. $C_S=7\%; C_T=3\%; C_F=4\%$
$$-7(C_S+C_T=100) \Rightarrow -7C_S-7C_T=-700$$
$$100(.07C_S+.03C_T=.04(100)) \Rightarrow \underline{7C_S+3C_T= 400}$$
$$-4C_T=-300$$
$$C_T= 75 \text{ gal}$$

10. $C_S+C_T=100 \Rightarrow C_S+(75)=100$
$$C_S=25 \text{ gal}$$

11. $W_T=93\%; W_V=97\%; W_X=96\%$
$$-97(W_T+W_V=100) \Rightarrow -97W_T-97W_V=-9,700$$
$$100(.93W_T+.97W_V=.96(100)) \Rightarrow \underline{93W_T+97W_V= 9,600}$$
$$-4W_T = -100$$
$$W_T = 25 \text{ gal}$$

12. $W_T+W_V=100 \Rightarrow (25)+W_V=100$
$$W_V=75 \text{ gal}$$

13. line and hyperbola

X	Y
-10	1
-1	10
-5	2
-2	5

14. see graph

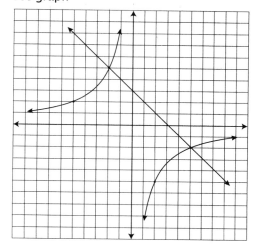

15. $XY = -10 \Rightarrow X(-X+3) = -10$

16.
$$-X^2 + 3X = -10$$
$$-X^2 + 3X + 10 = 0$$
$$(-1)(-X^2 + 3X + 10) = (-1)(0)$$
$$X^2 - 3X - 10 = 0$$
$$(X-5)(X+2) = 0$$
$$X - 5 = 0 \qquad X + 2 = 0$$
$$X = 5 \qquad X = -2$$

17.
$$Y = -X + 3 \Rightarrow Y = -(5) + 3$$
$$Y = -2$$
$$Y = -X + 3 \Rightarrow Y = -(-2) + 3$$
$$Y = 5$$

18. $(5, -2), (-2, 5)$

9. $R_S = 60\%; \ R_T = 20\%; \ R_F = 50\%$
$$-20(R_S + R_T = 20) \Rightarrow -20R_S - 20R_T = -400$$
$$100(.60R_S + .20R_T = .50(20)) \Rightarrow \underline{60R_S + 20R_T = 1{,}000}$$
$$40R_S = 600$$
$$R_S = 15$$

10. $R_S + R_T = 20 \Rightarrow (15) + R_T = 20$
$$R_T = 5$$

11. $M_F = 40\%; \ M_E = 80\%; \ M_H = 50\%$
$$-8(M_F + M_E = 20) \Rightarrow -8M_F - 8M_E = -160$$
$$10(.4M_F + .8M_E = .5(20)) \Rightarrow \underline{4M_F + 8M_E = 100}$$
$$-4M_F = -60$$
$$M_F = 15$$

12. $M_F + M_E = 20 \Rightarrow (15) + M_E = 20$
$$M_E = 5$$

Systematic Review 28E

1.
$$-10(D + Q = 14) \Rightarrow -10D - 10Q = -140$$
$$100(.10D + .25Q = 2.30) \Rightarrow \underline{10D + 25Q = 230}$$
$$15Q = 90$$
$$Q = 6$$

2. $D + Q = 14 \Rightarrow D + (6) = 14$
$$D = 8$$

3.
$$7(N) + 2(N+1) = 6(N+2) + 8$$
$$7N + 2N + 2 = 6N + 12 + 8$$
$$7N + 2N - 6N = 12 + 8 - 2$$
$$3N = 18$$
$$N = 6$$

4. 6, 7, 8

5.
$$3(N) - 7(N+2) = 2$$
$$3N - 7N - 14 = 2$$
$$3N - 7N = 2 + 14$$
$$-4N = 16$$
$$N = -4$$

6. $-4, -2, 0$

7.
$$5(N) - 3(N+4) = (N+2) + 1$$
$$5N - 3N - 12 = N + 3$$
$$5N - 3N - N = 3 + 12$$
$$N = 15$$

8. 15, 17, 19

13. circle and line

14. see graph

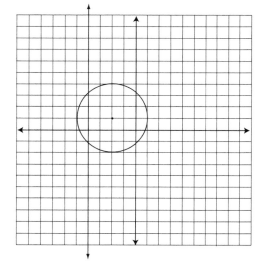

15. $2(X+2)^2 + 2(Y-1)^2 = 18 \Rightarrow 2((-4)+2)^2 + 2(Y-1)^2 = 18$

16. $2((-4)+2)^2 + 2(Y-1)^2 = 18$

$$2(-2)^2 + 2(Y-1)^2 = 18$$
$$2(4) + 2(Y-1)^2 = 18$$
$$8 + 2(Y-1)^2 = 18$$
$$2(Y-1)^2 = 10$$
$$(Y-1)^2 = 5$$
$$Y-1 = \pm\sqrt{5}$$
$$Y = 1 \pm \sqrt{5} \approx 1 \pm 2.24$$

17. $X = -4$

18. $\left(-4, 1+\sqrt{5}\right), \left(-4, 1-\sqrt{5}\right)$ or:

$(-4, 3.24), (-4, -1.24)$

Lesson Practice 29A

1. $(R+5) = 3(D+5)$ $\qquad$ $R = 5D$ $\qquad$ $R = 5D$

$R+5 = 3D+15$ $\qquad$ $(3D+10) = 5D$ $\qquad$ $R = 5(5)$

$R = 3D+10$ $\qquad$ $10 = 2D$ $\qquad$ $R = 25$

$\qquad\qquad\qquad$ $5 = D$

2. $J+1 = 6(N+1)$ $\qquad$ $\dfrac{1}{3}(J+7) = (N+7)$ $\qquad$ $(6N+5) = (3N+14)$

$J+1 = 6N+6$ $\qquad\qquad$ $(J+7) = 3(N+7)$ $\qquad$ $6N-3N = 14-5$

$J = 6N+5$ $\qquad\qquad$ $J+7 = 3N+21$ $\qquad\qquad$ $3N = 9$

$\qquad\qquad\qquad\qquad$ $J = 3N+14$ $\qquad\qquad\qquad$ $N = 3$

$J = 6N+5 \Rightarrow \ J = 6(3)+5$

$\qquad\qquad\qquad J = 18+5$

$\qquad\qquad\qquad J = 23$

3. $D-2 = 2(T-2)$ $\quad$ $\dfrac{2}{3}(D+23) = (T+23)$ $\qquad$ $2D = 3T+23$

$D-2 = 2T-4$ $\qquad\qquad$ $2(D+23) = 3(T+23)$ $\qquad$ $2(2T-2) = 3T+23$

$D = 2T-2$ $\qquad\qquad\quad$ $2D+46 = 3T+69$ $\qquad\quad$ $4T-4 = 3T+23$

$\qquad\qquad\qquad\qquad\qquad$ $2D = 3T+23$ $\qquad\qquad$ $4T-3T = 23+4$

$\qquad\qquad\qquad\qquad\qquad\qquad\qquad\qquad\qquad$ $T = 27$

$D = 2T-2 \Rightarrow \ D = 2(27)-2$

$\qquad\qquad\qquad D = 54-2$

$\qquad\qquad\qquad D = 52$

4. $(C+5) = 6(H+5)$ $\frac{1}{5}(C+7) = (H+7)$ $6H+25 = 5H+28$

$C+5 = 6H+30$ $(C+7) = 5(H+7)$ $H = 3$

$C = 6H+25$ $C+7 = 5H+35$

$C = 5H+28$

$C = 6H+25 \Rightarrow C = 6(3)+25$

$C = 18+25$

$C = 43$

5. $D_D = R_D T_D$ $D_U = R_U T_U$ $D_D = D_U$

$D_D = (B+W)T_D$ $D_U = (B-W)T_U$ $(12+W)2 = (12-W)6$

$D_D = (12+W)(2)$ $D_U = (12-W)(6)$ $24+2W = 72-6W$

$8W = 48$

$W = 6$

current $= 6$ mph

distance $= R_D T_D = (12+6)(2) = (18)(2) = 36$ mi

6. $D_D = R_D T_D$ $D_U = R_U T_U$ $36 = 3B+3W$

$D_D = (B+W)T_D$ $D_U = (B-W)T_U$ $36 = 3B+3(B-2)$

$36 = (B+W)3$ $2 = (B-W)1$ $36 = 3B+3B-6$

$36 = 3B+3W$ $2 = B-W$ $36 = 6B-6$

$W = B-2$ $42 = 6B$

$7 = B$

$W = B-2 \Rightarrow W = (7)-2 = 5$

wind speed $= 5$ mph

bird speed $= 7$ mph

7. $D_D = R_D T_D$ $D_U = R_U T_U$ $21 = BT+2T$

$D_D = (B+W)T$ $D_U = (B-W)T_U$ $\underline{-9 = -BT+2T}$ (result of column 2 multiplied by (-1))

$21 = (B+2)T$ $9 = (B-2)T$ $12 = 4T$

$21 = BT+2T$ $9 = BT-2T$ $3 = T$

$21 = BT+2T \Rightarrow 21 = B(3)+2(3)$

$21 = 3B+6$

$15 = 3B$

$5 = B$

time $= 3$ hours

rate of barge $= 5$ mph

8.　$D_D = R_D T_D$　　　$D_U = R_U T_U$　　　　　$D_D = D_U$

　　$D_D = (P + W)T_D$　　$D_U = (P - W)T_U$　　$1{,}800 + 9W = 2{,}200 - 11W$

　　$D_D = (200 + W)9$　　$D_U = (200 - W)11$　　　　$20W = 400$

　　$D_D = 1{,}800 + 9W$　　$D_U = 2{,}200 - 11W$　　　　$W = 20$

$D_D = 1{,}800 + 9W \Rightarrow D_D = 1{,}800 + 9(20)$

$\phantom{D_D = 1{,}800 + 9W \Rightarrow}\ D_D = 1{,}800 + 180$

$\phantom{D_D = 1{,}800 + 9W \Rightarrow}\ D_D = 1{,}980$

speed of wind = 20 mph

distance = 1,980 miles

Lesson Practice 29B

1.　$S - 1 = 2(J - 1)$　　　$\dfrac{2}{3}(S + 19) = (J + 19)$

　　$S - 1 = 2J - 2$　　　　$2(S + 19) = 3(J + 19)$

　　$S = 2J - 1$　　　　$2S + 38 = 3J + 57$

　　　　　　　　　$2S = 3J + 19 \Rightarrow 2(2J - 1) = 3J + 19$

　　　　　　　　　　　　　　　　$4J - 2 = 3J + 19$

　　　　　　　　　　　　　　　　　$J = 21$ years

$S = 2J - 1 \Rightarrow S = 2(21) - 1 = 42 - 1 = 41$ years

2.　$S = 3B$　　　　　$\dfrac{3}{5}(S + 30) = (B + 30)$

　　　　　　　　　$3(S + 30) = 5(B + 30)$

　　　　　　　　　$3S + 90 = 5B + 150$

　　　　　　　　　$3S = 5B + 60 \Rightarrow 3(3B) = 5B + 60$

　　　　　　　　　　　　　　　$9B = 5B + 60$

　　　　　　　　　　　　　　　$4B = 60$

　　　　　　　　　　　　　　　$B = 15$ years

$S = 3B \Rightarrow S = 3(15) = 45$ years

3.　$C = 4S$　　　$(S - 10) = \dfrac{1}{6}(C - 10)$

　　　　　　　$6(S - 10) = (C - 10)$

　　　　　　　$6S - 60 = C - 10$

　　　　　　　$6S = C + 50 \Rightarrow 6S = (4S) + 50$

　　　　　　　　　　　　　$2S = 50$

　　　　　　　　　　　　　$S = 25$　years

$C = 4S \Rightarrow C = 4(25) = 100$ years

4. $Je = \dfrac{1}{2} Ja$ $\dfrac{5}{7}(Ja + 15) = (Je + 15)$

$2Je = Ja$ $5(Ja + 15) = 7(Je + 15)$

$\qquad\qquad 5Ja + 75 = 7Je + 105$

$\qquad\qquad\quad 5Ja = 7Je + 30 \Rightarrow 5(2Je) = 7Je + 30$

$\qquad\qquad\qquad\qquad\qquad\qquad 10Je = 7Je + 30$

$\qquad\qquad\qquad\qquad\qquad\qquad\quad 3Je = 30$

$\qquad\qquad\qquad\qquad\qquad\qquad\quad\ Je = 10 \text{ years}$

$2Je = Ja \Rightarrow 2(10) = Ja = 20 \text{ years}$

5. $D_D = R_D T_D$ $D_U = R_U T_U$ $25 = \ FT_D + 2T_D$

$D_D = (F + W) T_D$ $D_U = (F - W) T_U$ $\underline{-5 = -FT_D + 2T_D}$

$25 = (F + 2) T_D$ $35 = (F - 2) 7T_D$ $20 = \qquad\ \ 4T_D$

$25 = FT_D + 2T_D$ $35 = 7FT_D - 14T_D$ $5 = \qquad\quad T_D$

$\qquad\qquad\qquad\qquad 5 = FT_D - 2T_D$

$25 = (F + 2) T_D \Rightarrow 25 = (F + 2)(5)$

$\qquad\qquad\qquad\qquad 25 = 5F + 10$

$\qquad\qquad\qquad\qquad 15 = 5F$

$\qquad\qquad\qquad\qquad\ \ 3 = F$

time downstream = 5 hours

speed of fish = 3 mph

6. $D_D = R_D T$ $D_U = R_U T$ $60 = \ BT + 4T$

$D_D = (B + W) T$ $D_U = (B - W) T$ $\underline{-12 = -BT + 4T}$

$60 = (B + 4) T$ $12 = (B - 4) T$ $48 = \qquad\ 8T$

$60 = BT + 4T$ $12 = BT - 4T$ $6 = \qquad\ T$

$12 = BT - 4T \Rightarrow 12 = B(6) - 4(6)$

$\qquad\qquad\qquad\qquad 12 = 6B - 24$

$\qquad\qquad\qquad\qquad 36 = 6B$

$\qquad\qquad\qquad\qquad\ \ 6 = B$

time each direction = 6 hours; total time is 12 hours

rate of paddling = 6 mph

7. $D_D = R_D T_D$ $D_U = R_U T_U$ $D_D = D_U$

$D_D = (P + W) T_D$ $D_U = (P - W) T_U$ $600 + 6W = 900 - 9W$

$D_D = (100 + W) 6$ $D_U = (100 - W) 9$ $15W = 300$

$D_D = 600 + 6W$ $D_U = 900 - 9W$ $W = 20 \text{ mph}$

$D_D = (100 + W) 6 \Rightarrow D_D = (100 + (20)) 6 = 120(6) = 720 \text{ miles}$

speed of wind = 20 mph

distance = 720 miles

8.
$$D_D = R_D T_D \qquad D_U = R_U T_U \qquad 240 = 4C + 4W$$
$$D_D = (C+W)T_D \qquad D_U = (C-W)T_U \qquad \underline{160 = 4C - 4W}$$
$$240 = (C+W)4 \qquad 160 = (C-W)4 \qquad 400 = 8C$$
$$240 = 4C + 4W \qquad 160 = 4C - 4W \qquad 50 = C$$

$$240 = 4C + 4W \Rightarrow 240 = 4(50) + 4W$$
$$240 = 200 + 4W$$
$$40 = 4W$$
$$10 = W$$

speed of car = 50 mph
speed of wind = 10 mph

Systematic Review 29C

1. $(P+2) = 2(E+2)$

2. $(P-16) = 4.25(E-16)$

3.
$$(P+2) = 2(E+2) \qquad (P-16) = 4.25(E-16)$$
$$P+2 = 2E+4 \qquad ((2E+2)-16) = 4.25(E-16)$$
$$P = 2E+2 \qquad 2E-14 = 4.25E - 68$$
$$-14+68 = 4.25E - 2E$$
$$54 = 2.25E$$
$$24 = E$$

4. $P = 2E+2 \Rightarrow P = 2(24)+2$
$$P = 48+2$$
$$P = 50$$

5. $K = 2.75M$

6. $(M-2) = \dfrac{1}{3}(K-2)$

7.
$$(M-2) = \dfrac{1}{3}((2.75M)-2)$$
$$3(M-2) = ((2.75M)-2)$$
$$3M-6 = 2.75M - 2$$
$$3M - 2.75M = -2 + 6$$
$$.25M = 4$$
$$M = 16$$

8. $K = 2.75M \Rightarrow K = 2.75(16)$
$$K = 44$$

9. $D_D = R_D T_D \Rightarrow 36 = R_D(4)$
$$9 = R_D \qquad B + W = 9$$

10. $D_U = R_U T_U \Rightarrow 15 = R_U(3)$
$$5 = R_U \qquad B - W = 5$$

11.
$$B + W = 9$$
$$\underline{+\quad B - W = 5}$$
$$2B \qquad = 14$$
$$B \qquad = 7$$

12. $B + W = 9 \Rightarrow (7) + W = 9$
$$W = 2$$

13.
$$-1(P + N = 24) \Rightarrow \quad -P \; -N = -24$$
$$100(.01P + .05N = .72) \Rightarrow \quad \underline{P + 5N = \quad 72}$$
$$4N = \quad 48$$
$$N = \quad 12$$

14. $P + N = 24 \Rightarrow P + (12) = 24$
$$P = 12$$

15. $4(N + 2) - 8(N + 1) = 9N$
$$4N + 8 - 8N - 8 = 9N$$
$$8 - 8 = 9N - 4N + 8N$$
$$0 = 13N$$
$$0 = N$$

16. 0, 1, 2

17. $(N) + (N + 2) + (N + 4) = 84$
$$3N + 6 = 84$$
$$3N = 78$$
$$N = 26$$

18. 26, 28, 30

19. $C_T = 20\%$; $C_F = 5\%$
$$-5(C_T + C_F = 12) \Rightarrow \quad -5C_T - 5C_F = -60$$
$$100(.20C_T + .05C_F = .15(12)) \Rightarrow \quad \underline{20C_T + 5C_F = 180}$$
$$15C_T \qquad = 120$$
$$C_T \qquad = 8 \text{ oz}$$

20. $C_T + C_F = 12 \Rightarrow (8) + C_F = 12$
$$C_F = 4 \text{ oz}$$

Systematic Review 29D

1. $(S - 1) = \dfrac{1}{2}(K - 1)$
$$2(S - 1) = (K - 1)$$
$$2S - 2 = K - 1$$
$$2S - 1 = K$$

2. $(K + 7) = 1.5(S + 7)$

3. $(K + 7) = 1.5(S + 7) \Rightarrow ((2S - 1) + 7) = 1.5(S + 7)$
$$2S + 6 = 1.5S + 10.5$$
$$2S - 1.5S = 10.5 - 6$$
$$.5S = 4.5$$
$$S = 9$$

4. $2S - 1 = K \Rightarrow 2(9) - 1 = K$
$$18 - 1 = K$$
$$17 = K$$

5. $(K - 1) = \dfrac{1}{2}(D - 1)$
$$2(K - 1) = (D - 1)$$
$$2K - 2 = D - 1$$
$$2K - 1 = D$$

6. $D + 11 = \dfrac{3}{2}(K + 11)$

7. $D + 11 = \dfrac{3}{2}(K + 11) \Rightarrow$
$$(2K - 1) + 11 = \frac{3}{2}(K + 11)$$
$$2K + 10 = \frac{3}{2}(K + 11)$$
$$2(2K + 10) = 3(K + 11)$$
$$4K + 20 = 3K + 33$$
$$K = 13$$

8. $2K - 1 = D \Rightarrow 2(13) - 1 = D$
$$26 - 1 = D$$
$$25 = D$$

9. $R_D = B + W \Rightarrow 48 = B + W$

10. $R_U = B - W \Rightarrow 24 = B - W$

11.
$$48 = \quad B + W$$
$$\underline{24 = \quad B - W}$$
$$72 = 2B$$
$$36 = \quad B$$

12. $48 = B + W \Rightarrow 48 = (36) + W$
$$12 = W$$

13.

$$-5\big(D+N=19\big) \Rightarrow \quad -5D-5N=-95$$
$$100\big(.10D+.05N=1.30\big) \Rightarrow \quad \underline{10D+5N=130}$$
$$5D = 35$$
$$D = 7$$

14.

$$D+N=19 \Rightarrow (7)+N=19$$
$$N=12$$

15.

$$3(N)-5(N+1)=3+2(N+2)$$
$$3N-5N-5=3+2N+4$$
$$3N-5N-2N=3+4+5$$
$$-4N=12$$
$$N=-3$$

16. $-3, -2, -1$

17.

$$8(N+4)=5(N)-28$$
$$8N+32=5N-28$$
$$8N-5N=-28-32$$
$$3N=-60$$
$$N=-20$$

18. $-20, -18, -16$

19. $B_F=25\%; \ B_T=10\%$

$$-10\big(B_F+B_T=90\big) \Rightarrow \quad -10B_F-10B_T=-900$$
$$100\big(.25B_F+.10B_T=.20(90)\big) \Rightarrow \quad \underline{25B_F+10B_T=1800}$$
$$15B_F \qquad = 900$$
$$B_F \qquad = 60 \ \text{ml}$$

20.

$$B_F+B_T=90 \Rightarrow (60)+B_T=90$$
$$B_T=30 \ \text{ml}$$

7.

$$(K-2)=\frac{1}{9}(L-2) \Rightarrow \quad (K-2)=\frac{1}{9}\big((5K)-2\big)$$
$$9(K-2)=(5K-2)$$
$$9K-18=5K-2$$
$$4K=16$$
$$K=4$$

8.

$$L=5K \Rightarrow L=5(4)$$
$$L=20$$

9.

$$D_D=R_D T_D \Rightarrow 65=R_D(5)$$
$$13=R_D \qquad B+W=13$$

10.

$$D_U=R_U T_U \qquad 24=R_U(8)$$
$$3=R_U \qquad B-W=3$$

11.

$$B+W=13$$
$$\underline{B-W=\ \ 3}$$
$$2B \quad =16$$
$$B \quad =8$$

12.

$$13=B+W \Rightarrow 13=(8)+W$$
$$5=W$$

13.

$$-10\big(D+Q=12\big) \Rightarrow \quad -10D-10Q=-120$$
$$100\big(.10D+.25Q=2.55\big) \Rightarrow \quad \underline{10D+25Q=\ \ 255}$$
$$15Q = 135$$
$$Q = 9$$

14.

$$D+Q=12 \Rightarrow D+(9)=12$$
$$D=3$$

15.

$$6(N+2)+8=8(N+1)-N$$
$$6N+12+8=8N+8-N$$
$$6N-8N+N=8-12-8$$
$$-1N=-12$$
$$N=12$$

16. $12, 13, 14$

17.

$$(N)+(N+2)+(N+4)=-51$$
$$3N+6=-51$$
$$3N=-57$$
$$N=-19$$

18. $-19, -17, -15$

Systematic Review 29E

1. $B=D+30 \ \text{or} \ B-30=D$

2. $(B+10)=2(D+10)$

3.

$$(B+10)=2(D+10) \Rightarrow \big((D+30)+10\big)=2(D+10)$$
$$D+40=2D+20$$
$$-D=-20$$
$$D=20$$

4.

$$B=D+30 \Rightarrow B=(20)+30$$
$$B=50$$

5. $L=5K$

6. $(K-2)=\frac{1}{9}(L-2)$

19.
$$-3\left(A_T + A_E = 80\right) \Rightarrow \quad -3A_T - 3A_E = -240$$
$$100\left(.03A_T + .08A_E = .06(80)\right) \Rightarrow \quad \underline{3A_T + 8A_E = 480}$$
$$5A_E = 240$$
$$A_E = 48 \text{ ml}$$

20. $A_T + A_E = 80 \Rightarrow A_T + (48) = 80$
$$A_T = 32 \text{ ml}$$

Lesson Practice 30A

1.   A $3X + 6Y - 4Z = 17$
      B $\underline{-X + 5Y + 4Z = 11}$
      D $2X + 11Y = 28$

2.   B $5(-X + 5Y + 4Z = 11) \Rightarrow \quad -5X + 25Y + 20Z = 55$
      C $2(2X + 2Y - 10Z = 0) \Rightarrow \quad \underline{4X + 4Y - 20Z = 0}$
      E $-X + 29Y = 55$

3.   D $2X + 11Y = 28$
      E $2(-X + 29Y = 55) \Rightarrow \underline{-2X + 58Y = 110}$
                              $69Y = 138$
                                $Y = 2$

4.   D $2X + 11Y = 28 \Rightarrow 2X + 11(2) = 28$
                          $2X + 22 = 28$
                              $2X = 6$
                                $X = 3$

5.   B $-X + 5Y + 4Z = 11 \Rightarrow -(3) + 5(2) + 4Z = 11$
                            $-3 + 10 + 4Z = 11$
                               $4Z = 11 + 3 - 10$
                               $4Z = 4$
                               $Z = 1$

6.   A $3X + 6Y - 4Z = 17 \Rightarrow 3(3) + 6(2) - 4(1) = 17$
                            $9 + 12 - 4 = 17$
                               $17 = 17$

      B $-X + 5Y + 4Z = 11 \Rightarrow -(3) + 5(2) + 4(1) = 11$
                            $-3 + 10 + 4 = 11$
                               $11 = 11$

      C $2X + 2Y - 10Z = 0 \Rightarrow 2(3) + 2(2) - 10(1) = 0$
                            $6 + 4 - 10 = 0$
                               $0 = 0$

7.   A $-3X - Y - 2Z = -13$
      B $2(2X + 2Y + Z = 16) \Rightarrow \underline{4X + 4Y + 2Z = 32}$
      D $X + 3Y = 19$

8. $\begin{matrix} B \\ C \\ E \end{matrix}$ $\quad -3(2X+2Y+Z=16) \Rightarrow \begin{array}{r} -6X-6Y-3Z=-48 \\ X+3Y+3Z=13 \\ \hline -5X-3Y=-35 \end{array}$

9. $\begin{matrix} D \\ E \end{matrix}$ $\quad \begin{array}{r} X+3Y=19 \\ -5X-3Y=-35 \\ \hline -4X=-16 \\ X=4 \end{array}$

10. D $\quad X+3Y=19 \Rightarrow \begin{array}{r} (4)+3Y=19 \\ 3Y=15 \\ Y=5 \end{array}$

11. B $\quad 2X+2Y+Z=16 \Rightarrow \begin{array}{r} 2(4)+2(5)+Z=16 \\ 8+10+Z=16 \\ Z=-2 \end{array}$

12. A $\quad -3X-Y-2Z=-13 \Rightarrow \begin{array}{r} -3(4)-(5)-2(-2)=-13 \\ -12-5+4=-13 \\ -13=-13 \end{array}$

    B $\quad 2X+2Y+Z=16 \Rightarrow \begin{array}{r} 2(4)+2(5)+(-2)=16 \\ 8+10-2=16 \\ 16=16 \end{array}$

    C $\quad X+3Y+3Z=13 \Rightarrow \begin{array}{r} (4)+3(5)+3(-2)=13 \\ 4+15-6=13 \\ 13=13 \end{array}$

13. $\begin{matrix} A \\ B \\ D \end{matrix}$ $\quad 2(4X+6Y+2Z=22) \Rightarrow \begin{matrix} A \\ B \\ D \end{matrix} \begin{array}{r} 8X+12Y+4Z=44 \\ -4X+3Y-4Z=-10 \\ \hline 4X+15Y=34 \end{array}$

14. $\begin{matrix} B \\ C \\ E \end{matrix}$ $\quad \begin{array}{l} 3(-4X+3Y-4Z=-10) \Rightarrow \\ -4(5X+4Y-3Z=4) \Rightarrow \end{array} \begin{array}{r} -12X+9Y-12Z=-30 \\ -20X-16Y+12Z=-16 \\ \hline -32X-7Y=-46 \end{array}$

15. $\begin{matrix} D \\ E \end{matrix}$ $\quad 8(4X+15Y=34) \Rightarrow \begin{array}{r} 32X+120Y=272 \\ -32X-7Y=-46 \\ \hline 113Y=226 \\ Y=2 \end{array}$

16. D $\quad 4X+15Y=34 \Rightarrow \begin{array}{r} 4X+15(2)=34 \\ 4X+30=34 \\ 4X=4 \\ X=1 \end{array}$

17. A $\quad 4X+6Y+2Z=22 \Rightarrow \begin{array}{r} 4(1)+6(2)+2Z=22 \\ 4+12+2Z=22 \\ 2Z=6 \\ Z=3 \end{array}$

18. A $\quad 4X + 6Y + 2Z = 22 \Rightarrow 4(1) + 6(2) + 2(3) = 22$
$$4 + 12 + 6 = 22$$
$$22 = 22$$

B $\quad -4X + 3Y - 4Z = -10 \Rightarrow -4(1) + 3(2) - 4(3) = -10$
$$-4 + 6 - 12 = -10$$
$$-10 = -10$$

C $\quad 5X + 4Y - 3Z = 4 \Rightarrow 5(1) + 4(2) - 3(3) = 4$
$$5 + 8 - 9 = 4$$
$$4 = 4$$

Lesson Practice 30B

1. A $\quad -1(4X - 4Y + 2Z = -2) \Rightarrow -4X + 4Y - 2Z = 2$
$\quad$ B $\qquad\qquad\qquad\qquad\qquad 5X + Y + 2Z = 1$
$\quad$ D $\qquad\qquad\qquad\qquad\qquad \overline{X + 5Y = 3}$

2. B $\quad 3(5X + Y + 2Z = 1) \Rightarrow 15X + 3Y + 6Z = 3$
$\quad$ C $\quad 2(X + 6Y - 3Z = -11) \Rightarrow \underline{2X + 12Y - 6Z = -22}$
$\quad$ E $\qquad\qquad\qquad\qquad\qquad 17X + 15Y = -19$

3. D $\quad -3(X + 5Y = 3) \Rightarrow -3X - 15Y = -9$
$\quad$ E $\qquad\qquad\qquad\qquad \underline{17X + 15Y = -19}$
$\qquad\qquad\qquad\qquad\qquad 14X = -28$
$\qquad\qquad\qquad\qquad\qquadX = -2$

4. D $\quad X + 5Y = 3 \Rightarrow (-2) + 5Y = 3$
$$5Y = 5$$
$$Y = 1$$

5. A $\quad 4X - 4Y + 2Z = -2 \Rightarrow 4(-2) - 4(1) + 2Z = -2$
$$-8 - 4 + 2Z = -2$$
$$2Z = 10$$
$$Z = 5$$

6. A $\quad 4X - 4Y + 2Z \Rightarrow 4(-2) - 4(1) + 2(5) = -2$
$$-8 - 4 + 10 = -2$$
$$-2 = -2$$

B $\quad 5X + Y + 2Z = 1 \Rightarrow 5(-2) + (1) + 2(5) = 1$
$$-10 + 1 + 10 = 1$$
$$1 = 1$$

C $\quad X + 6Y - 3Z = -11 \Rightarrow (-2) + 6(1) - 3(5) = -11$
$$-2 + 6 - 15 = -11$$
$$-11 = -11$$

7. A $\qquad\qquad\qquad\qquad\qquad X + 2Y + 3Z = 32$
$\quad$ B $\quad -3(4X - 3Y + Z = 7) \Rightarrow \underline{-12X + 9Y - 3Z = -21}$
$\quad$ D $\qquad\qquad\qquad\qquad\qquad -11X + 11Y = 11$

8. B $2(4X - 3Y + Z = 7) \Rightarrow$ $8X - 6Y + 2Z = 14$
    C $-2X + 6Y - 2Z = 4$
    E $\overline{6X = 18}$
                                       $X = 3$

9. $X = 3$ (already found)

10. D $-11X + 11Y = 11 \Rightarrow$ $-11(3) + 11Y = 11$
                                    $-33 + 11Y = 11$
                                       $11Y = 44$
                                          $Y = 4$

11. A $X + 2Y + 3Z = 32 \Rightarrow$ $(3) + 2(4) + 3Z = 32$
                                     $3 + 8 + 3Z = 32$
                                        $3Z = 21$
                                         $Z = 7$

12. A $X + 2Y + 3Z = 32 \Rightarrow$ $(3) + 2(4) + 3(7) = 32$
                                     $3 + 8 + 21 = 32$
                                        $32 = 32$

    B $4X - 3Y + Z = 7 \Rightarrow$ $4(3) - 3(4) + 7 = 7$
                                     $12 - 12 + 7 = 7$
                                        $7 = 7$

    C $-2X + 6Y - 2Z = 4 \Rightarrow$ $-2(3) + 6(4) - 2(7) = 4$
                                     $-6 + 24 - 14 = 4$
                                        $4 = 4$

13. A $X - 8Y + Z = 6$
    B $\underline{2X + 7Y - Z = 11}$
    D $3X - Y = 17$

14. A $3(X - 8Y + Z = 6) \Rightarrow$ $3X - 24Y + 3Z = 18$
    C $\underline{2X - 10Y - 3Z = -22}$
    E $5X - 34Y = -4$

15. D $-34(3X - Y = 17) \Rightarrow$ $-102X + 34Y = -578$
    E $\underline{5X - 34Y = -4}$
                                $-97X = -582$
                                  $X = 6$

16. D $3X - Y = 17 \Rightarrow$ $3(6) - Y = 17$
                               $18 - Y = 17$
                                  $-Y = -1$
                                   $Y = 1$

17. A $X - 8Y + Z = 6 \Rightarrow$ $(6) - 8(1) + Z = 6$
                                  $6 - 8 + Z = 6$
                                     $Z = 8$

18.   A $X - 8Y + Z = 6 \Rightarrow (6) - 8(1) + (8) = 6$
$$6 - 8 + 8 = 6$$
$$6 = 6$$

   B $2X + 7Y - Z = 11 \Rightarrow 2(6) + 7(1) - (8) = 11$
$$12 + 7 - 8 = 11$$
$$11 = 11$$

   C $2X - 10Y - 3Z = -22 \Rightarrow 2(6) - 10(1) - 3(8) = -22$
$$12 - 10 - 24 = -22$$
$$-22 = -22$$

Systematic Review 30C

1.   A $-2(5X - 3Y + 3Z = 3) \Rightarrow -10X + 6Y - 6Z = -6$
   B $ \underline{2X - 6Y - 4Z = 2}$
   D $ -8X - 10Z = -4$

2.   A $-5(5X - 3Y + 3Z = 3) \Rightarrow -25X + 15Y - 15Z = -15$
   C $3(3X - 5Y + Z = -3) \Rightarrow \underline{9X - 15Y + 3Z = -9}$
   E $ -16X - 12Z = -24$

3.   D $-2(-8X - 10Z = -4) \Rightarrow 16X + 20Z = 8$
   E $ \underline{-16X - 12Z = -24}$
$$8Z = -16$$
$$Z = -2$$

4.   D $-8X - 10Z = -4 \Rightarrow -8X - 10(-2) = -4$
$$-8X + 20 = -4$$
$$-8X = -24$$
$$X = 3$$

5.   A $5X - 3Y + 3Z = 3 \Rightarrow 5(3) - 3Y + 3(-2) = 3$
$$15 - 3Y - 6 = 3$$
$$-3Y = -6$$
$$Y = 2$$

6.   A $5X - 3Y + 3Z = 3 \Rightarrow 5(3) - 3(2) + 3(-2) = 3$
$$15 - 6 - 6 = 3$$
$$3 = 3$$

   B $2X - 6Y - 4Z = 2 \Rightarrow 2(3) - 6(2) - 4(-2) = 2$
$$6 - 12 + 8 = 2$$
$$2 = 2$$

   C $3X - 5Y + Z = -3 \Rightarrow 3(3) - 5(2) + (-2) = -3$
$$9 - 10 - 2 = -3$$
$$-3 = -3$$

7. A
    B $2(4X+3Y-2Z=6) \Rightarrow$
    D

$$3X+2Y+4Z = 9$$
$$8X+6Y-4Z = 12$$
$$\overline{11X+8Y = 21}$$

8. A $3(3X+2Y+4Z=9) \Rightarrow$
    C $4(5X+4Y-3Z=8) \Rightarrow$
    E

$$9X+6Y+12Z = 27$$
$$20X+16Y-12Z = 32$$
$$\overline{29X+22Y = 59}$$

9. D $29(11X+8Y=21) \Rightarrow$
    E $-11(29X+22Y=59) \Rightarrow$

$$319X+232Y = 609$$
$$-319X-242Y = -649$$
$$\overline{-10Y = -40}$$
$$Y = 4$$

10. D $11X+8Y=21 \Rightarrow 11X+8(4)=21$
$$11X+32 = 21$$
$$11X = -11$$
$$X = -1$$

11. A $3X+2Y+4Z=9 \Rightarrow 3(-1)+2(4)+4Z=9$
$$-3+8+4Z = 9$$
$$4Z = 4$$
$$Z = 1$$

12. A $3X+2Y+4Z=9 \Rightarrow 3(-1)+2(4)+4(1)=9$
$$-3+8+4 = 9$$
$$9 = 9$$

    B $4X+3Y-2Z=6 \Rightarrow 4(-1)+3(4)-2(1)=6$
$$-4+12-2 = 6$$
$$6 = 6$$

    C $5X+4Y-3Z=8 \Rightarrow 5(-1)+4(4)-3(1)=8$
$$-5+16-3 = 8$$
$$8 = 8$$

13. $(P-3) = 4(C-3)$
$$P-3 = 4C-12$$
$$P = 4C-9$$

14. $(P-33) = C$
$$P = C+33$$

15. $(4C-9) = (C+33)$
$$4C-C = 33+9$$
$$3C = 42$$
$$C = 14$$

16. $P = 4C-9 \Rightarrow P = 4(14)-9 = 56-9 = 47$

17. $D_D = R_D T_D \Rightarrow (34) = R_D(2)$
$$34 = (B+W)(2)$$
$$34 = 2B+2W$$
$$17 = B+W$$

18. $D_U = R_U T_U \Rightarrow 15 = R_U(3)$

$$15 = (B - W)(3)$$
$$5 = B - W$$

19. $17 = B + W$

$\underline{5 = B - W}$

$22 = 2B$

$11 = B$

20. $17 = B + W \Rightarrow 17 = (11) + W$

$$6 = W$$

Systematic Review 30D

1. A $\quad 6X + 3Y - 5Z = 5$

B $\quad \underline{-2X - 3Y - Z = -1}$

D $\quad 4X - 6Z = 4$

2. A $\quad -2(6X + 3Y - 5Z = 5) \Rightarrow -12X - 6Y + 10Z = -10$

C $\quad 3(4X + 2Y - 6Z = -2) \Rightarrow \underline{12X + 6Y - 18Z = -6}$

E $\quad -8Z = -16$

$$Z = 2$$

3. $Z = 2$ (already found)

4. D $\quad 4X - 6Z = 4 \Rightarrow 4X - 6(2) = 4$

$$4X - 12 = 4$$
$$4X = 16$$
$$X = 4$$

5. A $\quad 6X + 3Y - 5Z = 5 \Rightarrow 6(4) + 3Y - 5(2) = 5$

$$24 + 3Y - 10 = 5$$
$$3Y = -9$$
$$Y = -3$$

6. A $\quad 6X + 3Y - 5Z = 5 \Rightarrow 6(4) + 3(-3) - 5(2) = 5$

$$24 - 9 - 10 = 5$$
$$5 = 5$$

B $\quad -2X - 3Y - Z = -1 \Rightarrow -2(4) - 3(-3) - (2) = -1$

$$-8 + 9 - 2 = -1$$
$$-1 = -1$$

C $\quad 4X + 2Y - 6Z = -2 \Rightarrow 4(4) + 2(-3) - 6(2) = -2$

$$16 - 6 - 12 = -2$$
$$-2 = -2$$

7. A $\quad 2(X - 2Y + 4Z = -4) \Rightarrow 2X - 4Y + 8Z = -8$

B $\quad \underline{ 3X + 4Y - 5Z = 25}$

D $\quad 5X + 3Z = 17$

8. $\begin{array}{l} \text{A} \\ \text{C} \\ \text{E} \end{array}$ $\begin{array}{l} -3(X-2Y+4Z=-4) \Rightarrow \\ 2(5X-3Y+2Z=12) \Rightarrow \end{array}$ $\begin{array}{r} -3X+6Y-12Z=12 \\ 10X-6Y+4Z=24 \\ \hline 7X \qquad -8Z=36 \end{array}$

9. $\begin{array}{l} \text{D} \\ \text{E} \end{array}$ $\begin{array}{l} 8(5X+3Z=17) \Rightarrow \\ 3(7X-8Z=36) \Rightarrow \end{array}$ $\begin{array}{r} 40X+24Z=136 \\ 21X-24Z=108 \\ \hline 61X \qquad =244 \\ X \qquad =4 \end{array}$

10. D $\begin{array}{r} 5X+3Z=17 \Rightarrow 5(4)+3Z=17 \\ 20+3Z=17 \\ 3Z=-3 \\ Z=-1 \end{array}$

11. A $\begin{array}{r} X-2Y+4Z=-4 \Rightarrow (4)-2Y+4(-1)=-4 \\ 4-2Y-4=-4 \\ -2Y=-4 \\ Y=2 \end{array}$

12. A $\begin{array}{r} X-2Y+4Z=-4 \Rightarrow (4)-2(2)+4(-1)=-4 \\ 4-4-4=-4 \\ -4=-4 \end{array}$

    B $\begin{array}{r} 3X+4Y-5Z=25 \Rightarrow 3(4)+4(2)-5(-1)=25 \\ 12+8+5=25 \\ 25=25 \end{array}$

    C $\begin{array}{r} 5X-3Y+2Z=12 \Rightarrow 5(4)-3(2)+2(-1)=12 \\ 20-6-2=12 \\ 12=12 \end{array}$

13. $\begin{array}{l} (S+1)=2(G+1) \\ S+1=2G+2 \\ S=2G+1 \end{array}$

14. $\begin{array}{l} (G-5)=\dfrac{4}{9}(S-5) \\ 9(G-5)=4(S-5) \\ 9G-45=4S-20 \\ 9G-25=4S \end{array}$

15. $\begin{array}{r} 9G-25=4S \Rightarrow 9G-25=4(2G+1) \\ 9G-25=8G+4 \\ G=29 \end{array}$

16. $\begin{array}{r} S=2G+1 \Rightarrow S=2(29)+1 \\ S=58+1 \\ S=59 \end{array}$

17. $\begin{array}{r} D_D=R_D T_D \Rightarrow 42=R_D(3) \\ 14=R_D \\ 14=B+W \end{array}$

18. $D_U = R_U T_U \Rightarrow 30 = R_U(5)$

$\qquad\qquad\qquad\quad 6 = R_U$

$\qquad\qquad\qquad\quad 6 = B - W$

19. $\quad 14 = \; B + W$

$\qquad \underline{\; 6 = \; B - W}$

$\qquad 20 = 2B$

$\qquad 10 = \; B$

20. $\quad 14 = B + W \Rightarrow 14 = (10) + W$

$\qquad\qquad\qquad\qquad\qquad 4 = W$

Systematic Review 30E

1. A $\quad -3(-2X + 3Y + 5Z = -7) \Rightarrow \quad 6X - 9Y - 15Z = \;\; 21$

B

D $\qquad\qquad\qquad\qquad\qquad\qquad\quad \underline{-6X - 2Y \;\;\;\; - Z = -15}$

$\qquad\qquad\qquad\qquad\qquad\qquad\qquad\qquad -11Y - 16Z = \quad 6$

2. A $\quad -2(-2X + 3Y + 5Z = -7) \Rightarrow \quad 4X - 6Y - 10Z = \;\; 14$

C

E $\qquad\qquad\qquad\qquad\qquad\qquad\quad \underline{-4X + 4Y + \;\; 5Z = -15}$

$\qquad\qquad\qquad\qquad\qquad\qquad\qquad\qquad -2Y - \;\; 5Z = \;\; -1$

3. D $\quad -2(-11Y - 16Z = 6) \Rightarrow \quad 22Y + 32Z = -12$

E $\qquad 11(-2Y - 5Z = -1) \Rightarrow \underline{-22Y - 55Z = -11}$

$\qquad\qquad\qquad\qquad\qquad\qquad\qquad\quad -23Z = -23$

$\qquad\qquad\qquad\qquad\qquad\qquad\qquad\qquad\quad Z = \quad 1$

4. D $\quad -11Y - 16Z = 6 \Rightarrow \quad -11Y - 16(1) = 6$

$\qquad\qquad\qquad\qquad\qquad\qquad -11Y - 16 = 6$

$\qquad\qquad\qquad\qquad\qquad\qquad\qquad -11Y = 22$

$\qquad\qquad\qquad\qquad\qquad\qquad\qquad\qquad Y = -2$

5. A $\quad -2X + 3Y + 5Z = -7 \Rightarrow \quad -2X + 3(-2) + 5(1) = -7$

$\qquad\qquad\qquad\qquad\qquad\qquad\qquad\quad -2X - 6 + 5 = -7$

$\qquad\qquad\qquad\qquad\qquad\qquad\qquad\qquad\quad -2X = -6$

$\qquad\qquad\qquad\qquad\qquad\qquad\qquad\qquad\qquad X = 3$

6. A $\quad -2X + 3Y + 5Z = -7 \Rightarrow \quad -2(3) + 3(-2) + 5(1) = -7$

$\qquad\qquad\qquad\qquad\qquad\qquad\qquad\qquad -6 - 6 + 5 = -7$

$\qquad\qquad\qquad\qquad\qquad\qquad\qquad\qquad\qquad -7 = -7$

$\quad$ B $\quad -6X - 2Y - Z = -15 \Rightarrow \quad -6(3) - 2(-2) - (1) = -15$

$\qquad\qquad\qquad\qquad\qquad\qquad\qquad\qquad -18 + 4 - 1 = -15$

$\qquad\qquad\qquad\qquad\qquad\qquad\qquad\qquad\quad -15 = -15$

$\quad$ C $\quad -4X + 4Y + 5Z = -15 \Rightarrow \quad -4(3) + 4(-2) + 5(1) = -15$

$\qquad\qquad\qquad\qquad\qquad\qquad\qquad\qquad -12 - 8 + 5 = -15$

$\qquad\qquad\qquad\qquad\qquad\qquad\qquad\qquad\quad -15 = -15$

7. A $2(2X - 5Y + 2Z = -5) \Rightarrow$ $4X - 10Y + 4Z = -10$
    B $\underline{-3X + 4Y - 4Z = 6}$
    D $X - 6Y = -4$

8. A $2X - 5Y + 2Z = -5$
    C $2(5X + 6Y - Z = 18) \Rightarrow$ $\underline{10X + 12Y - 2Z = 36}$
    E $12X + 7Y = 31$

9. D $-12(X - 6Y = -4) \Rightarrow$ $-12X + 72Y = 48$
    E $\underline{12X + 7Y = 31}$
                                       $79Y = 79$
                                       $Y = 1$

10. $X - 6Y = -4 \Rightarrow X - 6(1) = -4$
    $X - 6 = -4$
    $X = 2$

11. $2X - 5Y + 2Z = -5 \Rightarrow 2(2) - 5(1) + 2Z = -5$
    $4 - 5 + 2Z = -5$
    $2Z = -4$
    $Z = -2$

12. A $2X - 5Y + 2Z = -5 \Rightarrow 2(2) - 5(1) + 2(-2) = -5$
    $4 - 5 - 4 = -5$
    $-5 = -5$

    B $-3X + 4Y - 4Z = 6 \Rightarrow -3(2) + 4(1) - 4(-2) = 6$
    $-6 + 4 + 8 = 6$
    $6 = 6$

    C $5X + 6Y - Z = 18 \Rightarrow 5(2) + 6(1) - (-2) = 18$
    $10 + 6 + 2 = 18$
    $18 = 18$

13. $(W + 15) = 1.4(C + 15)$
    $W + 15 = 1.4C + 21$
    $W = 1.4C + 6$

14. $(C - 5) = .6(W - 5)$
    $C - 5 = .6W - 3$
    $C = .6W + 2$

15. $W = 1.4C + 6 \Rightarrow W = 1.4(.6W + 2) + 6$
    $W = .84W + 2.8 + 6$
    $W - .84W = 8.8$
    $.16W = 8.8$
    $W = 55$

16. $C = .6W + 2 \Rightarrow C = .6(55) + 2$
    $C = 33 + 2$
    $C = 35$

17. $D_D = R_D T_D \Rightarrow 20 = R_D(2)$

$ 10 = R_D$

$ 10 = B + W$

18. $D_U = R_U T_U \Rightarrow 20 = R_U(5)$

$ 4 = R_U$

$ 4 = B - W$

19. $10 = B + W$

$\underline{ 4 = B - W}$

$ 14 = 2B$

$ 7 = B$

20. $10 = B + W \Rightarrow 10 = (7) + W$

$ 3 = W$

Lesson Practice 31A

1. $(+5) + (-8) = -3$

2. $(-6) + (+3) + (-4) = -7$

3. direction and magnitude

4. $H^2 = 6^2 + 8^2$

$H^2 = 36 + 64$

$H^2 = 100$

$H = \sqrt{100} = 10$

5. $\sin\theta = \dfrac{\text{opposite}}{\text{hypotenuse}} = \dfrac{5}{13} \approx .3846$

$\cos\theta = \dfrac{\text{adjacent}}{\text{hypotenuse}} = \dfrac{12}{13} \approx .9231$

$\tan\theta = \dfrac{\text{opposite}}{\text{adjacent}} = \dfrac{5}{12} \approx .4167$

6. $\sin\theta = .3846$

$\theta = 22.6°$

7. $H^2 = 5^2 + 10^2$

$H^2 = 25 + 100$

$H^2 = 125$

$H = \sqrt{125} \approx 11.18 \text{ mi}$

$\tan\theta = \dfrac{10}{5} = 2$

$\theta \approx 63.4°$

$11.18 \text{ mi}, 63.4°$

8. $H^2 = 100^2 + 40^2$

$H^2 = 10,000 + 1,600$

$H^2 = 11,600$

$H = \sqrt{11,600} \approx 107.7 \text{ mi}$

$\tan\theta = \dfrac{40}{100} = .4$

$\theta \approx 21.8°$

$107.7 \text{ mi}, 21.8°$

You could have used sin or cos and gotten the right answer, as long as you set the ratio up correctly.

Lesson Practice 31B

1. $\sin = \dfrac{\text{opposite}}{\text{hypotenuse}}$

$\cos = \dfrac{\text{adjacent}}{\text{hypotenuse}}$

$\tan = \dfrac{\text{opposite}}{\text{adjacent}}$

2. direction and magnitude

3. resultant

4. $H^2 = 3^2 + 11^2$

$H^2 = 9 + 121$

$H^2 = 130$

$H = \sqrt{130} \approx 11.4 \text{ mi}$

5. $\sin\theta = \dfrac{\text{opposite}}{\text{hypotenuse}} = \dfrac{11}{11.4} \approx .9649$

$\cos\theta = \dfrac{\text{adjacent}}{\text{hypotenuse}} = \dfrac{3}{11.4} \approx .2632$

$\tan\theta = \dfrac{\text{opposite}}{\text{adjacent}} = \dfrac{11}{3} \approx 3.6667$

6. $\tan\theta = 3.6667$

$\theta \approx 74.7°$

7. $H^2 = 100^2 + 10^2$

$H^2 = 10,000 + 100$

$H^2 = 10,100$

$H = \sqrt{10,100} \approx 100.50 \text{ mi}$

$\tan\theta = \dfrac{10}{100} = .1000$

$\theta \approx 5.7° \qquad 180° - 5.7° = 174.3°$

8. $(+4)+(+2) = +6$ on the X axis
$(+3)+(+1) = +4$ on the Y axis

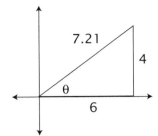

9. $H^2 = 6^2 + 4^2$
$H^2 = 36 + 16$
$H^2 = 52$
$H = \sqrt{52} = \sqrt{4}\sqrt{13} = 2\sqrt{13} \approx 7.21$
$\tan\theta = \dfrac{4}{6} \approx .6667$
$\theta \approx 33.7°$

Systematic Review 31C

1. $H^2 = 30^2 + 40^2$
$H^2 = 900 + 1,600$
$H^2 = 2,500$
$H = \sqrt{2,500} = 50$
$\tan\theta = \dfrac{40}{30} \approx 1.3333$
$\theta \approx 53.1°$
50 miles, 53.1°

2. $(+8)+(-5) = +3$ on the X axis
$(+2)+(+3) = +5$ on the Y axis
$H^2 = 3^2 + 5^2$
$H^2 = 9 + 25$
$H^2 = 34$
$H = \sqrt{34} \approx 5.83$
$\tan\theta = \dfrac{5}{3} \approx 1.6667$
$\theta \approx 59.0°$

3. $Y - X^2 = 0$
$Y = X^2$
see graph

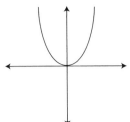

4. $2X + Y < 2$
$Y < -2X + 2$
test points: $(0, 0), (2, 2)$
$Y < -2X + 2 \Rightarrow (0) < -2(0) + 2$
$0 < 0 + 2$
$0 < 2 :$ true
$Y < -2X + 2 \Rightarrow (2) < -2(2) + 2$
$2 < -4 + 2$
$2 < -2 :$ false
see graph

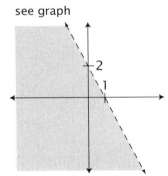

5. $XY = -4$
see graph

6. $m = \dfrac{(-5)-(-2)}{(-3)-(1)} = \dfrac{-3}{-4} = \dfrac{3}{4}$

7. $2\sqrt{-24} + 3\sqrt{8} =$

$2\sqrt{-1}\sqrt{4}\sqrt{6} + 3\sqrt{4}\sqrt{2} =$

$2i(2)\sqrt{6} + 3(2)\sqrt{2} =$

$4i\sqrt{6} + 6\sqrt{2}$

8. $X^2 - 16X + 64 =$

$(X-8)(X-8)$ or $(X-8)^2$

9. $\dfrac{X^2 - \dfrac{2}{X}}{X - \dfrac{3}{X}} = \dfrac{\dfrac{X^2(X)}{(X)} - \dfrac{2}{X}}{\dfrac{X(X)}{(X)} - \dfrac{3}{X}} = \dfrac{\dfrac{X^3-2}{X}}{\dfrac{X^2-3}{X}} =$

$\dfrac{X^3-2}{X} \times \dfrac{X}{X^2-3} = \dfrac{X^3-2}{X^2-3}$

10. $X^2 - 2X - 7 = 0$

$X = \dfrac{-(-2) \pm \sqrt{(-2)^2 - 4(1)(-7)}}{2(1)}$

$X = \dfrac{2 \pm \sqrt{4-(-28)}}{2} = \dfrac{2 \pm \sqrt{32}}{2} =$

$\dfrac{2 \pm \sqrt{16}\sqrt{2}}{2} = \dfrac{2 \pm 4\sqrt{2}}{2} = 1 \pm 2\sqrt{2}$

Systematic Review 31D

1. $H^2 = 75^2 + 125^2$

$H^2 = 5,625 + 15,625$

$H^2 = 21,250$

$H = \sqrt{21,250} \approx 145.77 \text{ mi}$

$\tan\theta = \dfrac{125}{75} \approx 1.6667$

$\theta \approx 59.0°$

2. $(-4) + (-1) = -5$ on the X axis

$(+2) + (+6) = +8$ on the Y axis

$H^2 = 5^2 + 8^2$

$H^2 = 25 + 64$

$H^2 = 89$

$H = \sqrt{89} \approx 9.43$

$\tan\alpha = \dfrac{8}{5} = 1.6$

$\alpha \approx 57.99°$ or $58°$ rounded to nearest tenth

$\theta = 180° - 58° = 122°$

(+5 is used instead of -5 in the Pythagorean theorem because distance or length is always positive. There will be more on this in PreCalculus.)

3. $X^2 + Y^2 = 16$

$\text{center} = (0, 0)$

$\text{radius} = \sqrt{16} = 4$

see graph

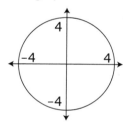

4. $X \geq -3 + Y$

$-Y \geq -X - 3$

$Y \leq X + 3$: sign is reversed because we multiplied by a negative number

test points: $(0, 0), (-3, 3)$

$X \geq -3 + Y \Rightarrow (0) \geq -3 + (0)$

$0 \geq -3$: true

$X \geq -3 + Y \Rightarrow (-3) \geq -3 + (3)$

$-3 \geq 0$: false

see graph

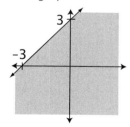

5. $X^2 - Y^2 = 4$

see graph

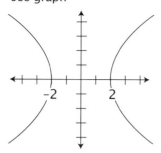

6. $m = \dfrac{1}{2}; (3, 4)$

$Y = mX + b \Rightarrow \qquad (4) = \left(\dfrac{1}{2}\right)(3) + b$

$4 = \dfrac{3}{2} + b$

$4 - \dfrac{3}{2} = b$

$\dfrac{8}{2} - \dfrac{3}{2} = b$

$\dfrac{5}{2} = b$

$Y = \dfrac{1}{2}X + \dfrac{5}{2}$ or $Y = \dfrac{1}{2}X + 2\dfrac{1}{2}$

7. $\dfrac{2\sqrt{12}}{\sqrt{10}} = \dfrac{2\sqrt{12}}{\sqrt{10}}\dfrac{\sqrt{10}}{\sqrt{10}} = \dfrac{2\sqrt{120}}{10} =$

$\dfrac{2\sqrt{4}\sqrt{30}}{10} = \dfrac{2(2)\sqrt{30}}{10} = \dfrac{2\sqrt{30}}{5}$

8. $6X^2 - X - 2 =$

$(3X - 2)(2X + 1)$

9. $\dfrac{\dfrac{1}{X} - 2}{\dfrac{4}{X}} = \dfrac{\dfrac{1}{X} - \dfrac{2(X)}{(X)}}{\dfrac{4}{X}} = \dfrac{\dfrac{1 - 2X}{X}}{\dfrac{4}{X}} =$

$\dfrac{1 - 2X}{X} \times \dfrac{X}{4} = \dfrac{1 - 2X}{4}$

10. $2X^2 - 3X = 2$

$2X^2 - 3X - 2 = 0$

$(2X + 1)(X - 2) = 0$

$\begin{array}{ll} 2X + 1 = 0 & X - 2 = 0 \\ \quad 2X = -1 & \quad X = 2 \\ \quad\ X = -\dfrac{1}{2} & \end{array}$

Systematic Review 31E

1. $H^2 = 15^2 + 30^2$

$H^2 = 225 + 900$

$H^2 = 1{,}125$

$H = \sqrt{1{,}125} \approx 33.54 \text{ ft}$

$\tan\theta = \dfrac{30}{15} = 2$

$\theta \approx 63.4°$

33.54 ft, 63.4°

2. $(-8) + (+2) = -6$ on the X axis

$(+6) + (-3) = +3$ on the Y axis

$H^2 = 6^2 + 3^2$

$H^2 = 36 + 9$

$H^2 = 45$

$H = \sqrt{45} \approx 6.71$

$\tan\alpha = \dfrac{3}{6} = .5$

$\alpha \approx 26.6°$

$\theta = 180° - 26.6° = 153.4°$

3. $X - Y^2 = 0$

$X = Y^2$

see graph

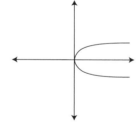

4. $\sqrt{X} = Y$

This is similar to $X = Y^2$, but shows only the part of the graph above the X-axis. If we had $\pm\sqrt{X} = Y$, it would show both the top and bottom half of the parabola.

see graph

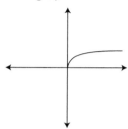

5. $4X^2 + 9Y^2 = 36 \Rightarrow 4(0)^2 + 9Y^2 = 36$

$$9Y^2 = 36$$
$$Y^2 = 4$$
$$Y = \pm 2$$

$4X^2 + 9Y^2 = 36 \Rightarrow 4X^2 + 9(0)^2 = 36$

$$4X^2 = 36$$
$$X^2 = 9$$
$$X = \pm 3$$

see graph

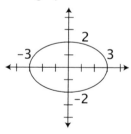

6. $D^2 = \left[(6) - (-2)\right]^2 + \left[(-3) - (1)\right]^2$

$D^2 = [8]^2 + [-4]^2 = 64 + 16 = 80$

$D = \sqrt{80} \approx 8.9$

7. $\dfrac{4}{\sqrt{3}} + \dfrac{6\sqrt{2}}{\sqrt{5}} = \dfrac{4\sqrt{3}}{\sqrt{3}\sqrt{3}} + \dfrac{6\sqrt{2}\sqrt{5}}{\sqrt{5}\sqrt{5}} =$

$\dfrac{4\sqrt{3}}{3} + \dfrac{6\sqrt{10}}{5} = \dfrac{4\sqrt{3}\,(5)}{3(5)} + \dfrac{6\sqrt{10}\,(3)}{5(3)} =$

$\dfrac{20\sqrt{3}}{15} + \dfrac{18\sqrt{10}}{15} = \dfrac{20\sqrt{3} + 18\sqrt{10}}{15}$

8. $10X^2 + 23X + 12 =$

$(5X + 4)(2X + 3)$

9. $3B^{-2}A^3C - \dfrac{2C}{B^2A^{-3}} + \dfrac{7CB^2}{A^{-3}} =$

$\dfrac{3A^3C}{B^2} - \dfrac{2A^3C}{B^2} + \dfrac{7A^3B^2C}{1} = \dfrac{A^3C}{B^2} + 7A^3B^2C$

10. $X^2 + X + 1 = 0$

$X = \dfrac{-(1) \pm \sqrt{(1)^2 - 4(1)(1)}}{2(1)} =$

$\dfrac{-1 \pm \sqrt{1 - 4}}{2} = \dfrac{-1 \pm \sqrt{-3}}{2} = \dfrac{-1 \pm i\sqrt{3}}{2}$

Honors Solutions

Honors Lesson 1

1. A. Multiplying by $\frac{1}{2}$:

$$1 \div \frac{3}{2} \times \frac{3}{4} = 1 \times \frac{2}{3} \times \frac{3}{4} = \frac{6}{12} = \frac{1}{2}$$

$$1 \times \frac{1}{2} = \frac{1}{2}$$

2. Substitute 3 for $(r - s)$:

$$3(3) + \frac{(3)}{18} - (3)^2 - 3 =$$

$$9 + \frac{1}{6} - 9 - 3 =$$

$$\frac{1}{6} - 3 = -2\frac{5}{6}$$

3. Since it is a square, we know all 4 sides are equal, therefore:

$$X + 9 = 4X$$
$$9 = 3X$$
$$3 = X$$

4. $A = (X + 9)(4X)$ square units
using $X = 3$ from #3
$$A = (3 + 9)(4 \cdot 3)$$
$$A = (12)(12) = 144 \text{ square units}$$

5. $$(X + 9)(4X) = 144$$
$$4X^2 + 36X = 144$$
$$4X^2 + 36X - 144 = 0$$
$$X^2 + 9X - 36 = 0$$
$$(X - 3)(X + 12) = 0$$
$$X = 3 \quad \text{same as #3}$$
$$X = -12 \quad \text{This solution does not make sense.}$$
$$\text{We say that it is invalid.}$$

6. $4 : 5$
$$\frac{4}{5} = \frac{8}{C}$$
$$4C = 40$$
$$C = 10$$

7. $\text{Slope} = \dfrac{\text{rise}}{\text{run}} = \dfrac{2}{1} = 2$

8. $180 - 35 = 145°$

9. $\dfrac{X^4Y^2 + X^2Y}{X^2Y} = X^2Y + 1$

10. Plug in values for X and Y:

$$(2)^2(3) + 1 = (4)(3) + 1 = 12 + 1 = 13 \text{ one side}$$
$$(2)^2(3) = (4)(3) = 12 \text{ other side}$$
$$\text{Area} = 13 \times 12 = 156 \text{ square units}$$

Honors Lesson 2

1. $$\frac{t}{8} + \frac{t}{12} = 1$$
$$24\left(\frac{t}{8} + \frac{t}{12}\right) = 24$$
$$3t + 2t = 24$$
$$5t = 24$$
$$t = 4\frac{4}{5} \text{ hours or 4 hours and 48 minutes}$$

2. $$\frac{t}{30} + \frac{t}{45} = 1$$
$$3t + 2t = 90 \quad \left(\begin{array}{l}\text{multiplied both} \\ \text{sides by 60}\end{array}\right)$$
$$5t = 90$$
$$t = 18 \text{ minutes}$$

3. $$\frac{t}{20} + \frac{t}{10} + \frac{t}{12} = 1$$
$$3t + 6t + 5t = 60 \quad \left(\begin{array}{l}\text{multiplied both} \\ \text{sides by 60}\end{array}\right)$$
$$14t = 60$$
$$t = 4\frac{2}{7} \text{ days}$$

4. Subtract this time, since the faucet and the drain are working against each other:

$$\frac{t}{15} - \frac{t}{20} = 1$$
$$4t - 3t = 60 \quad \left(\begin{array}{l}\text{multiplied both} \\ \text{sides by 60}\end{array}\right)$$
$$t = 60 \text{ minutes or 1 hour}$$

Honors Lesson 3

1.

rate of work	× time worked	= portion of job done
1/6	2 hours	1/3
1/10	6 2/3 hours	2/3

The rates have already been filled in. We are given the amount of time that the gardener worked, so we fill that in, and then figure out how much of the job he completed. If 1/3 of the job is done, then 2/3 of the job is left. Fill in that amount, and then figure the time worked by the helper by using the values and solving for time.

$RT = J$

$\left(\dfrac{1}{6}\right)(2) = J$

$\dfrac{1}{3} = J$

$RT = J$

$\left(\dfrac{1}{10}\right)(T) = \dfrac{2}{3}$

$T = \dfrac{20}{3}$

$T = 6\dfrac{2}{3}$ or 6 hours 40 minutes

2. 5

3. 1

4. 4

5. 2

6. 2

7. 2

8. 3

9. 4

10. 2.45×10^8 ft; 3 significant digits

11. 9×10^{-5} m; 1 significant digit

12. 1.304×10^3 tons; 4 significant digits

13. 1.50×10^0 g; 3 significant digits

Honors Lesson 4

1.

rate of work	×	time worked	= portion of job done
$\dfrac{1}{12}$		3 hours	$\dfrac{1}{4}$
$\dfrac{1}{20}$		15 hours	$\dfrac{3}{4}$

The mason works at the rate of 1/12 of the job per hour, and he worked for 3 hours. We also know his helper worked for a total of 15 hours. Using the formula, we find that the mason did 1/4 of the job. His helper, therefore, did 3/4 of the job. Use the formula again to find out the helper's rate:

$RT = J$

$R(15) = \dfrac{3}{4}$

$R = \dfrac{3}{60}$ or $\dfrac{1}{20}$

Working alone, the helper would have taken 20 hours to do the job.

2. $250 + 12.5 = 262.5$ ft; round to 260 ft

3. $.5 - .361 = .139$ in; round to .1 in

4. $\left(5.8 \times 0^4\right) + \left(1.2 \times 10^{-2}\right) = 58,000 + .012 = 58,000.012$ m

round to 58,000 m or 5.8×10^4 m

5. $650,000 - 3,400 = 646,600$ g; round to 650,000 or 6.5×10^5 g

6. $151 \times 6 = 906$ ft²; round to 900 ft²

7. $.0025 \div .10 = .025$ in; two significant digits

8. $\left(2.8 \times 10^2\right) \times \left(1.04 \times 10^2\right) = \left(2.912 \times 10^4\right)$ m²; round to $\left(2.9 \times 10^4\right)$ m²

9. $\left(3.6 \times 10^8\right) \div \left(1.2 \times 10^4\right) = 3.0 \times 10^4$ km; two significant digits

10. Area $= 19.1 \times 6 = 114.6$; round to 100 m² (one significant digit)
Perimeter $= 19.1 + 6 + 19.1 + 6 = 50.2$ m; round to 50 m

Honors Lesson 5

1. $d = rt$
   $d = (3)(40)$
   $d = 120$ miles

2. $d = (6)(40) = 240$ miles
   240 is twice 120, so increased by a factor of 2

3. $d = (5)(60) = 300$ miles
   $d = (10)(60) = 600$ miles
   600 is twice 300, so increased by a factor of 2

4. It will double, or increase by a factor of 2.

5. $5^2 = 25$

6. $10^2 = 100$
   100 is 4 times 25, so a factor of 4

7. $4^2 = 16$
   $8^2 = 64$
   64 is 4 times 16, so a factor of 4

8. The value of X should increase by a factor of 4.

9. $L = 12 \div 2$
   $L = 6$

10. $L = 12 \div 4$
    $L = 3$
    The length decreases as the width increases.
    Doubling the length decreases the width by a factor of 2.

11. The value of X should decrease by a factor of 2.

Honors Lesson 6

1. X

2. Use the pythagorean theorem:
   $A^2 + B^2 = C^2$
   substitute X for A, and 2X for B:
   $X^2 + (2X)^2 = C^2$
   $5X^2 = C^2$
   $\sqrt{5X^2} = C$
   $X\sqrt{5} = C$

3. $X\sqrt{5} + X$

4. $X\sqrt{5} - X$

5. $\dfrac{X\sqrt{5} + X}{2X}$

6. $\dfrac{\sqrt{5} + 1}{2}$

7. 1.618

8. answers will vary

9. $1 \div 1.618 = .618$

10. $\dfrac{5}{8}$ is close (.625), but you may have come up with something closer.

Honors Lesson 7

1. $A^{\frac{X}{Y}} = \left(\sqrt[Y]{A}\right)^X$

2. Q times itself R times

3. $\left(X^{\frac{a}{b}}\right)^{\frac{b}{a}} = X$

4. $\left(Y^{\frac{a}{b}}\right)^{\frac{c}{d}} = \left(\sqrt[bd]{Y}\right)^{ac}$

5. $\left(Y^F \cdot Y^G\right)^{\frac{1}{H}} = \left(\sqrt[H]{Y}\right)^{F+G}$

6. $\left(X^F \cdot Y^F\right)^G = X^{FG} Y^{FG}$

7. $\left(M^{\frac{x}{z}} \cdot M^{\frac{y}{z}}\right)^{\frac{z}{y}} = \left(M^{\frac{x+y}{z}}\right)^{\frac{z}{y}} = M^{\frac{x+y}{y}}$

8. $\left[\left(x^a\right)^b \cdot x^b\right]^{\frac{1}{c}} = \sqrt[c]{\left(x^{ab+b}\right)}$

9. $\left(p^a + p^a\right)^{\frac{a}{b}} = \left(\sqrt[b]{2p^a}\right)^a$

10. $\left(x^E \div x^F\right)^H = \left(x^{E-F}\right)^H$

Honors Lesson 8

1. 0

2. a negative number that is not a fraction or decimal, for example: –6

3. any fraction, for example: $\frac{3}{5}$

4. π, $\sqrt{2}$, $\sqrt{3}$, etc.

5. see chart below

6. Each number in the series is the sum of the previous two numbers

7. 8, 13, 21

# of months	drawing of pairs	# of pairs
1		1
2		1
3		2
4		3
5		5

Honors Lesson 9

1. Row 0 : 1
   Row 1 : 2
   Row 2 : 4
   Row 3 : 8
   Row 4 : 16
   Row 5 : 32
   Row 6 : 64
   The sum of each row is twice the previous row.

2.

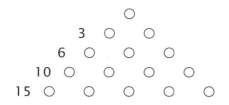

```
        3  ○  ○
      6  ○  ○  ○
   10  ○  ○  ○  ○
15  ○  ○  ○  ○  ○
```

3. 21

```
                1
             1     1
          1     2     1
       1    3    3    1
    1    4    6    4    1
  1   5   10   10   5   1
 1   6   15   20   15   6   1
1   7   21   35   35   21   7   1
```

4. $3+6=9$; $6+10=16$; $10+15=25$;
$15+21=36$ They are all perfect squares.

5. The Fibonacci
Sequence

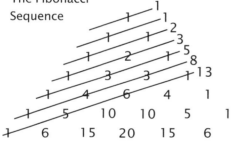

6. $4 \times 3 \times 2 \times 1 = 24$

7. $5 \times 4 \times 3 \times 2 \times 1 = 120$

8. $\dfrac{9!}{7!} = \dfrac{9 \times 8 \times 7 \times 6 \times 5 \times 4 \times 3 \times 2 \times 1}{7 \times 6 \times 5 \times 4 \times 3 \times 2 \times 1}$
$= 9 \times 8 = 72$

9. $\dfrac{6!}{3!3!} = \dfrac{6 \times 5 \times 4 \times 3 \times 2 \times 1}{3 \times 2 \times 1 \times 3 \times 2 \times 1}$

$\dfrac{6 \times 5 \times 4}{3 \times 2 \times 1} = 5 \times 4 = 20$

10. $\dfrac{201!}{200!} = \dfrac{201 \times 200!}{200!} = 201$

Honors Lesson 10

1. A, B, C
A, C, B
B, A, C
B, C, A
C, A, B
C, B, A
6 ways

2. $4! = 4 \cdot 3 \cdot 2 \cdot 1 = 24$

3. $6! = 6 \cdot 5 \cdot 4 \cdot 3 \cdot 2 \cdot 1 = 720$

4. $_9P_4 = \dfrac{9!}{(9-4)!} = \dfrac{9!}{5!} =$

$\dfrac{9 \times 8 \times 7 \times 6 \times 5 \times 4 \times 3 \times 2 \times 1}{5 \times 4 \times 3 \times 2 \times 1}$

$= 9 \times 8 \times 7 \times 6 = 3,024$

5. $_{20}P_5 = \dfrac{20!}{(20-5)!} = \dfrac{20!}{15!} =$
$20 \times 19 \times 18 \times 17 \times 16 =$
$1,860,480$

6. $_{21}P_6 = \dfrac{21!}{(21-6)!} = \dfrac{21!}{15!} =$
$21 \times 20 \times 19 \times 18 \times 17 \times 16 =$
$39,070,080$

Honors Lesson 11

1.

like	ilke	klie	elik
liek	ilek	klei	elki
leik	ikle	kiel	eilk
leki	ikel	kile	eikl
lkie	ielk	keli	ekli
lkei	iekl	keil	ekil

24 ways; yes

2.

look	olko	oklo
loko	ookl	kloo
lkoo	oolk	kool
olok	okol	kolo

12 ways; no

3. $P = \dfrac{5!}{2!} = \dfrac{5 \times 4 \times 3 \times 2 \times 1}{2 \times 1} =$
$5 \times 4 \times 3 = 60$

4. $P = \dfrac{6!}{3!} = \dfrac{6 \times 5 \times 4 \times \cancel{3 \times 2 \times 1}}{\cancel{3 \times 2 \times 1}} =$

$6 \times 5 \times 4 = 120$

5. $P = \dfrac{6!}{2!} = \dfrac{6 \times 5 \times 4 \times 3 \times \cancel{2 \times 1}}{\cancel{2 \times 1}} =$

$6 \times 5 \times 4 \times 3 = 360$

6. $P = \dfrac{6!}{3!2!} =$

$\dfrac{\overset{3}{\cancel{6}} \times 5 \times 4 \times \cancel{3 \times 2 \times 1}}{\cancel{3 \times 2 \times 1} \times \cancel{2 \times 1}} = 5 \times 4 \times 3 = 60$

7. m, a, and t each appear twice

$P = \dfrac{11!}{2!2!2!} =$

$\dfrac{11 \times 10 \times 9 \times 8 \times 7 \times 6 \times 5 \times \cancel{4} \times 3 \times \cancel{2 \times 1}}{\cancel{2} \times 1 \times \cancel{2} \times 1 \times \cancel{2 \times 1}} =$

$11 \times 10 \times 9 \times 8 \times 7 \times 6 \times 5 \times 3 = 4{,}989{,}600$

8. $P = \dfrac{20!}{15!3!2!} = \dfrac{20 \times 19 \times \overset{3}{\cancel{18}} \times 17 \times \overset{8}{\cancel{16}}}{\cancel{3 \times 2} \times 1 \times \cancel{2} \times 1} =$

$20 \times 19 \times 3 \times 17 \times 8 = 155{,}040$

Honors Lesson 12

1. $\dbinom{6}{5-1} X^{6-5+1} Y^{5-1} = \dbinom{6}{4} X^2 Y^4$

$\dfrac{6!}{2!4!} X^2 Y^4 = \dfrac{6 \times 5 \times \cancel{4!}}{2 \times \cancel{4!}} = 15 X^2 Y^4$

2. $\dbinom{4}{2-1} A^{4-2+1} 2^{2-1} = \dbinom{4}{1} A^3 2 = 8 A^3$

3. $\dbinom{5}{3-1} P^{5-3+1} Q^{3-1} = \dbinom{5}{2} P^3 Q^2$

$\dfrac{5!}{3!2!} P^3 Q^2 = \dfrac{5 \times 4 \times \cancel{3!}}{\cancel{3!} \times 2 \times 1} = 10 P^3 Q^2$

4. $\dbinom{7}{4-1}(2X)^{7-4+1}\left(-1^{4-1}\right) = \dbinom{7}{3}(2X)^4 \left(-1^3\right) =$

$\dfrac{7!}{4!3!}\left(-16X^4\right) = \dfrac{7 \times \cancel{6} \times 5 \times \cancel{4!}}{\cancel{4!} \times \cancel{3} \times 2 \times 1}\left(-16X^4\right)$

$= (35)\left(-16X^4\right) = -560X^4$

Honors Lesson 13

1. $\text{Area} = X(20 - 2X)$

$= 20X - 2X^2$

$48 = 20X - 2X^2$

$24 = 10X - X^2$

$X^2 - 10X + 24 = 0$

$(X - 6)(X - 4) = 0$

$X = 6,\ 4$

If $X = 6$ feet, then the long side would be:

$20 - 2(6) = 20 - 12 = 8$ ft

If $X = 4$ feet, then the long side would be:

$20 - 2(4) = 20 - 8 = 12$ ft

2. $\text{Area} = X\left(\dfrac{160 - 3X}{2}\right)$

$= \dfrac{160X - 3X^2}{2}$

$1{,}000 = \dfrac{160X - 3X^2}{2}$

$2000 = 160X - 3X^2$

$3X^2 - 160X + 2{,}000 = 0$

$(X - 20)(3X - 100) = 0$

$X = 20,\ 33\dfrac{1}{3}$

If $X = 20$, then the other side would be:

$(160 - 3(20)) \div 2 =$

$(160 - 60) \div 2 =$

$100 \div 2 = 50$ ft

$20 \times 50 = 1{,}000$ ft^2

If $X = 33\dfrac{1}{3}$, the other side would be:

$\left(160 - 3\left(33\dfrac{1}{3}\right)\right) \div 2 =$

$(160 - 100) \div 2 =$

$60 \div 2 = 30$ ft

$30 \times 33\dfrac{1}{3} = 1000$ ft^2

3. $\text{Area} = \dfrac{X(X+2)}{2}$

$24 = \dfrac{X^2 + 2X}{2}$

$48 = X^2 + 2X$

$0 = X^2 + 2X - 48$

$(X+8)(X-6) = 0$

$X = -8, 6$

$X = -8$ makes no sense

If $X = 6$, then the height is:

$(6) + 2 = 8$

$\dfrac{1}{2}(6)(8) = 24 \text{ in}^2$

4. $(x)(X+4) = 192$

$X^2 + 4X = 192$

$X^2 + 4X - 192 = 0$

$(X+16)(X-12) = 0$

$X = -16, 12$

$X = -16$ makes no sense

If $X = 12$, then the length is:

$(12) + 4 = 16$

$(12)(16) = 192 \text{ in}^2$

Honors Lesson 14

1. Done

2. $145,000 \times .26 = \$37,700$ increase

$145,000 + 37,700 = \$182,700$ now

3. $.25 \times 150 = \$37.50$ amount of decrease

$150 - 37.50 = \$112.50$ new price

4. $28.5 \times .29 = 8.265$ more bushels per acre

$4.15 \times 8.265 = \$34.30$ more per acre in sales

$34.30 - 25.00 = \$9.30$ benefit per acre

5. $9.30 \times 150 =$

$\$1,395.00$ more than without fertilizer

$4.15 \times 150 \times 28.5 = \$17,741.25$

$1,395 = WP \times 17,741.25$

$WP = .079$ or 7.9% (rounded)

6. $29,352 - 20,578 = 8,774$ increase

$8,774 = WP \times 20,578$

$WP = .426$ or 42.6% increase (rounded)

$.426 \times 29,352 =$

$\$12,503.95$ increase next year

$29,352 + 12,503.95 =$

$\$41,855.95$ in sales next year

if there is the same percentage increase

7. $4.7 - 4.1 = .6$ gallons saved

$.6 = WP \times 4.7$

$WP = .128$ or 12.8% (rounded)

8. $.6$ gallons saved per hundred miles driven, so

$.6 \times 6 = 3.6$ gallons saved

$3.6 \times 1.98 = \$7.13$ saved (rounded)

9. $20,567 \times 4.00 = 82,268$

10. $82,268 - 20,567 = 61,701$ increase

$61,701 = WP \times 20,567$

$WP = 3$ or 300%

Honors Lesson 15

1. $\dfrac{E}{h} = f$

2. $PA = F$

$A = \dfrac{F}{P}$

3. $P = 2L + 2W$

$P - 2L = 2W$

$\dfrac{P - 2L}{2} = W$

4. $kT = PV$

$P = \dfrac{kT}{V}$

5. $N = \dfrac{a+b}{2}$

$2N = a + b$

$2N - b = a$

6. $M = \dfrac{a+b}{c+d}$

$M(c+d) = a+b$

$c + d = \dfrac{a+b}{M}$

$c = \dfrac{a+b}{M} - d$

7. It will increase.

$t = 2, r = 40 : d = (2)(40) = 80$

$t = 2, r = 60 : d = (2)(60) = 120$

8. It will decrease.

$t = 2, r = 40 : d = (2)(40) = 80$

$t = 1, r = 40 : d = (1)(40) = 40$

9. R will increase as E increases.

10. R will decrease as i increases.

Honors Lesson 16

1. The smaller gear will move faster.

2. $RN = rn$

$120(12) = r(6)$

$1,440 = 6r$

$r = 240$ rpm

3. $RN = rn$

$\dfrac{RN}{r} = n$ Divide both sides by r.

$\dfrac{R}{r} = \dfrac{n}{N}$ Divide both sides by N.

4. $N = \dfrac{rn}{R}$ $R = \dfrac{rn}{N}$

$n = \dfrac{RN}{r}$ $r = \dfrac{RN}{n}$

5. $r = \dfrac{RN}{n}$

$r = \dfrac{300(40)}{30}$

$r = \dfrac{12,000}{30}$

$r = 400$ rpm

6. $N = \dfrac{rn}{R}$

$N = \dfrac{150(55)}{50}$

$N = \dfrac{8,250}{50}$

$N = 165$ teeth

7. $R = \dfrac{rn}{N}$

$R = \dfrac{600(60)}{90}$

$R = \dfrac{36,000}{90}$

$R = 400$ rpm

8. $2,000(10) = r(4,000)$

$20,000 = 4,000r$

$r = 5$ in

Honors Lesson 17

1. $X^4 + 3X^2 - 10$

$W^2 + 3W - 10$

$(W + 5)(W - 2)$

$(X^2 + 5)(X^2 - 2)$

2. $X^4 - 8X^2 + 12$

$W^2 - 8W + 12$

$(W - 2)(W - 6)$

$(X^2 - 2)(X^2 - 6)$

3. $X + 3\sqrt{X} + 2$

$W^2 + 3W + 2$

$(W + 1)(W + 2)$

$(\sqrt{X} + 1)(\sqrt{X} + 2)$

4. $\dfrac{X - 2}{-X^2 + 3X - 2} = \dfrac{X - 2}{(-1)(X^2 - 3X + 2)} =$

$\dfrac{\cancel{X - 2}}{(-1)(X - 1)\cancel{(X - 2)}} = \dfrac{1}{-X + 1}$ or $\dfrac{1}{1 - X}$

5. $\dfrac{3 - X}{X^2 - 9} = \dfrac{(-1)\cancel{(X - 3)}}{(X + 3)\cancel{(X - 3)}} = \dfrac{-1}{X + 3}$

6. $\dfrac{X^2 - 4}{2 - X} \cdot \dfrac{X + 3}{9 - X^2} =$

$\dfrac{(X + 2)\cancel{(X - 2)}}{(-1)\cancel{(X - 2)}} \cdot \dfrac{\cancel{X + 3}}{(-1)(X - 3)\cancel{(X + 3)}} = \dfrac{X + 2}{X - 3}$

Honors Lesson 18

1. center rectangle:

$$(X+3)\left[(X+1)+2\right] =$$
$$(X+3)(X+3) = X^2+6X+9$$

2 smaller rectangles:

$$2\left[(2)(X+1)\right] = 4X+4$$

together:

$$X^2+6X+9+4X+4 = X^2+10X+13 \text{ units}^2$$

2. $(3)^2+10(3)+13 =$
$9+30+13 = 52 \text{ ft}^2$

3. lower section:

$$(X)(2X+4)(X+3) =$$
$$\left(2X^2+4X\right)(X+3) =$$
$$2X^3+4X^2+6X^2+12X = 2X^3+10X^2+12X$$

top section:

$$(2X)(X+3)\left((2X+4)-2X\right) =$$
$$\left(2X^2+6X\right)(4) = 8X^2+24X$$

together :

$$2X^3+10X^2+12X+8X^2+24X =$$
$$2X^3+18X^2+36X \text{ units}^3$$

4. $\frac{4}{3}\pi(X+2)^3 = \frac{4}{3}\pi(X^3+6X^2+12X+8)$ or

$$\frac{4}{3}\pi X^3+8\pi X^2+16\pi X+\frac{32}{3}\pi$$

5. Answers may vary – choose a value for n and raise to the sixth power. For example, $3^6 = 729$. Other possibilities are: 1; 729; 4,096; 15,625

6. $n^{10} = n^5 \times n^5 = \left(n^5\right)^2$

$$n^{10} = n^2 \times n^2 \times n^2 \times n^2 \times n^2 = \left(n^2\right)^5$$

Honors Lesson 19

1. $\dfrac{N_P}{N_S} = \dfrac{E_P}{E_S}$

$$\frac{100}{20} = \frac{600}{E_S}$$

$$100E_S = 12,000$$

$$E_S = 120 \text{ volts}$$

2. $\dfrac{N_P}{N_S} = \dfrac{E_P}{E_S}$

$$\frac{480}{N_S} = \frac{7,200}{240}$$

$$7,200N_S = 480(240)$$

$$7,200N_S = 115,200$$

$$N_S = 16 \text{ turns}$$

3. $\dfrac{N_P}{N_S} = \dfrac{E_P}{E_S}$

$$\frac{500}{300} = \frac{E_P}{750}$$

$$300E_P = 500(750)$$

$$300E_P = 375,000$$

$$E_P = 1,250 \text{ volts}$$

Honors Lesson 20

1. $\rho = \dfrac{m}{V}$

$$m = V\rho$$

$$m = (10)(.009) = .09$$

2. $f = \dfrac{1}{T}$

$$T = \frac{1}{f}$$

$$T = \frac{1}{1.3} = .77 \text{ (rounded)}$$

3. $PE = mgh$

    $h = \dfrac{PE}{mg}$

    $h = \dfrac{1764}{(30)(9.8)} = 6$

4. $F = \dfrac{kq_1 q_2}{r^2}$

    $r^2 = \dfrac{kq_1 q_2}{F}$

    $r^2 = \dfrac{(9.0\times10^9)(4.0\times10^{-2})(2.0\times10^{-3})}{1.8\times10^5}$

    $r^2 = \dfrac{72\times10^4}{1.8\times10^5} = 40\times10^{-1} = 4$

    $r = 2$

5. $PV = nRT$

    $V = \dfrac{nRT}{P}$

    $V = \dfrac{(.5)(.0821)(293)}{.95} = 12.66$ (rounded)

Honors Lesson 21

1.

t(hours)	0	2	4	6	8	10	12	14	16
b (bacteria in thousands)	1	2	4	8	16	32	64	128	256

2.

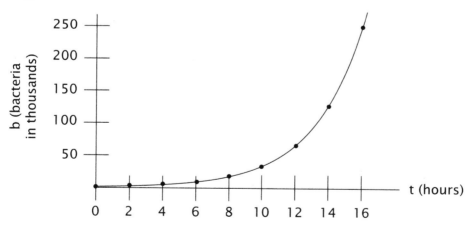

3.

X	Y
0	5
1	6
2	8
3	12
−1	4.5
−2	4.25
−3	4.125

4.

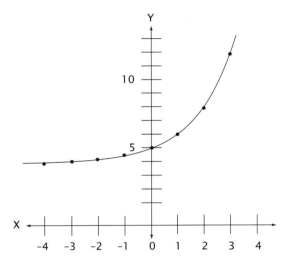

5. Y increases faster and faster.

2.

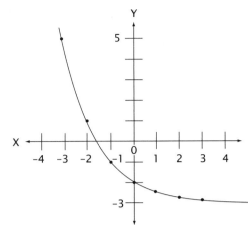

3. $M = A(1.1)^{t/d}$

$M = 100(1.1)^{4/2}$

$M = 100(1.1)^2 = 121$ students

4.

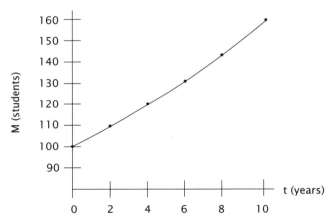

5. Since the number increases by a factor of 1.1 every 2 years, see how many times you have to multiply the original number by 1.1 to reach 133, and then multiply by 2:

$(100)(1.1) = 110$

$(110)(1.1) = 121$

$(121)(1.1) = 133.1$ so 6 years

Honors Lesson 22

1.

X	Y
0	−2
1	−2.5
2	−2.75
3	−2.875
−1	−1
−2	1
−3	5

Honors Lesson 23

1. $YX - YT = YZ$
   $X - T = Z$
   $X = Z + T$

2. $Q(X + B) = R(X + C)$
   $QX + QB = RX + RC$
   $QX - RX = RC - QB$
   $X(Q - R) = RC - QB$
   $X = \dfrac{RC - QB}{Q - R}$

3. $AX - BX - C = CX + X + E$
   $AX - BX - CX - X = E + C$
   $X(A - B - C - 1) = E + C$
   $X = \dfrac{E + C}{A - B - C - 1}$

4. $X(A + B + C) + Y - Z = A$
   $X(A + B + C) = A - Y + Z$
   $X = \dfrac{A - Y + Z}{A + B + C}$

5. $C(X - Y) + F = CAB - CY + F$
   $C(X - Y) = CAB - CY$
   $X - Y = AB - Y$
   $X = AB$

   Your work for the following problems may look different, depending on which equation was substituted into the other. You should have the same final answers.

6. $Y = R + 2X$
   $Y = S + X$
   Substitute S + X for Y in the first equation:
   $S + X = R + 2X$
   $S - R = 2X - X$
   $S - R = X$

7. $Y = EX$
   $Y + EX = Q$
   Substitute EX for Y in the second equation:
   $EX + EX = Q$
   $X(E + E) = Q$
   $X = \dfrac{Q}{2E}$

8. $X = Y + A$
   $X = BY - B$
   Solve first equation for Y: $X - A = Y$
   Substitute X – A for Y in the second equation:
   $X = B(X - A) - B$
   $X = BX - AB - B$
   $X - BX = -AB - B$
   $X(1 - B) = -AB - B$
   $X = \dfrac{-AB - B}{1 - B}$

9. $Y - X = Q$
   $Y + RX = T$
   Solve the first equation for Y:
   $Y = X + Q$
   Substitute X + Q for Y in the second equation:
   $(X + Q) + RX = T$
   $X + RX = T - Q$
   $X(1 + R) = T - Q$
   $X = \dfrac{T - Q}{1 + R}$

10. $Y - CX = C$
    $Y + DX = -D$
    Solve the first equation for Y:
    $Y = C + CX$
    Substitute C + CX for Y in the second equation:
    $C + CX + DX = -D$
    $CX + DX = -D - C$
    $X(C + D) = -D - C$
    $X = \dfrac{-D - C}{C + D}$

Honors Lesson 24

1.
$$AY + BX = C$$
$$-AY - DX = E$$
$$(B - D)X = C + E$$
$$X = \frac{C + E}{B - D}$$

2.
$$X - Y = R$$
$$AX + Y = T$$
$$AX + X = R + T$$
$$X(A + 1) = R + T$$
$$X = \frac{R + T}{A + 1}$$

$$\frac{R + T}{A + 1} - Y = R$$
$$\frac{R + T}{A + 1} - R = Y$$

3.
$$Y - QX = R$$
$$QY + QX = QT$$
$$(1 + Q)Y = R + QT$$
$$Y = \frac{R + QT}{1 + Q}$$
$$Y = \frac{(2) + (3)(6)}{1 + (3)}$$
$$Y = \frac{20}{4} = 5$$

Honors Lesson 25

1. minimum
2. maximum
3. minimum
4.
$$A = LW$$
$$L = 400 - 2W$$
$$A = W(400 - 2W) = -2W^2 + 400W$$
$$A = W^2 - 200W$$
$$\frac{-b}{2a} = \frac{-(-200)}{2(1)} = 100$$
If $W = 100$ then:
$$L = 400 - 2(100) = 200$$
100×200 encloses the largest area

5.
$$\frac{-b}{2a} = \frac{-(-2)}{2(.4)} =$$
2.5 thousand or 2,500 blouses

6. find maximum of $2X^2 - 60X - 36,000$:
$$\frac{-b}{2a} = \frac{-(-60)}{2(2)} = 15$$
$$X = 15 \text{ units of } 20,000$$
$$X = 300,000 \text{ gallons}$$

7. $\$1.20 + .15 = \$1.35$

8.
$$3,000,000 - 300,000 =$$
2,700,000 gallons sold
$\$1.35 \times 2,700,000 =$
$\$3,645,000$

Honors Lesson 26

1.
$$H = 0.2X^2 - 0.5X + 30$$
$$\text{minimum} = \frac{-(-.5)}{2(.2)} = 1.25$$

$$H = .2(1.25)^2 - .5(1.25) + 30$$
$$H = 29.6875$$
$$H = 29.7 \text{ ft (rounded)}$$

2.
$$P = -0.2X^2 + 50X$$
$$\text{maximum} = \frac{-50}{2(-.2)} = 125$$
$$P = -.2(125)^2 + 50(125) = \$3,125$$

3.
$$H = -0.002X^2 + .5X + 3$$
$$\frac{-.5}{2(-.002)} = 125 \text{ ft}$$
The ball will be at its highest point at 125 feet from where it was thrown, so after that, it will be descending.
Find the value of H when X is 200:

$$H = -0.002(200)^2 + .5(200) + 3$$
$$H = -0.002(40,000) + 100 + 3$$
$$H = -80 + 100 + 3 = 23 \text{ ft} \quad \text{yes}$$

4.
$$C = 0.3X^2 - 3X + 12$$
$$\frac{-(-3)}{2(.3)} = 5 \text{ tons}$$

5.
$$C = 0.3(5)^2 - 3(5) + 12$$
$$C = .3(25) - 15 + 12$$
$$C = 7.5 - 15 + 12 = \$4.50$$

Honors Lesson 27

1.

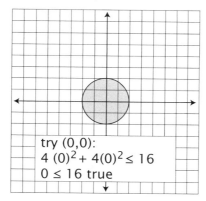

try (0,0):
$4(0)^2 + 4(0)^2 \leq 16$
$0 \leq 16$ true

2.

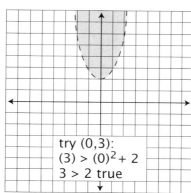

try (0,3):
$(3) > (0)^2 + 2$
$3 > 2$ true

3.

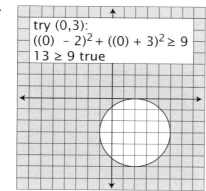

try (0,3):
$((0) - 2)^2 + ((0) + 3)^2 \geq 9$
$13 \geq 9$ true

4.

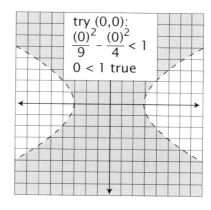

try (0,0):
$\dfrac{(0)^2}{9} - \dfrac{(0)^2}{4} < 1$
$0 < 1$ true

5.

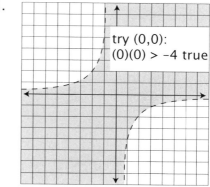

try (0,0):
$(0)(0) > -4$ true

6.

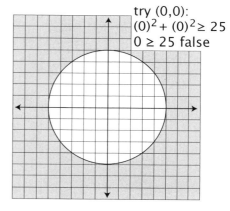

try (0,0):
$(0)^2 + (0)^2 \geq 25$
$0 \geq 25$ false

Honors Lesson 28

For clarity, only the final solutions are shaded.

1.

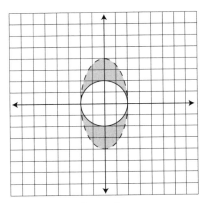

ellipse: $4(0)^2 + (0)^2 < 16$

$0 < 16$ true

circle: $(0)^2 + (0)^2 \geq 4$

$0 \geq 4$ false

2.

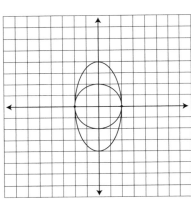

The intersection is in the points $(-2, 0)$ and $(2, 0)$

3.

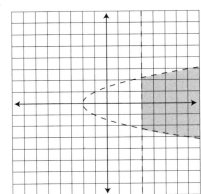

parabola: $(0) > (0)^2 - 2$

$0 > -2$ true

line: $(0) > 3$ false

4.

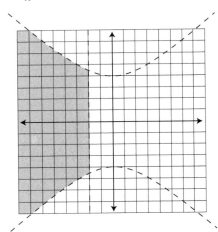

hyperbola: $(0)^2 - (0)^2 < 16$ true

line: $(0) < -2$ false

For #1, 3, and 4, it is a good idea to check additional points to accurately determine the shaded areas.

Honors Lesson 29

1.

	amount	cost per unit	value
low grade	X lb	2.00	2X
ground round	50 lb	3.75	187.50
mixture	X + 50 lb	2.50	2.50 (X + 50)

$$2X + 187.50 = 2.50(X + 50)$$
$$2X + 187.5 = 2.5X + 125$$
$$187.5 - 125 = 2.5X - 2X$$
$$62.5 = .5X$$
$$X = 125 \text{ lb}$$

2.

	amount	cost per unit	value
metal 1	12 lb	2.00	24.00
metal 2	12 lb	X	12X
mixture	24	1.25	30.00

$$24 + 12X = 30$$
$$12X = 6$$
$$X = \$0.50 \text{ per lb}$$

3. $4.75 \div 5 = \$0.95$ per lb

	amount	cost per unit	value
wheat	100	X	100X
corn	50	2X	100X
mixture	150	.95	142.50

$$100X + 100X = 142.50$$
$$200X = 142.50$$
$$X = \$.71 \text{ per lb for wheat}$$
$$X = \$1.43 \text{ per lb for corn}$$

Honors Lesson 30

1.
$$R = (20 + X)(80 - 2X)$$
$$R = -2X^2 + 40X + 1600$$
$$R = (-X^2 + 20X + 800)(2)$$
$$\frac{-b}{2a} = \frac{-20}{2(-1)} = 10$$
$$20 + 10 = 30 \text{ people}$$
$$R = -2(10)^2 + 40(10) + 1600$$
$$R = -200 + 400 + 1600 = \$1800$$

2. $R = (500 + 20W)(.90 - .03W)$

$R = -.6W^2 + 3W + 450$

$\dfrac{-b}{2a} = \dfrac{-3}{2(-.6)} = 2.5$ weeks

$R = -.6(2.5)^2 + 3(2.5) + 450$

$R = -3.75 + 7.5 + 450$

$R = \$453.75$

3. $R = (100,000 - 5,000X)(3 + .5X)$

$R = -2,500X^2 + 35,000X + 300,000$

$\dfrac{-b}{2a} = \dfrac{-35,000}{2(-2,500)} = 7$ units of .50

$\$3.00 + \$3.50 = \$6.50$

$R = -2,500(7)^2 + 35,000(7) + 300,000$

$R = -122,500 + 245,000 + 300,000$

$R = \$422,500$

4. $Y = (30 + X)\left(20 - \dfrac{1}{3}X\right)$ $\qquad X \le 18$

$Y = -\dfrac{1}{3}X^2 + 10X + 600$

$\dfrac{-b}{2a} = \dfrac{-10}{2\left(-\dfrac{1}{3}\right)} = 15$ more trees

$30 + 15 = 45$ trees per acre

Honors Lesson 31

1. $-4X + 3Y = 2; \quad -2X + Y = -6$

$X = \dfrac{\begin{vmatrix} 2 & 3 \\ -6 & 1 \end{vmatrix}}{\begin{vmatrix} -4 & 3 \\ -2 & 1 \end{vmatrix}} = \dfrac{2 - (-18)}{-4 - (-6)} = \dfrac{20}{2} = 10$

$Y = \dfrac{\begin{vmatrix} -4 & 2 \\ -2 & -6 \end{vmatrix}}{\begin{vmatrix} -4 & 3 \\ -2 & 1 \end{vmatrix}} = \dfrac{24 - (-4)}{-4 - (-6)} = \dfrac{28}{2} = 14$

2. $3X - Y = 1; \ 3X + 4Y = -19$

$$X = \frac{\begin{vmatrix} 1 & -1 \\ -19 & 4 \end{vmatrix}}{\begin{vmatrix} 3 & -1 \\ 3 & 4 \end{vmatrix}} = \frac{4 - (19)}{12 - (-3)} = \frac{-15}{15} = -1$$

$$Y = \frac{\begin{vmatrix} 3 & 1 \\ 3 & -19 \end{vmatrix}}{\begin{vmatrix} 3 & -1 \\ 3 & 4 \end{vmatrix}} = \frac{-57 - (3)}{12 - (-3)} = \frac{-60}{15} = -4$$

3. $X + 4Y = 11; \ -3X + 2Y = 9$

$$X = \frac{\begin{vmatrix} 11 & 4 \\ 9 & 2 \end{vmatrix}}{\begin{vmatrix} 1 & 4 \\ -3 & 2 \end{vmatrix}} = \frac{22 - (36)}{2 - (-12)} = \frac{-14}{14} = -1$$

$$Y = \frac{\begin{vmatrix} 1 & 11 \\ -3 & 9 \end{vmatrix}}{\begin{vmatrix} 1 & 4 \\ -3 & 2 \end{vmatrix}} = \frac{9 - (-33)}{2 - (-12)} = \frac{42}{14} = 3$$

4. $5X - 3Y + 3Z = 3; \ 2X - 6Y - 4Z = 2; \ 3X - 5Y + Z = -3$

$$X = \frac{\begin{vmatrix} 3 & -3 & 3 \\ 2 & -6 & -4 \\ -3 & -5 & 1 \end{vmatrix}}{\begin{vmatrix} 5 & -3 & 3 \\ 2 & -6 & -4 \\ 3 & -5 & 1 \end{vmatrix}} = \frac{(3)(-6)(1) + (-3)(-4)(-3) + (3)(2)(-5) - (-3)(-6)(3) - (-5)(-4)(3) - (1)(2)(-3)}{(5)(-6)(1) + (-3)(-4)(3) + (3)(2)(-5) - (3)(-6)(3) - (-5)(-4)(5) - (1)(2)(-3)}$$

$$\frac{(-18) + (-36) + (-30) - (54) - (60) - (-6)}{(-30) + (36) + (-30) - (-54) - (100) - (-6)} = \frac{-192}{-64} = 3$$

$$Y = \frac{\begin{vmatrix} 5 & 3 & 3 \\ 2 & 2 & -4 \\ 3 & -3 & 1 \end{vmatrix}}{\begin{vmatrix} 5 & -3 & 3 \\ 2 & -6 & -4 \\ 3 & -5 & 1 \end{vmatrix}} = \frac{(5)(2)(1) + (3)(-4)(3) + (3)(2)(-3) - (3)(2)(3) - (-3)(-4)(5) - (1)(2)(3)}{(5)(-6)(1) + (-3)(-4)(3) + (3)(2)(-5) - (3)(-6)(3) - (-5)(-4)(5) - (1)(2)(-3)} =$$

$$\frac{(10) + (-36) + (-18) - (18) - (60) - (6)}{(-30) + (36) + (-30) - (-54) - (100) - (-6)} = \frac{-128}{-64} = 2$$

$$Z = \frac{\begin{vmatrix} 5 & -3 & 3 \\ 2 & -6 & 2 \\ 3 & -5 & -3 \end{vmatrix}}{\begin{vmatrix} 5 & -3 & 3 \\ 2 & -6 & -4 \\ 3 & -5 & 1 \end{vmatrix}} = \frac{(5)(-6)(-3) + (-3)(2)(3) + (3)(2)(-5) - (3)(-6)(3) - (-5)(2)(5) - (-3)(2)(-3)}{(5)(-6)(1) + (-3)(-4)(3) + (3)(2)(-5) - (3)(-6)(3) - (-5)(-4)(5) - (1)(2)(-3)} =$$

$$\frac{(90) + (-18) + (-30) - (-54) - (-50) - (18)}{(-30) + (36) + (-30) - (-54) - (100) - (-6)} = \frac{128}{-64} = -2$$

Test Solutions

Test 1

1. B : $(3^0)(3^{-2})(3^2) = 3^{0+(-2)+2} = 3^0 = 1$

2. A : $Y^8 \div Y^2 = Y^{8-2} = Y^6$

3. D : $(3Q^2)^3 = 3^3 Q^{(2)(3)} = 27Q^6$

4. C : $\dfrac{P^3 N^{-2}}{N^2 P^4} = P^3 P^{-4} N^{-2} N^{-2} =$

   $P^{3+(-4)} N^{-2+(-2)} = P^{-1} N^{-4} = \dfrac{1}{PN^4}$

5. A : $3^{Y-1} = 81$
   $3^{Y-1} = 3^4$
   $Y - 1 = 4$
   $Y = 5$

6. C : Using trial and error
   A: $0^5 = 0$
   B: $(-1)^5 = -1$
   C: $(?)^5 = 16$
   (no whole number solution)
   D: $2^5 = 32$

7. A : $A^2 B^4 + B^3 A = AB^3(AB + 1)$
   AB^3 is the greatest common factor

8. D : $P^2 Q + P^4 Q^2 = P^2 Q(1 + P^2 Q)$ When the
   common factor of $P^2 Q$ is factored out
   $1 + P^2 Q$ is left.

9. C : $(-2 + 4)^{-2} = (2)^{-2} = \dfrac{1}{2^2} = \dfrac{1}{4}$

10. C : $\quad 3^6 = 9^X$
    $3^{(2)(3)} = 9^X$
    $\quad 9^3 = 9^X$
    $\quad 3 = X$

11. A : $X = \dfrac{1}{2} Y$

    $X + 2Y = 5 \Rightarrow \qquad \left(\dfrac{1}{2} Y\right) + 2Y = 5$

    $(2)\left(\dfrac{1}{2} Y\right) + (2)2Y = (2)5$

    $Y + 4Y = 10$
    $5Y = 10$
    $Y = 2$

12. B : In a rhombus, the sides are equal, so:
    $4a = a + 6$
    $3a = 6$
    $a = 2$

13. B : A straight line is 180°
    $180° - 20° = 160°$
    $b + a = 160°$

14. D : Let C = cups of flour needed:
    $\dfrac{3}{7} = \dfrac{15}{C}$
    $3C = 105$
    $C = 35$ cups

15. A : counting on graph:
    slope is $\dfrac{up}{over} = \dfrac{2}{3}$
    using formula:
    $(0, 0), (-3, -2)$
    $m = \dfrac{((0) - (-2))}{((0) - (-3))} = \dfrac{2}{3}$

Remember that questions 11-15 are
review questions. You may want to
grade them differently than questions 1-10.

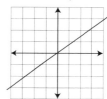

Test 2

1. C: $\dfrac{5X+2}{5X} = \dfrac{5X}{5X} + \dfrac{2}{5X} = 1 + \dfrac{2}{5X}$ $(X \neq 0)$

2. C

3. D: $\dfrac{2X}{X+2} - \dfrac{3X}{X-2} =$

$\dfrac{2X(X-2)}{(X+2)(X-2)} - \dfrac{3X(X+2)}{(X-2)(X+2)} =$

$\dfrac{(2X^2 - 4X) - (3X^2 + 6X)}{(X+2)(X-2)} =$

$\dfrac{2X^2 - 4X - 3X^2 - 6X}{(X+2)(X-2)} = \dfrac{-X^2 - 10X}{(X+2)(X-2)}$

$(X \neq 2, -2)$

4. A: $\dfrac{X^2 + 2X}{X} = \dfrac{X^2}{X} + \dfrac{2X}{X} = X + 2$

5. B: $\dfrac{9}{4X} - \dfrac{5}{4Y} =$

$\dfrac{9(Y)}{4X(Y)} - \dfrac{5(X)}{4Y(X)} = \dfrac{9Y - 5X}{4XY}$

$(X, Y \neq 0)$

6. D: $\dfrac{A}{A} + A^0 = 1 + 1 = 2$ $(A \neq 0)$

7. B: $\dfrac{18AB - 12A^2}{6A} = \dfrac{18AB}{6A} - \dfrac{12A^2}{6A} =$

$3B - 2A$ $(X \neq 0)$

8. C

9. C: $\dfrac{8Y}{X+1} + \dfrac{Y}{X-1} =$

$\dfrac{8Y(X-1)}{(X+1)(X-1)} + \dfrac{Y(X+1)}{(X-1)(X+1)} =$

$\dfrac{8XY - 8Y + XY + Y}{(X+1)(X-1)} = \dfrac{9XY - 7Y}{(X+1)(X-1)}$

$(X \neq 1, -1)$

10. A: $\dfrac{2}{Y} + \dfrac{5}{3Y} - Y^{-1} = \dfrac{2}{Y} + \dfrac{5}{3Y} - \dfrac{1}{Y} =$

$\dfrac{2(3)}{Y(3)} + \dfrac{5}{3Y} - \dfrac{1(3)}{Y(3)} =$

$\dfrac{6}{3Y} + \dfrac{5}{3Y} - \dfrac{3}{3Y} = \dfrac{8}{3Y}$

$(Y \neq 0)$

11. D: $N + 5 = 2N$

$5 = N$

12. A: $\left[\left(3^2 \right)^{-3} \right]^{-1} = 3^{(2)(-3)(-1)} = 3^6$

13. D: $-4^2 + 12 \div 4 - |6 - 8| =$

$-4^2 + 12 \div 4 - |-2| =$

$-(4)(4) + 12 \div 4 - 2 =$

$-16 + 3 - 2 = -15$

14. D: The area of a trapezoid is the average of the two bases, times the height.

$A = \left(\dfrac{3m + 5m}{2} \right)(m) = \left(\dfrac{8m}{2} \right)(m) =$

$(4m)(m) = 4m^2$

15. B

Test 3

1. C: $6{,}200 = 6.2 \times 10^3$

2. B: $.268 = 2.68 \times 10^{-1}$

3. C: $.000073 \times .0054 =$

$\left(7.3 \times 10^{-5} \right)\left(5.4 \times 10^{-3} \right) =$

$(7.3 \times 5.4)\left(10^{-5} \times 10^{-3} \right) =$

$39.42 \times 10^{-8} =$

$\left(3.942 \times 10^1 \right) \times 10^{-8} =$

3.942×10^{-7}

4. A: $32{,}000{,}000 \div 16{,}000 =$

$\left(3.2 \times 10^7 \right) \div \left(1.6 \times 10^4 \right) =$

$(3.2 \div 1.6)\left(10^7 \div 10^4 \right) = 2.0 \times 10^3$

5. B: $\dfrac{\left(2.3 \times 10^{-3} \right)\left(4 \times 10^4 \right)}{2 \times 10^{-2}} =$

$\dfrac{(2.3 \times 4.0)\left(10^{-3} \times 10^4 \right)}{2.0 \times 10^{-2}} =$

$\left(9.2 \times 10^1 \right) \div \left(2 \times 10^{-2} \right) =$

$(9.2 \div 2.0)\left(10^1 \div 10^{-2} \right) = 4.6 \times 10^3$

6. D: $2ab^{-1} + 3a^{-1}b - \dfrac{4b^{-1}}{a^{-1}} =$

$\dfrac{2a}{b} + \dfrac{3b}{a} - \dfrac{4a}{b} =$

$\dfrac{3b}{a} - \dfrac{2a}{b}$

$(a, b \neq 0)$

7. B : $\dfrac{A}{6} + \dfrac{3}{A^2} =$

$\dfrac{A(A^2)}{6(A^2)} + \dfrac{3(6)}{A^2(6)} = \dfrac{A^3 + 18}{6A^2}$

$(A \neq 0)$

8. C : $XXXY - YXXX + \dfrac{2}{Y^{-1}X^{-3}} =$

$X^3Y - X^3Y + 2X^3Y = 2X^3Y$

$(X, Y \neq 0)$

9. B : $3AABA^{-2} + 4AB - 6B =$

$3B + 4AB - 6B =$

$4AB - 3B$

10. A : $4X + \dfrac{3XY^2Y^{-1}}{Y^1} + 8XY =$

$4X + 3XY^2Y^{-2} + 8XY =$

$4X + 3X + 8XY =$

$7X + 8XY$

$(Y \neq 0)$

11. B : $\dfrac{1}{Y} - \dfrac{3Y}{Y} = 6$

$1 - 3Y = 6Y$

$1 = 9Y$

$Y = \dfrac{1}{9}$

12. D : $\dfrac{5}{8} = 5 \div 8 = .625$

13. A : $3 = 1 \times 3$

$6 = 1 \times 2 \times 3$

$8 = 1 \times 2 \times 2 \times 2$

$LCM = 1 \times 2 \times 2 \times 2 \times 3 = 24$

14. D : $P = 2(3b) + 2(a - b)$

$P = 6b + 2a - 2b$

$P = 4b + 2a$

15. C : 102°: An obtuse angle is greater than 90° and less than 180°.

Test 4

1. B : $(2\sqrt{3})(3\sqrt{2}) = 6\sqrt{6}$

2. C : $2\sqrt{3} + 3\sqrt{3} = 5\sqrt{3}$

3. A : $\dfrac{5}{\sqrt{3}} = \dfrac{5\sqrt{3}}{\sqrt{3}\sqrt{3}} = \dfrac{5\sqrt{3}}{\sqrt{9}} = \dfrac{5\sqrt{3}}{3}$

4. D : $2\sqrt{32} = 2\sqrt{16}\sqrt{2} = 2(4)\sqrt{2} = 8\sqrt{2}$

5. B : $\sqrt{3}\left(\sqrt{18Y} + 3\sqrt{3}\right) =$

$\sqrt{3}\sqrt{18Y} + 3\sqrt{3}\sqrt{3} = \sqrt{54Y} + 3\sqrt{9} =$

$\sqrt{9}\sqrt{6Y} + 3(3) = 3\sqrt{6Y} + 9$

6. C : $\dfrac{\sqrt{75}}{\sqrt{5}} = \dfrac{\sqrt{75}\sqrt{5}}{\sqrt{5}\sqrt{5}} = \dfrac{\sqrt{25}\sqrt{3}\sqrt{5}}{\sqrt{25}} =$

$\dfrac{5\sqrt{15}}{5} = \sqrt{15}$

7. D : $\dfrac{2}{\sqrt{5}} + \dfrac{5}{\sqrt{2}} = \dfrac{2\sqrt{5}}{\sqrt{5}\sqrt{5}} + \dfrac{5\sqrt{2}}{\sqrt{2}\sqrt{2}} =$

$\dfrac{2\sqrt{5}}{\sqrt{25}} + \dfrac{5\sqrt{2}}{\sqrt{4}} = \dfrac{2\sqrt{5}}{5} + \dfrac{5\sqrt{2}}{2} =$

$\dfrac{2\sqrt{5}(2)}{5(2)} + \dfrac{5\sqrt{2}(5)}{2(5)} =$

$\dfrac{4\sqrt{5}}{10} + \dfrac{25\sqrt{2}}{10} = \dfrac{4\sqrt{5} + 25\sqrt{2}}{10}$

8. A : $\dfrac{2}{3}\sqrt{63Y^{12}} = \dfrac{2}{3}\sqrt{9}\sqrt{7}\sqrt{Y^{12}} =$

$\dfrac{2}{3}(3)Y^6\sqrt{7} = 2Y^6\sqrt{7}$

9. A : $3\sqrt{\dfrac{9}{16}Y^4} = 3\left(\dfrac{\sqrt{9}}{\sqrt{16}}\right)\sqrt{Y^4} =$

$3\left(\dfrac{3}{4}\right)Y^2 = \dfrac{9}{4}Y^2$

10. A : $3\sqrt{50} - 2\sqrt{18} = 3\sqrt{25}\sqrt{2} - 2\sqrt{9}\sqrt{2} =$

$3(5)\sqrt{2} - 2(3)\sqrt{2} = 15\sqrt{2} - 6\sqrt{2} = 9\sqrt{2}$

11. C : complementary

12. B : In the slope-intercept form:

$Y = mX + b$, b indicates the Y intercept.

(reviewed in lesson 20 of Algebra 2)

13. D : V = area of base, times height

$V = (\pi b^2)6$ or $6\pi b^2$

14. A : $\dfrac{6(Y - 4)}{2} = \dfrac{3(Y - 4)}{1} = 3Y - 12$

15. B : $3^3 = 27$, and $3^4 = 81$, so four 3s

Remember that questions 11-15 are review questions. You may want to grade them differently than questions 1-10. Please read the introductory page of your Algebra 2 test booklet.

Test 5

1. B : $2X^2 - 9X + 9 = (2X - 3)(X - 3)$

2. C : $20 - 5X^2 =$
   $-5(X^2 - 4) =$
   $-5(X - 2)(X + 2)$

3. A : $2X^3 - X^2 - 3X =$
   $X(2X^2 - X - 3) =$
   $X(X + 1)(2X - 3)$

4. C : $Y^4 - 625 =$
   $(Y^2 - 25)(Y^2 + 25) =$
   $(Y - 5)(Y + 5)(Y^2 + 25)$

5. B : $2X^2 + 4X = 6$
   $X^2 + 2X = 3$
   $X^2 + 2X - 3 = 0$
   $(X + 3)(X - 1) = 0$

$X + 3 = 0$	$X - 1 = 0$
$X = -3$	$X = 1$

6. A : $-6X^2 = -27X + 12$
   $2X^2 = 9X - 4$
   $2X^2 - 9X + 4 = 0$
   $(2X - 1)(X - 4) = 0$

$2X - 1 = 0$	$X - 4 = 0$
$2X = 1$	$X = 4$
$X = \dfrac{1}{2}$	

7. C : $\dfrac{4}{X + 2} - \dfrac{2X}{2} = \dfrac{4(2)}{(X + 2)(2)} - \dfrac{2X(X + 2)}{2(X + 2)} =$

   $\dfrac{8 - (2X^2 + 4X)}{2(X + 2)} = \dfrac{-2X^2 - 4X + 8}{2(X + 2)} =$

   $\dfrac{-2(X^2 + 2X - 4)}{2(X + 2)} = \dfrac{-(X^2 + 2X - 4)}{X + 2}$

   $(X \neq -2)$

8. A : $\dfrac{3}{X + 4} - \dfrac{2X}{-X + 4} + \dfrac{X^2}{X^2 - 16} =$

   $\dfrac{3}{X + 4} - \dfrac{(-1)2X}{(-1)(-X + 4)} + \dfrac{X^2}{X^2 - 16} =$

   $\dfrac{3}{X + 4} - \dfrac{-2X}{X - 4} + \dfrac{X^2}{X^2 - 16} =$

   $\dfrac{3(X - 4)}{(X + 4)(X - 4)} - \dfrac{-2X(X + 4)}{(X + 4)(X - 4)} + \dfrac{X^2}{X^2 - 16} =$

   $\dfrac{(3X - 12) - (-2X^2 - 8X)}{X^2 - 16} + \dfrac{X^2}{X^2 - 16} =$

   $\dfrac{3X - 12 + 2X^2 + 8X + X^2}{X^2 - 16} = \dfrac{3X^2 + 11X - 12}{X^2 - 16}$

   $(X \neq 4, -4)$

9. D : $\dfrac{2 + \dfrac{6}{A}}{3 + \dfrac{12}{A - 1}} = \dfrac{\dfrac{2(A)}{(A)} + \dfrac{6}{A}}{\dfrac{3(A - 1)}{(A - 1)} + \dfrac{12}{A - 1}} = \dfrac{\dfrac{2A + 6}{A}}{\dfrac{3A - 3 + 12}{A - 1}} =$

   $\dfrac{\dfrac{2A + 6}{A}}{\dfrac{3A + 9}{A - 1}} = \dfrac{2A + 6}{A} \times \dfrac{A - 1}{3A + 9} =$

   $\dfrac{2(A + 3)}{A} \times \dfrac{A - 1}{3(A + 3)} = \dfrac{2}{A} \times \dfrac{A - 1}{3} = \dfrac{2(A - 1)}{3A}$

10. C : $\dfrac{\dfrac{X^2 + 9X + 20}{X^3 - 9X}}{\dfrac{X^2 + 8X + 16}{X^2 + X - 12}} = \dfrac{\dfrac{(X + 4)(X + 5)}{X(X - 3)(X + 3)}}{\dfrac{(X + 4)(X + 4)}{(X + 4)(X - 3)}} =$

    $\dfrac{(X + 4)(X + 5)}{X(X - 3)(X + 3)} \times \dfrac{(X + 4)(X - 3)}{(X + 4)(X + 4)} =$

    $\dfrac{(X + 5)}{X(X + 3)}$

    $(X \neq 0, -3, 3, -4)$

11. B : any number to the 0 power $= 1$

12. C : If $Y = mX + b$, m is the slope, and
    a perpendicular line has a slope of
    $-\dfrac{1}{m}$ or the negative reciprocal of m.

13. A :

    $$10(N + d = 7) \Rightarrow \quad 10N + 10D = \quad 70$$
    $$-100(.05N + .10D = .50) \Rightarrow \quad \underline{-5N - 10D = -50}$$
    $$5N \qquad\quad = \quad 20$$
    $$N = \quad 4$$

14. B: $\left(2^2+1\right)-\left(2^2-2\right)=$
$\left(4+1\right)-\left(4-2\right)=5-2=3$

15. D: $\left(B^2+1\right)+\left(B^2-2\right)=2B^2-1$

Test 6

1. C: $1{,}000^{\frac{2}{3}}=\left(\sqrt[3]{1{,}000}\right)^2=$
$\left(10\right)^2=100$

2. B: $\left(\frac{81}{25}\right)^{\frac{1}{2}}=$
$\sqrt{\frac{81}{25}}=\frac{\sqrt{81}}{\sqrt{25}}=\frac{9}{5}$

3. C: $\left(32^{\frac{3}{5}}\right)^2=\left[\left(\sqrt[5]{32}\right)^3\right]^2=$
$\left[\left(2\right)^3\right]^2=2^6=64$

4. D: $\left(\frac{R^{\frac{1}{3}}}{2}\right)^2=\frac{R^{\frac{2}{3}}}{4}\text{ or }\frac{1}{4}R^{\frac{2}{3}}$

5. D: $\left(\frac{-1}{\sqrt{25}}\right)^{-3}=\left(\frac{5}{-1}\right)^3=\frac{5^3}{\left(-1\right)^3}=$
$\frac{125}{-1}=-125$

6. A: $\sqrt{\sqrt{81}}=\left(81^{\frac{1}{2}}\right)^{\frac{1}{2}}=81^{\frac{1}{4}}=3$

7. A: $\sqrt[3]{B^6}=\left(B^6\right)^{\frac{1}{3}}=B^{\left(\frac{6}{3}\right)}=B^2$

8. D: $\left(\sqrt[3]{27}\right)^4=27^{\frac{4}{3}}=24^{\left(\frac{1}{3}\right)\left(4\right)}=\left(27^{\frac{1}{3}}\right)^4=$
$3^4=81$

9. B: $\left(\sqrt[3]{64}\right)^{-2}=64^{-\frac{2}{3}}=64^{\left(\frac{1}{3}\right)\left(-2\right)}=$
$4^{-2}=\frac{1}{4^2}=\frac{1}{16}$

10. B: $\left(\frac{3}{\sqrt{4}}\right)^{-2}=\left(\frac{3}{2}\right)^{-2}=$
$\left(\frac{2}{3}\right)^2=\frac{2^2}{3^2}=\frac{4}{9}$

11. B: $A+B+C=180°$ True for all triangles

12. D: Equations with 3 variables produce a three-dimensional figure.

13. C: Large rectangle:
$A=LW=\left(6\right)\left(5\right)=30\text{ ft}^2$
Shaded square:
$A=LW=\left(2\right)\left(2\right)=4\text{ ft}^2$
Shaded triangle:
$A=\frac{1}{2}bh=\frac{1}{2}\left(1\right)\left(2\right)=1\text{ ft}^2$
Unshaded:
$30-\left(4+1\right)=30-5=25\text{ ft}^2$

14. A: Sam missed $2\left(30\%\right)$ or 60%
$\left(.60\right)\left(60\right)=36$ wrong answers
$60-36=24$ correct answers

15. C: $93{,}000{,}000\div.3=$
$\left(9.3\times10^7\right)\div\left(3.0\times10^{-1}\right)=$
$\left(9.3\div3.0\right)\left(10^7\div10^{-1}\right)=3.1\times10^8$

Test 7

1. B: $\sqrt{-121}=\sqrt{121}\sqrt{-1}=11i$

2. A: $\sqrt{\frac{-81}{100}}=\frac{\sqrt{-81}}{\sqrt{100}}=$
$\frac{\sqrt{81}\sqrt{-1}}{\sqrt{100}}=\frac{9i}{10}\text{ or }\frac{9}{10}i$

3. C: $\sqrt{\frac{-16}{7}}=\frac{\sqrt{-16}}{\sqrt{7}}=\frac{\sqrt{16}\sqrt{-1}}{\sqrt{7}}=$
$\frac{4i}{\sqrt{7}}=\frac{4i\sqrt{7}}{\sqrt{7}\sqrt{7}}=\frac{4i\sqrt{7}}{7}$

4. C: $\sqrt{-4}+\sqrt{-8}=$
$\sqrt{4}\sqrt{-1}+\sqrt{4}\sqrt{-1}\sqrt{2}=2i+2i\sqrt{2}$

5. D: $\left(3\sqrt{-6}\right)\left(5\sqrt{-15}\right)=$
$\left(3\right)\left(5\right)\sqrt{6}\sqrt{-1}\sqrt{15}\sqrt{-1}=15\sqrt{90}i^2=$
$-15\sqrt{90}=-15\sqrt{9}\sqrt{10}=$
$-15\left(3\right)\sqrt{10}=-45\sqrt{10}$

6. B: $\left(i^3\right)^2=i^{\left(3\right)\left(2\right)}=i^6=\left(-1\right)^3=-1$

7. A: $\sqrt{81}-\sqrt{-4}=9-\sqrt{4}\sqrt{-1}=9-2i$

8. B: $\left(7i\right)\left(-3i\right)=\left(7\right)\left(-3\right)\left(i\right)\left(i\right)=-21\left(-1\right)=21$

9. C: $\left(2\sqrt{-4}\right)\left(5\sqrt{-9}\right)=(2)(5)\sqrt{-1}\sqrt{4}\sqrt{-1}\sqrt{9}=$
$10(2)(3)i^2=60(-1)=-60$

10. C: $\left[(3i)(4i)\right]^2=\left[(3)(4)(i)(i)\right]^2=$
$\left[12(-1)\right]^2=\left[-12\right]^2=144$

11. A: The line has a negative slope

12. D: Slope is -2, Y intercept is 2, so equation is $Y=-2X+2$

13. C: $6X^2\div2X=3X$

14. A: By definition

15. B: She must travel south. Latitude measures distance north and south, starting with 0° at the equator. Lines of longitude run through the north and south poles.

Test 8

1. B: $X+2i$

2. C: $3-\sqrt{2}$

3. B: $7+i$

4. A: $2-\sqrt{A}$

5. D: $\dfrac{Y}{4-3i}=\dfrac{Y(4+3i)}{(4-3i)(4+3i)}=$
$\dfrac{4Y+3Yi}{4^2-(3i)^2}=\dfrac{4Y+3Yi}{16-(-9)}=\dfrac{4Y+3Yi}{25}$

6. C: $\dfrac{5Q}{2+\sqrt{7}}=\dfrac{5Q(2-\sqrt{7})}{(2+\sqrt{7})(2-\sqrt{7})}=$
$\dfrac{10Q-5Q\sqrt{7}}{2^2-\sqrt{7}^2}=\dfrac{10Q-5Q\sqrt{7}}{4-7}=$
$\dfrac{10Q-5Q\sqrt{7}}{-3}$

7. B: $\dfrac{-6}{9-3\sqrt{3}}=\dfrac{-2}{3-\sqrt{3}}=$
$\dfrac{-2(3+\sqrt{3})}{(3-\sqrt{3})(3+\sqrt{3})}=\dfrac{-6-2\sqrt{3}}{3^2-\sqrt{3}^2}=$
$\dfrac{-6-2\sqrt{3}}{9-3}=\dfrac{-6-2\sqrt{3}}{6}=\dfrac{-3-\sqrt{3}}{3}$

8. C: $\dfrac{2X+1}{i}=\dfrac{(2X+1)(i)}{i(i)}=\dfrac{2Xi+i}{-1}=-2Xi-i$

9. D: $\dfrac{i}{1+\sqrt{2}}=\dfrac{i(1-\sqrt{2})}{(1+\sqrt{2})(1-\sqrt{2})}=\dfrac{i-i\sqrt{2}}{1^2-\sqrt{2}^2}=$
$\dfrac{i-i\sqrt{2}}{1-2}=\dfrac{i-i\sqrt{2}}{-1}=-i+i\sqrt{2}=i\sqrt{2}-i$

10. D: $\dfrac{3}{2-\sqrt{Y}}=\dfrac{3(2+\sqrt{Y})}{(2-\sqrt{Y})(2+\sqrt{Y})}=$
$\dfrac{6+3\sqrt{Y}}{2^2-\sqrt{Y^2}}=\dfrac{6+3\sqrt{Y}}{4-Y}$

11. C: By definition

12. A

13. D: $X-Y=3\Rightarrow X=Y+3$
$3X-Y=13\Rightarrow 3(Y+3)-Y=13$
$3Y+9-Y=13$
$2Y=4$
$Y=2$

$X=Y+3\Rightarrow X=(2)+3$
$X=5$

$(5,2)$

14. B: $180°(4)=720°$

15. B: $720°\div6=120°$

Test 9

1. C: $X^2+10X+25=(X+5)(X+5)$

2. A: $4A^2+16A+16=(2A+4)(2A+4)$

3. B: $(X+4)(X+4)=X^2+8X+16$

4. A: $(2X-2)(2X-2)=4X^2-8X+4$

5. D: $(X+2)^3=$
$(1)X^3 2^0+(3)X^2 2^1+(3)X^1 2^2+(1)X^0 2^3=$
$X^3+(3)(2)X^2+(3)(4)X+(1)(8)(1)=$
$X^3+6X^2+12X+8$

6. B: $(A-B)^3=$
$(1)A^3(-B)^0+(3)A^2(-B)^1+(3)A^1(-B)^2+(1)A^0(-B)^3=$
$A^3-3A^2B+3AB^2-B^3$

7. B

8. C: Y^3

9. A: $-3(3X)^2(1)=-27X^2$

10. D : $\left(\dfrac{1}{2}\right)^3 = \dfrac{1}{8}$

11. A : $\dfrac{864\ \cancel{in^2}}{1} \times \dfrac{1\ ft}{12\ \cancel{in}} \times \dfrac{1\ ft}{12\ \cancel{in}} =$

    $\dfrac{864\ ft^2}{144} = 6\ ft^2$

12. D : $m\angle ABC = 180° - 129° = 51°$
    (supplementary angles)

13. C : $m\angle BCA = 180° - 115° = 65°$
    $m\angle A = 180° - (m\angle ABC + m\angle BCA) =$
    $180° - (51° + 65°) = 180° - 116° = 64°$

14. C : base 10

15. B : $3 = 1 \times 2^1 + 1 \times 2^0 = 11_2$

Test 10

1. A
2. B
3. A
4. C
5. D
6. B : $6 + 1 = 7$
7. A : Coefficient has 2 factors $(3-1)$
    $\dfrac{6 \cdot 5}{1 \cdot 2} = 15$
    Y-exponent $= 3 - 1 = 2$
    X-exponent $= 6 - 2 = 4$
    $15X^4Y^2$
8. C : Coefficient has 3 factors $(4-1)$
    $\dfrac{5 \cdot 4 \cdot 3}{1 \cdot 2 \cdot 3} = 10$
    B-exponent $= 4 - 1 = 3$
    2A-exponent $= 5 - 3 = 2$
    $(10)(2A)^2 B^3 = 40A^2 B^3$
9. D : Coefficient has 4 factors $(5-1)$
    $\dfrac{4 \cdot 3 \cdot 2 \cdot 1}{1 \cdot 2 \cdot 3 \cdot 4} = 1$
    exponent of $\left(-\dfrac{1}{2}\right)$ is 4
    X-exponent $= 4 - 4 = 0$
    $(1)(X)^0\left(-\dfrac{1}{2}\right)^4 = (1)(1)\left(\dfrac{1}{16}\right) = \dfrac{1}{16}$

10. D : Coefficient has 1 factor $(2\text{-}1)$
    $\dfrac{7}{1} = 7$
    2Y-exponent $= 2 - 1 = 1$
    X-exponent $= 7 - 1 = 6$
    $7X^6(2Y)^1 = 14X^6Y$

11. A : $A^2 + (2A)^2 = H^2$
    $A^2 + 4A^2 = H^2$
    $5A^2 = H^2$
    $H = A\sqrt{5}$

12. C : 1 triangle:
    $A = \dfrac{1}{2}bh = \dfrac{1}{2}(4)(2\sqrt{3}) = 4\sqrt{3}\ in^2$
    6 triangles:
    $A = (6)4\sqrt{3} = 24\sqrt{3}\ in^2$

13. C : $b^2 + b^2 = H^2$
    $2b^2 = H^2$
    $H = b\sqrt{2}$, so multiply b by $\sqrt{2}$
    The legs of a $45° - 45° - 90°$
    triangle are congruent.

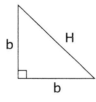

14. B : A kilogram is a little over 2 pounds

15. C : area of 1st rectangle $= XY\ ft^2$
    area of 2nd rectangle $= (2X)(2Y) = 4XY\ ft^2$
    four times as great

Unit Test 1

1. $(2^0)(2^{-3})(2^3) = 2^{0+(-3)+3} = 2^0 = 1$
2. $X^9 \div X^3 = X^{9-3} = X^6$
3. $(4Q^3)^2 = 4^2 Q^{(3)(2)} = 16Q^6$

4. $\dfrac{X^3Y^{-2}}{Y^2X^4} = \dfrac{X^3Y^{-2}Y^{-2}X^{-4}}{1} =$

$X^{3+(-4)}Y^{-2+(-2)} =$

$X^{-1}Y^{-4}$ or $\dfrac{1}{XY^4}$

5. $\dfrac{3}{4A} - \dfrac{8}{4B} = \dfrac{3(B)}{4A(B)} - \dfrac{8(A)}{4B(A)} = \dfrac{3B - 8A}{4AB}$

6. $\dfrac{R}{R} + R^0 = 1 + 1 = 2$

7. $\dfrac{2Y}{X+Y} + \dfrac{Y}{X-Y} =$

$\dfrac{2Y(X-Y)}{(X+Y)(X-Y)} + \dfrac{Y(X+Y)}{(X-Y)(X+Y)} =$

$\dfrac{2XY - 2Y^2}{X^2 - Y^2} + \dfrac{XY + Y^2}{X^2 - Y^2} = \dfrac{3XY - Y^2}{X^2 - Y^2}$

8. $4R^6TR^{-2} + 5R^4T - 2T =$

$4R^{6+(-2)}T + 5R^4T - 2T =$

$4R^4T + 5R^4T - 2T =$

$9R^4T - 2T$

9. $(.0056)(.034) =$

$\left(5.6 \times 10^{-3}\right)\left(3.4 \times 10^{-2}\right) =$

$(5.6 \times 3.4)\left(10^{-3} \times 10^{-2}\right) =$

$19.04 \times 10^{-5} = 1.904 \times 10^{-4}$

or 1.9×10^{-4} with SD

10. $(45,500)(21,000,000) =$

$\left(4.55 \times 10^4\right)\left(2.1 \times 10^7\right) =$

$(4.55 \times 2.1)\left(10^4 \times 10^7\right) =$

9.555×10^{11}

or 9.6×10^{11} with SD

11. $(32,000) \div (.00016) =$

$\left(3.2 \times 10^4\right) \div \left(1.6 \times 10^{-4}\right) =$

$(3.2 \div 1.6)\left(10^4 \div 10^{-4}\right) = 2.0 \times 10^8$

12. $\dfrac{(.00023)(160)}{.002} =$

$\left(2.3 \times 10^{-4}\right)\left(1.6 \times 10^2\right) \div \left(2.0 \times 10^{-3}\right) =$

$(2.3 \times 1.6 \div 2.0)\left(10^{-4} \times 10^2 \div 10^{-3}\right) =$

1.84×10^1

or 2.0×10^1 with SD

13. $\left(4\sqrt{5}\right)\left(5\sqrt{3}\right) = (4)(5)\sqrt{5}\sqrt{3} = 20\sqrt{15}$

14. $5\sqrt{6} + 2\sqrt{6} = 7\sqrt{6}$

15. $\dfrac{6}{\sqrt{2}} + \dfrac{1}{\sqrt{3}} = \dfrac{6\sqrt{2}}{\sqrt{2}\sqrt{2}} + \dfrac{1\sqrt{3}}{\sqrt{3}\sqrt{3}} =$

$\dfrac{6\sqrt{2}}{\sqrt{4}} + \dfrac{\sqrt{3}}{\sqrt{9}} = \dfrac{6\sqrt{2}}{2} + \dfrac{\sqrt{3}}{3} =$

$\dfrac{6\sqrt{2}\,(3)}{2(3)} + \dfrac{\sqrt{3}\,(2)}{3(2)} = \dfrac{18\sqrt{2}}{6} + \dfrac{2\sqrt{3}}{6} =$

$\dfrac{18\sqrt{2} + 2\sqrt{3}}{6} = \dfrac{2\left(9\sqrt{2} + \sqrt{3}\right)}{2(3)} = \dfrac{9\sqrt{2} + \sqrt{3}}{3}$

16. $\sqrt{36X^4} = \sqrt{36}\sqrt{X^4} = 6X^2$

17. $\left(\dfrac{16}{25}\right)^{\frac{1}{2}} = \sqrt{\dfrac{16}{25}} = \dfrac{\sqrt{16}}{\sqrt{25}} = \dfrac{4}{5}$

18. $\sqrt{\sqrt{16}} = \sqrt{4} = 2$

19. $3X^2 + 17X + 10 = (X+5)(3X+2)$

20. $3X^2 - 9X + 6 =$

$3\left(X^2 - 3X + 2\right) =$

$3(X-1)(X-2)$

21. $X^4 - 1 = \left(X^2 - 1\right)\left(X^2 + 1\right) =$

$(X-1)(X+1)\left(X^2+1\right)$

22. $2X^2 + 3X - 2 = (2X-1)(X+2)$

23. $\quad X^2 - 10X = -18 - X$

$\quad X^2 - 9X + 18 = 0$

$\quad (X-3)(X-6) = 0$

$X - 3 = 0 \qquad X - 6 = 0$

$\quad X = 3 \qquad\quad X = 6$

24. $\quad 2X^2 + 2X + 14 = 32 + 2X$

$\qquad\quad 2X^2 - 18 = 0$

$\qquad\quad 2\left(X^2 - 9\right) = 0$

$\quad 2(X-3)(X+3) = 0$

$X - 3 = 0 \qquad X + 3 = 0$

$\quad X = 3 \qquad\quad X = -3$

25. $\qquad 2X + 15 = X^2$

$\qquad\qquad 0 = X^2 - 2X - 15$

$(X+3)(X-5) = 0$

$X + 3 = 0 \qquad X - 5 = 0$

$\quad X = -3 \qquad\quad X = 5$

26.
$$X^3 = 16X$$
$$X^3 - 16X = 0$$
$$X(X^2 - 16) = 0$$
$$X(X-4)(X+4) = 0$$
$$X = 0 \quad X - 4 = 0 \quad X + 4 = 0$$
$$\qquad\qquad X = 4 \qquad X = -4$$

27. $\sqrt{-144} = \sqrt{144}\sqrt{-1} = 12i$

28. $\sqrt{-8} + \sqrt{-4} =$
$\sqrt{4}\sqrt{-1}\sqrt{2} + \sqrt{4}\sqrt{-1} = 2i\sqrt{2} + 2i$

29. $\left(4\sqrt{-5}\right)\left(2\sqrt{-6}\right) = (4)(2)\sqrt{-1}\sqrt{5}\sqrt{-1}\sqrt{6} =$
$8(-1)\sqrt{30} = -8\sqrt{30}$

30. $\left(i^3\right)^2 = i^6 = i \cdot i \cdot i \cdot i \cdot i \cdot i = (-1)^3 = -1$

31. $\dfrac{X}{8+2i} = \dfrac{X(8-2i)}{(8+2i)(8-2i)} = \dfrac{8X - 2Xi}{64 - (2i)^2} =$

$\dfrac{8X - 2Xi}{64 - 4(-1)} = \dfrac{8X - 2Xi}{64 + 4} = \dfrac{8X - 2Xi}{68} =$

$\dfrac{4X - Xi}{34}$

32. $\dfrac{2}{1+\sqrt{2}} = \dfrac{2\left(1-\sqrt{2}\right)}{\left(1+\sqrt{2}\right)\left(1-\sqrt{2}\right)} = \dfrac{2 - 2\sqrt{2}}{1-2} =$

$\dfrac{2 - 2\sqrt{2}}{-1} = -2 + 2\sqrt{2}$

33. $\dfrac{5 \cdot 4}{1 \cdot 2} X^3 Y^2 = 10X^3 Y^2$

34. $\dfrac{4}{1} A^3 B^1 = 4A^3 B$

35. $\dfrac{6 \cdot 5 \cdot 4}{1 \cdot 2 \cdot 3} D^3 \left(-\dfrac{1}{2}\right)^3 = 20D^3 \left(-\dfrac{1}{8}\right) =$
$-\dfrac{20}{8} D^3 = -\dfrac{5}{2} D^3$

Test 11

1. D : $\left(\dfrac{1}{2}\right)(16) = 8; \; 8^2 = 64$

2. A : $\left(\dfrac{1}{2}\right)(-7) = -\dfrac{7}{2}; \left(-\dfrac{7}{2}\right)^2 = \dfrac{49}{4}$

3. A : $\left(\dfrac{1}{2}\right)\left(\dfrac{3}{4}\right) = \dfrac{3}{8}; \left(\dfrac{3}{8}\right)^2 = \dfrac{9}{64}$

4. B : $\sqrt{81} = 9; \; 9(2) = 18;$
middle term is 18X

5. C : $\sqrt{144} = 12; \; 12(2) = 24;$
middle term is 24X

6. A : $\sqrt{\dfrac{16}{25}} = \dfrac{4}{5}; \dfrac{4}{5}(2) = \dfrac{8}{5};$
middle term is $\dfrac{8}{5}X$

7. D : $X^2 - 14X + 49 = (X-7)(X-7)$

8. B :
$$X^2 + 8X = 4$$
$$X^2 + 8X + 16 = 4 + 16$$
$$X^2 + 8X + 16 = 20$$
$$(X+4)^2 = 20$$
$$X + 4 = \pm\sqrt{20}$$
$$X + 4 = \pm 2\sqrt{5}$$
$$X = -4 \pm 2\sqrt{5}$$

9. D :
$$X^2 - 10X = -3$$
$$X^2 - 10X + 25 = -3 + 25$$
$$(X-5)^2 = 22$$
$$X - 5 = \pm\sqrt{22}$$
$$X = 5 \pm \sqrt{22}$$

10. C :
$$X^2 + 4X = -8$$
$$X^2 + 4X + 4 = -8 + 4$$
$$(X+2)^2 = -4$$
$$X + 2 = \pm\sqrt{-4}$$
$$X + 2 = \pm 2i$$
$$X = -2 \pm 2i$$

11. D :
$$4^2 + L^2 = 8^2$$
$$16 + L^2 = 64$$
$$L^2 = 48$$
$$L = \sqrt{48}$$
$$L = \sqrt{16}\sqrt{3}$$
$$L = 4\sqrt{3}$$

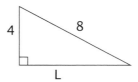

12. A : 1 liter ≈ 1.06 quarts

13. B : by definition

14. C : The sum of the measures of the interior angles of a triangle is 180°.

15. B : by definition

Test 12

1. B : A, C and D are equations whose highest exponent is 2. The quadratic equation works for equations of this nature only.

2. B : $\dfrac{-B \pm \sqrt{B^2 - 4AC}}{2A}$

3. C : All quadratic equations have constants A, B and C, which can be substituted into the quadratic formula. Not every quadratic equation can be factored.

4. C : Standard form provides A, B and C with the proper sign

5. A : $7X^2 + 2X - 1 = 0$
$A = 7; B = 2; C = -1$

6. A :
$$X^2 - 36 = 0$$
$$(X - 6)(X + 6) = 0$$
$$X - 6 = 0 \quad X + 6 = 0$$
$$X = 6 \qquad X = -6$$

7. D : $X^2 + 3X + 3 = 0$
$A = 1; B = 3; C = 3$
$$X = \frac{-(3) \pm \sqrt{(3)^2 - 4(1)(3)}}{2(1)} =$$
$$\frac{-3 \pm \sqrt{9 - 12}}{2} = \frac{-3 \pm \sqrt{-3}}{2} = \frac{-3 \pm i\sqrt{3}}{2}$$

8. B : $5X^2 + 2X - 1 = 0$
$A = 5; B = 2; C = -1$
$$X = \frac{-(2) \pm \sqrt{(2)^2 - 4(5)(-1)}}{2(5)} =$$
$$\frac{-2 \pm \sqrt{4 - (-20)}}{10} = \frac{-2 \pm \sqrt{4 + 20}}{10} =$$
$$\frac{-2 \pm \sqrt{24}}{10} = \frac{-2 \pm \sqrt{4}\sqrt{6}}{10} =$$
$$\frac{-2 \pm 2\sqrt{6}}{10} = \frac{-1 \pm \sqrt{6}}{5}$$

9. D : $4X^2 + 20X + 25 = 0$
$$(2X + 5)(2X + 5) = 0$$
$$2X + 5 = 0$$
$$2X = -5$$
$$X = -\frac{5}{2}$$

10. C : $4X^2 + 4X - 10 = 0$
$A = 4; B = 4; C = -10$
$$X = \frac{-(4) \pm \sqrt{(4)^2 - 4(4)(-10)}}{2(4)} =$$
$$\frac{-4 \pm \sqrt{16 - (-160)}}{8} = \frac{-4 \pm \sqrt{16 + 160}}{8} =$$
$$\frac{-4 \pm \sqrt{176}}{8} = \frac{-4 \pm \sqrt{16}\sqrt{11}}{8} =$$
$$\frac{-4 \pm 4\sqrt{11}}{8} = \frac{-1 \pm \sqrt{11}}{2}$$

11. C : $\overline{ED}$

12. B : A rhombus and a parallelogram have 2 pairs of parallel sides. A regular polygon may have any number of sides.

13. C : SAS stands for side angle side.

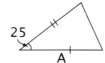 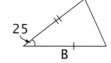

25   A   25   B

14. D: Knowing that the angles of one triangle are the same as the angles of another triangle proves similarity, not congruence.

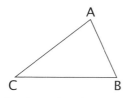

15. A: 1 yard ≈ .9 meters

Test 13

1. B: real, rational, equal

2. A: real, rational, unequal

3. D: imaginary

4. C: real, irrational, unequal

5. C: $X^2 + X = X + 9$
$X^2 - 9 = 0$
$A = 1; B = 0; C = -9$

$B^2 - 4AC \Rightarrow (0)^2 - 4(1)(-9) =$
$0 - (-36) =$
$0 + 36 = 36$

6. D: $X^2 + 5 = 2X$
$X^2 - 2X + 5 = 0$
$A = 1; B = -2; C = 5$

$B^2 - 4AC \Rightarrow (-2)^2 - 4(1)(5) =$
$4 - 20 = -16$

7. B: $X^2 + 9 = -6X$
$X^2 + 6X + 9 = 0$
$A = 1; B = 6; C = 9$

$B^2 - 4AC \Rightarrow (6)^2 - 4(1)(9) =$
$36 - 36 = 0$

8. A: $X^2 - 32 = -4X$
$X^2 + 4X - 32 = 0$

$X = \dfrac{-(4) \pm \sqrt{(4)^2 - 4(1)(-32)}}{2(1)} =$

$\dfrac{-4 \pm \sqrt{16 - (-128)}}{2} = \dfrac{-4 \pm \sqrt{144}}{2} =$

$\dfrac{-4 \pm 12}{2} = \dfrac{-2 \pm 6}{1} = -2 \pm 6$

$X = 4; X = -8$

9. A: $X^2 + 3X - 6 = 0$

$X = \dfrac{-(3) \pm \sqrt{(3)^2 - 4(1)(-6)}}{2(1)} =$

$\dfrac{-3 \pm \sqrt{9 - (-24)}}{2} = \dfrac{-3 \pm \sqrt{9 + 24}}{2} =$

$\dfrac{-3 \pm \sqrt{33}}{2}$

10. C: $X^2 - 5X = -8$
$X^2 - 5X + 8 = 0$

$X = \dfrac{-(-5) \pm \sqrt{(-5)^2 - 4(1)(8)}}{2(1)} =$

$\dfrac{5 \pm \sqrt{25 - 32}}{2} = \dfrac{5 \pm \sqrt{-7}}{2} = \dfrac{5 \pm i\sqrt{7}}{2}$

11. D: A and B are not true, C is true, but does not prove triangles congruent.

12. B: $62,000 \times .75 =$
$(6.2 \times 10^4)(7.5 \times 10^{-1}) =$
$(6.2 \times 7.5)(10^4 \times 10^{-1}) =$
$46.5 \times 10^3 = 4.65 \times 10^4$

13. A: similar

14. D: They are mirror images of each other.

15. A: The figure has been translated or moved over 2 and down 6.

Test 14

1. C: $250 - 200 = \$50$ saved
$WP \times 250 = 50$

$WP = \dfrac{50}{250}$

$WP = \dfrac{1}{5} = .20 = 20\%$

2. A : $24 - 12 = \$12$ markup

$WP \times 12 = 12$

$W = \dfrac{12}{12} = 1 = 100\%$

3. C : $7.83 - 7.25 = .58$ raise

$WP \times 7.25 = .58$

$WP = \dfrac{.58}{7.25} = .08 = 8\%$ raise

4. D : $P =$ the amount paid for the house

$P + 25\%$ of $P = \$100,000$

$P + .25P = 100,000$

$P(1 + .25) = 100,000$

$1.25P = 100,000$

$P = \dfrac{100,000}{1.25} = 80,000$

5. A : $38.95 \times .05 \approx 1.95$ tax

$38.95 \times .20 = 7.79$ tip

$38.95 + 1.95 + 7.79 = \$48.69$

6. A : $.45 \times 75 = 33.75$ off

$75 - 33.75 = \$41.25$

7. B : $.40 \times 32 = \$12.80$ markup

$32.00 + 12.80 = \$44.80$

8. D : $\dfrac{35}{23 + 35} = \dfrac{35}{58} \approx .60 = 60\%$

9. D : $\dfrac{2(16)}{12 + 32} = \dfrac{32}{44} \approx .73 = 73\%$

10. C : $\dfrac{23}{23 + 16 + 1} = \dfrac{23}{40} = .575 \approx 58\%$

11. B : $\left(9^{-2}\right) = \dfrac{1}{9^2} = \dfrac{1}{81}$

12. A : $\dfrac{Y}{X} + \dfrac{4Y}{X+2} = \dfrac{Y(X+2)}{X(X+2)} + \dfrac{4Y(X)}{(X+2)(X)} =$

$\dfrac{XY + 2Y}{X(X+2)} + \dfrac{4XY}{X(X+2)} = \dfrac{5XY + 2Y}{X(X+2)}$

$(X \neq 0, -2)$

13. B : Area of circle $= \pi r^2 \approx$

$3.14(3)^2 = 28.26$ units^2

unshaded portion is

$100\% - 15\% = 85\%$ of whole circle

$85\% \times 28.26 = .85 \times 28.26 \approx$

24.02 units^2

14. C : the quadratic formula

15. D : $\left(2\sqrt{5}\right)\left(5\sqrt{12}\right) = (2)(5)\sqrt{5}\sqrt{12} = 10\sqrt{60} = 10\sqrt{4}\sqrt{15} = 10(2)\sqrt{15} = 20\sqrt{15}$

Test 15

In this test, all unknowns are such that denominators are not equal to 0.

1. D : $ABC = D$

$A = \dfrac{D}{BC}$

2. C : $\dfrac{YZ}{B} = \dfrac{A}{X}$

$AB = XYZ$

$B = \dfrac{XYZ}{A}$

3. A : $\dfrac{Q}{P} - R = 0$

$\dfrac{Q}{P} = R$

$Q = RP$

4. D : $X(Y - Z) + D = 4$

$X(Y - Z) = 4 - D$

$X = \dfrac{4 - D}{Y - Z}$

5. B : $\dfrac{1}{B} = \dfrac{1}{C}$

$B = C$

6. B : $\dfrac{X}{YZ} = \dfrac{S}{T}$

$YZS = XT$

$Y = \dfrac{TX}{SZ}$

7. C : $\dfrac{RS}{T} = \dfrac{B}{A}$

$RSA = BT$

$A = \dfrac{BT}{RS}$

8. A : $X - Z = Y + 5$

$X - Z - 5 = Y$

$Y = X - Z - 5$

9. D : $A(B+C) - D = X$

$A(B+C) = X + D$

$B + C = \dfrac{X+D}{A}$

$B = \dfrac{X+D}{A} - C$

10. B : $-X + Y - 4 = A + B$

$X - Y + 4 = -A - B$

$X = -A - B + Y - 4$

11. B : $\dfrac{3}{2\sqrt{8}} = \dfrac{3\sqrt{2}}{2\sqrt{8}\sqrt{2}} = \dfrac{3\sqrt{2}}{2\sqrt{16}} = \dfrac{3\sqrt{2}}{2(4)} = \dfrac{3\sqrt{2}}{8}$

12. A : An example would be

$X^2 + 5X + 6 = (X+3)(X+2)$

13. B : $6 + (-8) - 4 + 3 = 6 - 8 - 4 + 3 = -3$

14. A : $C = \pi d \approx 3.14(10) = 31.4$ ft

15. C : $\dfrac{a}{A} \div \dfrac{b}{B} = \dfrac{a}{A} \times \dfrac{B}{b}$ (invert and multiply)

To prove :

$\dfrac{\dfrac{a}{A}}{\dfrac{b}{B}} \times \dfrac{\dfrac{B}{b}}{\dfrac{B}{b}} = \dfrac{aB}{Ab}$

$\left(\dfrac{B}{b} \text{ is the reciprocal of } \dfrac{b}{B} \right)$

Test 16

1. D : 3 wins + 5 losses = 8 total

$\dfrac{\text{losses}}{\text{total}} = \dfrac{5}{8}$, not $\dfrac{5}{15}$

2. B : 3 successes + 1 failure = 4 attempts

$\dfrac{\text{successes}}{52} = \dfrac{3}{4}$

3. D : $\dfrac{R}{P} = \dfrac{2}{7}; \dfrac{R}{21} = \dfrac{2}{7}$

$7R = 2(21)$

$7R = 42$

$R = 6$

4. A : $\dfrac{F}{243} = \dfrac{5}{9}$

$F = \dfrac{5(243)}{9} = \dfrac{1,215}{9} = 135$

5. A : $\dfrac{L}{540} = \dfrac{3}{5}$

$L = \dfrac{3(540)}{5} = \dfrac{1,620}{5} = 324$ only looked

6. C : $\dfrac{R}{S} = \dfrac{4}{5}; \dfrac{R}{100} = \dfrac{4}{5}$

$R = \dfrac{4(100)}{5} = \dfrac{400}{5} = 80$ rainy days

7. B : $\dfrac{S_2}{CS_2} = \dfrac{64}{76}; \dfrac{S_2}{798} = \dfrac{64}{76}$

$S_2 = \dfrac{798(64)}{76} = \dfrac{51,072}{76} = 672$ g

8. B : $\dfrac{H_2}{H_2O} = \dfrac{2}{18}; \dfrac{H_2}{720} = \dfrac{2}{18}$

$H_2 = \dfrac{2(720)}{18} = \dfrac{1,440}{18} = 80$ g

9. C : $\dfrac{O}{H_2O} = \dfrac{16}{18}; \dfrac{O}{1,440} = \dfrac{16}{18}$

$O = \dfrac{16(1,440)}{18} = \dfrac{23,040}{18} = 1,280$ g

10. A : $K = 39; C = 12; N = 14$

$39 + 12 + 14 = 65$

11. D : $\sqrt{\sqrt[3]{X}} = \sqrt{X^{\frac{1}{3}}} = \left(X^{\frac{1}{3}}\right)^{\frac{1}{2}} = X^{\left(\frac{1}{3}\right)\left(\frac{1}{2}\right)} = X^{\frac{1}{6}}$

12. C : imaginary: real numbers always yield positive results when squared.

13. A : $\left(\dfrac{1}{2}\right)(-18) = -9; (-9)^2 = 81$

14. B : The easiest way to find the answer is to substitute the number given in each possible answer into the second equation, and try each resulting value of X in the first equation.

15. D : Volume of cube:

$10 \times 10 \times 10 = 1,000$ cm^3

Volume of tube:

$\pi r^2 h \approx 3.14(1)^2(10) = 31.4$ cm^3

Total:

$1,000 - 31.4 = 968.6$ cm^3

Test 17

1. D : multiplying by a fraction equal to one
2. B : 2 multipliers needed for square units
3. D : $\dfrac{80\ \cancel{oz}}{1} \times \dfrac{1\ lb}{16\ \cancel{oz}} = 5\ lb$
4. A : $\dfrac{6\ \cancel{yd}}{1} \times \dfrac{3\ ft}{1\ \cancel{yd}} = 18\ ft$
5. B :

$$\dfrac{360\ \cancel{cm^3}}{1} \times \dfrac{1\ m}{100\ \cancel{cm}} \times \dfrac{1\ m}{100\ \cancel{cm}} \times \dfrac{1\ m}{100\ \cancel{cm}}$$

$$= \dfrac{360\ m}{1{,}000{,}000} = .00036\ m^3$$

6. C : $\dfrac{3\ \cancel{mi^2}}{1} \times \dfrac{5{,}280\ ft}{1\ \cancel{mi}} \times \dfrac{5{,}280\ ft}{1\ \cancel{mi}} =$

83,635,200 ft^2

7. D : $\dfrac{9\ \cancel{km}}{1} \times \dfrac{.62\ mi}{1\ \cancel{km}} = 5.58\ mi$
8. C : $\dfrac{56\ \cancel{oz}}{1} \times \dfrac{28\ g}{1\ \cancel{oz}} = 1{,}568\ g$
9. A : $\dfrac{10\ \cancel{qt}}{1} \times \dfrac{.95\ liters}{1\ \cancel{qt}} = 9.5\ liters$
10. B : $\dfrac{2\ \cancel{greens^2}}{1} \times \dfrac{5\ \cancel{blue}}{1\ \cancel{green}} \times \dfrac{5\ \cancel{blue}}{1\ \cancel{green}} \times$

$\dfrac{3\ reds}{1\ \cancel{blue}} \times \dfrac{3\ reds}{1\ \cancel{blue}} = 450\ reds^2$

11. A : $4 - \sqrt{10}$
12. A : $\dfrac{X}{6+i} = \dfrac{X(6-i)}{(6+i)(6-i)} = \dfrac{6X-iX}{36-i^2} =$

$\dfrac{6X-iX}{36-(-1)} = \dfrac{6X-iX}{36+1} = \dfrac{6X-iX}{37}$

13. C : $\left|3^2 - 8^2\right| = |9 - 64| = |-55| = 55$
14. A : a line has 1 dimension: length
15. B : area $= LW = \left(2(2A)\right)\left(\dfrac{1}{2}B\right) =$

$(2)(2)\left(\dfrac{1}{2}\right)AB = 2AB\ units^2$

Test 18

1. A : D = RT can be manipulated to arrive at the answers in B and D.
2. B : $D = RT \Rightarrow D = (60)(4) = 240$ miles
3. C : $T = \dfrac{D}{R} \Rightarrow T = \dfrac{270}{3} = 90$ minutes
4. D : $R = \dfrac{D}{T} \Rightarrow R = \dfrac{400}{1.75} \approx 229$ mph
5. B : distance
6. A : $D_P = D_H$

$R_P T_P = R_H T_H$

$(9)T_P = (6)(6) \Rightarrow \begin{cases} R_P = 9 \\ R_H = 6 \\ T_H = 6 \end{cases}$

$9T_P = 36$

$T_P = 4$ hours to the park

7. D : $D = RT \Rightarrow D = (9)(4) = 36$ mi
8. C : $D_S = D_J$

$R_S T_S = R_J T_J$

$R_S(4) = (R_S + 2)(3) \Rightarrow \begin{cases} T_S = 4 \\ R_J = R_S + 2 \\ T_J = T_S - 1 = 3 \end{cases}$

9. C : $4R_S = 3R_S + 6$

$R_S = 6$ mph

10. A : $D = RT \Rightarrow D = (6)(4) = 24$ miles
11. B : $4 - 1 = 3$ factors for coefficient

exponent of Y term is $4 - 1 = 3$

exponent of 2X term is $5 - 3 = 2$

$\dfrac{5 \cdot 4 \cdot 3}{1 \cdot 2 \cdot 3}(2X)^2 Y^3 = (10)4X^2 Y^3 = 40X^2 Y^3$

12. B : $m\angle 2 = 180° - 132° = 48°$

(supplementary angles)

$m\angle 7 = m\angle 2 = 48°$

(alternate exterior angles)

13. D : obtuse; see example below

14. C : $V = \pi r^2 h = \pi\left(\dfrac{4X}{2}\right)^2 H = \pi(2X)^2 H = 4\pi X^2 H$

15. C : SA = 2 times the area of the base, plus the area of the side:

$SA = 2\pi r^2 + \pi dh =$

$2\pi(2X)^2 + \pi(4X)(H) =$

$2\pi 4X^2 + 4\pi XH =$

$8\pi X^2 + 4\pi XH$

Test 19

1. C : figure 3
2. A : time
3. A : $D_A + D_J = 24$

$R_A T_A + R_J T_J = 24$

$(4)T + (8)T = 24 \Rightarrow \begin{cases} R_A = 4 \\ R_J = 8 \\ T_A = T_J \end{cases}$

$12T = 24$

$T = 2$ hours

4. A : figure 1
5. B : $D_R + D_S = 39$

$R_R T_R + R_S T_S = 39$

$(5)(2T_S) + (3)(T_S) = 39 \Rightarrow \begin{cases} R_R = 5 \\ T_R = 2T_S \\ R_S = 3 \end{cases}$

$10T_S + 3T_S = 39$

$13T_S = 39$

$T_S = 3$ hours

6. D : $T_R = 2T_S \Rightarrow T_R = 2(3) = 6$ hours
7. B : figure 2
8. D : $D = RT$

$(130) = R(5) \Rightarrow \begin{cases} D = 130 \\ T = 5 \end{cases}$

$130 = 5R$

$R = 26$ mph

9. C : $D_A + D_V = 130$

$R_A T_A + R_V T_V = 130$

$(26)T_A + (26)(T_A - 1) = 130 \Rightarrow \begin{cases} R_A = R_V = 26 \\ T_V = T_A - 1 \end{cases}$

$26T_A + 26T_A - 26 = 130$

$52T_A - 26 = 130$

$52T_A = 156$

$T_A = 3$ hours

$D_A = R_A T_A \Rightarrow D_A = (26)(3) = 78$ miles

10. B : $D_V = R_V T_V \Rightarrow D_V = (26)(2) = 52$ miles

or $D_A + D_V = 130 \Rightarrow D_V + (78) = 130$ miles

$D_V = 52$ miles

11. A : $\left.\begin{array}{l} 4 = 2 \times 2 \\ 16 = 2 \times 2 \times 2 \times 2 \\ 18 = 2 \times 3 \times 3 \end{array}\right\}$ GCF is 2

12. B : $(3^3)(3^2) = (3)(3)(3) \times (3)(3) = 3^5 = 243$

13. B : $2X + 3 \overline{\smash{\big)}\, 2X^2 + 11X + 12}$ with quotient $X + 4$

$\underline{-2X^2 - 3X}$

$8X + 12$

$\underline{-8X - 12}$

0

14. C : used in the decimal system
15. B : $6^2 + 7^2 = H^2$

$36 + 49 = H^2$

$85 = H^2$

$H = \sqrt{85} \approx 9.22$ miles

Unit Test II

1. $X^2 + 6X - 6 = 0$

$X^2 + 6X = 6$

$X^2 + 6X + 9 = 6 + 9$

$(X + 3)^2 = 15$

$X + 3 = \pm\sqrt{15}$

$X = -3 \pm \sqrt{15}$

2. $X^2 + 4X + 1 = 0$

$X^2 + 4X = -1$

$X^2 + 4X + 4 = -1 + 4$

$(X + 2)^2 = 3$

$X + 2 = \pm\sqrt{3}$

$X = -2 \pm \sqrt{3}$

3. $X^2 + 8X - 5 = 0$

$$X = \frac{-(8) \pm \sqrt{(8)^2 - 4(1)(-5)}}{2(1)} =$$

$$\frac{-8 \pm \sqrt{64 - (-20)}}{2} = \frac{-8 \pm \sqrt{64 + 20}}{2} =$$

$$\frac{-8 \pm \sqrt{84}}{2} = \frac{-8 \pm \sqrt{4}\sqrt{21}}{2} = \frac{-8 \pm 2\sqrt{21}}{2} =$$

$$\frac{2(-4 \pm \sqrt{21})}{2} = -4 \pm \sqrt{21}$$

4. $2X^2 + 3X + 6 = 0$

$$X = \frac{-(3) \pm \sqrt{(3)^2 - 4(2)(6)}}{2(2)} = \frac{-3 \pm \sqrt{9 - 48}}{4} =$$

$$\frac{-3 \pm \sqrt{-39}}{4} = \frac{-3 \pm i\sqrt{39}}{4}$$

5. $X^2 + 3X = 10$

$X^2 + 3X - 10 = 0$

$a = 1; b = 3; c = -10$

$b^2 - 4ac \Rightarrow (3)^2 - 4(1)(-10) =$

$9 - (-40) =$

$9 + 40 = 49$

real, rational, unequal

6. $X^2 + 12X + 36 = 0$

$a = 1; b = 12; c = 36$

$b^2 - 4ac \Rightarrow (12)^2 - 4(1)(36) =$

$144 - 144 = 0$

real, rational, equal

7. $4X^2 - 8X = -20$

$4X^2 - 8X + 20 = 0$

$a = 4; b = -8; c = 20$

$b^2 - 4ac \Rightarrow (-8)^2 - 4(4)(20) =$

$64 - 320 = -256$

imaginary

8. $X^2 - 5X + 3 = 0$

$a = 1; b = -5; c = 3$

$b^2 - 4ac \Rightarrow (-5)^2 - 4(1)(3) =$

$25 - 12 = 13$

real, irrational, unequal

9. $WXY = Z$

$W = \dfrac{Z}{XY}$

10. $\dfrac{TR}{A} = \dfrac{X}{B}$

$TRB = AX$

$A = \dfrac{TRB}{X}$

11. $\dfrac{X}{Y} - A = 0$

$\dfrac{X}{Y} = A$

$X = AY$

12. $\dfrac{1}{Y} = \dfrac{1}{T}$

$Y = T$

13. $20\% \times \$2,345 =$

$.20 \times \$2,345 = \$469$ discount

$\$2,345 - \$469 = \$1,876$ new price

14. $6\% \times \$1,876 =$

$.06 \times \$1,876 = \$112.56$ tax

$\$1,876 + \$112.56 = \$1,988.56$ total

15. Weight $= 12 + 2(16) = 12 + 32 = 44$

$12 = WP \times 44$

$\dfrac{12}{44} = WP$

$WP \approx .27 = 27\%$

16. $\dfrac{B}{G} = \dfrac{6}{8} \Rightarrow \dfrac{36}{G} = \dfrac{6}{8}$

$6G = 36(8)$

$G = 6(8)$

$G = 48$ greens

17. $\dfrac{B}{T} = \dfrac{5}{5+9} \Rightarrow \dfrac{B}{56} = \dfrac{5}{14}$

$14B = 56(5)$

$B = 4(5)$

$B = 20 \text{ baskets}$

18. $\dfrac{400 \ \cancel{oz}}{1} \times \dfrac{1 \ lb}{16 \ \cancel{oz}} = 25 \ lb$

19. $\dfrac{8 \ \cancel{km}}{1} \times \dfrac{.62 \ mi}{1 \ \cancel{km}} = 4.96 \ mi$

20. $D = RT \Rightarrow D = (50)(6)$

$\qquad\qquad D = 300 \text{ miles}$

21. $D = RT \Rightarrow (280) = (20)T$

$\qquad\qquad\qquad T = 14 \text{ hours}$

22. $D_C = R_C T_C \Rightarrow D_C = (60)(1)$

$\qquad\qquad\qquad D_C = 60 \text{ miles}$

$R_H = \dfrac{1}{3}\left(R_C\right) \Rightarrow R_H = \dfrac{1}{3}(60) = 20 \text{ mph}$

$D_H = R_H T_H \Rightarrow (60) = (20) T_H$

$\qquad\qquad\qquad\qquad T_H = 3$

23. $D_L = R_L T_L$

$D_C = R_C T_C$

$D_L = D_C \Rightarrow$

$\qquad R_L T_L = R_C T_C$

$R_L (5) = (R_L + 2)(4) \begin{cases} T_L = 5 \\ R_C = R_L + 2 \\ T_C = T_L - 1 = (5) - 1 = 4 \end{cases}$

$\qquad 5R_L = 4R_L + 8$

$\qquad\quad R_L = 8 \text{ mph}$

$R_C = R_L + 2 \Rightarrow R_C = (8) + 2$

$\qquad\qquad\qquad\quad R_C = 10 \text{ mph}$

24. $D_J = R_J T_J$

$D_D = R_D T_D$

$D_J + D_D = 230 \Rightarrow$

$\qquad R_J T_J + R_D T_D = 230$

$(35)T_J + (45)\left(T_J - 2\right) = 230 \begin{cases} R_J = 35 \\ R_D = R_J + 10 = (35) + 10 = 45 \\ T_D = T_J - 2 \end{cases}$

$35T_J + 45T_J - 90 = 230$

$\qquad\quad 80T_J - 90 = 230$

$\qquad\qquad\quad 80T_J = 320$

$\qquad\qquad\qquad T_J = 4 \text{ hours}$

$1:00 \text{ pm} + 4 \text{ hours} = 5:00 \text{ pm}$

25. $D_J = R_J T_J \Rightarrow D_J = (35)(4)$

$\qquad\qquad\qquad D_J = 140 \text{ miles}$

$T_D = T_J - 2; \ T_D = 4 - 2 = 2 \text{ hours}$

$D_D = R_D T_D \Rightarrow D_D = (45)(2)$

$\qquad\qquad\qquad D_D = 90 \text{ miles}$

Test 20

1. A : Y is by itself on the left side of the equation.

2. C : slopes down to the right

3. D : $Y = 5X - 3$

Y-intercept is negative

steep positive slope

4. B : $Y = -2X$

Y-intercept is 0

negative slope

5. C : $Y = 3X + 2$

Y-intercept is positive

moderate positive slope

6. C : $Y = mX + b \Rightarrow (2) = (3)(1) + b$

$\qquad\qquad\qquad\qquad 2 = 3 + b$

$\qquad\qquad\qquad\qquad b = -1$

7. A : $\dfrac{Y_2 - Y_1}{X_2 - X_1}$ (change in Y ÷ change in X)

8. D : $m = \dfrac{Y_2 - Y_1}{X_2 - X_1} = \dfrac{(5) - (1)}{(-1) - (1)} = \dfrac{4}{-2} = -2$

$Y = mX + b \Rightarrow (1) = (-2)(1) + b$

$\qquad\qquad\qquad\qquad 1 = -2 + b$

$\qquad\qquad\qquad\qquad b = 3$

$Y = -2X + 3$

9. B : $Y = -2X + 3$
    $2X + Y = 3$
    (may also be written as $2X + Y - 3 = 0$)

10. A : $m = \dfrac{(1) - (-2)}{(-1) - (3)} = \dfrac{3}{-4} = -\dfrac{3}{4}$

    $Y = mX + b \Rightarrow \qquad (1) = \left(-\dfrac{3}{4}\right)(-1) + b$

    $1 = \dfrac{3}{4} + b$

    $1 - \dfrac{3}{4} = b$

    $\dfrac{4}{4} - \dfrac{3}{4} = b$

    $b = \dfrac{1}{4}$

    $Y = -\dfrac{3}{4}X + \dfrac{1}{4}$

11. A : quadratic formula

12. D : Since the discriminant is under a square root sign, a negative number will always yield an imaginary result.

13. B : Let X = original price
    $X + 15\%$ of $X + 5\%$ of $X = \$68.15$
    $X + .15X + .05X = \$68.15$
    $1.2X = \$68.15$
    $X = \dfrac{\$68.15}{1.2} \approx \$56.79$

14. A : $93{,}000{,}000 \div 30{,}000 =$
    $\left(9.3 \times 10^7\right) \div \left(3.0 \times 10^4\right) =$
    $(9.3 \div 3.0)\left(10^7 \div 10^4\right) = 3.1 \times 10^3$ days
    If significant digits were taken into account, this would be rounded to 3.0×10^3 days.

15. D : $3.1 \times 10^3 \div 365 =$
    $\left(3.1 \times 10^3\right) \div \left(3.65 \times 10^2\right) =$
    $(3.1 \div 3.65)\left(10^3 \div 10^2\right) \approx$
    $.85 \times 10^1 = 8.5$ years

Test 21

1. B : parallel

2. D : negative reciprocals

3. A : $2Y = 4X + 3 \Rightarrow Y = 2X + \dfrac{3}{2}$
    same slope as given line

4. D : new slope should be -4

5. C : slope needs to be 2
    $Y = mX + b \Rightarrow (-2) = (2)(2) + b$
    $-2 = 4 + b$
    $b = -6$
    $Y = 2X - 6$

6. B : slope needs to be $-\dfrac{1}{2}$
    $Y = mX + b \Rightarrow (-2) = \left(-\dfrac{1}{2}\right)(2) + b$
    $-2 = -1 + b$
    $b = -1$
    $Y = -\dfrac{1}{2}X - 1$

7. A : the inequality sign changes direction

8. D : check point $(0, 0)$
    $Y > 2X - 3 \Rightarrow (0) > 2(0) - 3$
    $0 > 0 - 3$
    $0 > -3$ true
    $>$ sign indicates dotted line
    Figure 4 has a dotted line, and the side of the line containing the point $(0, 0)$ is shaded.

9. C : Same as number 8, but $\geq$ indicates that the line must be solid. Figure 3 has a solid line, and the side of the line containing the point $(0, 0)$ is shaded.

10. B : check point $(0, 0)$
    $Y < 2X - 3 \Rightarrow (0) < 2(0) - 3$
    $0 < 0 - 3$
    $0 < -3$ false
    $<$ sign indicates dotted line. Figure 2 has a dotted line, and the side of the line containing the point $(0, 0)$ is unshaded.

11. C : trapezoid; by definition all of the other three figures have two pairs of parallel sides.

12. C : $(x^4 - 1) = (x^2 - 1)(x^2 + 1)$
     $(x^4 - 1) \div (x^2 - 1) = (x^2 + 1)$

13. D : By definition, supplementary angles add to 180°. They may or may not be congruent.

14. B : $3|-4 - 2| + 1 - 2(3 - 6)^2 =$
     $3|-6| + 1 - 2(-3)^2 = 3(6) + 1 - 2(9) =$
     $18 + 1 - 18 = 1$

15. A : $8^{-\frac{1}{3}} = \dfrac{1}{8^{\frac{1}{3}}} = \dfrac{1}{\sqrt[3]{8}} = \dfrac{1}{2}$

Test 22

1. A : the Pythagorean theorem
2. C
3. B : $\left(\dfrac{X_1 + X_2}{2}, \dfrac{Y_1 + Y_2}{2} \right)$
4. A : distance is always positive
5. D : $A = (-2, 1); C = (3, 2)$
     $D = \sqrt{[(3) - (-2)]^2 + [(2) - (1)]^2} =$
     $\sqrt{[5]^2 + [1]^2} = \sqrt{25 + 1} = \sqrt{26}$
6. B : $B = (-1, 5); C = (3, 2)$
     $D = \sqrt{[(3) - (-1)]^2 + [(2) - (5)]^2} =$
     $\sqrt{[4]^2 + [-3]^2} = \sqrt{16 + 9} = \sqrt{25} = 5$
7. A : $A = (-2, 1); D = (-1, -2)$
     $D = \sqrt{[(-1) - (-2)]^2 + [(-2) - (1)]^2} =$
     $\sqrt{[1]^2 + [-3]^2} = \sqrt{1 + 9} = \sqrt{10}$
8. C : $B = (-1, 5); D = (-1, -2)$
     $M = \left(\dfrac{(-1 + (-1))}{2}, \dfrac{(5) + (-2)}{2} \right) =$
     $\left(\dfrac{-2}{2}, \dfrac{3}{2} \right) = \left(-1, \dfrac{3}{2} \right)$

9. B : $D = (-1, -2); C = (3, 2)$
     $M = \left(\dfrac{(-1) + (3)}{2}, \dfrac{(-2) + (2)}{2} \right) =$
     $\left(\dfrac{2}{2}, \dfrac{0}{2} \right) = (1, 0)$

10. C : $A = (-2, 1); B = (-1, 5)$
     $M = \left(\dfrac{(-2) + (-1)}{2}, \dfrac{(1) + (5)}{2} \right) =$
     $\left(\dfrac{-3}{2}, \dfrac{6}{2} \right) = \left(-\dfrac{3}{2}, 3 \right)$

11. B : $\dfrac{AB}{C} = \dfrac{Y}{X}$
     $CY = ABX$
     $C = \dfrac{ABX}{Y}$

12. D : T = total customers = 3 + 4 = 7
     $\dfrac{51}{T} = \dfrac{3}{7}$
     $3T = 51(7)$
     $T = 17(7)$
     $T = 119$

13. A : area of a trapezoid $= \left(\dfrac{b_1 + b_2}{2} \right)(h)$

14. C : $\dfrac{192.5 \text{ cm}}{1} \times \dfrac{.4 \text{ in}}{1 \text{ cm}} \times \dfrac{1 \text{ ft}}{12 \text{ in}} \approx 6.4 \text{ ft}$

15. D : $8^2 + 10^2 = H^2$
     $64 + 100 = H^2$
     $164 = H^2$
     $H = \sqrt{164}$
     $H = \sqrt{4}\sqrt{41}$
     $H = 2\sqrt{41}$

Test 23

1. **D :** The equation of a circle consists of the sum of two squares equal to some number.

2. **A :** The equation of an ellipse is similar to that of a circle, but the squared terms have coefficients that are different from one another.

3. **C :** In an equation of the form $X^2 + Y^2 = Z^2$, the radius of the circle is the square root of the number on the right side of the equals sign.

4. **C :** Complete the two squares on the left side of the equation:

$$X^2 - 6X + Y^2 + 10Y = -18$$
$$X^2 - 6X + 9 + Y^2 + 10Y + 25 = -18 + 9 + 25$$
$$(X^2 - 6X + 9) + (Y^2 + 10Y + 25) = 16$$
$$(X - 3)^2 + (Y + 5)^2 = 4^2$$

5. **A :** The coordinates of the center are the opposite of what is in the parentheses with X and Y respectively.

6. **B**

7. **A :** The coefficients are equal, so the equation represents a circle. The center of the circle is at $(-3, -3)$, so figure E.

8. **C :** The coefficients are unequal, so the equation represents an ellipse. The center of the ellipse is at $(0, 0)$, and the coefficient of X is greater than the coefficient of Y, so the ellipse is longer in the vertical direction, so figure G.

9. **D :** $(1, -1)$

10. **D :** Make each term equal to 0 in turn, and solve for the opposite variable.

11. **C :** Check all four:

A $\begin{aligned}-3^2 \div 3 + 6 &= \\ -9 \div 3 + 6 &= \\ -3 + 6 &= 3\end{aligned}$ B $\begin{aligned}(3)^2 \div 3 + 5 &= \\ 9 \div 3 + 5 &= \\ 3 + 5 &= 8\end{aligned}$

C $\begin{aligned}-(3)^2 \div 3 + 5 &= \\ -9 \div 3 + 5 &= \\ -3 + 5 &= 2\end{aligned}$ D $\begin{aligned}(-3)^2 \div 3 + 4 &= \\ 9 \div 3 + 4 &= \\ 3 + 4 &= 7\end{aligned}$

12. **B :** $\dfrac{2}{X} + \dfrac{6}{3X} - X^{-1} = \dfrac{2}{X} + \dfrac{6}{3X} - \dfrac{1}{X} =$

$$\frac{2(3)}{X(3)} + \frac{6}{3X} - \frac{1(3)}{X(3)} =$$
$$\frac{6}{3X} + \frac{6}{3X} - \frac{3}{3X} = \frac{9}{3X} = \frac{3}{X}$$

13. **B**

14. **A :**
$$D_{Dav} = R_{Dav}T_{Dav}$$
$$D_{Dan} = R_{Dan}T_{Dan}$$
$$D_{Dav} = D_{Dan} \quad R_{Dav}T_{Dav} = R_{Dan}T_{Dan}$$
$$(4)T_{Dav} = (6)(T_{Dav} - 1) \left\{ \begin{array}{l} R_{Dav} = 4 \\ R_{Dan} = 6 \\ T_{Dan} = T_{Dav} - 1 \end{array} \right.$$
$$4T_{Dav} = 6T_{Dav} - 6$$
$$-2T_{Dav} = -6$$
$$T_{Dav} = 3 \text{ hours}$$
$$D_{Dav} = R_{Dav}T_{Dav} \Rightarrow D_{Dav} = (4)(3)$$
$$D_{Dav} = 12 \text{ mi}$$

15. **C :** Area of original rectangle = XY units^2
Area of new rectangle =
$(3X)(3Y) = 9XY$ units^2
$9XY \div XY = 9$ times as great

Test 24

1. **A :** The equation of a parabola always has one term squared.

2. **A :** The coefficient of the X term is positive, and X is the squared term.

3. **D :** The coefficient of the Y term is negative, and Y is the squared term.

4. **B :** The coefficient of the squared term has the largest absolute value.

5. C: When X is 0:

$$Y = X^2 + 2 \Rightarrow Y = (0)^2 + 2$$
$$Y = 0 + 2$$
$$Y = 2$$

vertex $= (0, 2)$

6. A: The Y-intercept is 1; the graph opens in an upward direction and is moderately narrow.

7. C: The X-intercept is -2; the graph opens toward the left and is wide.

8. A: The Y-intercept is 1; the graph opens in a downward direction and is wide.

9. D: The X-intercept is 1; the graph opens toward the right and is of normal width.

10. B: The Y-intercept is -1; the graph opens in an upward direction and is of normal width.

11. B: The point naming the vertex is always the middle of the three points naming an angle.

$$\vdash \overset{\longleftrightarrow}{\underset{44\ km}{\qquad}} \dashv$$

12. D: $D_1 = R_1 T_1$

$$D_2 = R_2 T_2$$
$$D_1 + D_2 = 44 \Rightarrow R_1 T_1 + R_2 D_2 = 44$$

$$(4)T_1 + (9)(6 - T_1) = 44 \begin{cases} R_1 = 4 \text{ kph} \\ R_2 = 9 \text{ kph} \\ T_2 = 6 - T_1 \end{cases}$$

$$4T_1 + 54 - 9T_1 = 44$$
$$-5T_1 = -10$$
$$T_1 = 2 \text{ hours}$$

$$T_2 = 6 - T_1 \Rightarrow T_2 = 6 - (2)$$
$$T_2 = 4 \text{ hours}$$

13. C: $P = 2(X + 1) + 2(3X) =$

$$2X + 2 + 6X = 8X + 2$$

14. C: Add up the area of the 6 faces:

$$2\big[(3X)(X + 1)\big] + 2\big[(3X)(X)\big] + 2\big[(X)(X + 1)\big] =$$
$$2\big[3X^2 + 3X\big] + 2\big[3X^2\big] + 2\big[X^2 + X\big] =$$
$$6X^2 + 6X + 6X^2 + 2X^2 + 2X =$$
$$14X^2 + 8X \text{ units}^2$$

15. A: $V = lwh = (X)(3X)(X + 1) =$

$$(3X^2)(X + 1) =$$
$$3X^3 + 3X^2 \text{ units}^3$$

Test 25

1. D
2. D: minima
3. A: maxima
4. C: divides the parabola into symmetrical halves
5. B: $\dfrac{-B}{2A}$
6. B: axis of symmetry:

$$X = \frac{-B}{2A} \Rightarrow X = \frac{-(-8)}{2(1)} = \frac{8}{2} = 4$$

y-coordinate of vertex:

$$Y = X^2 - 8X + 1 \Rightarrow Y = (4)^2 - 8(4) + 1$$
$$Y = 16 - 32 + 1$$
$$Y = -15$$

vertex is $(4, -15)$

7. C: axis of symmetry:

$$X = \frac{-B}{2A} \Rightarrow X = \frac{-(6)}{2(-3)} = \frac{-6}{-6} = 1$$

y-coordinate of vertex:

$$Y = -3X^2 + 6X \Rightarrow Y = -3(1)^2 + 6(1)$$
$$Y = -3(1) + 6$$
$$Y = -3 + 6$$
$$Y = 3$$

vertex is $(1, 3)$

8. A : axis of symmetry:

$$X = \frac{-B}{2A} \Rightarrow X = \frac{-(-4)}{2\left(\frac{1}{2}\right)} = \frac{4}{1} = 4$$

y-coordinate of vertex:

$$Y = \frac{1}{2}X^2 - 4X \Rightarrow Y = \frac{1}{2}(4)^2 - 4(4)$$
$$Y = \frac{1}{2}(16) - 16$$
$$Y = 8 - 16$$
$$Y = -8$$

vertex is $(4, -8)$

9. D : A = lw

Let X represent length of rectangle. This means that 2X feet of fencing will be used in two opposite sides of the rectangle, leaving 240 – 2X to be divided between the other two sides. Thus, $w = \frac{240 - 2X}{2}$, and the area of the rectangle, represented by Y, is:

$$Y = X\left(\frac{240 - 2X}{2}\right) = \frac{X(240 - 2X)}{2} =$$
$$\frac{240X - 2X^2}{2} = \frac{2(120X - X^2)}{2} =$$
$$120X - X^2 = -X^2 + 120X$$

axis of symmetry:

$$X = \frac{-B}{2A} \Rightarrow X = \frac{-(120)}{2(-1)} = \frac{-120}{-2} = 60$$

vertex:

$$Y = -X^2 + 120X \Rightarrow Y = -(60)^2 + 120(60)$$
$$Y = -3,600 + 7,200$$
$$Y = 3,600 \text{ ft}^2$$

$$\frac{240 - 2X}{2}$$

X [rectangle] X

$$\frac{240 - 2X}{2}$$

10. B : A = lw

If X = l, 96 – 2X = w

Area, or Y = X(96 – 2X)

$$Y = 96X - 2X^2$$
$$Y = -2X^2 + 96X$$

axis of symmetry:

$$X = \frac{-B}{2A} \Rightarrow X = \frac{-(96)}{2(-2)} = \frac{-96}{-4} = 24 \text{ ft}$$

$$w = 96 - 2X \Rightarrow w = 96 - 2(24)$$
$$w = 96 - 48$$
$$w = 48 \text{ ft}$$

$$96 - 2X$$

X [rectangle] X

$$96 - 2X$$

11. B : $3\sqrt{-27} - 4\sqrt{-8} = 3\sqrt{9}\sqrt{-1}\sqrt{3} - 4\sqrt{4}\sqrt{-1}\sqrt{2} =$
$3(3)i\sqrt{3} - 4(2)i\sqrt{2} = 9i\sqrt{3} - 8i\sqrt{2}$

12. A : $9i\sqrt{-64} = 9i\sqrt{-1}\sqrt{64} = 9i(i)(8) = 72i^2 = -72$

13. A : $\frac{2i}{8 + 5i} = \frac{2i(8 - 5i)}{(8 + 5i)(8 - 5i)} = \frac{2i(8) - 2i(5i)}{64 - 25i^2} =$

$$\frac{16i - 10i^2}{64 - (25)(-1)} = \frac{16i - (10)(-1)}{64 - (-25)} =$$

$$\frac{16i - (-10)}{64 + 25} = \frac{16i + 10}{89}$$

14. D : Using quadratic formula:

$$2X^2 + 2X - 5 = 0$$

$$X = \frac{-(2) \pm \sqrt{(2)^2 - 4(2)(-5)}}{2(2)} =$$

$$\frac{-2 \pm \sqrt{4 - (-40)}}{4} = \frac{-2 \pm \sqrt{44}}{4} =$$

$$\frac{-2 \pm \sqrt{4}\sqrt{11}}{4} = \frac{-2 \pm 2\sqrt{11}}{4} = \frac{-1 \pm \sqrt{11}}{2}$$

15. B : the square root of a negative number is imaginary

Test 26

1. D : A) is a hyperbola

   B) $X = \dfrac{16}{Y} \Rightarrow XY = 16$

   C) The difference of two squares is a hyperbola.

   D) The sum of two squares is a circle or an ellipse.

2. A : One variable increases while the other decreases.

3. D : The lines approach but do not touch the axis.

4. C : $XY = 6 \Rightarrow (-1)(Y) = 6$
   $$Y = -6$$

5. C : $X^2 - Y^2 = 8 \Rightarrow X^2 - (-1)^2 = 8$
   $$X^2 - 1 = 8$$
   $$X^2 = 9$$
   $$X = \pm 3$$

6. D : II and IV

   Choose a negative and a positive value for X and find the corresponding values for Y. The results will show what quadrants the graph lies in:

   $X = 4$:

   $XY = -1 \Rightarrow (4)Y = -1$
   $$Y = -\frac{1}{4}$$

   $X = -4$:

   $XY = -1 \Rightarrow (-4)Y = -1$
   $$Y = \frac{1}{4}$$

   Points $\left(4, -\dfrac{1}{4}\right)$ and $\left(-4, \dfrac{1}{4}\right)$ are on the hyperbola, and lie in quadrants IV and II, respectively.

7. B : a hyperbola in the difference of two squares form

8. A : a hyperbola in quadrants I and III

9. C : a hyperbola in quadrants II and IV

10. D : a hyperbola in the difference of two squares form

11. C : $Y = \dfrac{1}{2}X + 4 \Rightarrow m = \dfrac{1}{2}$

   negative reciprocal of $\dfrac{1}{2}$ is -2

   $Y = mX + b \Rightarrow (0) = (-2)(0) + b$
   $$0 = 0 + b$$
   $$b = 0$$

   $Y = mX + b \Rightarrow Y = -2X$

12. A : $\text{midpoint} = \left(\dfrac{X_1 + X_2}{2}, \dfrac{Y_1 + Y_2}{2}\right) =$

   $\left(\dfrac{(-5) + (7)}{2}, \dfrac{(-3) + (-6)}{2}\right) =$

   $\left(\dfrac{2}{2}, \dfrac{-9}{2}\right) = \left(1, -\dfrac{9}{2}\right)$

13. B : a square

14. D : $\$299.00 = \$254.00 = \$45.00$ markup
   $$WP \times 299 = 45$$
   $$WP = \frac{45}{299} \approx .15 = 15\%$$

15. D : $\dfrac{Na}{NaCl} = \dfrac{23}{23 + 35} = \dfrac{23}{58} \approx .40 = 40\%$

Test 27

1. C : I, III, and IV only: translation is a type of geometric transformation, not a method for solving equations.

2. A : Since Y and $2X - 1$ have the same value, one can be substituted for the other.

3. D: $3X + 2Y = 12$

$2Y = -3X + 12$

$Y = -\dfrac{3}{2}X + 6$

$XY = 6 \Rightarrow \qquad (X)\left(-\dfrac{3}{2}X + 6\right) = 6$

$-\dfrac{3}{2}(X^2) + 6X = 6$

$\left(-\dfrac{2}{3}\right)\left(-\dfrac{3}{2}\right)(X^2) + \left(-\dfrac{2}{3}\right)(6X) = \left(-\dfrac{2}{3}\right)(6)$

$X^2 - 4X = -4$

$X^2 - 4X + 4 = 0$

$(X - 2)(X - 2) = 0$

$X - 2 = 0$

$X = 2$

$3X + 2Y = 12 \Rightarrow 3(2) + 2Y = 12$

$6 + 2Y = 12$

$2Y = 6$

$Y = 3$

solution $= (2, 3)$

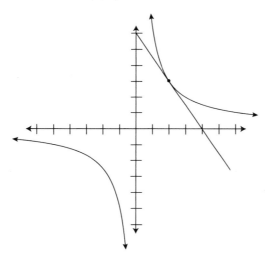

4. B: circle and line

5. C: $Y = X - 3$

$X^2 + Y^2 = 9 \Rightarrow \qquad X^2 + (X - 3)^2 = 9$

$X^2 + X^2 - 6X + 9 = 9$

$2X^2 - 6X = 0$

$(2X)(X - 3) = 0$

$2X = 0$

$X = 0$

$Y = X - 3 \Rightarrow Y = (0) - 3$

$Y = -3$

solution 1: $(0, -3)$

$X - 3 = 0$

$X = 3$

$Y = X - 3 \Rightarrow Y = (3) - 3$

$Y = 0$

solution 2: $(3, 0)$

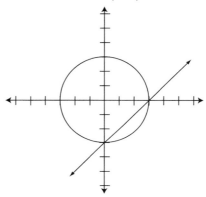

6. D: see illustration for problem 8

7. B : $Y = X^2 + 2$

$X^2 + Y^2 = 4 \Rightarrow \quad X^2 + \left(X^2 + 2\right)^2 = 4$

$\qquad\qquad\qquad X^2 + X^4 + 4X^2 + 4 = 4$

$\qquad\qquad\qquad\qquad\qquad X^4 + 5X^2 = 0$

$\qquad\qquad\qquad\qquad \left(X^2\right)\left(X^2 + 5\right) = 0$

$X^2 = 0$

$X = 0$

$Y = X^2 + 2 \Rightarrow \quad Y = \left(0\right)^2 + 2$

$\qquad\qquad\qquad\qquad Y = 0 + 2$

$\qquad\qquad\qquad\qquad Y = 2$

solution 1: $\left(0, 2\right)$

$X^2 + 5 = 0$

$\quad X^2 = -5$

$\qquad X = \pm\sqrt{-5}$: not a real solution

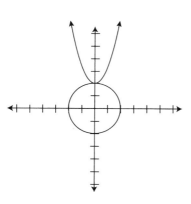

8. A : sketch hyperbola and circle:

work solution:

$X^2 + Y^2 = 34$

$\qquad X^2 + Y^2 = 34$

$\qquad \underline{X^2 - Y^2 = 16}$

$\qquad 2X^2 \qquad = 50$

$\qquad\qquad X^2 = 25$

$\qquad\qquad X = \pm 5$

$X^2 + Y^2 = 34 \Rightarrow \left(\pm 5\right)^2 + Y^2 = 34$

$\qquad\qquad\qquad\qquad 25 + Y^2 = 34$

$\qquad\qquad\qquad\qquad\qquad Y^2 = 9$

$\qquad\qquad\qquad\qquad\qquad Y = \pm 3$

solutions are:

$\left(5, 3\right); \left(5, -3\right); \left(-5, 3\right); \left(-5, -3\right)$

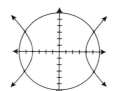

9. C : some roots may require imaginary numbers to solve the equations

10. C : gives a good estimate of the answer

11. A : $A = lw = \left(X\right)\left(\dfrac{176 - 2X}{2}\right) =$

$\dfrac{176X - 2X^2}{2} = 88X - X^2 = -X^2 + 88X$

axis of symmetry:

$\dfrac{-B}{2A} \Rightarrow \dfrac{-\left(88\right)}{2\left(-1\right)} = 44$

$Y = $ area

$Y = -X^2 + 88X \Rightarrow \quad Y = -\left(44\right)^2 + 88\left(44\right)$

$\qquad\qquad\qquad\qquad\qquad Y = -1{,}936 + 3{,}872$

$\qquad\qquad\qquad\qquad\qquad Y = 1{,}936 \text{ ft}^2$

12. A : $\dfrac{A}{BC} = \dfrac{D}{E}$

$AE = BCD$

$\dfrac{AE}{CD} = B$

$B = \dfrac{AE}{CD}$

13. C : $1,200,000,000 \times 3,000,000 =$

$\left(1.2 \times 10^9\right)\left(3.0 \times 10^6\right) =$

$\left(1.2 \times 3.0\right)\left(10^9 \times 10^6\right) = 3.6 \times 10^{15}$

If significant digits had been taken into account, the answer would have been 4×10^{15}

14. B : $D = RT \Rightarrow \ (18) = (15 - 3)T$

$18 = (12)T$

$\dfrac{18}{12} = T$

$T = \dfrac{3}{2} = 1.5$ hours

15. D : $3 \times 1.6 = 4.8$ km

$2 \times 1.6 = 3.2$ km

$a = (4.8)(3.2) = 15.36$ km^2

Test 28

1. B : $\begin{aligned} N + D &= 11 \ \Rightarrow \ {-5N} - \ 5D = -55 \\ .05N + .10D &= .80 \ \Rightarrow \ \underline{\ \ 5N + 10D = \ \ 80} \\ &\qquad\qquad\qquad\quad 5D = \ \ 25 \\ &\qquad\qquad\qquad\quad\ \ D = 5 \end{aligned}$

2. A : $\begin{aligned} D + Q &= 7 \ \Rightarrow \ {-10D} - 10Q = -70 \\ .10D + .25Q &= 1.15 \ \Rightarrow \ \underline{\ \ 10D + 25Q = 115} \\ &\qquad\qquad\qquad\qquad 15Q = \ \ 45 \\ &\qquad\qquad\qquad\qquad\ \ Q = 3 \end{aligned}$

3. C : $\begin{aligned} P + N &= 25 \ \Rightarrow \ {-P} - \ N = -25 \\ .01P + .05N &= .57 \ \Rightarrow \ \underline{\ \ P + 5N = \ \ 57} \\ &\qquad\qquad\qquad\qquad 4N = \ \ 32 \\ &\qquad\qquad\qquad\qquad\ N = 8 \\ &\qquad\qquad P = 25 - 8 = 17 \end{aligned}$

4. C : N; N + 2; N + 4 are the three integers

$3(N) + (N + 2) + 2 = 3(N + 4)$

$3N + N + 2 + 2 = 3N + 12$

$4N + 4 = 3N + 12$

$N = 8$

integers are 8, 10, and 12

5. D : each odd integer is 2 more than the previous odd integer

6. A : N; N + 1; N + 2 are the three integers

$4(N) + 2(N + 1) = 4(N + 2)$

$4(N) + 2(N) + 2 = 4(N) + 8$

$6N + 2 = 4N + 8$

$2N = 6$

$N = 3$

integers are 3, 4, and 5

7. D : N; N + 2; N + 4 are the three integers

$10(N) + 10(N + 2) = 10 + 10(N + 4)$

$10N + 10N + 20 = 10 + 10N + 40$

$20N + 20 = 10N + 50$

$10N = 30$

$N = 3$

integers are 3, 5 and 7

third integer is 7

8. B : $A_T = 20\%$ solution; $A_E = 8\%$ solution

$A_T + A_E = 90$ ml

$.20A_T + .08A_E = .10(90)$

Multiply first equation by -8 and second equation by 100:

$\begin{aligned} {-8A_T} - 8A_E &= -720 \\ \underline{20A_T + 8A_E} &= \ \ 900 \\ 12A_T &= \ \ 180 \\ A_T &= 15 \text{ ml} \end{aligned}$

9. B : $F_1 = 50\%$ solution; $F_2 = 5\%$ solution

$F_1 + F_2 = 150$ lb

$.50F_1 + .05F_2 = .12(150)$

Multiply first equation by -50 and second equation by 100:

$\begin{aligned} {-50F_1} - 50F_2 &= -7500 \\ \underline{50F_1 + \ 5F_2} &= \ \ 1800 \\ -45F_2 &= -5700 \\ F_2 &\approx 126.7 \text{ lb} \end{aligned}$

10. C: $150 \text{ lb} - 126.7 \text{ lb} = 23.3 \text{ lb}$

11. D: $D_C + D_B = 20$

D $\overbrace{}$

$$R_C T_C + R_B T_B = 20$$

$$(4)(T_C) + (6)(T_C) = 20 \Rightarrow \begin{cases} R_C = 4 \\ R_B = 6 \\ T_B = T_C \end{cases}$$

$$4T_C + 6T_C = 20$$

$$10T_C = 20$$

$$T_C = 2 \text{ hours}$$

$$12\!:\!00 \text{ noon} + 2 \text{ hours} = 2\!:\!00 \text{ pm}$$

12. A: B is a line, C is a hyperbola, and D is a circle

13. C: 6 triangles $\times 180° = 1,080°$

14. C: width $= X$; length $= 720 - 2X$

area, or $Y = (X)(720 - 2X)$

$720 - 2X$

$$Y = 720X - 2X^2$$

$$Y = -2X^2 + 720X$$

axis of symmetry:

$$\frac{-B}{2A} \Rightarrow \frac{-(720)}{2(-2)} = \frac{-720}{-4} = 180 \text{ ft}$$

length $= 720 - 2X = 720 - 2(180) =$

$720 - 360 = 360 \text{ ft}$

dimensions: $180 \text{ ft} \times 360 \text{ ft}$

15. B: $\dfrac{\frac{3}{X}}{\frac{2}{X+1}} = \frac{3}{X} \times \frac{X+1}{2} = \frac{3(X+1)}{(X)(2)} = \frac{3X+3}{2X}$

Test 29

1. A: $R + 2 = \left(3\frac{1}{2}\right)(S + 2)$

$$R + 2 = \left(\frac{7}{2}\right)S + 7$$

$$R = \frac{7}{2}S + 5$$

$$R - 3 = 6(S - 3) \Rightarrow \left(\frac{7}{2}S + 5\right) - 3 = 6(S - 3)$$

$$\frac{7}{2}S + 2 = 6S - 18$$

$$7S + 4 = 12S - 36$$

$$40 = 5S$$

$$S = 8 \text{ years}$$

2. B: $P - 8 = \frac{1}{3}(K - 8)$

$$P - \frac{24}{3} = \frac{1}{3}K - \frac{8}{3}$$

$$P = \frac{1}{3}K - \frac{8}{3} + \frac{24}{3}$$

$$P = \frac{1}{3}K + \frac{16}{3}$$

$$2(P + 2) = K + 2$$

$$2P + 4 = K + 2$$

$$2P = K - 2 \Rightarrow 2\left(\frac{1}{3}K + \frac{16}{3}\right) = K - 2$$

$$\frac{2}{3}K + \frac{32}{3} = K - 2$$

$$2K + 32 = 3K - 6$$

$$38 = K$$

$$K = 38 \text{ years}$$

3. C: $C - 5 = \frac{1}{2}(D - 5)$

$$C - \frac{10}{2} = \frac{1}{2}D - \frac{5}{2}$$

$$C = \frac{1}{2}D + \frac{5}{2}$$

$$C + 3 = \frac{5}{6}(D + 3) \Rightarrow \left(\frac{1}{2}D + \frac{5}{2}\right) + 3 = \frac{5}{6}(D + 3)$$

$$\frac{1}{2}D + \frac{5}{2} + \frac{6}{2} = \frac{5}{6}D + \frac{15}{6}$$

$$3D + 15 + 18 = 5D + 15$$

$$3D + 33 = 5D + 15$$

$$18 = 2D$$

$$D = 9 \text{ years}$$

4. B : $D - 5 = 4(S - 5)$

$D - 5 = 4S - 20$

$D = 4S - 15$

$S + 30 = \dfrac{3}{5}(D + 30)$

$5S + 150 = 3(D + 30)$

$5S + 150 = 3D + 90$

$5S + 60 = 3D \Rightarrow 5S + 60 = 3(4S - 15)$

$5S + 60 = 12S - 45$

$105 = 7S$

$S = 15$ years

5. D : $R = 10C$

$R + 10 = 4(C + 10)$

$R + 10 = 4C + 40$

$R = 4C + 30 \Rightarrow (10C) = 4C + 30$

$6C = 30$

$C = 5$

$C - 2 = 3$ years

6. C : With this formula, the rate upstream would decrease when the rate of the boat increased, which is the opposite of the truth.

7. D : $D_D = (B + W)T$

$(42) = (B + (4))T$

$42 = BT + 4T$

$D_U = (B - W)T$

$(18) = (B - (4))T$

$18 = BT - 4T \Rightarrow \quad -18 = -BT + 4T$

$ \underline{42 = BT + 4T}$

$ 24 = \phantom{-BT + {}}8T$

$ T = 3$ hours

$42 = (B + 4)T \Rightarrow 42 = (B + 4)(3)$

$42 = 3B + 12$

$30 = 3B$

$B = 10$ mph

8. D : $D_D = (B + W)T$

$(60) = (B + (2))T$

$60 = BT + 2T$

$D_U = (B - W)T$

$(12) = (B - (2))T$

$12 = BT - 2T \qquad -12 = -BT + 2T$

$ \underline{60 = BT + 2T}$

$ 48 = \phantom{-BT + {}}4T$

$T = 12$ hours each way

$12 + 12 = 24$ hours total

9. A : $D_D = (B + W)T$

$D_D = ((10) + W)(2)$

$D_D = 20 + 2W$

$D_U = (B - W)T$

$D_U = ((10) - W)(5)$

$D_U = 50 - 5W$

$D_D = D_U \Rightarrow (20 + 2W) = (50 - 5W)$

$7W = 30$

$W \approx 4.3$ mph

10. C : $D = (B + W)T \Rightarrow D = ((10) + (4.3))(2)$

$D = (14.3)(2)$

$D = 28.6$ miles

11. B : While corresponding sides MAY be equal, we can't say with certainty that they are.

12. D : A is a line with slope of 1: its equation can be expressed as $Y = X - 4$. B and C are not lines. D is a line whose equation is $Y = -X - 4$.

13. B : Divide the coefficient of the second term by 2, and square the result.

$(3 \div 2)^2 = \left(\dfrac{3}{2}\right)^2 = \dfrac{9}{4}$

14. C : $ N + Q = 14 \quad \Rightarrow \quad -5N -5Q = -70$

$.05N + .25Q = 1.90 \Rightarrow \quad \underline{5N + 25Q = 190}$

$ 20Q = 120$

$ Q = 6$

$N + Q = 14 \Rightarrow N + (6) = 14$

$N = 8$ nickels at first

$8 - 3 = 5$ nickels now

15. A : The length of a side of a square is equal to the square root of its area.

$$\sqrt{9X^2Y} = \sqrt{9}\sqrt{X^2}\sqrt{Y} = 3X\sqrt{Y}$$

Test 30

1. C : 3

2. C : Eliminate one variable, creating two new equations using those variables.

3. B : Substitution and elimination are the main tools for solving them.

4. A : Multiplying equation A by 2 changes the Z term to −2Z, which is the additive inverse of the Z term in equation B.

5. D :

A $4X + 2Y + 3Z = 10$
B $\underline{2X - 2Y + \ \ Z = -6}$
D $6X \ \ \ \ \ \ \ + 4Z = \ \ 4$

B $3(2X - 2Y + Z = -6) \Rightarrow \ \ 6X - 6Y + 3Z = -18$
C $2(X + 3Y - 4Z = 22) \Rightarrow \ \ \underline{2X + 6Y - 8Z = \ \ 44}$
E $8X \ \ \ \ \ \ \ - 5Z = \ \ 26$

D $5(6X + 4Z = 4) \Rightarrow \ \ 30X + 20Z = \ \ 20$
E $4(8X - 5Z = 26) \Rightarrow \ \ \underline{32X - 20Z = 104}$
      $62X \ \ \ \ \ \ \ \ = 124$
             $X = 2$

6. D:

D $6X + 4Z = 4 \Rightarrow \ \ 6(2) + 4Z = 4$
            $12 + 4Z = 4$
               $4Z = -8$
               $Z = -2$

A $4X + 2Y + 3Z = 10 \Rightarrow \ \ 4(2) + 2Y + 3(-2) = 10$
                      $8 + 2Y - 6 = 10$
                        $2Y + 2 = 10$
                          $2Y = 8$
                          $Y = 4$

7. C : see first part of solution for number 6

8. A:

A $3(2X - Y + 4Z = 5) \Rightarrow \ \ 6X - 3Y + 12Z = 15$
B
D $\underline{X + 3Y - \ \ 2Z = \ \ 8}$
                  $7X \ \ \ \ \ \ \ + 10Z = 23$

B $-1(X + 3Y - 2Z = 8) \Rightarrow \ \ -X - 3Y + 2Z = -8$
C
E $\underline{3X + 3Y - \ \ Z = \ \ 9}$
                  $2X \ \ \ \ \ \ \ + \ \ Z = \ \ 1$

D $7X + 10Z = \ \ 23$
E $-10(2X + Z = 1) \ \ \underline{-20X - 10Z = -10}$
              $-13X \ \ \ \ \ \ \ \ = 13$
                  $X = -1$

9. D : E $2X + Z = 1 \Rightarrow \ \ 2(-1) + Z = 1$
                  $-2 + Z = 1$
                     $Z = 3$

    B $X + 3Y - 2Z = 8 \Rightarrow \ \ (-1) + 3Y - 2(3) = 8$
                       $-1 + 3Y - 6 = 8$
                         $3Y - 7 = 8$
                           $3Y = 15$
                           $Y = 5$

10. B : see first part of solution for number 9

11. D : $X = \dfrac{3}{Y} \Rightarrow XY = 3$

12. B : The first equation is a line, and the second is a circle. A line can intersect a circle in at most two places.

13. B : $\left(\sqrt[3]{27}\right)^{-1} = (3)^{-1} = \dfrac{1}{3}$

14. A : $CS_2 = 12 + 32 + 32 = 76$

$$\frac{12}{76} = \frac{C}{1,368}$$

$$C = \frac{12(1,368)}{76} = 216 \text{ grams}$$

15. C : $2[(2 + X) - (-3)]^2 =$
$2[2 + X + 3]^2 =$
$2[X + 5]^2 =$
$2[X^2 + 10X + 25] =$
$2X^2 + 20X + 50$

Test 31

1. B : vectors have direction and magnitude
2. C : resultant
3. A : opposite over hypotenuse
4. C : using pythagorean theorem:

$$H^2 = 60^2 + 80^2$$
$$H^2 = 3,600 + 6,400$$
$$H^2 = 10,000$$
$$H = 100$$

5. A : $\tan\theta = \dfrac{80}{60}$; $\theta \approx 53.1°$

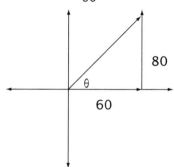

6. C : opposite over adjacent
7. D : $H^2 = 50^2 + 70^2$

$$H^2 = 7,400$$
$$H = \sqrt{7,400}$$
$$H \approx 86 \text{ miles}$$

8. B : $\tan\theta = \dfrac{70}{50} = 1.4$

$$\theta \approx 54.5°$$

9. C : $(+2) + (+3) = +5$ on the X-axis

$$(+5) + (-1) = +4 \text{ on the Y-axis}$$

10. D : $H^2 = 5^2 + 4^2$

$$H^2 = 25 + 16$$
$$H^2 = 41$$
$$H = \sqrt{41} \approx 6.4$$
$$\tan\theta = \dfrac{4}{5} = .8$$
$$\theta = 38.7°$$

11. C : $4i - 3i^2 = 4i - 3(-1) = 4i - (-3) = 4i + 3$
12. C : a square is a rectangle with four equal sides

13. D : $\dfrac{16^{-\frac{1}{2}} \times 3^{-2}}{2^{-2}} = \dfrac{2^2}{16^{\frac{1}{2}} \times 3^2} =$

$$\dfrac{4}{4 \times 9} = \dfrac{4}{36} = \dfrac{1}{9}$$

14. C : $i^{401} = \left(i^{400}\right)\left(i^1\right) = (-1)^{200}(i) = 1^{100}(i) = i$

15. C : $\dfrac{B^{\frac{1}{3}}B^4}{B^{-2}} = \dfrac{B^{\frac{1}{3}+\frac{12}{3}}}{B^{-2}} = \dfrac{B^{\frac{13}{3}}}{B^{-2}} =$

$$B^{\frac{13}{3}}B^2 = B^{\frac{13}{3}+\frac{6}{3}} = B^{\frac{19}{3}}$$

Unit Test III

1. $m = \dfrac{Y_2 - Y_1}{X_2 - X_1} = \dfrac{(1) - (3)}{(-1) - (3)} = \dfrac{-2}{-4} = \dfrac{1}{2}$

$$Y = mX + b \Rightarrow \qquad (1) = \dfrac{1}{2}(-1) + b$$
$$1 = -\dfrac{1}{2} + b$$
$$\dfrac{2}{2} + \dfrac{1}{2} = b$$
$$\dfrac{3}{2} = b = 1\dfrac{1}{2}$$

$$Y = \dfrac{1}{2}X + 1\dfrac{1}{2}$$

2. They are parallel.
3. They are perpendicular.
4. The direction of the sign is reversed.
5. $D = \sqrt{\Delta X^2 + \Delta Y^2}$

$$D = \sqrt{4^2 + 3^2}$$
$$D = \sqrt{16 + 9}$$
$$D = \sqrt{25}$$
$$D = 5$$

6. $\text{midpoint} = \left(\dfrac{X_1 + X_2}{2}, \dfrac{Y_1 + Y_2}{2} \right) =$

$$\left(\dfrac{(-2) + (2)}{2}, \dfrac{(5) + (2)}{2} \right) = \left(\dfrac{0}{2}, \dfrac{7}{2} \right) = \left(0, 3\dfrac{1}{2} \right)$$

7. $X^2 + Y^2 = 4$

circle: center $= (0, 0)$; radius $= 2$

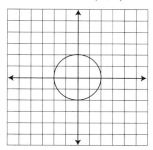

8. ellipse: center $= (0, 0)$; find extremities by making each term 0 in turn:

$$4X^2 + Y^2 = 16 \qquad\qquad 4X^2 + Y^2 = 16$$
$$4X^2 + (0)^2 = 16 \qquad\qquad 4(0)^2 + Y^2 = 16$$
$$4X^2 = 16 \qquad\qquad\qquad Y^2 = 16$$
$$X^2 = 4 \qquad\qquad\qquad\quad Y = \pm 4$$
$$X = \pm 2$$

extremities: $(2, 0)$; $(-2, 0)$; $(0, 4)$; $(0, -4)$

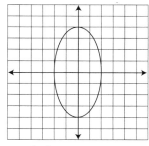

9. parabola:

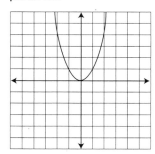

10. hyperbola:

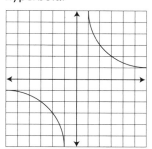

11. $Y = X^2 - 4X + 2$

axis of symmetry $= \dfrac{-B}{2A} = \dfrac{-(-4)}{2(1)} = \dfrac{4}{2} = 2$

$Y = X^2 - 4X + 2 \Rightarrow \quad Y = (2)^2 - 4(2) + 2$
$$Y = 4 - 8 + 2$$
$$Y = -2$$

vertex $= (2, -2)$

12. sketch circle:

$X^2 + Y^2 = 9$

center is $(0, 0)$ and radius $= \sqrt{9} = 3$

sketch line:

$Y = X + 3$

intercept is 3 and slope is 1

solve by substitution:

$X^2 + Y^2 = 9 \Rightarrow \qquad X^2 + (X + 3)^2 = 9$
$$X^2 + X^2 + 6X + 9 = 9$$
$$2X^2 + 6X = 0$$
$$(2X)(X + 3) = 0$$

$2X = 0$
$\quad X = 0$
$\quad Y = X + 3 \Rightarrow \quad Y = (0) + 3$
$$Y = 3$$

solution 1: $(0, 3)$

$X + 3 = 0$
$\quad X = -3$
$\quad Y = X + 3 \Rightarrow \quad Y = (-3) + 3$
$$Y = 0$$

solution 2: $(-3, 0)$

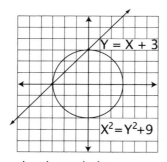

13. sketch parabola:

$Y = X^2$

vertex is $(0, 0)$ opens upward

sketch hyperbola:

$XY = 8$

curves in quadrants I and III

make a chart if needed

solve by substitution:

$XY = 8 \Rightarrow X(X^2) = 8$

$X^3 = 8$

$X = 2$

$XY = 8 \Rightarrow (2)Y = 8$

$Y = 4$

solution: $(2, 4)$

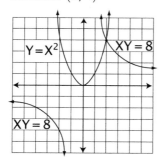

14.

$$10(D + N = 12) \Rightarrow 10D + 10N = 120$$
$$-100(.10D + .05N = .85) \Rightarrow \underline{-10D - 5N = -85}$$
$$5N = 35$$
$$N = 7 \text{ nickels}$$

$D + N = 12 \Rightarrow D + (7) = 12$

$D = 5 \text{ dimes}$

15. integers are N, $N+1$, and $N+2$

$2(N) + (N + 1) = (N + 2) + 9$

$2N + N + 1 = N + 2 + 9$

$3N + 1 = N + 11$

$2N = 10$

$N = 5$

integers are 5, 6 and 7

16. $S_T = $ amount of 20% solution

$S_F = $ amount of 50% solution

$$20(S_T + S_F = 3) \Rightarrow 20S_T + 20S_F = 60$$
$$-100(.20S_T + .50S_F = .30(3)) \Rightarrow \underline{-20S_T - 50S_F = -90}$$
$$-30S_F = -30$$
$$S_F = 1 \text{ liter}$$

$S_T + S_F = 3 \Rightarrow S_T + (1) = 3$

$S_T = 2 \text{ liters}$

17. $K + 2 = 3(L + 2)$

$K + 2 = 3L + 6$

$K = 3L + 4$

$6(L - 1) = K - 1$

$6L - 6 = K - 1$

$6L = K + 5 \Rightarrow 6L = (3L + 4) + 5$

$6L = 3L + 9$

$3L = 9$

$L = 3 \text{ years old}$

$K = 3L + 4 \Rightarrow K = 3(3) + 4$

$K = 9 + 4$

$K = 13 \text{ years old}$

18. $D_D = R_D T_D \Rightarrow (36) = R_D(3)$

$R_D = 12$

$D_U = R_U T_U \Rightarrow (30) = R_U(5)$

$$R_U = 6$$

$R_D = B + W \Rightarrow \quad 12 = \quad B + W$

$R_U = B - W \Rightarrow \quad \underline{6 = \quad B - W}$

$$18 = 2B$$

$$B = 9 \text{ mph}$$

$R_D = B + W \Rightarrow \quad (12) = (9) + W$

$$12 - 9 = W$$

$$W = 3 \text{ mph}$$

19.

A

B $\quad 3(4X + Y - 2Z = 0) \Rightarrow$

D

$$\begin{array}{r} 2X - 3Y + 3Z = 9 \\ \underline{12X + 3Y - 6Z = 0} \\ 14X \qquad - 3Z = 9 \end{array}$$

B $\quad 2(4X + Y - 2Z = 0) \Rightarrow$

C

E

$$\begin{array}{r} 8X + 2Y - 4Z = 0 \\ \underline{-6X - 2Y + Z = 0} \\ 2X \qquad - 3Z = 0 \end{array}$$

D

E $\quad -1(2X - 3Z = 0) \Rightarrow$

$$\begin{array}{r} 14X - 3Z = 9 \\ \underline{-2X + 3Z = 0} \\ 12X = 9 \end{array}$$

$$X = \frac{3}{4}$$

D $\quad 14X - 3Z = 9 \Rightarrow 14\left(\frac{3}{4}\right) - 3Z = 9$

$$\frac{42}{4} - 3Z = 9$$

$$-3Z = \frac{36}{4} - \frac{42}{4}$$

$$-3Z = -\frac{6}{4}$$

$$Z = \frac{2}{4} = \frac{1}{2}$$

B $\quad 4X + Y - 2Z = 0 \Rightarrow 4\left(\frac{3}{4}\right) + Y - 2\left(\frac{1}{2}\right) = 0$

$$\frac{12}{4} + Y - \frac{2}{2} = 0$$

$$3 + Y - 1 = 0$$

$$Y = -3 + 1$$

$$Y = -2$$

20. direction and magnitude

Final Exam

1. $\left(X^7 \div X^3\right) + \left(X^2 X^2\right) = \left(X^{7-3}\right) + \left(X^{2+2}\right) =$

$X^4 + X^4 = 2X^4$

2. $\dfrac{A^5 B^{-3}}{B^3 A^2} = A^5 B^{-3} A^{-2} B^{-3} =$

$A^{5+(-2)} B^{-3+(-3)} = A^3 B^{-6}$ or $\dfrac{A^3}{B^6}$

3. $\left(\dfrac{8}{27}\right)^{-\frac{1}{3}} = \left(\dfrac{27}{8}\right)^{\frac{1}{3}} = \dfrac{27^{\frac{1}{3}}}{8^{\frac{1}{3}}} = \dfrac{3}{2}$ or $1\dfrac{1}{2}$

4. $2\sqrt{5} + 7\sqrt{5} = (2 + 7)\sqrt{5} = 9\sqrt{5}$

5. $\dfrac{X}{3+i} = \dfrac{X(3-i)}{(3+i)(3-i)} = \dfrac{3X - Xi}{9 - i^2} =$

$\dfrac{3X - Xi}{9 - (-1)} = \dfrac{3X - Xi}{10}$ or $\dfrac{X(3-i)}{10}$

6. $\dfrac{3}{1+\sqrt{3}} = \dfrac{3\left(1-\sqrt{3}\right)}{\left(1+\sqrt{3}\right)\left(1-\sqrt{3}\right)}$

$= \dfrac{3 - 3\sqrt{3}}{1 - 3} = \dfrac{3 - 3\sqrt{3}}{-2}$

7. $\dfrac{5}{6X} + \dfrac{4}{3Y} = \dfrac{5(Y)}{6X(Y)} + \dfrac{4(2X)}{3Y(2X)} = \dfrac{5Y + 8X}{6XY}$

8. $5Q^{-1}RQ^2 + 3QR - R = 5Q^{-1+2}R + 3QR - R =$

$5Q^1R + 3QR - R = 5QR + 3QR - R = 8QR - R$

9. $(.0009)(.027) =$

$\left(9.0 \times 10^{-4}\right)\left(2.7 \times 10^{-2}\right) =$

$(9.0 \times 2.7)\left(10^{-4} \times 10^{-2}\right) =$

$24.3 \times 10^{-6} = 2.43 \times 10^{-5}$

If significant digits are taken into account:

2.0×10^{-5} (either answer is correct)

10. $\dfrac{3,700,000}{.002} = \dfrac{3.7 \times 10^6}{2.0 \times 10^{-3}} =$

$\left(3.7 \times 10^6\right) \div \left(2.0 \times 10^{-3}\right) =$

$(3.7 \div 2.0)\left(10^6 \div 10^{-3}\right) =$

1.85×10^9

2.0×10^9 with significant digits

11. $2X^2 - 9X = 35$

$2X^2 - 9X - 35 = 0$

$X = \dfrac{-(-9) \pm \sqrt{(-9)^2 - 4(2)(-35)}}{2(2)} =$

$\dfrac{9 \pm \sqrt{81 - (-280)}}{4} = \dfrac{9 \pm \sqrt{361}}{4} = \dfrac{9 \pm 19}{4}$

$X = \dfrac{9 + 19}{4} \qquad X = \dfrac{9 - 19}{4}$

$X = \dfrac{28}{4} \qquad X = \dfrac{-10}{4}$

$X = 7 \qquad X = -\dfrac{5}{2}$

$\qquad\qquad\qquad X = -2\dfrac{1}{2}$

12. $X^2 + 4X - 4 = -3X$

$X^2 + 7X - 4 = 0$

$X = \dfrac{-(7) \pm \sqrt{(7)^2 - 4(1)(-4)}}{2(1)} =$

$\dfrac{-7 \pm \sqrt{49 - (-16)}}{2} = \dfrac{-7 \pm \sqrt{65}}{2}$

13. sketch parabola and line
(see graph on next page)

$Y = X^2 + 2$

$Y = X + 2 \Rightarrow \quad (X^2 + 2) = X + 2$

$\qquad\qquad\qquad X^2 - X = 0$

$\qquad\qquad (X)(X - 1) = 0$

$X = 0$

$Y = X + 2 \Rightarrow \quad Y = (0) + 2$

$\qquad\qquad\qquad Y = 2$

solution 1: $(0, 2)$

$X - 1 = 0$

$\qquad X = 1$

$Y = X + 2 \Rightarrow \quad Y = (1) + 2$

$\qquad\qquad\qquad Y = 3$

solution 2: $(1, 3)$

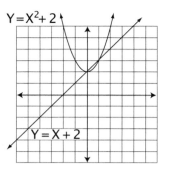

14. sketch circle and hyperbola

$X^2 + Y^2 = 1$

$\dfrac{X^2 - Y^2 = 1}{2X^2 \qquad = 2}$

$X^2 = 1$

$X = \pm 1$

$X = 1:$

$X^2 + Y^2 = 1 \Rightarrow (1)^2 + Y^2 = 1$

$\qquad\qquad\qquad 1 + Y^2 = 1$

$\qquad\qquad\qquad Y^2 = 0$

$\qquad\qquad\qquad Y = 0$

solution 1: $(1, 0)$

$X = -1:$

$X^2 + Y^2 = 1 \Rightarrow (-1)^2 + Y^2 = 1$

$\qquad\qquad\qquad 1 + Y^2 = 1$

$\qquad\qquad\qquad Y^2 = 0$

$\qquad\qquad\qquad Y = 0$

solution 2: $(-1, 0)$

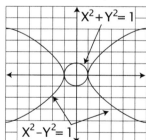

15. $15\% \times \$1,565 = .15 \times \$1,565 = \$234.75$ off
$\$1,565 - \$234.75 = \$1,330.25$

16. Weight of NaCl $= 23 + 35 = 58$

$\dfrac{Na}{NaCl} = \dfrac{23}{58} \approx .40 = 40\%$

17. $\dfrac{C}{D} = \dfrac{5}{18} \Rightarrow \dfrac{(10)}{D} = \dfrac{5}{18}$
$10(18) = 5D$
$2(18) = D$
$D = 36$ dogs

18. $\dfrac{10 \ \cancel{km}}{1} \times \dfrac{.62 \ mi}{1 \ \cancel{km}} = (10)(.62 \ mi) = 6.2 \ mi$

19. $D_M = R_M T_M \Rightarrow D_M = (55)(9)$
$D_M = 495$ mi

$D_A = R_A T_A$

$(495) = (45)T_A \Rightarrow \begin{cases} D_A = D_M \\ R_A = 45 \end{cases}$

$T_A = 11$ hours; arrived at 7:00 PM

20.

$\begin{array}{l} -10(D+Q=15) \Rightarrow \ -10D - 10Q = -150 \\ 100(.10D+.25Q = 3.15) \Rightarrow \ \underline{10D + 25Q = \ \ \ 315} \\ \hspace{4.5cm} 15Q = \ \ \ 165 \\ \hspace{4.5cm} Q = 11 \ \text{quarters} \end{array}$

$D + Q = 15 \Rightarrow D + (11) = 15$
$D = 4$ dimes

21. integers are N, N + 2, and N + 4

$3(N) + 2(N+2) - (N+4) = 16$
$3N + 2N + 4 - N - 4 = 16$
$4N = 16$
$N = 4$

*Note that in this line, the entire quantity $(N+4)$ is subtracted. Think of it as distributing −1 across the two terms inside the parentheses.

Integers are 4, 6 and 8.

22. $M_T = 10\%$ mixture; $M_S = 60\%$ mixture

$\begin{array}{l} -10(M_T + M_S = 100) \Rightarrow \ -10M_T - 10M_S = -1,000 \\ 100(.10M_T + .60M_S = .45(100)) \Rightarrow \ \underline{10M_T + 60M_S = \ \ 4,500} \\ \hspace{5.5cm} 50M_S = \ \ 3,500 \\ \hspace{5.5cm} M_S = 70 \ \text{lb} \end{array}$

$M_T + M_S = 100 \Rightarrow M_T + (70) = 100$
$M_T = 30$ lb

23. $R + 6 = 2(A + 6)$
$R + 6 = 2A + 12$
$R = 2A + 6$
$(A - 4)(3) = R - 4$
$3A - 12 = R - 4$
$3A - 8 = R \Rightarrow 3A - 8 = (2A + 6)$
$A = 14$ years old

$R = 2A + 6 \Rightarrow R = 2(14) + 6$
$R = 28 + 6$
$R = 34$ years old

24. $D_D = R_D T_D$
$D_D = (B + W)T_D$

$(26) = (B + (5))(T_D) \Rightarrow \begin{cases} D_D = 26 \\ W = 5 \end{cases}$

$T_D = \dfrac{26}{B + 5}$

$D_U = R_U T_U$
$D_U = (B - W)T_U$

$(6) = (B - (5))(T_U) \Rightarrow \begin{cases} D_U = 6 \\ W = 5 \end{cases}$

$T_U = \dfrac{6}{B - 5}$

$T_D = T_U \Rightarrow \dfrac{26}{B + 5} = \dfrac{6}{B - 5}$
$26(B - 5) = 6(B + 5)$
$26B - 130 = 6B + 30$
$20B = 160$
$B = 8$ mph

Symbols & Tables

SYMBOLS

$<$	less than
$>$	greater than
$\leq$	less than or equal to
$\geq$	greater than or equal to
$=$	equal in numerical value
$\neq$	not equal to
$\approx$	approximately equal
%	percent
| |	absolute value
$\sqrt{}$	square root
Δ	change in

FACTORING

$(A + B)^2 = A^2 + 2AB + B^2$

$(A - B)^2 = A^2 - 2AB + B^2$

$(A + B)^3 = A^3 + 3A^2B + 3AB^2 + B^3$

$(A - B)^3 = A^3 - 3A^2B + 3AB^2 - B^3$

QUADRATIC FORMULA

$$X = \frac{-B \pm \sqrt{B^2 - 4AC}}{2A}$$

MIDPOINT FORMULA

$$\left(\frac{X_1 + X_2}{2}, \frac{Y_1 + Y_2}{2} \right)$$

PYTHAGOREAN THEOREM

$L^2 + L^2 = H^2$

DISTANCE FORMULA

$D^2 = \Delta X^2 + \Delta Y^2$ or $D = \sqrt{\Delta X^2 + \Delta Y^2}$

IMPERIAL TO METRIC

1 inch $\approx$ 2.5 centimeters

1 yard (36") $\approx$ .9 meters

1 mile $\approx$ 1.6 kilometers

1 ounce $\approx$ 28 grams

1 pound $\approx$ .45 kilograms

1 quart $\approx$ .95 liters

METRIC TO IMPERIAL

1 centimeter $\approx$ .4 inches

1 meter (39.37") $\approx$ 1.1 yards

1 kilometer $\approx$ .62 miles

1 gram $\approx$ .035 ounces

1 kilogram $\approx$ 2.2 pounds

1 liter $\approx$ 1.06 quarts

SLOPE OF A LINE

$$\text{slope} = \frac{\text{up}}{\text{over}} = \frac{Y_2 - Y_1}{X_2 - X_1}$$

DISCRIMINANT

$b^2 - 4ac$

AXIS OF SYMMETRY OF A PARABOLA

$$X = \frac{-B}{2A}$$

SLOPE-INTERCEPT FORM

$Y = mX + b$

EQUATION OF A:

line	$AX + BY = C$
circle	$AX^2 + AY^2 = C$
ellipse	$AX^2 + BY^2 = C$
parabola	$Y = X^2$
hyperbola	$XY = C$ and $X^2 - Y^2 = C$

ORDER OF OPERATIONS

Parachute Expert My Dear Aunt Sally

1. Parentheses
2. Exponents
3. Multiplication and Division
4. Addition and Subtraction

BINOMIAL THEOREM

$$(A - B)^N = A^N B^0 + \frac{N}{1} A^{N-1} B^1 + \frac{N(N-1)}{1(2)} A^{N-2} B^2 + \frac{N(N-1)(N-2)}{1(2)(3)} A^{N-3} B^3 \ldots A^0 B^N$$

ATOMIC WEIGHT TABLE

Symbol	Element	Atomic Weight	Symbol	Element	Atomic Weight
H	Hydrogen	1	Mg	Magnesium	24
Li	Lithium	7	Si	Silicon	28
Be	Beryllium	9	P	Phosphorus	31
B	Boron	11	S	Sulfur	32
C	Carbon	12	Cl	Chlorine	35
N	Nitrogen	14	K	Potassium	39
O	Oxygen	16	Ca	Calcium	40
F	Fluorine	19	Cr	Chromium	52
Na	Sodium	23	Fe	Iron	56

Glossary

absolute value - the value of a number without its sign, or the difference between a number and zero expressed as a positive number

algebra - a branch of mathematics that deals with numbers, which may be represented by letters or symbols

asymptote - a line that is continually approached by a given curve but is never met by that curve

base - a particular side or face of a geometric figure used to calculate area or volume; a number that is raised to a power; the number that is the foundation in a given number system

binomial - an algebraic expression with two terms

binomial theorem - a formula for finding the complete expansion of any positive power of a binomial

coefficient - a quantity placed before and multiplying the variable in an algebraic expression

completing the square - a technique for solving a quadratic equation that involves rewriting it as a perfect square plus a constant

complex number - a combination of a real and an imaginary number in the form $a + bi$

cone - a solid with a circular base and a curved surface that rises to a point

conic section - a curve that results when a cone is intersected by a plane

conjugate - a binomial formed by negating the second term of a given binomial

constant - a fixed, unchanging value

difference of two squares - an expression in which one squared number is subtracted from another squared number

E–I

ellipse - a regular oval created by moving a point around two foci

empty set - a set having no elements; also called *null set*

hyperbola - a conic section that forms two congruent open curves facing in opposite directions on a graph

imaginary number - a number that, when squared, gives a negative product; generally written in the form bi, where i equals the square root of -1

integer - a non-fractional number that can be positive, negative, or zero

irrational numbers - numbers that cannot be written as fractions and form non-repeating, non-terminating decimals

L–O

linear equation - an equation that creates a straight line when graphed

magnitude - length of a vector

maximum - the greatest value of a function at a particular point in its domain; plural is *maxima*

minimum - the least value of a function at a particular point in its domain; plural is *minima*

multiplicative inverse - the number that, when multiplied by a given number, has a product of 1; also called *reciprocal*

natural numbers - whole numbers from 1 to infinity; also called *counting numbers*

origin - on a coordinate grid, the point at the intersection of the axes, generally identified by the ordered pair (0, 0)

P–Q

parabola - a conic section that forms a symmetrical curve on a graph

parallel lines - lines in the same plane that do not intersect

Pascal's triangle - a triangular array of numbers that has a variety of mathematical applications

perfect cube - a number that has a whole number as its cube root

perfect square - a number that has a whole number as its square root

polynomial - an algebraic expression with more than one term

quadratic equation - an equation where the highest power of the variable is 2

quadratic expression - an expression where the highest power of the variable is 2

R–S

radical - an expression containing a root

ratio - the relationship between two values; can be written in fractional form

rational expression - an expression that is the ratio of two polynomials

rational numbers - numbers that can be written as ratios or fractions, including decimals

real numbers - numbers that can be written as decimals, including rational and irrational numbers

reciprocal - the number that, when multiplied by a given number, has a product of 1; also called *multiplicative inverse*

resultant vector - the combination of two or more vectors

scientific notation - a way to write numbers using the product of a base and a power of ten

significant digits - digits that indicate the accuracy of a measurement

simultaneous equations - a pair of equations with two unknown variables that must be solved at the same time

T–Z

trinomial - an algebraic expression with three terms

unknown - a specific quantity that has not yet been determined

variable - a value that is not fixed or determined, often representing a range of possible values

vector - a quantity with both direction and magnitude

vertex - the highest or lowest point of a parabola; the endpoint shared by two rays, line segments, or edges; plural is *vertices*

500

Secondary Levels Master Index

This index lists the levels at which main topics are presented in the instruction manuals for Pre-Algebra through PreCalculus. For more detail, see the description of each level at www.mathusee.com.

Algebra 2 Index